W9-ABG-579

DATE DUE

BRODART, CO.   Cat. No. 23-221-003

Centennial College
P.O. Box 631, Station A,
Scarborough, Ont.
M1K 5E9

DRAFTING FOR ELECTRONICS
Third Edition

LOUIS GARY LAMIT
DeAnza College

SANDRA J. LLOYD
Grays Harbor College

Merrill
Prentice Hall

Prentice Hall
Upper Saddle River, New Jersey   *Columbus, Ohio*

Library of Congress Cataloging-in-Publication Data

Lamit, Louis Gary
    Drafting for electronics / Louis Gary Lamit, Sandra J.
Lloyd–3d ed.
      p.   cm.
    Includes index.
    ISBN 0-13-602137-9
    1. Electronic drafting.   I. Lloyd, S. J. II. Title.
TK7866.L36   1998
621.381′022′1—dc21   98-17353
                                  CIP

Cover photo: Courtesy of Parametric Technology Corp.
Editor: Stephen Helba
Production Editor: Rex Davidson
Editorial/Production Supervision: Gretchen K. Chenenko
Design Coordinator: Julia Zonneveld Van Hook
Cover Designer: Rod Harris
Production Manager: Laura Messerly
Marketing Manager: Frank Mortimer, Jr.

This book was set in Times Roman by Bi-Comp, Inc. and was
printed and bound by Courier/Kendallville, Inc. The cover
was printed by Courier/Kendallville, Inc.

Merrill Prentice Hall

©1998 by Prentice-Hall, Inc.

Upper Saddle River, New Jersey 07458

Earlier editions ©1993 by Macmillan Publishing Company and
©1985 by Merrill Publishing Company.

Printed in the United States of America

10 9 8 7 6 5 4

ISBN: 0-13-602137-9

Prentice-Hall International (UK) Limited, *London*
Prentice-Hall of Australia Pty. Limited, *Sydney*
Prentice-Hall Canada Inc., *Toronto*
Prentice-Hall Hispanoamericana, S.A., *Mexico*
Prentice-Hall of India Private Limited, *New Delhi*
Prentice-Hall of Japan, Inc., *Tokyo*
Pearson Education Asia Pte. Ltd., *Singapore*
Editora Prentice-Hall do Brasil, Ltda., *Rio de Janeiro*

TRADEMARK INFORMATION

Prentice Hall has compiled the following list of trademarks
for company names, products, and services mentioned in this
book from various sources and cannot attest to the accuracy
or completeness of the information.

Accufilm is a registered trademark of Bishop Graphics, Inc.
AMES Lettering Guide is a registered trademark of Olson
    Manufacturing
Apple II is a registered trademark of Apple Computer, Inc.
Art Gum is a registered trademark of Faber Castell Corp.
AutoCAD is a registered trademark of Autodesk, Inc.
Autoplacement is a trademark of Prime-Computervision
Autoroute is a trademark of Prime-Computervision
Braddock-Rowe is a registered trademark of Braddock Instru-
    ment Co.
CADAM is a registered trademark of Lockheed-California
    Co.
CADKEY is a registered trademark of Micro Control Systems
CALMA is a registered trademark of CALMA Corporation
Ground Plane Grid Strips is a registered trademark of Bishop
    Graphics, Inc.
IBM is a registered trademark of International Business Ma-
    chines Corporation
Koh-I-Noor Rapidometric Guide is a registered trademark of
    Koh-I-Noor, Inc.
Lin CMOS is a registered trademark of Texas Instruments
Macintosh II is a registered trademark of Apple Computer,
    Inc.
Modified Crosshair is a registered trademark of Bishop Graph-
    ics, Inc.
P-CAD is a registered trademark of Personal CAD Sys-
    tems, Inc.
Personal Designer is a registered trademark of Prime-Comput-
    ervision Corporation
Pink Pearl is a registered trademark of Eberhard Faber, Inc.
Precut Stick On is a registered trademark of Bishop Graph-
    ics, Inc.
Puppets is a trademark of Bishop Graphics, Inc.
Rubylith is a registered trademark of Ulano Corp.
The Grabber is a registered trademark of Triplett Corp.
Universal Target is a registered trademark of Bishop Graph-
    ics, Inc.
VersaCAD is a registered trademark of The VersaCAD Cor-
    poration
Versatec is a registered trademark of American Microsystems
X-ACTO is a registered trademark of Hunt Manufacturing
    Co.

To my parents, Norma and Frank Lloyd, for their continued guidance and support, and to my daughter, Teresa and grandsons, Luke and Lynden, for making my life complete.

S. J. L.

In loving memory of my parents, Frances and Louis Lamit.

L. G. L.

Preface

This book is intended to be a text first and a practical guide second. Conventional practices and current standards have been adhered to, making the text comprehensive and up-to-date. More than 250 photos and 600 drawings have also been included. Students in three categories will find this text especially appealing—those with drafting skills who need to apply drafting to electronics; those with electronics skills who need introduction to drafting principles; and those who need exposure to both electronics applications and drafting fundamentals.

The chapters in this edition of the text have been re-ordered to keep similar materials together. The first five chapters of the text cover drafting fundamentals, particularly how they relate to the field of electronics and electromechanical design and drafting. Chapter 1 introduces the specialized equipment used by the electronic designer and drafter, concentrating on manual tools and equipment. Chapter 2 covers lettering, including manual and machine lettering as well as CAD variations and techniques. Chapter 3 overviews the types of lines found on electronic diagrams and electromechanical drawings for packaging. Chapter 4 introduces variations of projection and dimensioning that are found throughout industry and, in particular, the various forms utilized by the electronics industry. Chapter 5 covers the use of pictorial representations and how they are an integral part of electronic documentation and literature. Chapter 6 has been rewritten to include not only an overview of CAD and its use in the electronics field, but also a fundamental introduction to AutoCAD. This is meant only as an introduction to the use of the software and is not meant to replace intensive study of the software and its capabilities and use.

After the first six chapters, the primary emphasis of the text is a detailed analysis of electronics and the types and variations of graphic documentation required to bring an electronic product to market. Chapters 7–10 overview electronics and its specialized language. Chapter 7 covers the basics of electronics for those students who have not had an introductory electronics class. Chapters 8 and 9 are important to the electronics drafter and designer, since they are the foundation for all drawings used in this field. In Chapter 8, electronic components and symbols are introduced. Chapter 9 discusses electronic standards, designations, and abbreviations used on electronics diagrams and drawings. Chapter 10 finishes this section of material with an introduction to microcircuits.

Chapters 11–13 cover the general types of diagrams found in electronics. Schematic (Chapter 11), block (Chapter 12), and wiring (Chapter 13) diagrams are used in almost every kind of electronics system design. These chapters are the most important drafting-oriented chapters in the text and provide the foundation for more specialized drawings introduced later in the text.

Chapters 14–17 introduce and overview the various specialized areas found in the electrical or industrial areas of the electronics industry. Motors and Control Circuits (Chapter 14) was expanded in the second edition at the request of a variety of schools. Chapters 15 and 16 introduce logic circuits—the symbols and theory—and then expand that theory to programmable controllers and robotics. The material in Chapter 16 was new to the second edition, as is Chapter 17, which introduces power distribution.

The final two chapters include very specialized areas of electronics. Chapter 18 covers printed circuit board design and development. Chapter 19, this text's only mechanical drafting chapter, covers the essentials of electromechanical design.

By selecting those chapters appropriate to the students' background, the instructor can use this text in a single-term course. For students having little or no background in either drafting or electronics, the authors recommend that this material be separated and offered in a two-term sequence. The first course should include the first eleven chapters. In the second course, Chapter 6 might be included for review, along with the remaining chapters of the book. The material could be arranged to cover a three-course sequence by including the first six chapters in the initial, or drafting fundamentals, term. Chapters 7–13 and 15 would be used in the second term, or electronics drafting course. A third course would include Chapters 6 and 14–19 and would be an applications, or production drawing course.

The extensive appendices include two glossaries, quick reference symbols, and many tables. Though the text includes numerous questions and problems, supplementing these with some real-world work projects would enhance the information presented in the text.

The third edition reviewers deserve recognition for their contributions: Judith Wooderson, San Juan College; David Dillon, North Carolina A & T State University; and Dave Larue, ITT Technical Institute.

L. G. Lamit
S. J. Lloyd

Contents

1

EQUIPMENT, INSTRUMENTS, AND MATERIALS

INTRODUCTION

Drafting tools used in electronic drafting and design are the same as those for all other fields of engineering and design work. If a difference exists, it is in the level and complexity of drafting technology and the variety of drafting and design techniques available, ranging from plastic taping and preprinted transfer materials to **computer-aided design (CAD)** systems. The traditional manual drafting techniques and tools are still used throughout the industry, however. Therefore, as an aspiring electronic technician, drafter, designer, or engineer, you need to have a firm grasp of procedures and be familiar with these tools. Every engineering office uses the equipment described in this chapter, with varying degrees of sophistication in the design process. Even where CAD systems have replaced manual methods, the simple and essential drafting tools can still be found—pencils, compasses, dividers, templates, drafting machines, drafting tables, and drawing storage cabinets. The simple lead holder and the complex electric light pen are of equal importance to the total drafting and design process.

This chapter is meant to be used as an overview of drafting equipment, instruments, and materials. *Equipment* includes drafting boards, T squares, triangles, templates, and computer-aided design hardware. *Instruments* are precision manufactured drawing tools, such as compasses and dividers in all their variations. Drawing mediums and related support items (drafting paper, grid underlays, preprinted title blocks, and transfer drafting aids) are drafting *materials*.

Drafters and designers must know how to use their equipment, instruments, and materials to be able to communicate effectively by lettering and linework. Both the user and the originator of a drawing must understand the procedures, conventions, and concepts used in the drawing. In electronic drafting and design, symbols, linework, projection procedures, and notation must be assembled according to standardized drafting conventions.

EQUIPMENT

The most important and conspicuous piece of equipment found in any drafting room is the **drafting table.** Originally, all drafting was done on simple flat-surfaced wooden drawing boards. Normally, one or more edges were cut as straight and square as possible, making a straight edge for the drafter to guide a T square. Today board sizes range from the hand-carried versions of 9 ×

FIGURE 1–1
Modern steel drafting table. (Courtesy Hamilton Industries)

FIGURE 1–2
Large-surface, high-quality metal frame drafting board. (Courtesy Hamilton Industries)

A grid pattern can greatly aid the engineer, designer, or drafter in preparing block, logic, schematic, or wiring diagrams. A grid system in electronic drafting and design is especially beneficial since most electronic drawings are two-dimensional diagrams composed almost exclusively of vertical and horizontal lines. Drawings and sketches to be digitized on a CAD system

FIGURE 1–3
Light table. (Courtesy Hamilton Industries)

12 in. to the large-format, stand-alone tables commonly found in industry and the classroom. Figure 1–1 shows a modern drafting table. This table is vertically adjustable and can be tilted to any comfortable angle. Figure 1–2 shows a large 3 ½ × 8 ft automatically adjustable drafting board. Whatever the size or material used for a drafting table, the table surface must have a pliable cover. This surface can be a plastic or vinyl covering that permits drafting without destroying the table surface or marring the drawing medium (paper, plastic film, cloth).

Light tables are also used throughout the electronics field to prepare printed circuit artwork, draw pictorial illustrations, and trace projects. A light table is shown in Fig. 1–3. Normally the drawing surface is an opaque glass or plastic sheet that scatters the rays from the light source. Figure 1–4 shows a modern drafting station and reference desk. Figure 1–5 shows a small portable table with a light for tracing and doing artwork. This table has a built-in grid pattern as part of its drawing surface. A grid pattern covering can also be applied to tables without lights.

FIGURE 1–4
Modern drafting stations, reference desk, and drawing storage. (Courtesy Hamilton Industries)

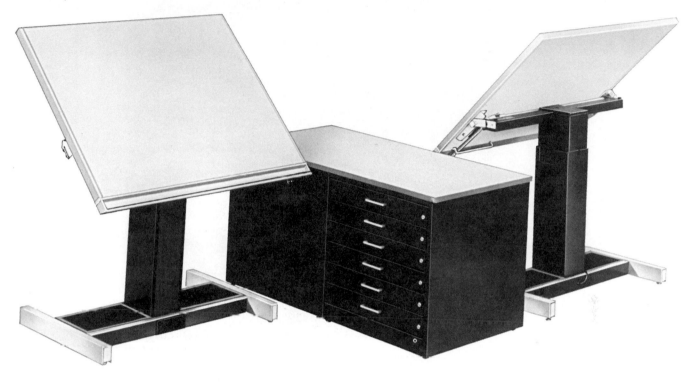

require the use of a grid system. Grid patterns are also essential for the proper alignment and layout of printed circuit boards.

CAD Equipment

CAD equipment is normally referred to as hardware. Hardware includes all types of computer and drafting equipment associated with a computer-aided design: computer screen, keyboard, electronic pen or mouse, digitizer tablet, mainframe or local processor, mass stor-

FIGURE 1–5
Grid-surfaced portable light table. (Courtesy Kroy, Inc.)

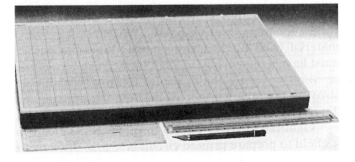

age components (such as a hard disk), and hard-copy device (printer or plotter) (Fig. 1–6).

The CRT, or cathode-ray tube, is the television type of monitor used to view a project as it is *drawn*. The software (program) is used to relate commands from the operator to the systems hardware. Commands can be entered on the keyboard. The **digitizer** tablet or table is used to convert existing graphic data into a digital form understood by the computer. A graphics/data tablet is shown in Fig. 1–6.

The pen **plotter** converts the computer's digital data into graphic data by plotting the drawing on paper or drafting film. In Fig. 1–7, the pen plotter illustrated is a drum-type plotter.

The two most important pieces of computer hardware in a CAD system are the processor and the mass storage system. In a PC-based drafting system, the processor is built into the workstation. In more sophisticated systems, the processor is a stand-alone centralized computer unit capable of handling multiple stations as well as performing non-CAD duties, such as business and word processing applications. In CAD engineering workstations (Fig. 1–6), the processor is separate from the keyboard and monitor. The storage medium of a CAD system today is

FIGURE 1–6
Electronic DesignCAD system.
(Courtesy Prime-
Computervision)

generally a hard disk, often with one or more gigabytes of space.

Storage and Reproduction Equipment

After a project is drawn, regardless of the method, it must be stored and reproduced. CAD electronic draw-

FIGURE 1–7
High-speed 14-pen plotter. (Courtesy Houston Instruments)

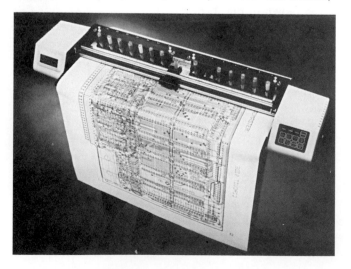

ings are plotted on pen or photoplotters. Reprographics, computer-stored design data for quick reproduction, often replaces cabinet storage of original drawings.

Storage

Frequently, drawings are stored as paper prints and originals. Multiple-drawer cabinets, like the one shown in Fig. 1–8, as well as tube storage systems, all require expensive, cumbersome equipment that takes up valuable office space. The graphic data must be cataloged and physically available to several departments. This method of storage is time consuming and requires a good deal of office space.

Reprographic systems can eliminate the need for bulky storage of original drawings. For example, drawings can be stored on **microfilm.** Computer graphics systems enable you to reproduce design data stored on disk or tape. Another form, reprographics, uses 35-mm **micrographic aperture cards** (design data cards).

The design data system can be used with traditional manual drafting techniques or computerized procedures. Design data card systems such as the one developed by the 3M Corp. allow access to 1,300 design drawings in less than 7 ½ in. of space in a desk drawer. When a new or revised drawing is checked and ready for release, it is taken to a processor camera, or film

FIGURE 1–8
Hamilton Unit System File drawing storage cabinet.
(Courtesy Hamilton Industries)

plotter (Fig. 1–9). In seconds, a master data card—an accurately reduced version of the original—is produced. This card is correct in every detail but is smaller and easier to use and reproduce than an original drawing. Multiple copies of the data card are then made from the original for distribution to the users. The users can review the drawing with a display device at their workstations. Many of the industry-provided drawings in the text were reproduced from design data cards.

Reproduction

Traditionally, the *blueprint* machine was used to make multiple prints of drawings. The term blueprint is not quite accurate today since *blueprints* are actually *whiteprints*, or what may be called *blue-line* prints. A whiteprint machine, shown in Fig. 1–10, is a more accurate description because most, if not all, reproduction with this method involves developing a print with blue lines and white background, not the opposite.

For a drawing completed on a CAD system, the user must be able either to reproduce the drawing from a hard-copy device like a photocopier or to *plot* the drawing with a pen plotter, shown in Fig. 1–11 and Fig. 1–7, or a photoplotter, shown in Fig. 1–12. The pen plotting method allows the production of an accurate original every time the drawing is replotted. Multiple copies can then be made from a whiteprinter or from input to a data card system. The beauty of the CAD system lies in the ease of reducing or enlarging the originals and reproducing originals on a wide variety of drafting paper.

Aperture data cards enable the user to make fast multiple photoprints with several reduction and enlargement printout options. The manually operated, low-volume photoprinter shown in Fig. 1–13 allows the operator to print a drawing on various kinds of paper and instantly switch enlargement scales. The print paper is manually fed into the front of the printer, and the viewing screen allows easy monitoring.

T Squares, Parallel Bars, and Drafting Machines

Three methods of drafting—CAD, drafting machine, and parallel bar—are used in industry.

Originally, the primary straightedge device used in drafting was the T square, shown in Fig. 1–14. This piece of equipment is still used in some drafting classes and for personal drafting. It is said that "if you can draw with a T square you can draw with anything" because the T square is the most difficult of all straightedge drawing devices. If you must learn on a T square, then be comforted that once you master it other straightedges

FIGURE 1–9
Film plotter. Converts CAD-generated engineering drawings directly to 35-mm aperture cards. (Courtesy 3M Corp.)

FIGURE 1–10
Nonammonia whiteprinter.
(Courtesy Bruning)

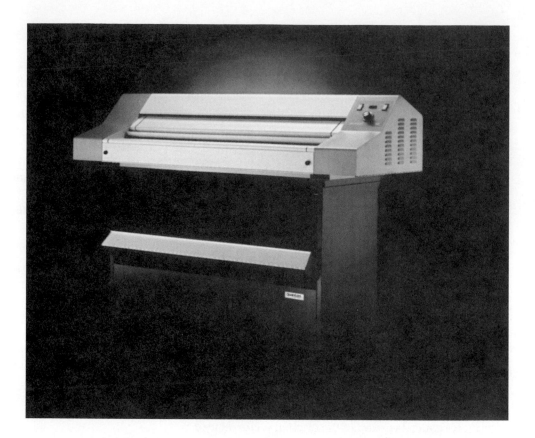

FIGURE 1–11
Check plotter—low cost, large area, on-line check-plotting
capabilities. (Courtesy Gerber Scientific Instrument Co.)

FIGURE 1–12
GSI photoplotter. (Courtesy Gerber Scientific Instrument
Co.)

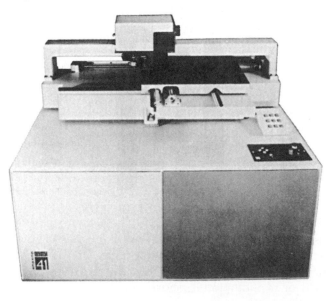

FIGURE 1–13
3M enlarger–printer. Makes prints from aperture cards on vellum, paper, or printing plates. (Courtesy 3M Corp.)

will be easier. The T square itself is not so difficult to use, but it is the easiest to misalign of all straightedge devices. The bar portion of the T is placed along the edge of a drafting board or table. Parallel horizontal lines are drawn with the length of the T square, and parallel vertical lines are drawn with a triangle placed on the T square. Obviously, if the T square and the table edge are not aligned properly, the linework will be inconsistent.

FIGURE 1–14
Wood and plastic-edged T squares. (Courtesy Pickett)

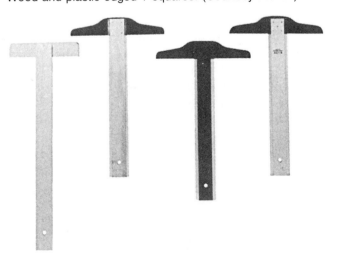

The parallel bar, or straightedge, is found throughout industry, since it is an excellent tool for drawing long horizontal lines. The parallel straightedge is attached to the drafting table by a series of cables and pulleys. It remains parallel or at a preset angle to the drafting table as it is slid up or down the table surface. Parallel straightedges are excellent for electronic drawings, where long, straight parallel horizontal and vertical lines make up most of a drawing.

The **drafting machine** comes in two standard versions, the drafting arm type, shown in Fig. 1–15, and the track type, shown in Fig. 1–16. The track type is the most accurate and costly. Drafting machines are mounted to the drafting tables, as shown in these figures. Drafting machines take the place of triangles, protractors, and scales. The control head can be rotated to any angle and set by pushing a button or locking the head. Most drafting heads automatically lock in 15° increments and must be hand-locked for intermediate angles. When you use the drafting machine—or, for that matter, any drafting straightedges—avoid dragging the equipment across the drawing. Always lift the equipment above the paper to avoid smearing the drawing.

Proper lighting is essential for relaxed, unstrained work with manual drafting techniques. Figure 1–15 shows an arm-type drafting machine with a lamp attached to the board. Lighting requirements are consid-

FIGURE 1–15
Arm-type drafting machine. (Courtesy Picket Co.)

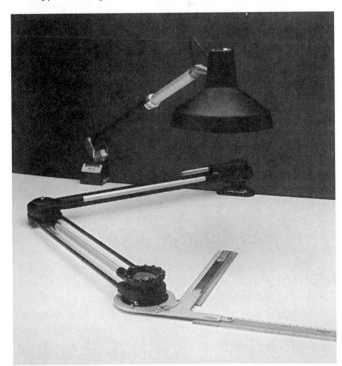

FIGURE 1–16
VariTilt drafting table with track-type drafting machine. (Courtesy Keuffel & Esser/Kratos)

FIGURE 1–17
Drafting tools. (Courtesy Teledyne Post)

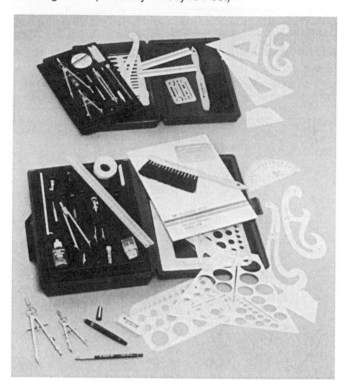

erably different when a CAD system is used because the CRT screen is easier to read if it is shaded from external light sources.

General Equipment

Traditional manual drafting requires a variety of small tools and equipment, which are shown in Fig. 1–17. Special templates, inking pens, stick-on symbols, and so on, are used throughout the electronics industry. This section provides an overview of the wide range of equipment on the market. The quality of your drafting will be directly influenced by the range and quality of the equipment you use. This is not to say that expensive, high-quality tools *draw* the project, but good-quality tools are beneficial for fast, efficient, and precise linework and projection.

The skills required to use drafting equipment come only through constant practice and repetition. Drafting is a skill that must be cultivated throughout your career.

Drafting kits are available through a variety of reputable companies. These kits are sufficient for most classes in drafting. In general, however, precision high-quality tools and instruments should be purchased indi-

vidually, either at a drafting supply store or through a drafting equipment catalog. Table 1–1 lists standard drafting tools you can buy. The purchase of all items listed would be quite expensive. Therefore, essential items are distinguished from optional items, which can be added as needed.

Pencils

The choice of pencil, mechanical lead holder, or thin-line mechanical pencil depends on the preference of the user. All forms of drawing pencils can be used to do excellent linework and lettering. In many cases the choice of pencil is determined by cost. Traditional wooden pencils are the cheapest but are also the hardest to keep sharpened with a consistent conical point.

The **lead holder,** shown in Fig. 1–18, has replaced the wooden pencil. It holds a long single stick of lead.

FIGURE 1–18
Drafting lead holders. (Courtesy Koh-I-Noor Rapidograph, Inc.)

TABLE 1–1
Manual Drafting Tools

Essential Items	Optional Items
Pencils (Grades 3H, 2H, H, F)	Lead holder
Sandpaper block	Thin-line mechanical pencil
Pencil sharpener	Lead (3H, 2H, H, F)
Eraser	Electric eraser
Erasing shield	Adjustable triangle
Drafting tape	Lettering guide
Dusting pad or powder	Lettering template
Drafting brush	Ames or Braddock–Rowe lettering guides
Scales (metric, mechanical, architectural)	Drop compass
Protractor	Beam compass
30/60° Triangle	Compass inking attachment
45° Triangle	Proportional dividers
French curve	Inking pens (3 × 0, 00, 1, 2, 2½, 3)
Templates (circle, ellipse, logic symbols, block diagram, schematic diagram, wiring diagram, PC component)	Ink
	Ink eraser
	Leroy set
	Transfer materials (lettering, symbols, dolls, slit tape)
Bow compass	Grid paper (nonreproducible lines)
Dividers	Drafting table
Paper (vellum, plastic film)	
Drafting board	

FIGURE 1–19
Ultra-thin mechanical pencil. (Courtesy Berol RapiDesign)

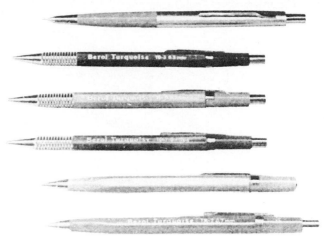

Having more than one lead holder ready with various leads ensures that a variety of lead grades is quickly available. This drawing tool is easily sharpened and increases the speed, consistency, and ease of drafting.

The **thin-line mechanical pencil** is an excellent tool for drawing consistent-width lines and letters. It never requires sharpening. As shown in Fig. 1–19, thin-line mechanical pencils come in metric sizes, including .3, .5, .7, and .9 mm. These sizes are used to draw the typical lineweights: .35 to .45 mm (centerlines, dimension lines, construction lines), .5 to .7 mm (object lines, diagram flow lines, hidden lines), and .7 to .9 mm (cutting plane lines, borders, emphasized diagram flow lines). The trouble with thin-line pencils is the simple fact that each pencil holds only one thickness of lead. The purchase of multiple thin-line pencils (or mechanical lead holders) may be prohibitively expensive. Of course, when using a smaller pencil, you can thicken the line to any width desired. The thinner the lead, however, the more frequently it breaks. Another drawback to thin-line mechanical pencils is that they are difficult to work with when using a lettering guide or a template.

To gain experience with the three types of drawing instruments, purchase two types of wooden pencils (H and F), one lead holder with three or four lead grades (F, H, 2H), and a .5- or .7-mm thin-line mechanical pencil with a range of lead grades. The thin-line pencil should be used for lettering and the lead holder, for most darkened linework. The wooden pencils can be used for lettering and all types of lines.

Regardless of the type of drawing pencil that you choose, purchase a variety of leads. Standard graphite leads used on vellum and film drafting mediums range from soft and dark (6B, 5B, 4B, 2B, B, HB, and F) to the medium hard and dark (H, 2H, 3H). The hardest and lightest types are 4H, 5H, 6H, 7H, 8H, and 9H. In general, only the medium leads are used: 2H and 3H for construction lines and blocking in a drawing, H for darkened finished lines. Some drafters prefer HB for finished lines, but great care must be taken to ensure that the drawing is kept clean and unsmudged. To be adequately reproduced, all lines must *block light* if a whiteprinter is used, or they must be dark and thick enough to be recorded by a camera if a photocopier or micrographics aperture card machine is used.

Plastic leads are used on plastic film drawing mediums, as shown in Fig. 1–20. Film leads come in a variety of grades (E, K, CF, and so on).

Lead holders and wooden pencils require frequent sharpening. Figure 1–21 shows an electric sharpener. A sharp conical point is needed to make thin, erasable lines required for the construction stage of the project. For darkened, finished linework, slightly dull the pencil point on scrap paper to avoid frequent breaking. To maintain the line thickness, rotate the pencil or lead holder as the line is drawn. A sharpened but slightly dulled lead point is also required for lettering. The

FIGURE 1–20
Wooden pencils and plastic leads (grade E1) for drafting film. (Courtesy Berol RapiDesign)

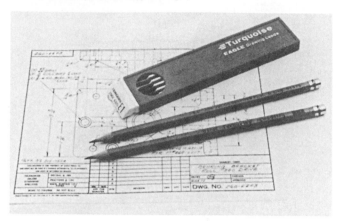

advantage of using a thin-line mechanical pencil lies in never needing to sharpen the lead, since it is a consistent thickness.

One sharpening device that is necessary for wooden pencils and lead holders is the **sandpaper pad** shown in

FIGURE 1–21
Lead holder and pencil sharpener. (Courtesy Berol RapiDesign)

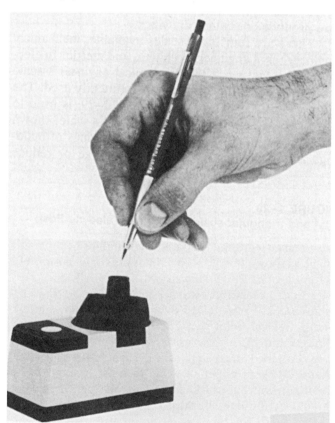

FIGURE 1–22
Sandpaper pencil pointer. (Courtesy Hearlihy & Co.)

Fig. 1–22. A sandpaper pad is used to sharpen lead points on compasses and pencils. Compass leads are sharpened as wedge shapes instead of conical points in order to slow down the dulling process. After sanding the lead, wipe it clean with a soft cloth or tissue.

Erasing Tools

Erasers come in many shapes and sizes, from hand held, as shown in Fig. 1–23, to electric, shown in Fig. 1–24. Regardless of the expertise of the drafter, erasing is unavoidable. Manual and electric erasers come in many grades. Pink Pearl® and Art Gum® erasers are used for both paper (vellum) and drafting film. Special vinyl erasers (Fig. 1–23) are available for erasing inked drawings on drafting film. Note that ink is extremely hard to erase when used on paper. An electric eraser is essential in this situation, but great care is required to avoid rubbing holes in the paper. An electric eraser should not be used on drafting film, since it tends to destroy the *tooth*, or surface, of the drafting film. For ink drawings completed on film, a small amount of moisture applied to a vinyl eraser is the best technique.

To save the linework around the erasing area, an **erasing shield,** shown in Fig. 1–25, is used. Erasing shields are thin metal plates perforated with different

FIGURE 1–23
Vinyl eraser for drafting film. (Courtesy Keuffel & Esser/ Kratos)

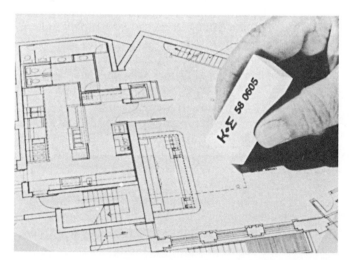

FIGURE 1–24
Electric erasing system. (Courtesy Koh-I-Noor Rapidograph, Inc.)

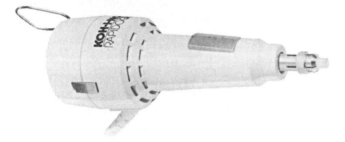

FIGURE 1–26
Professional drafting brush and student drafting brush. (Courtesy Hearlihy & Co.)

holes and slots that can be positioned over only the portion of the line to be removed.

Two other items are essential for clean, unsmudged drawings: a **dry cleaning pad** (an erasing dust pad) and a **drafting brush.** Dry cleaning pads contain ground eraser pieces and powder. They are used to remove dirt and leftover crumbled graphite deposited from drawing and lettering. When you use a dry cleaning pad, don't drag it across the drawing; instead, lightly pat the linework and lettering after a small portion of the drawing is complete. Then use a drafting brush, shown in Fig. 1–26, to sweep the drawing clean of powder and dirt. Frequent patting and dusting ensure a higher-quality drawing, although you must be careful not to lighten the lines and lettering by too much dusting.

Scales

Instrument drawings show each aspect of an object or graphical form in proper proportion. The size of the

FIGURE 1–25
Metal erasing shield, drafting pencil, and lead holder. (Courtesy Hearlihy & Co.)

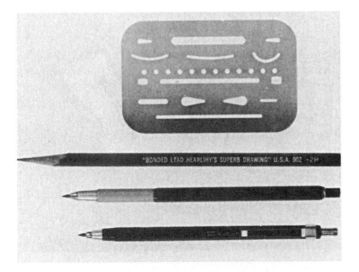

drawing may be the full size of the shape or it may be reduced or enlarged. For electronic diagrams, the choice of drawing size is normally determined by the complexity of the diagram and the sheet size, since electronic drawings are normally without a particular *scale.* For sheet metal drawings (panels, enclosures, cabinets, chassis) on electronic equipment, the choice of *scale* depends on the size of the object, as in traditional mechanical drafting. A drawing with accurate proportions is said to be drawn to *scale.* **Scales** are measuring instruments that are made to accurately represent specific units of measurement. The scale is a precision instrument and, with proper use, will help you produce consistent drawings.

Four basic scales are available: mechanical engineer, civil engineer, architect, and metric. In electromechanical work the mechanical engineer's scale and the metric scale are the most frequently used. The civil engineer's scale, shown in Fig. 1–27, is used to draw very large projects. The architect's scale, shown in Fig. 1–28 (top), is used to make drawings of buildings and structures, as well as to enlarge or reduce drawings in electromechanical work. A metric scale, shown in Fig. 1–28 (bottom), is used for all types of projects and is easily adaptable to every form of technical and engineering work.

The civil engineer, architect, and metric scales are triangular and are about 12 in. long. The triangular shape makes six surfaces available for the different-sized scales.

The markings on the scales are arranged in two ways, fully divided and open divided. A fully divided scale has each main unit completely divided into specific units, as in the metric, civil engineer's, and mechanical engineer's scales. An open-divided scale has each main unit of the scale undivided, and an extra main unit is fully divided at the 0 end of the scale. The architect's scale is open divided.

FIGURE 1–27
Flat and triangular scales. (Courtesy Teledyne Post)

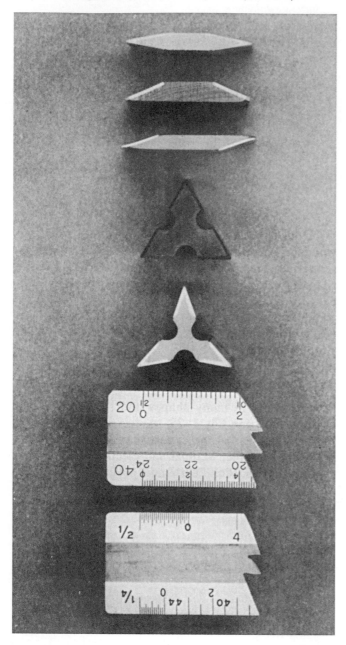

The Mechanical Engineer's Scale.   The mechanical engineer's scale, which is two sided, is flat. One side is the full-inch scale, which is divided into either decimal units of .10 in. or as many as 50 divisions (every .2 in.). The opposite side is at half-scale (1:2).

The Civil Engineer's Scale.   The civil engineer's scale is either the two-sided type [Fig. 1–27 (top)] or the three-sided type, which has six scales that are equally divided. These scales have 10, 20, 30, 40, 50, and 60 divisions per inch and are numbered at each tenth division along the length of each scale. The number of divisions per inch is marked at the 0 end of each scale. The user can assign any units needed to the divisions of these scales, although the usual usage is that each division equals 1 ft.

The Architect's Scale.   The architect's scale has a *foot ruler* full-sized scale (Fig. 1–28, top) and ten reduced, open-divided scales. The open-divided scales are used to permit the reduced scales to be placed on the remaining five sides. All scales on an architect's scale are divided in feet and inches. The open-divided scale has only full, 1-ft units reading in one direction from the 0 as well as a fully divided 1-ft unit reading in the opposite direction. Therefore, the number of feet is read along the length of the scale and the number of inches is read in the fully divided unit at the 0 end of that same scale, both numbers becoming larger as the distance from the 0 becomes greater. Each scale is identified by the number or fraction at the 0 end, which indicates the unit of length in inches that represents 1 ft of real size.

The Metric Scale.   Metric units, also called SI units (Système Internationale d'Unités), are measured with a metric scale. The full-sized metric scale (Fig. 1–28, bottom) is divided into major units of centimeters and smaller units of millimeters. There are 10 millimeters in each centimeter. (See Appendix A for U.S. customary and metric equivalents.) Metric units in full scale as well as in reductions and enlargements are used in all forms of electronic work.

This text has been designed to be used with all types of scales. Many problems in the text are undimensioned, allowing your instructor to choose the unit of measurement. Use of different scales is suggested. Table 1–2 compares the four basic scales.

Protractor

The protractor, shown in Fig. 1–29, is a tool that measures angles instead of lines. Note that the drafting machine can replace not only the straightedge, triangle, and scale, but also the protractor. A circular protractor is still an excellent investment, whether you have access to a drafting machine or not. A circular 360° protractor is the easiest to use. A 180° protractor is shown in Fig. 1–29.

Triangles and French Curves

Three basic types of triangles are normally found in drafting work, 30/60, 45, and adjustable. The 30/60 triangle is made up of 30°, 60°, and 90° angles, and the 45 triangle is composed of 45°, 45°, and 90° angles (Fig.

FIGURE 1–28
Architect's scale and metric scale. (Courtesy Teledyne Post)

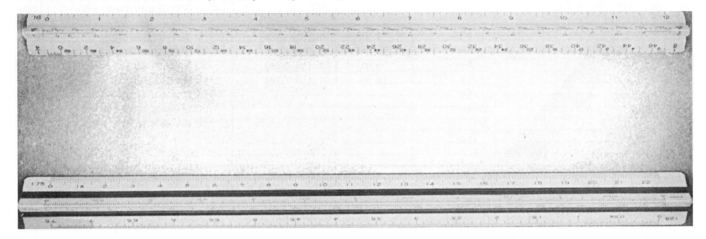

FIGURE 1–29
Protractor and French curves. (Courtesy Hearlihy & Co.)

1–17). The **adjustable triangle,** shown in Fig. 1–30, is very useful although more expensive. Adjustable triangles can be set at any angle between 0° and 90°. The higher cost of an adjustable triangle is partly overcome by the fact that it can take the place of both the 45 and 30/60 triangles. Triangles are used to draw all straight lines that are not horizontal when a T square or parallel bar is used.

For curved lines that are not circles or circular arcs, a French curve (Fig. 1–29) is required. Triangles and French curves are made of a hard, flexible, clear plastic.

Templates

A **template** is a tool for drafting shapes of all sizes. Standard templates, shown in Fig. 1–31, are essential for the quick, easy construction of circular, square, rectangular, triangular, and elliptical shapes. Electronic templates, shown in Fig. 1–32, are one of the most important pieces of equipment in electronic and electrical

TABLE 1–2
Basic Drafting Scales

Architect's Scale	Mechanical Engineer Flat	Civil Engineer Triangle	Metric Triangle
3/32	1 in. = 1 in. (full size)	10 divisions/unit	1:10
1/8	1/2 in. = 1 in. (1/2 size)	20 divisions/unit	1:20
3/16	1/4 in. = 1 in. (1/4 size)	30 divisions/unit	1:25
1/4	1/8 in. = 1 in. (1/8 size)	40 divisions/unit	1:33.3
1/2		50 divisions/unit	1:50
3/4		60 divisions/unit	1:75
1		80 divisions/unit	1:80
1 1/2			1:100
3			1:150

FIGURE 1–30
Rapid rule, architect's scale, and adjustable triangle.
(Courtesy Hearlihy & Co.)

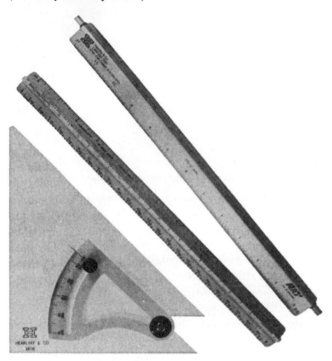

drafting and design. Seldom are any of the vast array
of electronic symbols ever hand-constructed except in
the sketching stage. After you master the essentials of
linework and compasswork, use templates for all standard shape construction. Templates are presented
throughout this text when electronic and drawing applications, pictorials, electronic diagram construction, and
printed circuit board layout are discussed.

INSTRUMENTS

Instruments include all forms of **compasses** and **dividers.**
The drafting instrument set, shown in Fig. 1–33, is commonly composed of one or two sizes of compasses, a
divider, and accessories. The compass is used to draw
circles and circular arcs. Although a wide variety of
drafting sets is available, they normally contain such
useless and obsolete items as the ruling pen (Fig. 1–33).
This item has been totally replaced by the technical
pen. As a beginning drafter you need only purchase a
medium-sized, high-quality **bow compass** and a divider.

Three expensive but extremely useful tools are
available for special situations: the **drop compass,** shown
in Fig. 1–34, for very small accurate circles; the **beam
compass,** Fig. 1–35, for very large circles and arcs; and
proportional dividers, Fig. 1–36, for reductions and enlargements. Most electronic drafters and designers need
only a good set of dividers and a medium-sized bow
compass.

Most compasses can have inking pen or **technical
pen** attachments, as shown in Fig. 1–37. When purchasing a compass set, make sure that it can be equipped
with an attachment for holding a technical pen, not just
a ruling pen.

Although many drafting sets contain ruling pens,
our discussion is limited to technical pens, since the
ruling pens are not used in industry. Technical pens like
the one in Fig. 1–38, although expensive, have totally
replaced all other forms of inking tools. Technical pens
come in a wide range of pen widths (diameters), as
shown in Fig. 1–39. Each pen size corresponds to a
metric thickness. Electronic drafters and illustrators are
frequently called on to complete a project in ink, including diagrams and pictorials for technical manuals, sales

FIGURE 1–31
Templates. (Courtesy Koh-I-Noor Rapidograph, Inc.)

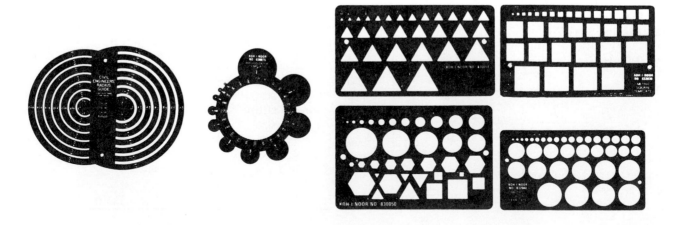

FIGURE 1–32
Electronic and electrical templates. (Courtesy Koh-I-Noor Rapidograph, Inc.)

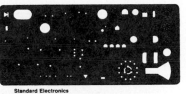
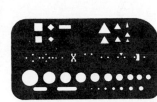
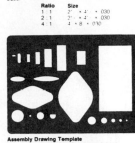

Electronics Jr. Template
An updated template with symbols of universal application. Contours lend themselves to combination usage, thus allowing you to create the majority of symbols in the electronics field. Size 4˝ × 8˝ × .030

Standard Electronics
Template contains the most current electronic symbols for use by designers and engineers. Size 5˝ × 11˝ × .030

Wiring
Electrical and electronic symbols arranged and grouped for drawing ease. Size 5˝ × 9˝ × .030

Dual In-Line Pattern Template
A must template for laying out integrated circuit patterns utilizing either the 14 or 16 lead TO-116 package. Contains the three most popular pad sizes.

Ratio	Size
1:1	2˝ × 4˝ × .030
2:1	2˝ × 4˝ × .030
4:1	4 × 8˝ × .030

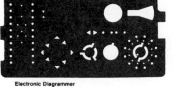
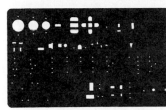

Electronic Diagrammer
Template contains the latest revisions for the professional diagrammer. Size 6˝ × 10˝ × .030

Public Utilities
Template can be combined in various ways to represent complete electrical devices. Contains component parts and elements, such as coils, connections, transformers, instruments, etc. Size 4˝ × 8˝ × .030

Electric Components
Standard symbols used to depict electrical circuitry for machine operation and production processes. Size 6˝ × 10˝ × .030

Assembly Drawing Template
Designed to aid in the detailing of printed circuit assembly drawings. Contains the most commonly used components in the electronics industry.

Ratio	Size
1:1	3 × 5 × .030
2:1	6 × 8 × .030
4:1	8 × 12 × .030

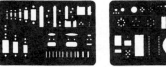

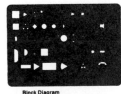

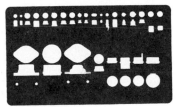

Design Component Master
A comprehensive template set designed for printed circuit layouts. Available in three ratios 1:1, 2:1 and 4:1 in sets of two and four templates. Templates contain body outlines and layout patterns, composition resistors, axial and radial lead capacitors, potentiometers and connector strips. Also included are lead patterns on both .200˝ and .400˝ diameter lead circles. TO-3 and TO-66 outlines.

Electronic Components
A general electronic symbol template. Size 4 × 6˝ × .030

Standard Logic Symbols
For digital processing. Contains pinhole location points. Conforms to ANSI Y32.14

Size	Scale
3⅜˝ × 7½˝ × .060˝	Full
4˝ × 6˝ × .030	3/4
3˝ × 4½˝ × .030˝	1/2
3˝ × 5˝ × .030	3/8
2¼˝ × 4˝ × .030	1/4

Block Diagram
Combination of symbols used in block diagrams. Symbols are in IEC117 and BS3939 and 108. Size 5˝ × 7˝ × .030

Computer Diagrammer
Flow chart symbols in accordance with ANSI X3.5. Contains card volume scale, plus scales in 6ths and 10ths of an inch. Scale: Three-quarter size. Size: 4˝ × 8˝ × .030

Transistor Outlines
A basic tool for designers and engineers. Template conforms to JEDEC, contains both sizes and designating numbers. Size 6 × 10 × .030

brochures, and graph and chart presentations. In general, technical pens in .25-, .35-, .45-, .50-, and .70-mm sizes are required for such drafting. (See Chapter 3 for ANSI line-width standards.)

MATERIALS

Drafting materials include drawing papers (vellum, film, grid, preprinted) and preprinted transfer items (title blocks, lettering, electronic symbols, and printed circuit artwork aids).

An extensive variety of electronic drafting aids is available; in fact, electronic drafting has been at the forefront of all technological advances concerning pin graphics, appliqués, transfer symbols, and other time-saving items.

This text does not discuss title block construction or borders, since industry typically uses preprinted vellum or film standard drafting paper sizes with company or standard title blocks and borders. Your instructor can assign title block format with specific problems.

The completion of the title block is one of the most important parts of a project. The title block normally includes spaces for the following information:

FIGURE 1–33
Drawing instruments: ruling pen, compass, and dividers.
(Courtesy VEMCO Corp.)

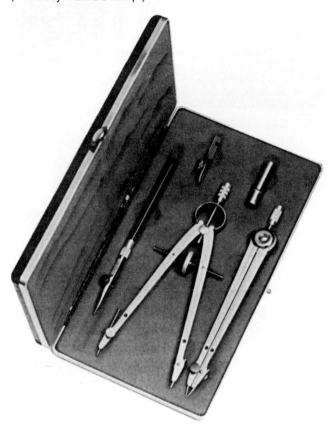

FIGURE 1–34
Drop bow pen and lead compass. (Courtesy Keuffel &
Esser/Kratos)

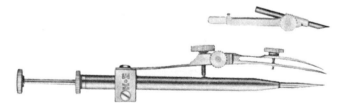

FIGURE 1–35
Beam compass. (Courtesy Keuffel & Esser/Kratos)

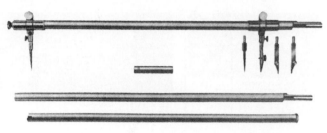

FIGURE 1–36
Proportional dividers. (Courtesy Keuffel & Esser/Kratos)

FIGURE 1–37
Bow compass for technical pen. (Courtesy Koh-I-Noor
Rapidograph, Inc.)

FIGURE 1–38
Technical pen. (Courtesy Koh-I-Noor Rapidograph, Inc.)

FIGURE 1–39
Pen sizes and tip widths.

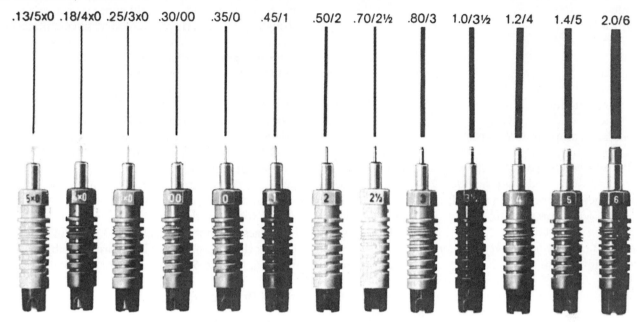

FIGURE 1–40
ANSI standard title block.

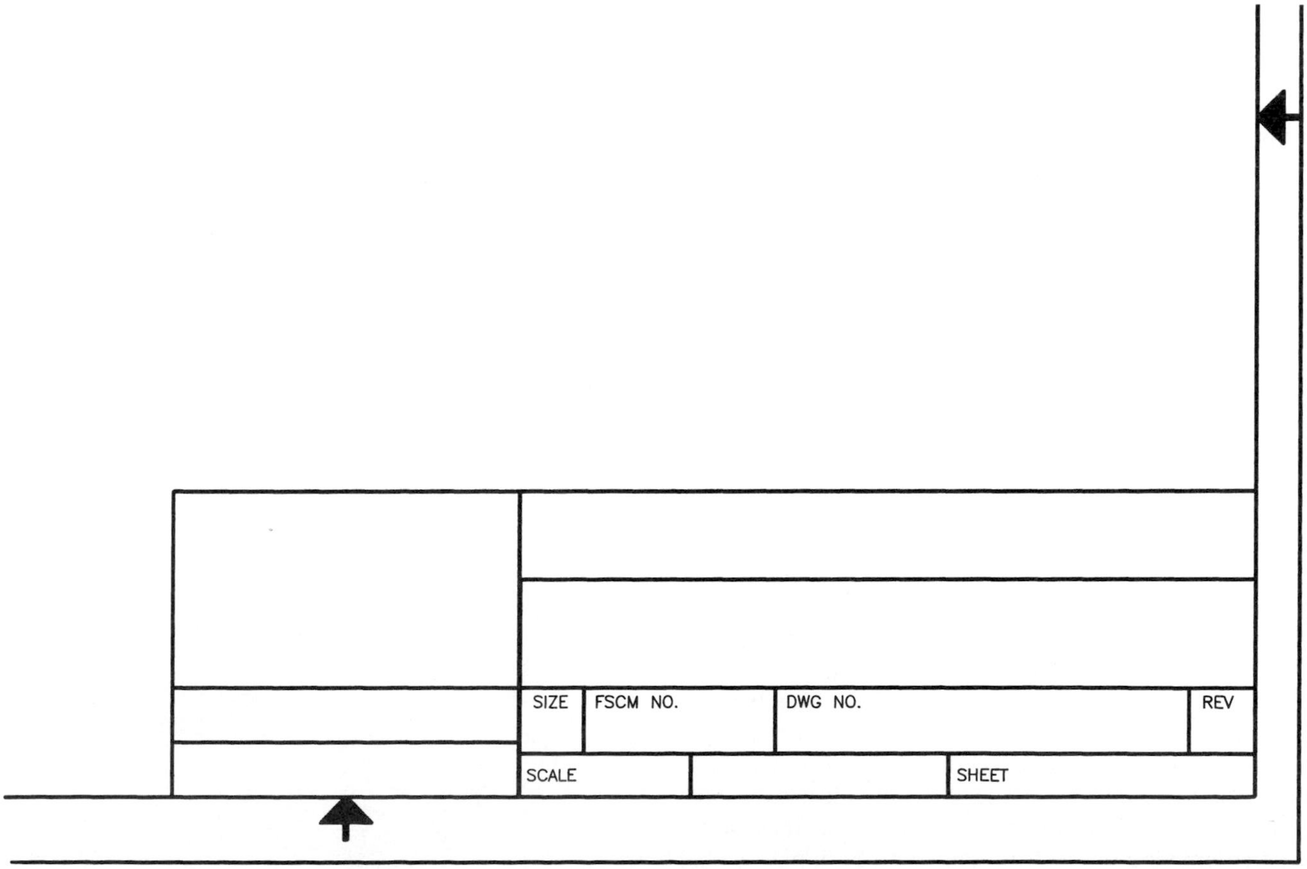

	SIZE	FSCM NO.	DWG NO.		REV
	SCALE			SHEET	

- ☐ Company or school name
- ☐ Project title or part name
- ☐ Scale
- ☐ Drawn by
- ☐ Material specification
- ☐ Date
- ☐ Checked by
- ☐ Sheet number
- ☐ Drawing number
- ☐ Revision
- ☐ Standard company tolerances

Preprinted company title blocks need only be filled in by the drafter or designer. Title blocks will need to be constructed according to school, company, or ANSI standards when preprinted blocks are not available. Since many instructors have their own title block formats, this text leaves format selection to the instructor. Border widths also need to be specified. For general drawing practice, ½ in. (12.7 mm) can be used for size C sheets and larger. For small format paper, A and B sizes, ¼ in. (6.35 mm) can be used. Check with your instructor before using these border widths.

Title blocks and paper or film are covered by the ANSI standard *Drawing Sheet Size and Format,* ANSI Y14.1–1980. Your instructor can assign paper types, sizes, and format as necessary. An ANSI standard title block is shown in Fig. 1–40.

Drafting Mediums

Traditional drafting mediums that are transparent enough to be whiteprinted include vellum, drafting cloth, and drafting film. Vellum and plastic film are the most widely used throughout industry.

All drafting mediums are secured to the drafting table with **drafting tape.** Drafting tape is a high quality version of masking tape, but is designed not to pull the finish off the paper surface. Note that many drafters have tried staples and tacks without success, since they tear the drawing medium and destroy the drafting table surface.

Drawing paper, sometimes referred to as **vellum,** is a high-quality translucent drawing paper. Vellum is an excellent drawing medium, since it is translucent enough to be reproduced in a whiteprinter. Vellum is easily erased when pencil is used, and it takes ink well.

Drafting film is made of durable, high-quality polyester sheets. This drafting medium is excellent for ink and plastic lead alike. Special drafting-film-grade leads and erasers are also available. Although it is expensive compared to vellum, every student should have some practice drawing on drafting film, with both plastic lead and ink.

Both drawing paper (vellum) and drafting film are available plain and with a fine nonreproducible or *fade-out* grid pattern. Fade-out **grid paper** is used extensively throughout the electronics industry, since a majority of electronic drawing involves two-dimensional diagrams composed primarily of horizontal and vertical lines. United States customary and metric unit grid rulings are available. Since the introduction of drafting film, the use of drafting cloth has diminished. Cloth is a closely woven muslin and is used primarily for special projects where durability and a long life are essential.

Other types of paper used frequently by the drafter, designer, or engineer include sketch pads, cross-section paper, and special graph paper, such as log–log, semilog, linear, and polar, as shown in Fig. 1–41.

All forms of drafting mediums come in standard sheet sizes, as shown in Table 1–3, and rolls.

Roll drafting paper and film are available in widths of 30, 36, 42, and 54 in. in 25-ft lengths and up.

Printed Drafting Transfer Aids

The electronics industry makes extensive use of rub-on and transfer preprinted letters, numbers, symbols, lines,

FIGURE 1–41
Graph paper. (Courtesy Keuffel & Esser/Kratos)

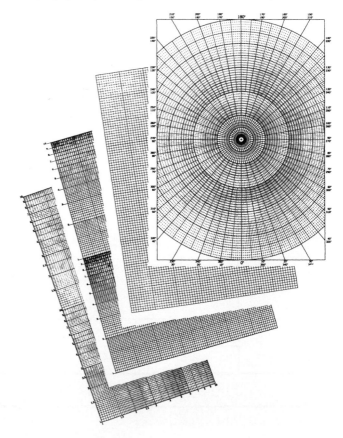

TABLE 1–3
Standard Paper Sizes

American National Standard ANSI Y14.1–1980, in.	International Standard, mm
A–8 ½ × 11	A 4–210 × 249
B–11 × 17	A 3–297 × 420
C–17 × 22	A 2–420 × 594
D–22 × 34	A 1–594 × 841
E–34 × 44	A 0–841 × 1189
F–28 × 40	

FIGURE 1–42
Precut StickOn® transfer symbols. (Provided courtesy of Bishop Graphics, Inc.)

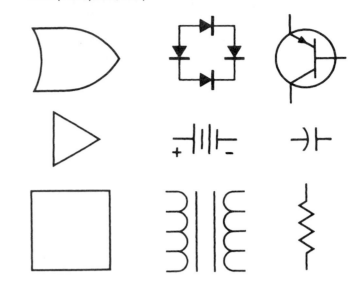

and standard shapes. Printed drafting aids are available in many shapes and sizes, as shown in Fig. 1–42. Electronic and electrical symbols, conductor strips, component shapes, and slit tape have replaced hand-drawn lines and symbols in many instances, especially for printed circuit board artwork, as shown in Fig. 1–43.

FIGURE 1–43
Tapes for printed circuit board artwork, where consistent precision line widths are required. (Courtesy Graphic Products)

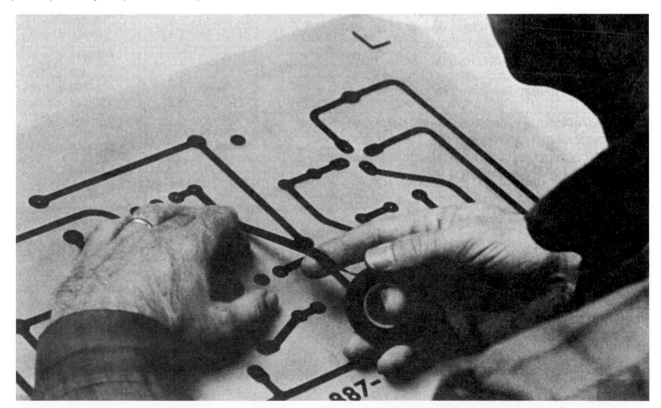

REVIEW QUESTIONS

Many of the technical terms discussed in this and other chapters may be unfamiliar to you. Consult the glossaries at the back of the book for definitions of terms when necessary.

1–1 Which type of drafting board is used for tracing and artwork?
    a. digitizer   c. wooden table
    b. light table   d. metal drafting table

1–2 Dry-transfer drafting aids are used to ( ).
    a. replace drafting   c. increase quality of art
    b. increase speed   d. save lead

1–3 CAD hard-copy devices provide instant copies of ( ).
    a. permanent design drawings   c. electronic artwork
    b. drafting layouts   d. CRT-displayed graphics

1–4 Which item is the least used and least accurate straightedge device?
    a. parallel bar   c. T square
    b. CAD digitizer   d. drafting machine

1–5 Light pens, wooden pencils, mechanical lead holders, and ( ) are used to draw or draft.
    a. electronic pens   c. thin lines
    b. cursors   d. compasses

1–6 A sandpaper pad is used to ( ).
    a. file   c. sand
    b. shave   d. sharpen

1–7 Dry cleaning pads are used to (and) a drawing.
    a. dust   c. clean
    b. remove graphite   d. rub

1–8 An architect's scale is ( ) divided.
    a. partially   c. fully
    b. open   d. half

1–9 An adjustable triangle cannot be used as a ( ).
    a. 30/60 triangle   c. foot scale
    b. protractor   d. 45 triangle

1–10 Which two instruments are obsolete?
    a. ruling pen   c. divider
    b. cursor   d. T square

1–11 Define the following terms:

CAD	scale
micrographics	architect's scale
vellum	sandpaper pad
lead holder	dry cleaning pad
thin-line mechanical pencil	erasing shield
plotter	bow compass
printer	beam compass
template	metric scale
light table	light pen
grid paper	drafting film
CRT	drafting brush
floppy disk	drafting tape
magnetic tape	adjustable triangle
microfilm	dividers
digitizer	drop compass
drafting machine	technical pen
drafting table	

2

LETTERING

INTRODUCTION

Line drawings are never complete until they are explained by callouts, dimensions, notations, and titles. This information is often carefully hand lettered with a freehand, single-stroke gothic alphabet. The gothic alphabet does not have short bars or serifs at the ends of strokes as does the roman alphabet.

Legibility is the first requirement for any lettering, followed by ease of printing and speed. All drafters, designers, and engineers must master the art of freehand lettering through studying letter forms and the direction of strokes through consistently practicing lettering styles. The importance of good lettering cannot be overemphasized. The lettering can *make or break* an otherwise excellent drawing. In electronic drafting and design, as in all engineering disciplines, sloppy or misplaced lettering can cause misconceptions and inaccurate communication of data.

Lettering can be divided into three separate methods: (1) manual or freehand, (2) mechanical, and (3) machine. Table 2–1 lists the three categories along with some of the available equipment or techniques associated with each group.

A wide variety of lettering styles, or fonts, is available. A font is an assortment of type all of one size and style. For most drafting disciplines, the single-stroke, commercial, uppercase gothic alphabet is required. Figure 2–1 shows a few of the many fonts commercially available in phototypesetting and printing processes and in dry transfer letters. Note that vertical uppercase DRAFTING STANDARD is available.

Figure 2–2 is an example of a CAD title block using a variety of lettering styles. CAD lettering fonts and styles are limited only by the capabilities of the system and the imagination of the operator. The drawing in Fig. 2–3 was made with freehand, single-stroke, vertical gothic lettering. A template was used for Fig. 2–4. Each of these examples represents a different level of technology: manual (Fig. 2–3), mechanical (Fig. 2–4), and machine (Fig. 2–2).

MANUAL LETTERING

Many electronic drawings are still made with freehand lettering techniques. Both vertical and inclined lettering are found throughout industry. Vertical lettering is slightly preferred since it reduces and microfilms better than inclined lettering. Still, inclined lettering is easier to master and normally is faster to complete. In general, only uppercase lettering is used on drawings. However,

21

TABLE 2–1
Methods of Lettering

Types of Letters	Manual	Mechanical	Machine
Vertical Inclined Uppercase Lowercase	Freehand Lettering aid (slot guide)	Template WRICO Leroy Letterguide Varigraph	Typewriter Printer Dry transfer Phototypesetter CAD

FIGURE 2–1
Examples of typefaces.

since long columns of uppercase characters are not pleasing to the eye and are harder to read, lowercase lettering may be used for drawing *notes*. The use of lowercase lettering is specified in company standards when acceptable.

Guidelines

Freehand lettering, whether vertical or inclined, has guidelines at the top and bottom of the letters unless grid underlays or fade-out grid paper is used. Avoid using grid underlays until you have gained some experience with lettering from hand-drawn guidelines. Guidelines should be very thin, sharp, and gray, made with 3H or 4H grade lead, as shown in Fig. 2–5 (also see Fig. 2–9). Figure 2–5 has two callouts specifying the required markings on a piece of electronic equipment. The equipment is upside down. The object and note were completed first and then the drawing was turned and lettered as required.

Since most lettering is done with capital letters and whole numbers, only two lines are necessary. Guidelines can be drawn with a straightedge or with the aid of a line spacing guide, such as the AMES® lettering guide or the Braddock–Rowe® triangle. Consult Table 2–2 for the spacing of lettering guidelines. For dimensions, notes, and callouts, most lettering is 5/32 in. high in U.S. customary units or 5 mm high in SI units. For all problems in the text, use the standards for lettering heights and guideline spacing shown in Table 2–2. This table corresponds to *Conventions and Lettering* from the American National Standards Institute (ANSI). The units are from ANSI Y14.2M–1979 for all U.S. unit measurements and ISO 3098/1–1974 for all metric measurements. (Metric sizes are not U.S. customary conversions.)

The distance between lines of lettering, such as that for notes and callouts, should be equal to the full height of the letter being used. This spacing is best for reproducible, legible letters for reduction and blow back (when a microfilmed drawing is returned to its original size). When upper- and lowercase lettering is used on D size sheets and larger, the minimum uppercase height

FIGURE 2–2
CAD lettering.

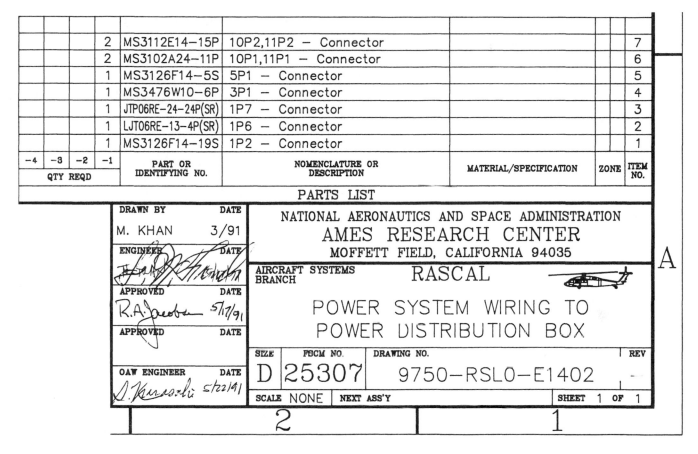

		2	MS3112E14–15P	10P2,11P2 — Connector			7
		2	MS3102A24–11P	10P1,11P1 — Connector			6
		1	MS3126F14–5S	5P1 — Connector			5
		1	MS3476W10–6P	3P1 — Connector			4
		1	JTP06RE–24–24P(SR)	1P7 — Connector			3
		1	LJT06RE–13–4P(SR)	1P6 — Connector			2
		1	MS3126F14–19S	1P2 — Connector			1
−4	−3	−2	−1	PART OR IDENTIFYING NO.	NOMENCLATURE OR DESCRIPTION	MATERIAL/SPECIFICATION	ZONE	ITEM NO.
	QTY REQD							

PARTS LIST

DRAWN BY M. KHAN **DATE** 3/91
ENGINEER **DATE**
APPROVED R.A. Jacobs 5/17/91 **DATE**
APPROVED **DATE**
OAW ENGINEER S. Kurosaki 5/22/91 **DATE**

NATIONAL AERONAUTICS AND SPACE ADMINISTRATION
AMES RESEARCH CENTER
MOFFETT FIELD, CALIFORNIA 94035

A

AIRCRAFT SYSTEMS BRANCH **RASCAL**

POWER SYSTEM WIRING TO
POWER DISTRIBUTION BOX

SIZE D **FBCM NO.** 25307 **DRAWING NO.** 9750–RSL0–E1402 **REV** -

SCALE NONE **NEXT ASS'Y** **SHEET** 1 **OF** 1

2 | 1

should be no less than .180 in. to allow for legible enlargement of microfilmed drawings.

The freehand lettering aid shown in Fig. 2–6 was used to letter the diagram in Fig. 2–7. This device eliminates the need for guidelines, since it limits the height of the lettering to the space within the slots. Note that this form of lettering tends to flatten the upper and lower portions of some letters. Do not use lettering guides while attempting to learn and perfect freehand lettering.

FIGURE 2–3
Vertical lettering.

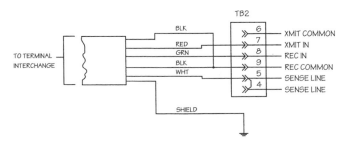

Guidelines are unnecessary when you use vellum, or drafting film, with nonreproducible grid lines. The grid spacing can be used as guidelines for the lettering. Guidelines are also unnecessary when you use a lettering guide or template.

Vertical and inclined lettering should never be mixed on one drawing. Many times a drafter is called on to complete or revise an existing drawing. If you are given such a project, try to match the existing lettering style. The lettering in Fig. 2–8 is both vertical and inclined, and guidelines were obviously not used.

Pencil Technique

Freehand lettering places a requirement on linework that is different from what is possible with instrument lines. Instrument lines are made more dense when the line is retraced. It is impossible to consistently retrace freehand lines; therefore, they must be drawn with the proper density in only one stroke. To help get the proper density, a soft lead is used. We prefer the H lead. A .5-, .7-, or .9-mm F lead works well for some drafters.

FIGURE 2–4
Template lettering.

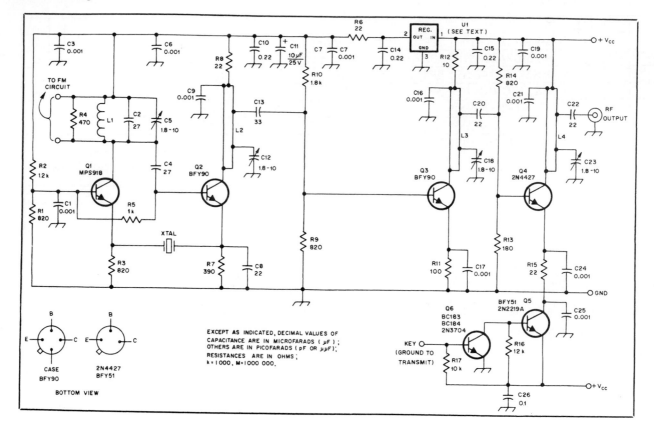

FIGURE 2–5
Freehand lettering with guidelines. The lettering in this figure is slightly back-slanted.
(Courtesy Simpson Electric Co.)

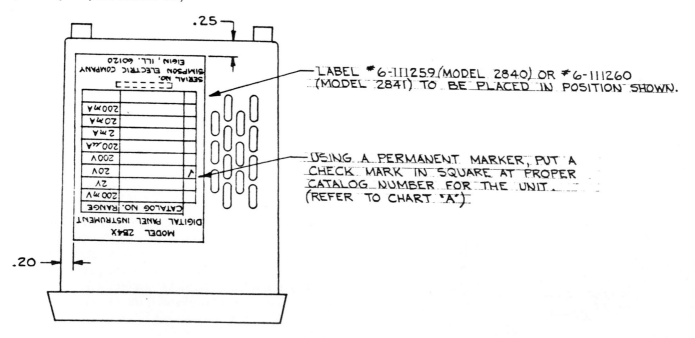

TABLE 2–2
Recommended Lettering Heights for Manual and Mechanical Lettering (Uppercase Letters)

Project	Size of Drawing	Height of Manual Letters, U.S. (Metric) Units	Height of Mechanical Letters, U.S. (Metric) Units
Numbers in	A–C[a]	.250 in., ¼ in. (7 mm)	.240 in. (7 mm)
a title block	D and above[a]	.312 in., ⁵⁄₁₆ in. (7 mm)	.290 in. (7 mm)
Title, section	A–F	.250 in., ¼ in. (7 mm)	.240 in. (7 mm)
lettering			
Zone letters and	A–F	.188 in., ³⁄₁₆ in. (5 mm)	.175 in. (5 mm)
numerals in			
borders			
Lettering in dimensions,	A–C	.125 in., ⅛ in. (3.5 mm)	.120 in. (3.5 mm)
tolerances, notes,	D and above	.156 in., ⁵⁄₃₂ in. (5 mm)	.140 in. (5 mm)
tables, limits			

[a] Drawing sizes: A, 8 ½ × 11 in.; B, 11 × 17 in.; C, 17 × 22 in.; D, 22 × 34 in.

FIGURE 2–6
Lettering guide. (Courtesy Koh-I-Noor Rapidograph, Inc.)

Soft lead contributes to the *dirt* on the drawing because it chalks more easily. Frequent use of the drafting brush is necessary. Due to its tendency to smear, lettering is usually the last step in the completion of a drawing. When lettering, do not let your hand come in contact with the drawing surface. Always place a sheet of clean paper between your hand and the drawing medium. This will help keep the drawing free of body oils and dirt, as well as prevent smearing of the linework.

The thin-line pencil is an excellent lettering device, since it requires no sharpening. A .5- or .7-mm thin line

FIGURE 2–7
Inclined lettering drawn with a lettering guide.

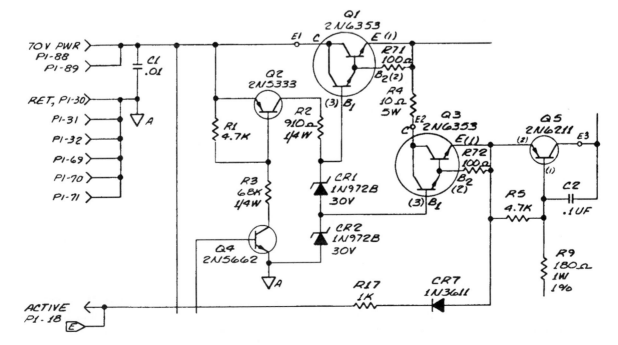

FIGURE 2–8
Mixed vertical and inclined lettering.

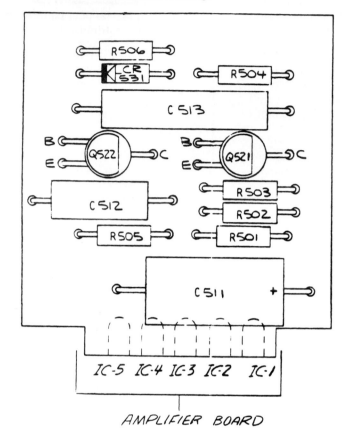

should be used for the lettering when available. Rotating a wooden pencil or a lead holder minimizes depletion of the point and helps maintain consistency of character width. A .7- or .9-mm thin-line mechanical pencil may also need to be rotated.

Stroke, Sequence, Form, and Spacing

The strokes of letters must be kept consistent in both width and density. The density of the lettering should approach the density of the linework. Obvious variation in the densities between lettering and linework should be avoided. Strive for consistent, uniform, well-spaced letters. The stroke sequence suggested in Fig. 2–9 is the same for both inclined and vertical lettering. Note that this suggested stroke sequence is meant only to be a general guide. Your lettering style, manner of holding the pencil, and whether you are right- or left-handed all affect your choice of stroke sequence.

Develop a lettering style that is comfortable for you, a style that communicates the necessary engineering data without confusion and mistakes. It is very important to catch any bad habits early in order not to ingrain them in your lettering style.

In the beginning it is important to eliminate any individualized style, until your lettering becomes clear, concise, dark, and well formed. Through practice, a more attractive personal style will emerge and become obviously *yours.*

In Fig. 2–9 an alphabet of vertical and inclined letters and numerals is shown along with a suggested stroke sequence. This figure highlights typical problems in forming letters. Lettering *examples* are also provided below each *comment.* The examples used are reference designations found on electronic drawings (see Chapter 9). Note that guidelines were used throughout.

The grid pattern shown in Fig. 2–9 gives the ideal width and height relationship for single-stroke gothic lettering. All characters are six units in height and vary in width from the *1* and *I* to the *W* and *M.* Most of the characters are 6 units wide.

Spacing for letters and words should correspond to the following specifications:

☐ Background areas between letters in words should be separated by approximately equal spaces.
☐ Numerals separated by a decimal point (5.375, 2.54 mm, etc.) should be a minimum of two-thirds of the character height used.
☐ Spaces between words should be approximately equal and a minimum of .06 in. (1.5 mm). A full character width for word spacing is suggested.
☐ The horizontal space between lines of lettering should be at least one-half of the height of the characters, but preferably one full character of space should be left between lines.
☐ Sentences should be separated by at least one full character height and preferably two character heights if space permits.

In Fig. 2–10, the typical slant angle used for inclined lettering is shown. Any angle between 90° (vertical) and 65° is acceptable unless an individual company has a preferred practice. Left-handed drafters can slant their lettering backward 1° to 10° if it is more convenient. Inclined lettering is ideally slanted at approximately 67 ½° for right-handed drafters. (Figure 2–5 showed freehand lettering that was slightly back-slanted.)

Lowercase lettering as shown in Fig. 2–11 is seldom used on engineering work except for some architectural and structural drawings. *Notes* on other kinds of engineering drawings sometimes have both uppercase and lowercase letters, since mixed lettering is more pleasing to the eye and is easier to read. Lowercase lettering, whether inclined or vertical, requires extra guidelines. Guidelines for the waist line (top of main body of letter) and base line (bottom of main body of letter), as well as for ascender and descender lines, should be added to the drawing before the letters are drawn.

FIGURE 2-9
Stroke sequence for lettering.

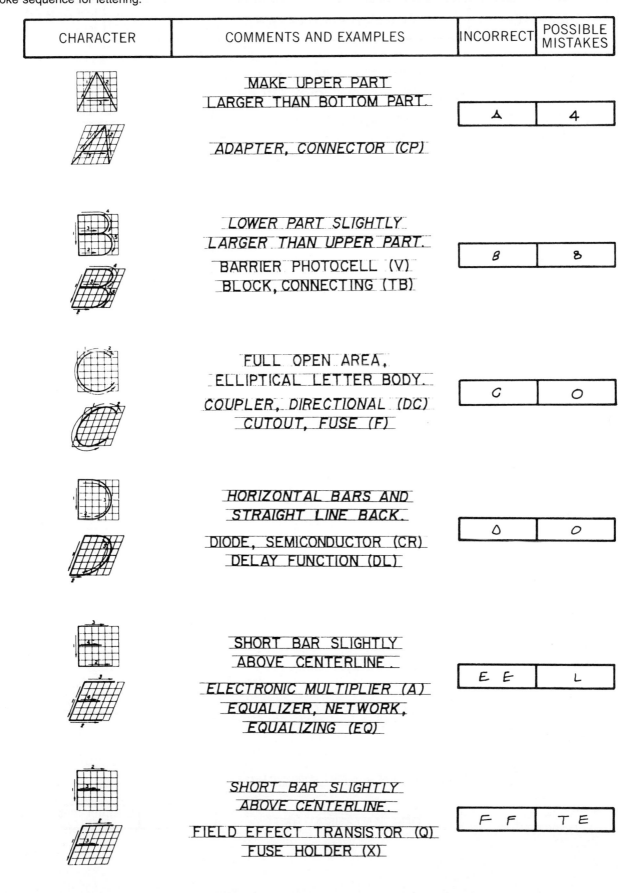

FIGURE 2–9 (cont.)

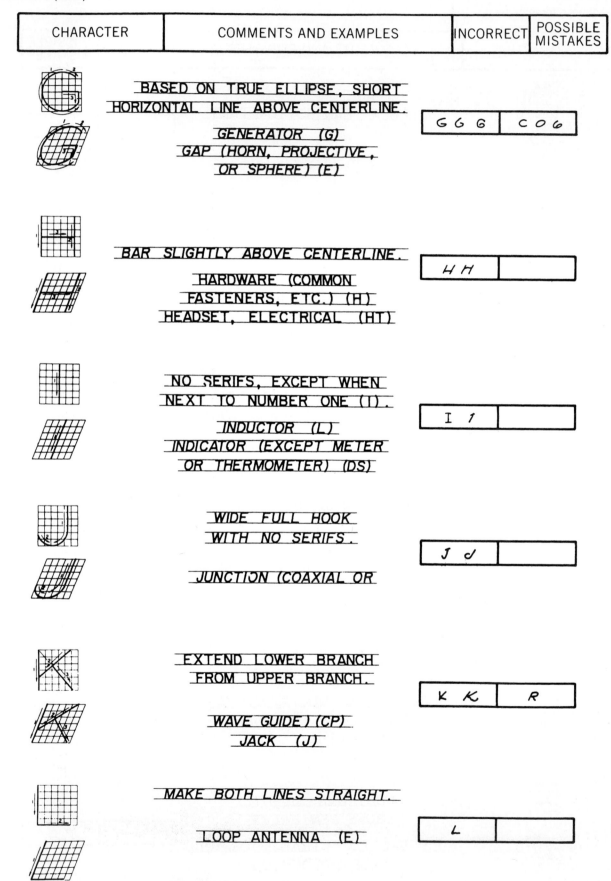

CHARACTER	COMMENTS AND EXAMPLES	INCORRECT	POSSIBLE MISTAKES
	BASED ON TRUE ELLIPSE, SHORT HORIZONTAL LINE ABOVE CENTERLINE. GENERATOR (G) GAP (HORN, PROJECTIVE, OR SPHERE) (E)	G G G	C O G
	BAR SLIGHTLY ABOVE CENTERLINE. HARDWARE (COMMON FASTENERS, ETC.) (H) HEADSET, ELECTRICAL (HT)	H H	
	NO SERIFS, EXCEPT WHEN NEXT TO NUMBER ONE (I). INDUCTOR (L) INDICATOR (EXCEPT METER OR THERMOMETER) (DS)	I 1	
	WIDE FULL HOOK WITH NO SERIFS. JUNCTION (COAXIAL OR	J J	
	EXTEND LOWER BRANCH FROM UPPER BRANCH. WAVE GUIDE) (CP) JACK (J)	K K	R
	MAKE BOTH LINES STRAIGHT. LOOP ANTENNA (E)	L	

FIGURE 2–9 (cont.)

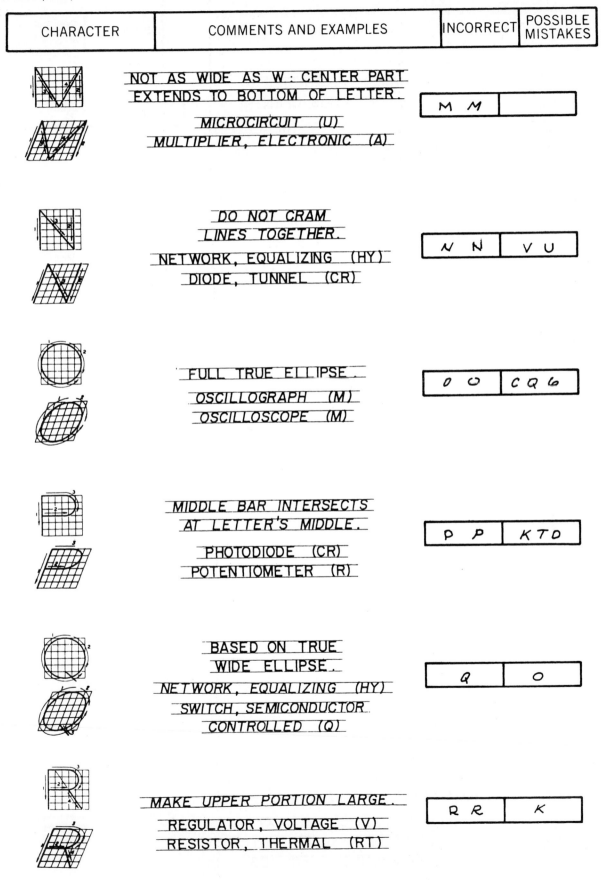

CHARACTER	COMMENTS AND EXAMPLES	INCORRECT	POSSIBLE MISTAKES
	NOT AS WIDE AS W: CENTER PART EXTENDS TO BOTTOM OF LETTER. MICROCIRCUIT (U) MULTIPLIER, ELECTRONIC (A)	M M	
	DO NOT CRAM LINES TOGETHER. NETWORK, EQUALIZING (HY) DIODE, TUNNEL (CR)	N N	V U
	FULL TRUE ELLIPSE. OSCILLOGRAPH (M) OSCILLOSCOPE (M)	O O	C Q 6
	MIDDLE BAR INTERSECTS AT LETTER'S MIDDLE. PHOTODIODE (CR) POTENTIOMETER (R)	P P	K T D
	BASED ON TRUE WIDE ELLIPSE. NETWORK, EQUALIZING (HY) SWITCH, SEMICONDUCTOR CONTROLLED (Q)	Q	O
	MAKE UPPER PORTION LARGE. REGULATOR, VOLTAGE (V) RESISTOR, THERMAL (RT)	R R	K

FIGURE 2–9 (cont.)

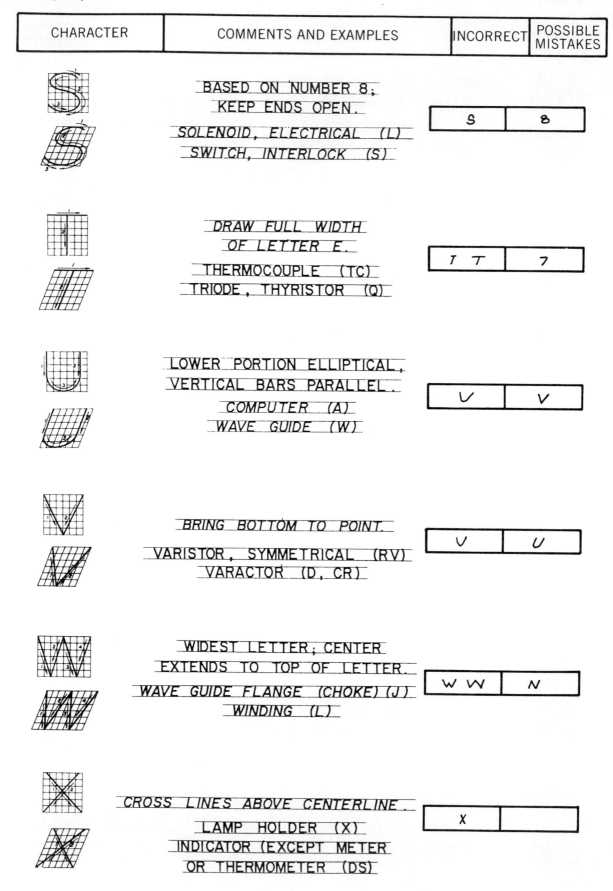

CHARACTER	COMMENTS AND EXAMPLES	INCORRECT	POSSIBLE MISTAKES
	BASED ON NUMBER 8; KEEP ENDS OPEN. SOLENOID, ELECTRICAL (L) SWITCH, INTERLOCK (S)	S	8
	DRAW FULL WIDTH OF LETTER E. THERMOCOUPLE (TC) TRIODE, THYRISTOR (Q)	T T	7
	LOWER PORTION ELLIPTICAL, VERTICAL BARS PARALLEL. COMPUTER (A) WAVE GUIDE (W)	U	U
	BRING BOTTOM TO POINT. VARISTOR, SYMMETRICAL (RV) VARACTOR (D, CR)	V	U
	WIDEST LETTER; CENTER EXTENDS TO TOP OF LETTER. WAVE GUIDE FLANGE (CHOKE) (J) WINDING (L)	W W	N
	CROSS LINES ABOVE CENTERLINE. LAMP HOLDER (X) INDICATOR (EXCEPT METER OR THERMOMETER (DS)	X	

FIGURE 2–9 (cont.)

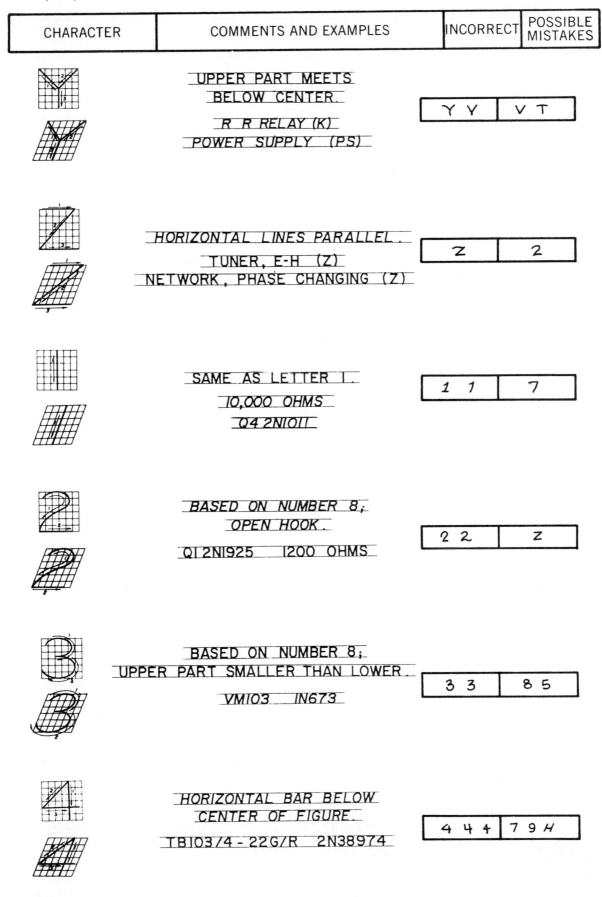

CHARACTER	COMMENTS AND EXAMPLES	INCORRECT	POSSIBLE MISTAKES
	UPPER PART MEETS BELOW CENTER. R R RELAY (K) POWER SUPPLY (PS)	Y Y	V T
	HORIZONTAL LINES PARALLEL. TUNER, E-H (Z) NETWORK, PHASE CHANGING (Z)	Z	2
	SAME AS LETTER I. 10,000 OHMS Q4 2N1011	1 1	7
	BASED ON NUMBER 8; OPEN HOOK. Q1 2N1925 1200 OHMS	2 2	Z
	BASED ON NUMBER 8; UPPER PART SMALLER THAN LOWER. VM103 1N673	3 3	8 5
	HORIZONTAL BAR BELOW CENTER OF FIGURE. TB1037/4 - 22G/R 2N38974	4 4	7 9 H

FIGURE 2–9 (cont.)

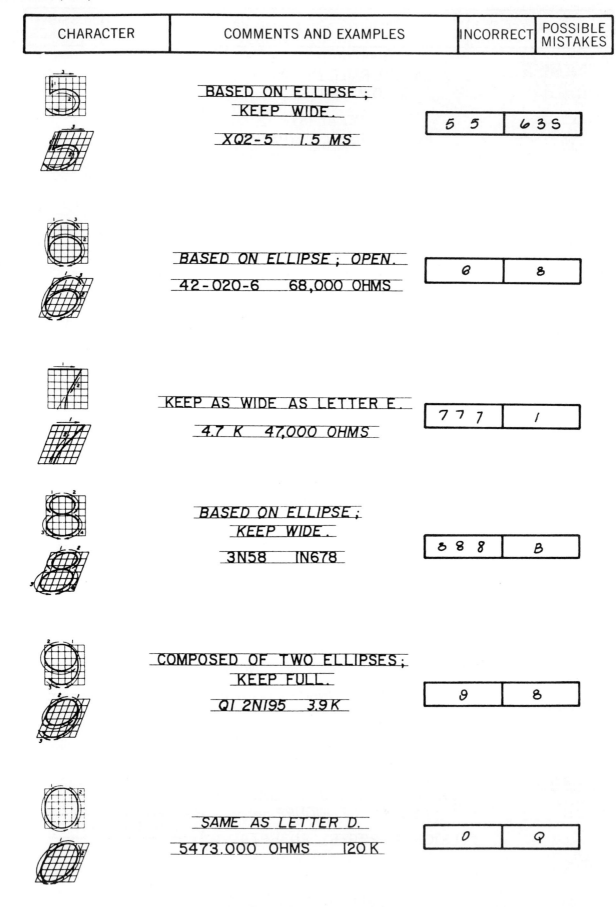

CHARACTER	COMMENTS AND EXAMPLES	INCORRECT	POSSIBLE MISTAKES
	BASED ON ELLIPSE ; KEEP WIDE. XQ2-5 1.5 MS	5 5	6 3 S
	BASED ON ELLIPSE ; OPEN. 42 - 020 - 6 68,000 OHMS	6	8
	KEEP AS WIDE AS LETTER E. 4.7 K 47,000 OHMS	7 7 7	1
	BASED ON ELLIPSE ; KEEP WIDE. 3N58 1N678	8 8 8	B
	COMPOSED OF TWO ELLIPSES ; KEEP FULL. Q1 2N195 3.9 K	9	8
	SAME AS LETTER D. 5473.000 OHMS 120 K	0	Q

FIGURE 2–10
Guidelines and inclined lettering.

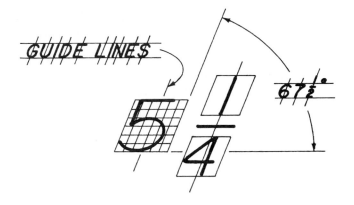

FIGURE 2–11
Lowercase lettering.

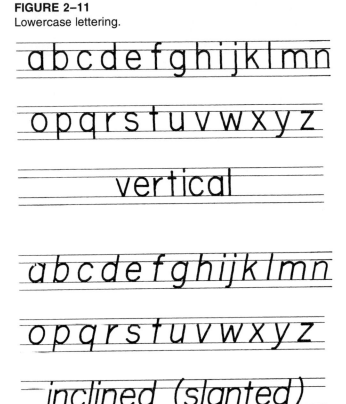

Ascender lines or *cap lines* designate the top of strokes for letters that extend above the waist line, as in *b*, *d*, *f*, *h*, *k*, and *l*. Descender lines or *drop lines* designate the bottom of strokes for letters below the base line, as in *g*, *j*, *p*, and *y*.

Most electronic drawings are dimensioned with decimal or metric units. These methods are easier, more accurate, and quicker to draw, since all numbers are placed between two equally spaced guidelines. For some cabinets, panels, chassis, and other types of sheet metal enclosures, the tolerance and accuracy required for manufacturing and construction are loose enough to permit fraction dimensioning. In Fig. 2–12 the height ratio of fraction number to whole number is provided. When a drawing is to be reduced, the size of lettering may need to be larger than normal for accurate enlargement from reduction size. Some drafters prefer to make the fraction number size almost exactly the same as the whole number to avoid clarity problems. The ANSI standard on lettering states that the height of the fraction number should be the same as that for the whole number. Most drafting books, however, suggest the relationship shown in Fig. 2–12. Note that the numbers must not be placed too close to the fraction division bar. The division bar is normally drawn horizontally between the

numbers, not at an angle, except in *Notes,* when the angled division bar is acceptable. Some company standards require the angled division bar.

Lettering Composition

A variety of special circumstances affects the composition and placement of letters on an electronic drawing. Expanded (extended), compressed, stopline, centered, and symmetrical lettering are found on electronic diagrams and drawings. In Fig. 2–13 the block diagram has been phototypeset in compressed lettering to fit within the blocks. Centered and symmetrical composition is used on this diagram.

FIGURE 2–12
Fractions.

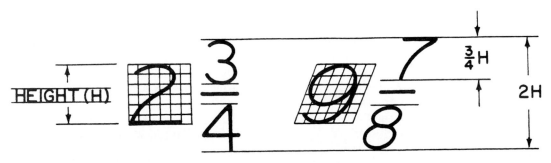

FIGURE 2–13
Symmetrical and centered lettering for block diagrams. (Courtesy TRW LSI Products)

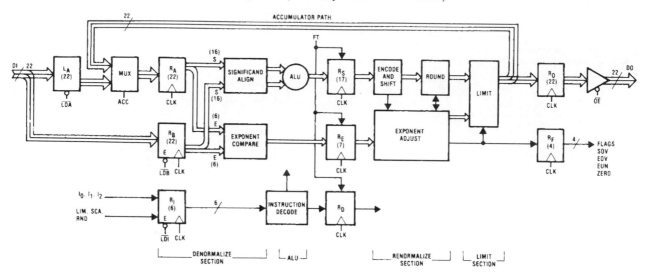

Stopline lettering is used on many electronic drawings. In most cases, lettering is *left-justified* (aligned on the left in a column). When lettering must stop along a given line or at a specific point, it is called *right-justified*. Right-justified freehand lettering is called stopline lettering. Many of the drawings throughout the text show this type of lettering composition. Stopline lettering is somewhat difficult since the letters are normally drawn from the left toward the right. The drafter may have to letter from the right toward the left to complete stopline lettering properly.

Expanded lettering is the opposite of compressed lettering since the individual letters are drawn wider than the standard widths. Other variations include **boldface**, *italics*, and ***boldface italics.***

The beauty of machine lettering lies in its ability to do all the preceding variations automatically. The hand-lettered lists and notes in Figs. 2–14 and 2–15 are time consuming and tiring. The notes in Fig. 2–15 are standard preprinted company notes; 7 and 8 were added in freehand. Notes and lists should always be placed horizontally on the drawing. The vertical distance between two separately numbered notes should be at least two times the character height used for the letters so that the identity of each note is maintained. Traditionally, notes have been placed above the title block area on the drawing on the far right. The newest ANSI standards have reversed the placement of notes. *Notes are now to be placed on the lower left or the upper left of the drawing.* Many companies still follow the older practice of placing notes above the title block on the right side of the drawing.

MECHANICAL LETTERING AIDS

One of the most common lettering devices found in any drafting room is the template. Although freehand lettering is the rule rather than the exception, templates are used for many types of electronic and electrical drawings and for portions of hand-lettered drawings. Templates are available for almost any size and style of lettering, as shown in Fig. 2–16, and can be adapted for ink as well as pencil use. The beauty of a template lies in its ability to produce repeatable uniform letters and numerals. Template lettering, however, takes considerably more time than freehand lettering.

With a template, guidelines are unnecessary. A template must rest on a straightedge while in use so that all the letters are aligned properly (Fig. 2–17). In general, electrical drawings are inked when a lettering template is used. The major drawback of templates is that it is hard to create perfectly formed letters without a great deal of practice. The pencil or inking pen must be kept almost perpendicular to the paper while the letters are drawn, as shown in Fig. 2–18.

The Koh-I-Noor Rapidometric Guide® template in Fig. 2–18 is an excellent example of a template designed to be used for inking. Note that the inset drawing shows how the template shelf does not come in contact with the drawing surface. Thin stick-on pads may also be fastened to the bottom of the template. This eliminates smearing of the ink so common with traditional flat templates.

Technical pens tend to dry up and clog at the pen tip. The special Leroy technical pen set in Fig. 2–19 is

FIGURE 2–14
Hand-lettered parts list.

REF NO	COMPONENT	PART NO
R-401	33K	216480
R-402	24K	216477
R-403	9.1K	216467
R-404	33	549978
R-405	100K	216491
R-406	430K	216731
R-407	7.5K	216465
R-408	100	595359
R-409	1K	216445
R-410	5.1K	216461
R-411	15K	216472
R-412	47K	216484
R-413	100K	216491
R-414	680	216442
J	JUMPER	1207833
C-421	.15/35 MFD	491255
C-422	150 PF DISC	1207587
C-423	3.3/35 MFD	1207585
C-424	.47/35 MFD	1208599
C-425	.33/35 MFD	1208591
C-426	2.2/35 MFD	1208601
C-427	.0068/100 MFD	492500
C-428	.0027/100 MFD	491309
C-429	150 PF DISC	1207587
Q-441	GREEN	1207577
Q-442	GREEN	1207577
Q-443	BLACK	1207601

FIGURE 2–15
Hand-lettered notes of vertical characters added to preprinted standard company notes. (Courtesy Fairchild Industries, Inc.)

NOTES:

△1 MARK PER MIL-STD-130 APPROXIMATELY WHERE SHOWN, .093 HIGH CHARACTERS USING ITEM 48

2 SOLDER IN ACCORDANCE WITH NHB5300.4 (3A-1)

3 PARTIAL REFERENCE DESIGNATIONS ARE SHOWN FOR COMPLETE DESIGNATIONS PREFIX WITH UNIT NUMBER AND SUBASSEMBLY DESIGNATIONS

△4 ELECTROSTATIC DEVICE, HANDLE PER DOD-STD-1686

△5 TORQUE 2-2.5 INCH LBS

△6 FINISH: CONFORMAL COAT PER GEN-PS5205 EXCEPT CONNECTOR AND DESIGNATED AREAS SHOWN

△7 BOND ITEM 67 TO ITEM 1 PRIOR TO POPULATION OF CARD ASSEMBLY PER GEN-PS5402 CLASS 7

△8 APPLY FILLET TO COMPONENTS INDICATED AFTER CONFORMAL COATING PER GEN-PS5402 CL II

FIGURE 2–16
Lettering template and samples. (Courtesy Wrico, Wood–Regan)

ABCDEFGHIJKLMN
Guide No. SC 3-16P 7

abcdefghijklmnopqrs
Guide No. SL 3-16P 7

1234567890$¢&+#%
Guide No. SN 3-16P 7

SET NO. NS 3-16 ·····

ABCDEFGHIJKLMN
Guide No. VC 3-16P 7

abcdefghijklmnopqrst
Guide No. VL 3-16P 7

1234567890$¢&#%
Guide No. VN 3-16P 7

SET NO. NV 3-16 ·····

FIGURE 2–17
Vertical lettering from a template.

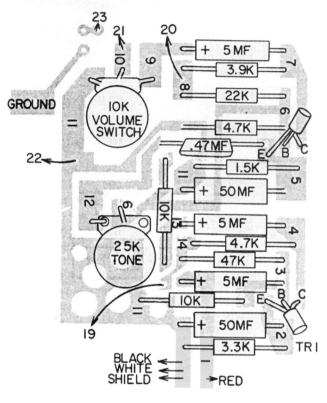

designed to keep the pen tips free flowing by encasing them in a small humidifier holder when they are not in use. The holder shown in this figure eliminates or slows down the drying process.

The Leroy lettering set shown in Fig. 2–20 uses a scriber, shown in Fig. 2–21, and a template with slot

FIGURE 2–18
Lettering guide template. (Courtesy Koh-I-Noor Rapidograph, Inc.)

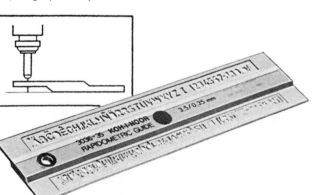

FIGURE 2–19
Leroy technical pen set. (Courtesy Keuffel & Esser/Kratos)

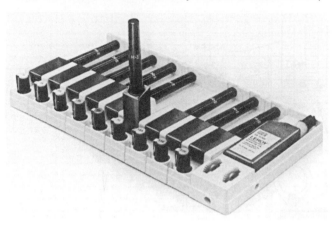

guide to assist in making close-to-perfect letters, shown in Fig. 2–22. The scriber can be adjusted to alter the slant of the lettering. Other similar scribers and templates are available for use on WRICO templates and templates from other companies. Skill in using a lettering template and scriber can be accomplished only through practice. The most difficult part of lettering systems is mastering the spacing of the characters.

Hand-operated lettering systems are expensive and take more time than traditional freehand lettering. Mechanical lettering devices and the inking of drawings are usually limited to the drawings for publication. Manuals, catalogs, and sales literature all require more precise lettering and linework than design, detailing, and assembly drawings for electronic equipment.

FIGURE 2–20
Leroy lettering set. (Courtesy Keuffel & Esser/Kratos)

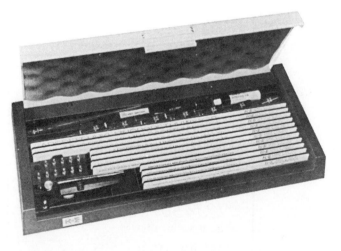

FIGURE 2–21
Rapidograph scriber. (Courtesy Hearlihy & Co.)

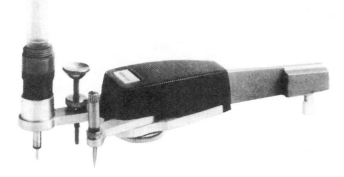

FIGURE 2–22
Leroy lettering set in use. (The template guide should be against a straightedge, not as in the photograph.) (Courtesy Keuffel & Esser/Kratos)

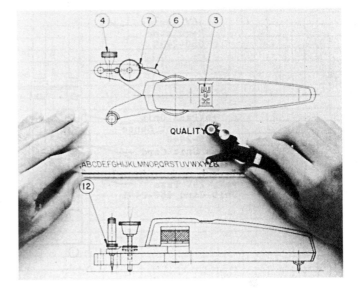

Mechanical lettering devices enable the user to make slightly smaller letters than do manual techniques. (See Table 2–2 to compare sizes.) The variation in recommended minimum standard letter heights between freehand and mechanical devices is needed because freehand lettering does not reduce and blow back as accurately as mechanically drawn characters.

MACHINE LETTERING DEVICES

Typewriters with specially designed carriages and gothic typefaces can be used on A, B, and C size sheets. Figure 2–23 shows a panel drawing where the callouts have been typed on the drawing. Hand lettering was used for markings on the panel itself. A special inking ribbon must be used on the typewriter so that the characters are not smeared.

Dry transfer lettering and appliqués, shown in Fig. 2–24, are normally confined to artwork or headings. It is time consuming to apply each letter or number separately. The Kroy lettering system, shown in Fig. 2–25, is an example of dry transfer lettering where the user can dial a sequence of letters or numbers as required. The result is a dry, adhesive-backed strip for easy attachment to the drawing. Notes are easy to apply with this system, as shown in Fig. 2–26.

Phototypesetting and printing, shown in Fig. 2–27, are used for publication-level artwork and drawings, when the quality is extremely important. Many of the instructional figures throughout this text were photo-

typeset. Figure 2–28 is an example of phototypeset lettering on an electronic diagram.

The speed of lettering with a CAD system is limited only by the efficiency and speed of the user who enters the data on the terminal keyboard. CAD systems allow almost unlimited lettering fonts and sizes without mistakes. Figure 2–29 shows a few examples of CAD fonts.

CAD lettering can be left justified, right justified, or center justified (Fig. 2–30). The size, height, width, angle, and slant can be operator determined (Fig. 2–31). A wide variety of altered lettering is possible, including mirrored, backward, and curved styles (Fig. 2–32). The wiring diagram shown in Fig. 2–33 was drawn on a CAD system and plotted with a pen plotter.

Whether you use a CAD system or one of the many different lettering aids described in this chapter, you will still need to master freehand lettering. Engineering and design sketches and other types of written communication involve the mastery of freehand lettering to ensure proper and correct transferring of data. Regardless of future innovations in technology, handwritten communication will always be necessary in engineering work.

FIGURE 2–23
Typing used on panel drawing.

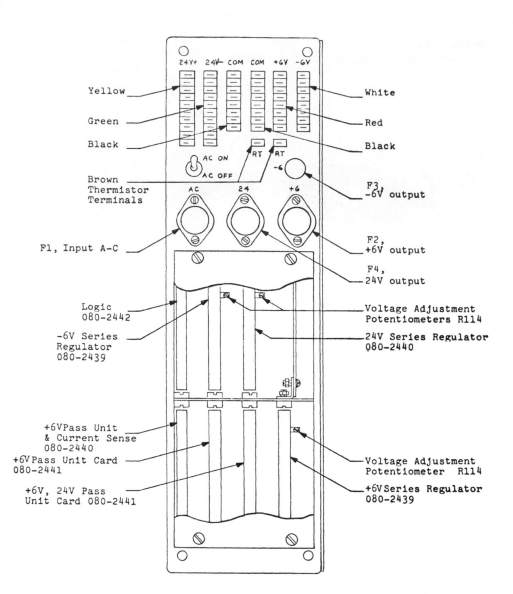

Yellow
Green
Black
Brown
Thermistor
Terminals

White
Red
Black

F1, Input A-C

F3,
-6V output

F2,
+6V output

F4,
24V output

Logic
080-2442

-6V Series
Regulator
080-2439

Voltage Adjustment
Potentiometers R114

24V Series Regulator
080-2440

+6V Pass Unit
& Current Sense
080-2440

+6V Pass Unit Card
080-2441

+6V, 24V Pass
Unit Card 080-2441

Voltage Adjustment
Potentiometer R114

+6V Series Regulator
080-2439

FIGURE 2–24
Transfer lettering. (Courtesy Graphic Products)

FIGURE 2–25
Kroy lettering system. (Courtesy Kroy, Inc.)

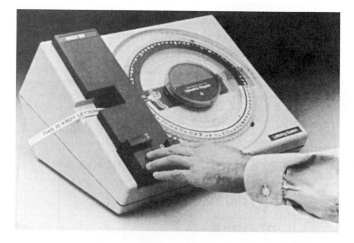

FIGURE 2–26
Kroy lettered notes. (Courtesy Kroy, Inc.)

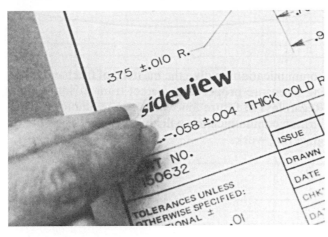

FIGURE 2–27
Electronic lettering. (Courtesy Xerox Corp.)

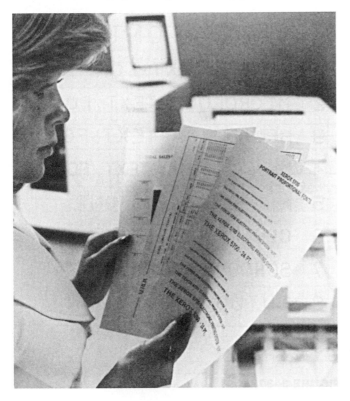

FIGURE 2–28
Typeset lettering. (Courtesy TRW LSI Products)

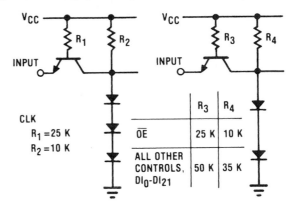

FIGURE 2–29
Text fonts.

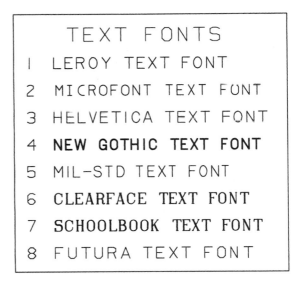

FIGURE 2–30
Text justification.

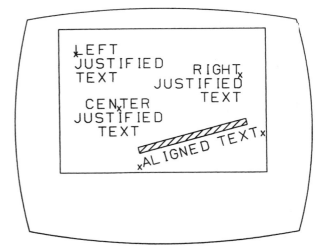

FIGURE 2–31
Text variations.

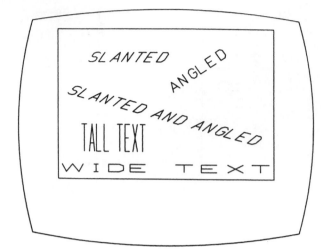

FIGURE 2–32
Text capabilities.

FIGURE 2–33
Wiring diagram.

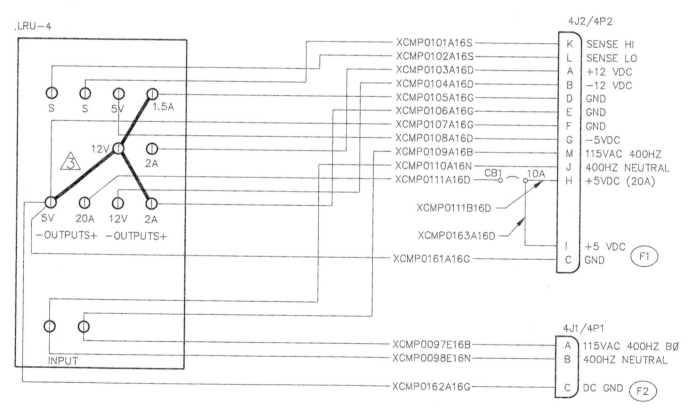

REVIEW QUESTIONS

2–1 Which of the following is not considered *machine* lettering?
  a. CAD lettering   c. Leroy
  b. dry transfer   d. phototypesetting

2–2 (   ) lettering is used on most engineering drawings.
  a. Emboss light   c. Folio light
  b. Single-stroke gothic capitals   d. Drafting standard

2–3 *True or false:* Vertical and inclined lettering should be used on the same drawing.

2–4 *True or false:* Lowercase lettering is used on all engineering drawings.

2–5 Inclined or vertical (   ) lettering is standard practice for electronic drawings.
  a. gothic   c. uppercase
  b. lowercase   d. upper- and lowercase

2–6 *True or false:* Guidelines are needed for all types of manual lettering.

2–7 What is the typical slope angle used for inclined lettering?
  a. 72°   c. 30/60
  b. 67 ½°   d. 78°

2–8 Notes and dimensions should be at least (   ) on D size sheets and larger.
  a. .250 in.   c. .156 in.
  b. .125 in.   d. 5 mm

2–9 What type of lettering is most frequently found on drawings used in manuals?
  a. freehand   c. CAD
  b. phototypeset   d. template

2–10 *True or false:* High-quality freehand lettering skills are not necessary for the aspiring drafter or designer, since automated methods will eventually replace freehand lettering.

PROBLEMS

The problems in this chapter are meant to provide lettering practice. Confine the width of each *Standard* to 4.5 in. (115 mm). Your instructor will assign the height of lettering, whether it is to be lead or ink, and whether you are to use freehand, template, or Leroy lettering. Example: ⅛ in., ink, template.

2–1 NOTES:
1. Printed circuit board assembly should be drawn viewed from the component side. If components are mounted on both sides, then both front and back views should be shown.
2. Use auxiliary views to clarify any details such as the precise location and orientation of mechanical parts.
3. Maximum component height or overall board thickness should be included.
4. Electronic components should be identified within the component outline with the reference designation used on the schematic diagram.
5. All parts other than electronic components should be identified by ballooned item numbers.

2–2 MIL-STD-429 Printed Wiring and Printed Circuits Terms and Definitions
MIL-STD-1313 Microelectronic Terms and Definitions
IPC-T-50 Terms and Definitions
EIA RS-208 Definition and Register, Printed Wiring

2–3 Documentation & Drafting
MIL-STD-100 Engineering Drawing Practices
MIL-D-1000 Drawings, Engineering and Associated Lists
MIL-D-1000/3 Drawings, Installation Plans and Preliminary Data for Electronic and Related Equipment (EC)
MIL-D-8510 Drawings, Undimensioned, Reproducibles, Photographic and Contact
ANSI Y14.5-1966 Dimensioning and Tolerancing for Engineering Drawings

2–4 Documentation & Drafting
ANSI Y14.5–1970 Electrical and Electronics Diagrams
ANSI Y32.–1970 Graphic Symbols for Electrical and Electronics Diagrams. (IEEE Std. #315)
ANSI Y32.14–1973 Graphic Symbols for Logic Diagrams (Two-State Devices) (IEEE Std. #91)
ANSI Y32.16–1970 Reference Designations for Electrical and Electronic Parts and Equipment

2–5 Printed Circuit Design
Single-sided & Two-sided Boards
MIL-STD-275D, Printed Wiring
MIL-P-55110 Printed Wiring Boards
IEC 326 General Requirements and Measuring Methods for Printed Wiring Boards
IPC-D-300 Printed Wiring Board Dimensions and Tolerances
Single and Two-Sided Rigid Boards
IPC-CM-770 Guidelines for Printed Circuit Board Component Mounting

2–6 Sources
Specifications and Standards listed herein may be obtained from the following sources:
ANSI American National Standards Institute, Inc., 1430 Broadway, New York, NY 10018
EIA Electronic Industries Association, 2001 Eye Street, NW, Washington, DC 20006
IEC International Electrotechnical Commission, 1, rue de Varembe, Geneve, Suisse, Europe
IPC Institute for Interconnecting & Packaging Electronic Circuits, 3451 Church Street, Evanston, IL 60203
Armed Services Electro Standards Agency (ASESA), Fort Monmouth, Red Bank, NJ 07703
General Services Administration (GSA), Washington, DC 20405
National Bureau of Standards (NBS), Washington, DC 20234
Office of Technical Services (OTS), 5285 Port Royal Road, Springfield, VA 22171

2–7 Letter the following note freehand using ³⁄₁₆-in. (4.7-mm) height characters and confining the width for notes to 5 in. (127 mm). Use H lead and guidelines.
Common circuit drawing conventions:
1. Circuit signal flow left to right with inputs on the left and outputs on the right.
2. The various functional stages of the circuit located in the same sequence as the signal flow.
3. Voltage potentials located with the highest voltage at the top of the sheet and the lowest at the bottom.
4. Auxiliary circuits that are included, but are not a main part of the signal flow, such as oscillators and power supplies, on the lower half of the drawing.
5. Frequently, a repetition of identical circuits is encountered in schematic diagrams. It is essential that these like circuits be drawn in the same form so that they are easily recognized.
6. Interconnections between components should be carefully planned for a minimum of right-angle bends and crossovers.

2–8 Using a mechanical lettering device (template or Leroy), ink the following note using ⅛-in. (3.0-mm) height lettering.
Electronic diagrams are normally drawn using drafting templates. Logic and block diagrams use templates that have been designed in accordance with MIL-STD-806 and are available in five scales: ¼, ⅜, ½, ¾, and full size. Schematic templates are generally designed to be in accordance with ANSI-Y32.2.

2–9 Your instructor can assign all or any part of the following designation list as a freehand lettering exercise. Specify lettering style (vertical, inclined), height (⅛ in. or ³⁄₁₆ in.), and composition parameters (normal, compressed, or expanded).

A

accelerometer
(1) assembly, separable or repairable (2)
circuit element, general
computer
divider, electronic
generator, electronic function
integrator
modulator
multiplier, electronic
recorder, sound
recording unit

reproducer, sound
servomechanism, positional
sensor (transducer to electric power)
subassembly, separable or repairable
telephone set
telephone station
teleprinter
teletypewriter

AR

amplifier (magnetic, operational, or summing)   repeater, telephone

AT

attenuator (fixed or variable)   isolator (nonreciprocal device)
capacitive termination   pad
inductive termination   resistive termination

B

blower   motor
fan   synchro

BT

barrier photocell   cell, battery
battery   cell, solar
blocking layer cell   transducer, photovoltaic

C

capacitor   capacitor bushing

CB

circuit breaker

CP

adapter, connector   junction (coaxial or waveguide)
coupling (aperture, loop, or probe)

2–10 Using stopline lettering (right justified), copy the following terms, which use Q as a reference designation symbol. Use 5/32-in. (for SI units drawings, use 5 mm) inclined lettering.

Q

transistor
rectifier, semiconductor controlled
switch, semiconductor controlled
thyraton (semiconductor device)
thyristor (semiconductor triode)

2–11 Letter the following note using .156-in. (3.96-mm) vertical lettering with a template in ink. Confine the note to a width of 4 in. (100 mm). Center justify (symmetrically) as shown. Your instructor may assign this problem to be completed with Leroy lettering instead of with a template.

Graphic Symbols for Electrical and Electronics Diagrams
(IEEE No. 315, 1971)/(ANSI Y32.2–1970) and
Graphic Symbols for Logic Diagrams (Two-State Devices)
(IEEE Std. 91, 1973)/(ANSI Y32.14–1973)

3

LINEWORK

INTRODUCTION

High-quality drafting skills associated with all forms of engineering are essential when graphics are used to communicate design, production, and manufacturing data. An engineer, designer, or drafter's graphical calculations must be neat and accurate to convey the proper message. In many cases the drawing will be reproduced by such processes as whiteprinting, photocopying, or reducing onto microfilm. Lettering and linework must be dark and of high quality. This chapter, an overview of the basics of linework, provides guidelines, procedures, and techniques for developing high-quality drawing skills.

Drafting pencils are graded according to the hardness of their lead, which determines the quality and shade of the line that can be drawn. A hard lead can make a very sharp, accurate, and thin line, which is quite beneficial for design calculations that need not be reproduced. Hard leads do not reproduce well in most cases. A soft lead will make dark lines, but their leads are very difficult to keep sharp. There are 11 grades of lead, from hardest to softest: 9H, 8H, 7H, 6H, 5H, 4H, 3H, 2H, H, F, and HB. For electronic drawings on a good grade of drafting paper, 3H or 4H is recommended for layout and construction of linework; H or 2H is

suggested for reproducible lines and for lettering. These leads, when used properly, produce excellent drawings having sharp, dense lines and dark, readable letters that will make good prints. A hard lead like 3H or 4H is sufficient for sketches used to lay out electronic diagrams and other electronic drawings before they are digitized on a CAD system. A digitizer is used with a menu (with electronic symbols and lines) to construct the diagram on the CRT. Figure 3–1 shows a CAD operator digitizing an electronic symbol. In Fig. 3–2, each of the four CRTs has a different type of electronic drawing displayed.

LINEWORK AND TECHNIQUE

Knowledge of lines and their symbolic meanings is essential to the electronic drafter. The most important aspect of an electronic drafter's job is the understanding of the process, intent, and content of an electronic drawing. In electronics, most drawings are made up of single-line diagrams, primarily composed of straight vertical and horizontal lines. Related drawings used for manufacturing electronic equipment include multiview, dimensioned drawings: equipment drawings, sheet metal enclosure layouts and developments, and component

FIGURE 3–1
Electronic diagram menu. (Courtesy Bausch & Lomb Interactive Graphics Division)

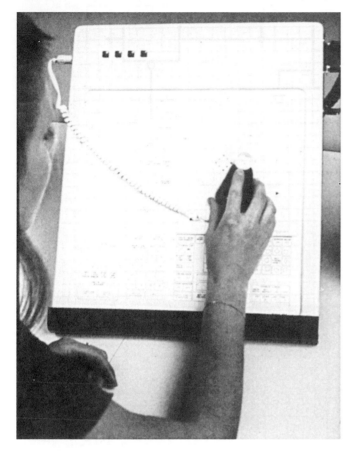

Line Technique

A properly drawn line is uniform for its entire length. You can make a line consistent in two ways:

1. Incline the pencil or lead holder so that it makes an angle of about 60° with the surface of the paper and then pull in the direction that it is leaning.
2. Rotate the pencil or lead holder slowly as the line is drawn and maintain a semisharp point. This will enable you to control the thickness and quality of the line.

These techniques take practice but will soon become automatic, and lines will be uniform from end to end and from one line to another. Most diagrams are drawn with one lineweight both for the symbols and for the vertical and horizontal lines representing the flow of the system. Figure 3–7 is a wiring diagram with three lineweights. Different lineweights can be used to emphasize a particular portion of the circuit, as illustrated in Fig. 3–8.

Since most drawings are reduced or enlarged at times, correct lineweights and line techniques are essen-

FIGURE 3–2
Four types of electronic diagrams displayed on CRTs. (Courtesy Prime-Computervision Corp.)

drawings. Graphs and charts are frequently used to represent tabular and related data.

Most electronic drawings are used to describe a process. The single-line diagram is the most frequently used to convey this information. Therefore, unlike mechanical drawings, where the shape description and projection technique are most important, electronic drawings require mastery of the symbol meaning, as well as of the actual symbols and lines used.

Although the types of drawings, linework, and weights are varied in the field of electronics, the most common drawing is the single-line diagram. Block diagrams, shown in Fig. 3–3, logic diagrams, Fig. 3–4, schematic diagrams, Fig. 3–5, and wiring diagrams, Fig. 3–6, are all single-line diagrams.

All drawings are made of lines that convey an idea from the drafter to the user. The control the drafter has over the pencil or pen and the techniques determine the quality of the drawings produced. The serious drafter, designer, or engineer constantly strives to improve technique through practice and attention to detail.

FIGURE 3–3
Block diagram. (Courtesy American Microsystems, Inc.)

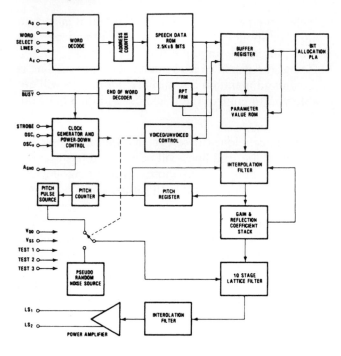

FIGURE 3–4
Logic diagram. (Courtesy TRW LSI Products)

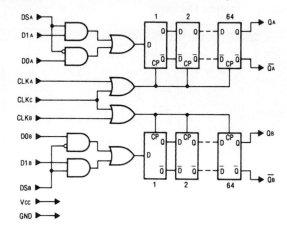

FIGURE 3–5
Hand-drawn schematic diagram.

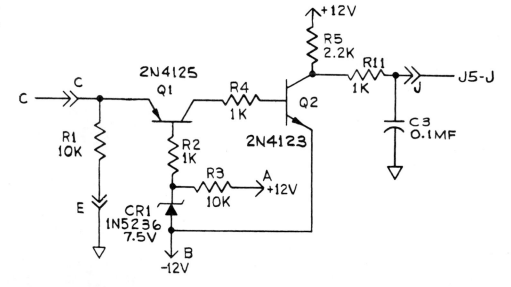

FIGURE 3–6
Portion of a control panel layout and interconnection wiring diagram. (Courtesy California Computer Products, Inc.)

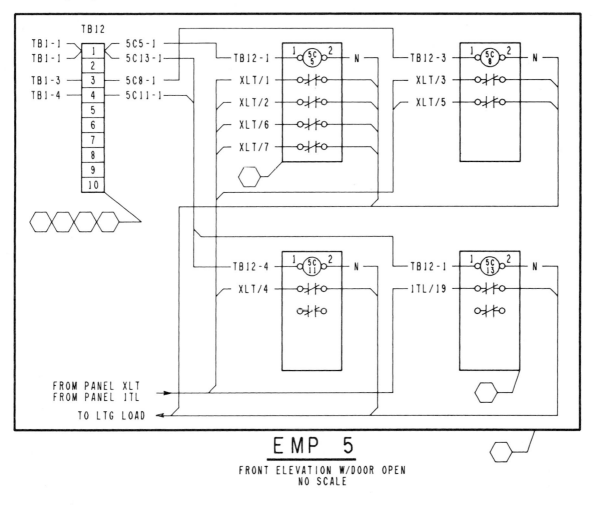

EMP 5
FRONT ELEVATION W/DOOR OPEN
NO SCALE

tial. For drawings that are to be enlarged, special care must be taken since even the smallest mistake tends to *mushroom*. In other words, the mistake will also be enlarged. For drawings that are to be reduced—for instance, printed circuit artwork drawings that are normally drawn at 2:1 or 4:1 enlargements—the corresponding reduction will *clean up* and *reduce* any small problems. It is important to draw the project accurately, however, at whatever scale is used and not to rely on a subsequent reduction to hide any poorly constructed areas of the drawing.

For CAD-generated drawings the lineweights are drawn and altered to any desired thickness, depending on the project, company standard, or special requirements for the drawing. CAD systems allow flexibility in the construction and revision of lines and symbols, as shown in Fig. 3–9. Lineweights can be exactly reproduced every time the drawing is plotted.

How well lines will print is determined by their density—that is, by how dark they are. Density is controlled by the hardness of the lead and by the pressure applied while the line is drawn. The width and sharpness of the line are determined by the size of the point touching the paper. A pencil point must always be smoothed and rounded on scratch paper after being repointed. It can also be resharpened on scratch paper. Uniform lines require uniform point preparation.

Thin-line pencils are widely used. These excellent lead holders are available in different lead thicknesses. A .5- and a .7-mm lead holder with H or 2H leads work well for lettering and linework. Construction lines can be prepared with .3- and .4-mm fine-line pencils with 3H or 4H leads. These instruments require no sharpening and help maintain a high-quality, consistently uniform line. However, a different pencil must be purchased for each line thickness required, and the thinner

FIGURE 3–7
CAD-drawn wiring diagram.

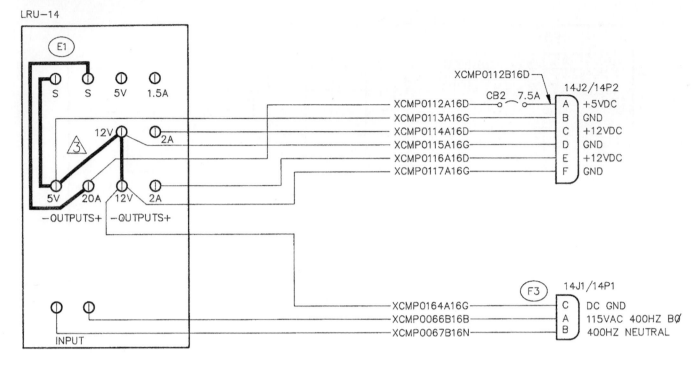

FIGURE 3–8
Schematic diagram showing dashed, emphasized, and regular solid lines.

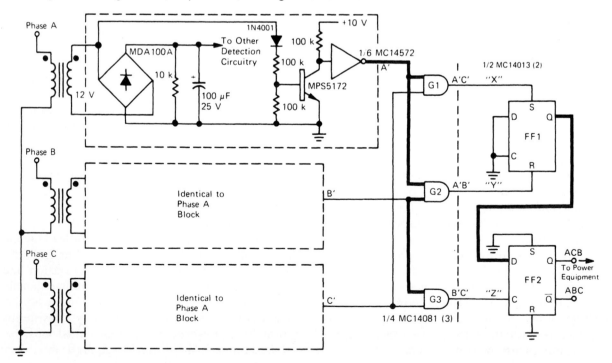

FIGURE 3–9
Logic diagram displayed on a CRT.

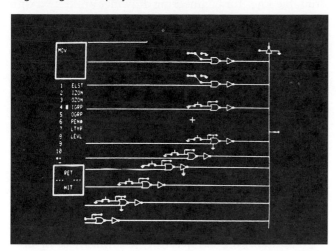

leads tend to break more often when soft grades of lead are used. Holding a thin-line pencil almost vertically will reduce lead breakage.

Using a dust pad or erasing powder helps keep linework and paper clean. Brushing the paper frequently and using an extra piece of paper to rest the hand on while drawing will also help reduce smudges.

Instrument Drawings

All drafters of electronic drawings use triangles and some form of straightedge (T square, drafting machine, parallel straightedge). As in mechanical drafting, electronic drawings are created as instrument drawings and are not to be constructed with anything except high-quality drafting instruments. Vertical lines are to be constructed with a straightedge and triangle. Horizontal lines are to be constructed with a straightedge that will give consistent parallel lines. Curved lines are to be plotted and then drawn with a compass, template, or French or irregular curve. No lines are to be formed without instruments. Only lettering can be formed freehand.

The drawing of an instrument line is a two-step process. First, determine the position and length of the line using dividers or a scale, and then draw the line with correct width and density. This process requires that you draw two completely different lines. The first line is for positioning and is thin and gray, drawn with 3H or 4H lead. This is called a construction line. It is suggested that you complete each electronic drawing problem totally using construction lines and then go over all lines to make them the proper thickness and density. This procedure ensures that all dimensions and

measurements are taken from thin, sharp, accurate points and lines, creating a more precise and correct drawing. Draw the second line uniform, thicker, and more dense using H or 2H lead. Erase the lines used for construction purposes before you darken the drawing, unless company or school practice allows construction lines to remain on the finished drawing. Ask your instructor which method should be followed before completing your project.

Erasing and Keeping the Drawing Clean

Erasing is a necessary part of drafting and, when done properly, improvements to drawings are easily made. The eraser should have good pickup power without smudging, such as a Faber Castell Pink Pearl®. To protect adjacent areas that are to remain, all erasing is done through the perforations of a stainless steel erasing shield. Firmly hold the erasing shield in place on the drawing with one hand while erasing through a particular slot or hole with the other. Be sure not to erase surrounding areas through adjacent openings. After each erasing, brush the drawing so that erasing particles will not be ground into the drawing. Electric erasers like the one in Fig. 3–10 are also available.

All drawings attract dirt. Cleanliness does not just happen; it must be consistently cultivated. The following procedures help to keep drawings clean:

1. *Clean hands:* Periodically wash your hands to remove accumulations of graphite, dirt, perspiration, and body oils.
2. *Equipment:* Periodically clean with soap and water all tools that come in contact with the drawing. Clean

FIGURE 3–10
Electric eraser. (Courtesy Keuffel & Esser/Kratos)

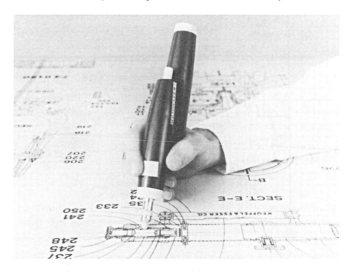

tools that contain wood with a damp sponge. Scrub down drawing boards when they become soiled.

3. *Graphite:* Most *dirt* on a drawing is actually graphite dust (lead particles) resulting from drawing and from the lines themselves. Use the brush to remove this graphite dust before other tools smear it. This will help tremendously to keep drawings clean.

4. *Pencil pointer:* The pencil pointer leaves dust clinging to the lead after sharpening. After sharpening the lead, wipe it with tissue or poke it into a dry cleaning pad or a piece of scrap Styrofoam to remove clinging lead dust. Do not use the same dry cleaning pad to clean the drawing!

5. *Equipment use:* Proper use of the straightedge and triangles will always place these instruments between the hands and the paper, except for lettering. While lettering, always rest the hands on a clean sheet of scratch paper. Even clean hands put body oils onto the paper, and this oil has a magnetic effect on dirt.

Ink Drawings

Electronic drawings are frequently inked on drafting film or vellum. Electronic drawings used in sales literature, technical manuals, and pictorial illustrations are normally drawn in ink, since they need to be of photographic quality.

Ink drawings must be laid out in pencil and then traced or inked over the pencil lines. It is impossible to ink a drawing as it is laid out. Light tables are excellent for inking and tracing drawings. Triangles and templates must be raised from the drawing when ink is used because ink will tend to flow between the two surfaces and cause smeared lines. Specially designed equipment with a ledge or with inking risers prevents the equipment from being flush with the paper or film.

Ink drawings (Fig. 3–10) should be prepared with **technical pens,** not ruling pens. Keeping the technical pen almost vertical helps prevent uneven and ragged linework. If possible, avoid having to take more than one pass for thin and medium lines. Extremely thick lines can be drawn with an appropriate pen size, although better results may be attained if a thinner pen is used to thicken the line in stages.

Allow all ink to dry completely before moving to another portion of the drawing. Some drafters prefer to ink all horizontal lines from the top of the project downward and then from left to right. Erase ink lines very carefully, especially when drawing on vellum or another type of paper. You can easily erase ink from drafting film using the proper type of eraser and applying a small amount of moisture. As in pencil drawings, protect the surrounding lines while erasing.

Line Types

Printable lines for engineering drawings are drawn with different widths to provide specific information. In reality, each line type is a symbol and expresses an idea or communicates a special situation. It is important to understand that the thickness of any line is determined by its intended use and the smallest size to which it will be reduced. Lines representing the same type must be a consistent thickness throughout a single drawing to avoid confusion. The minimum spacing between parallel lines is determined by how much the drawing will be reduced. Two parallel lines that are placed too close together will merge if reduced; this process is called *fill-in* and must be avoided. Normally, .06-in. (1.5-mm) parallel spacing meets reduction requirements for most drawings.

The following list describes the thickness of lines on drawings.

1. *Fine lines:* Thin black lines used to provide information about the drawing or to construct the drawing. These include dimension lines, construction elements (permanent construction lines such as development elements), leader lines, extension lines, and centerlines.

2. *Medium lines:* Intermediate, black lines used to outline planes, lines, and solid shapes (components and so on). Medium lines are also used for hidden and dashed lines and for symbols and flow lines on electronic diagrams.

3. *Heavy lines:* Solid, thick black lines used for the border, cutting plane lines, and break lines. Emphasized diagram lines are also heavy lines.

The one thing that all lines have in common is that they are *black,* clean-cut, precise, and opaque, with sufficient contrast in thickness between different types of lines.

Precedence of Lines

Whenever lines of different kinds coincide in a view, certain ones take precedence. Since the visible features of a part (object lines) are represented by thick solid lines, they take precedence over all other lines. If a centerline and cutting plane coincide, the more important one should take precedence, normally the cutting-plane line, since it is drawn with a thicker lineweight. The following list gives the preferred precedence of lines on mechanical and nondiagramic electronic drawings:

- Visible or object lines
- Hidden lines
- Cutting planes
- Centerlines

□ Break lines
□ Dimension and extension lines
□ Section lines

Line Conventions

This section provides a general overview of the accepted standard line conventions for mechanical drawings. The suggested standard line thicknesses are shown on the line key in Fig. 3–11. Examples are provided in Fig. 3–12. These lines are divided by thickness and format. Line types and conventions are covered in ANSI Standard Y14.2M for mechanical drawings. Electronic drafting includes many of these lines for production, assembly, manufacturing, and pictorial representation of electronic equipment. Suggested lineweights for specific electronic drawings and diagrams are discussed later.

Visible or Object Lines

Visible lines or **object lines** are thick lines used to represent the visible edges and contours of an object so that the views stand out clearly. This type of line is used to draw object outlines on mechanical drawings, component outlines for electronic drawings, and object outlines on sheet metal details. Visible lines must stand out from all other secondary lines on the drawing since they are the most important. In mechanical drawing, visible lines are normally drawn about .032 in. thick (between .5 and .7 mm).

Hidden Lines

Hidden lines consist of short, thin dashes, approximately .12 in. (3.0 mm) long, spaced about .03 to .06 in. apart to show the hidden features of an object. Hidden lines should always begin and end with a dash, except when a dash would form a continuation of a visible line.

Dashes always meet at corners, and an arc should start with dashes at the tangent points. When an arc is small, the length of the dash may be modified to maintain a uniform and neat appearance. Show only lines or features that add to the clearness and conciseness of the drawing. Eliminate confusing and conflicting hidden lines. Excessive hidden lines are difficult to follow. Where hidden lines do not adequately define an object's configuration, a **section** should be taken. Whenever possible, hidden lines are normally eliminated from the sectioned portion of a drawing. Hidden lines should be drawn approximately .016 in. (.35 to .45 mm) thick.

Centerlines

Centerlines are composed of thin, long and short dashes, alternately and evenly spaced, with the long dash at each end. The long dash normally varies in length from .75 to 2 in. (19.0 to 50.8 mm), depending on the size of the drawing. Short dashes should be approximately .06 to .12 in. (1.52 to 3.0 mm), depending on the length of the required centerline. Very short centerlines may be unbroken, with dashes at either end.

Centerlines are used to indicate the axes of symmetrical parts or features, bolt circles, paths of motion, and pitch circles. They should extend about .12 in. beyond the outline of symmetry unless they are used as extension lines for dimensioning. Every circle should have two centerlines that intersect at its center on the short dashes. Extended centerlines are frequently used as extension lines. Centerlines are usually drawn about .012 in. (between .25 and .35 mm) thick.

Dimension Lines

Dimension lines are thin lines used to show the extent and direction of dimensions. Space for a single line of numerals is provided by a break in the dimension line. However, when two lines of numerals are used in the form of limits, one may be placed above and the other below an unbroken dimension line, when this is an acceptable company practice.

Dimension lines should be aligned if possible and grouped for uniform appearance and ease of reading. Parallel dimension lines should be spaced not less than .25 in. (6.35 mm) apart. No dimension line should be closer than .38 in. (9.65 mm) to the outline of a view; .50 in. (12.7 mm) is the preferred distance.

When you draw several parallel dimension lines, stagger the numerals to make them easier to read. When possible, avoid crossing dimension lines and extension lines. Place the shorter dimension lines closer to the view, inside the longer ones, to avoid crossing extension lines. Do not use a centerline, an extension line, or an object line as a dimension line.

All dimension lines terminate with an arrowhead, a slash, or a dot. The preferred ending is the arrowhead. **Arrowheads** should be drawn with a ratio of 3:1, length to thickness. Arrowheads are normally .0625 in. wide and .1875 in. long (1.58 × 4.76 mm). The size used, however, is determined by the drawing scale, the total drawing size and area used, and the reduction requirements. Large, elaborate arrowheads should be avoided. Dimension lines are drawn the same thickness as centerlines, about .012 in. (between .25 and .35 mm).

Extension Lines

Extension lines are used to indicate the termination of a dimension. An extension line should not touch the feature from which it extends but should start approximately .04 to .06 in. (1.52 mm) away and terminate approximately .12 in. (3.0 mm) beyond the dimension line. Where extension lines cross other extension lines,

FIGURE 3–11
ANSI Y14.2M, *Width and Type of Lines.*

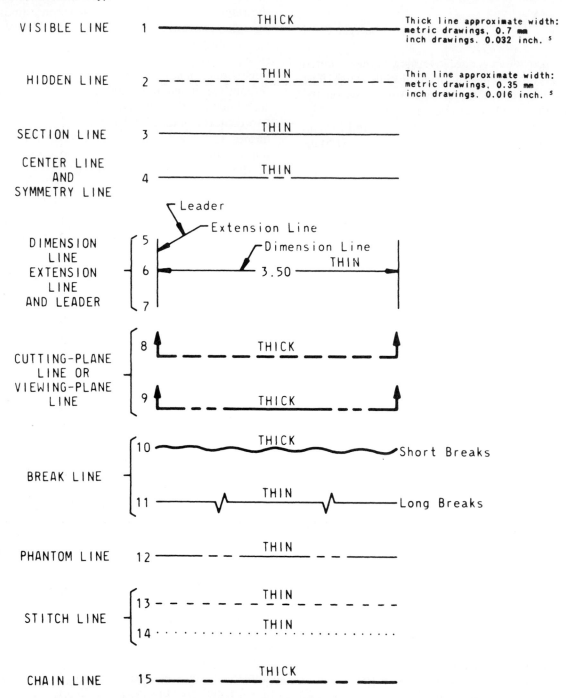

5 The metric line widths agree with ISO/DIS/128 (June 1977) and are not a soft metric conversion of the inch value.

These approximate line widths are intended to differentiate between THICK and THIN lines and are not values for control of acceptance or rejection of the drawings.

FIGURE 3–12
ANSI Y14.2M, *Application of Lines.*

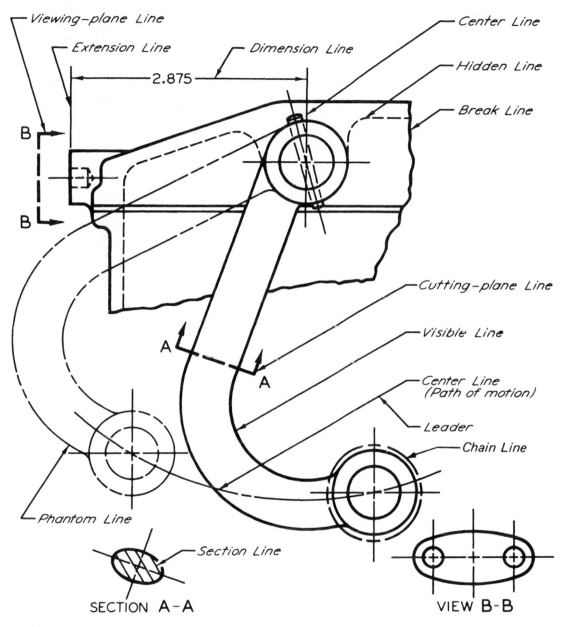

dimension lines, leader lines, or object lines, they usually are not broken unless this is a company's accepted practice. Where extension lines cross dimension lines close to an arrowhead, a break in the extension line at the arrowhead is recommended for clarity. Extension lines are drawn the same thickness as dimension lines and centerlines.

Leader Lines

A **leader** is a continuous straight line that extends at an angle from a note, dimension, or other reference to a feature where attention is directed. An arrowhead at one end touches the feature. A horizontal bar .25 in. (6.35 mm) long at the note end terminates approximately .12 in. (3.0 mm) away from the midheight of the lettering, either at the beginning of the first line or at the end of the last line, never from a point in between. Leaders should not be bent to underline the lettering or part numbers. Unless its unavoidable, they should not be bent in any way except to form the horizontal bar at the lettering.

Leaders usually do not cross. Leaders or extension lines may cross an outline of an object or other extension lines if necessary, but the lines usually are continuous and unbroken at the point of intersection. Where a leader is directed to a circle or circular arc, its direction should be radial. Leader lines are drawn the same thickness as centerlines, dimension lines, and extension lines.

Cross-Section Lines

Cross-section lines are uniformly spaced thin lines used to indicate the exposed cut surfaces of a part in a sectional view. Spacing should be between .06 in. (1.5 mm) and .12 in. (3.0 mm) depending on the possible reduction of the drawing. Cross-section lines are drawn slightly thinner than centerlines and dimension lines.

Phantom Lines

Phantom lines are made up of thin long and short dashes. They are used to indicate alternative positions of moving parts, adjacent positions of related parts, and repeated details. They also show the cast or rough shape of a part before machining. The line starts and ends with the long dash .60 in. (15.2 mm) or longer, and the short dashes are approximately .12 in. (3.0 mm) long, with about .06 in. (1.52 mm) space between. A phantom line is drawn approximately as thick as a hidden line.

Cutting-Plane and Viewing-Plane Lines

Cutting-plane and **viewing-plane lines** are made up of thick, long dashes separated by two short dashes. They are used to indicate the location of cutting planes for sectional views and the viewing positions for removed views. These lines start and stop with long dashes .60 in. (15.24 mm) or longer, and the short dashes are approximately .12 in. (3.0 mm) long, with about .06 in. (1.5 mm) space between. An alternative method uses dashed lines for the total cutting plane. Both methods are acceptable. Cutting-plane lines are normally drawn with a thickness of about .032 in. (.70 mm).

Break Lines

Short **break lines** are thick, freehand, continuous lines that are used to limit a broken view, partial view, or broken section. For long breaks where space is limited, a neat break may be made with long, thin, ruled dashes joined by freehand zigzags. Break lines are drawn as thick as cutting-plane lines when the ragged method is used and about as thick as hidden lines when the long-break method is used.

Electronic and Electrical Line Conventions

Specific lineweights (thicknesses of lines) are not standardized for electronic drawings. ANSI Standard Y32.2, *Graphic Symbols for Electrical and Electronic Diagrams,* allows the drafter to decide the lineweights:

> A4.3 Line Width. The width of a line does not affect the meaning of the symbol. In specific cases, a wider (heavier) line may be used for emphasis.

Although this standard refers to symbols, it can also be applied to the lines connecting the symbols, since they are normally drawn with the same thickness. Note that electronic and electrical drawings do not define the thickness of lines as meaning anything in particular, as do mechanical line conventions.

All electrical and electronic lineweights should conform with ANSI Y14.2. In 15–3.5, *Line Conventions and Lettering* of ANSI Standard USAS Y14.15–1966, *Electrical and Electronic Diagrams* makes the following recommendations:

> The selection of line thickness as well as letter size should take into account size reduction or enlargement, when it is felt that legibility will be affected. A line of medium thickness is recommended for general use on diagrams. A thin line may be used for brackets, leader lines, etc. When emphasis of special features such as main or transmission paths is essential, a line thickness sufficient to provide the desired contrast may be used.

Fig. 3–13 illustrates recommended lineweights for electronic drawings and diagrams. The labels on this figure are explained in the following list:

A. **Dimension Line, Fig. 3–13 (1):** Same as for mechanical conventions; used on dimensioned production drawings, pictorial drawings (1), component multiview drawings (6), and sheet metal details and layouts.

B. **Object Line/Visible Line, Fig. 3–13 (6):** Same as for mechanical drawings; found on the same type of drawings as dimension lines.

C. **Centerline, Fig. 3–13 (6):** Same as for mechanical drawings.

D. **Component Outline, Fig. 3–13 (7):** Can be drawn as thick as an object line, although a less thick (medium to thin) line is preferred, depending on the reduction amount; used on PC board layouts (7) and a variety of other electronic drawings.

E. **Phantom Line, Fig. 3–13 (2):** Used to represent future or alternative arrangements (as on the wiring diagram shown in No. 2).

F, G, H, I. **Diagram Lines, Fig. 3–13 (4, 8, 9):** The majority of all lines used in electrical and electronic drawing, since they represent all diagram flow lines and symbols; not standardized; used on block diagrams (3), logic diagrams (8), schematic diagrams (4, 9), and wiring diagrams (2). (The thickness of a

FIGURE 3–13
Line key for electronic drawings.

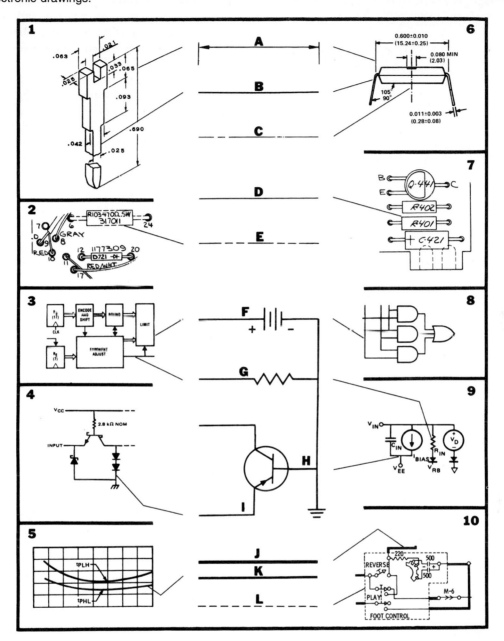

diagram line is determined by the company standard practice, desired emphasis of a particular circuit, and the reduction amount.)

J, K. **Emphasis Line, Fig. 3–13 (5, 10):** A thicker line (10), used when a particular flow line (or symbol) on a diagram needs to be emphasized; also found on graphs (5).

L. **Dashed Line, Fig. 3–13 (10):** Used on electronic diagrams (10); also used to indicate mechanical linkage or connection between components (10, foot control switch). JIC Electronic Standard EL–1–71

states that "Discrete items or the equivalent circuit contained within a packaged (plotted) unit shall be shown within dashed lines on the elementary diagram. For clarity, individual circuits within the package may be separated and each enclosed within dashed lines."

The dimensioned IC package in Fig. 3–14 uses a variety of conventional mechanical lines, including visible lines, object lines, centerlines, dimensioning lines, extension lines, and leader lines.

FIGURE 3–14
Dimensioned orthographic drawing of integrated circuit package. (Courtesy Texas Instruments, Inc.)

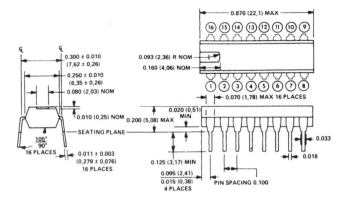

Sketching

Sketching is the primary means of graphic communication among engineering personnel. The ability to sketch is an essential skill for all drafters, designers, and technicians. Sketching is used to convey original design ideas from the designer to the drafter and to clarify design alternatives. Sketches are also used to lay out diagrams so that they can be digitized on a CAD system. In Fig. 3–15 the CAD operator is digitizing a schematic diagram sketch.

In general there are three types of sketches: pictorial, orthographic with or without dimensions, and diagramic. The use of graph and grid paper speeds the construction of any sketch and enables it to be used as a digitizing layout since all digitizers are based on a grid pattern.

FIGURE 3–15
Drafter digitizing sketch. (Courtesy Bausch & Lomb Interactive Graphics Division)

As a designer, sketching allows you freedom to try alternative positions of components and trial layouts for electronic diagramming. Sketch with thin, light construction lines, starting with box shapes. Darken the shapes only after the design is complete. Sketch centerlines and lines establishing symmetry in the early stages to locate important features of the design or layout. Block in circles and circular arcs before drawing the curve, using the diameter as the controlling dimension.

Isometric sketching is used for clarifying the design of three-dimensional parts and objects and can aid in establishing the proper view orientation for a pictorial or multiview drawing.

Figure 3–16 shows the original sketch and the finished electronic diagram. The sketch was completed on vellum with a *fade-out* grid.

Lines

Lines and their relationships are the most important parts of all engineering disciplines. This is particularly true of electronic drafting and design, where a vast majority of drawings are two-dimensional diagrams primarily made up of vertical and horizontal lines and symbols. Block diagrams are composed of straight, perpendicular vertical and horizontal lines. Logic, schematic, and wiring diagrams are primarily constructed with straight lines except for accompanying symbols and components.

A line is considered to have length but no width. A straight line is the shortest distance between two points and is implied when you speak of a *line*. A line that bends is called a *curve*.

Parallel lines are equally spaced along their entire length, neither becoming closer together nor further apart. The symbol for parallel lines is / /.

Perpendicular lines lie at an angle of 90° to one another and can be intersecting or nonintersecting.

Horizontal and vertical lines are easily constructed with the straightedge and a triangle or a drafting machine.

Curved Lines

Arcs, circles, and other curved lines require special linework techniques. The compass lead is fixed in the compass and cannot be rotated. Noncircular curves are drawn with an irregular curve as a guide, but the guide only fits the curve for a short distance. Moreover, a curve must be drawn equal in width to the straight lines in order to produce a uniform drawing.

The use of the compass and the irregular curve to create dark, consistent linework is typically one of the most frustrating aspects of mastering drafting. Circle, ellipse, and other curved templates are available in most

FIGURE 3–16
Original sketch and completed schematic.

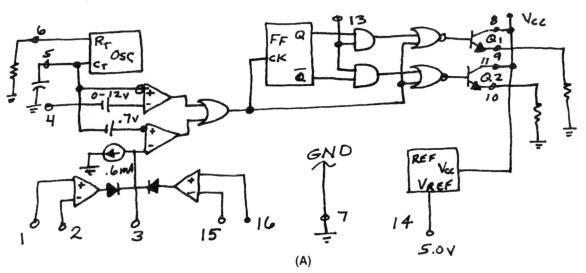

(A)

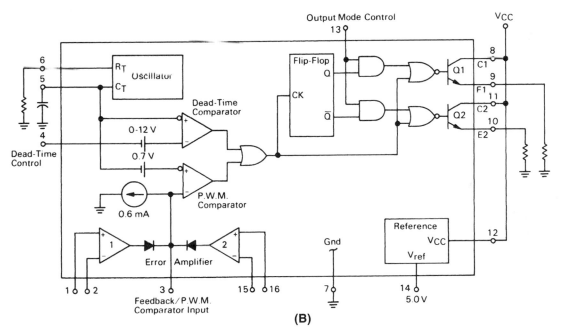

(B)

standard sizes. These excellent tools can be applied to many of the constructions. Unfortunately, they are limited in sizes and shapes and are relatively expensive. It is suggested that you wait to practice curves with templates until you have mastered the compass and irregular curve.

The Bow Compass and Dividers

A good bow compass and dividers are essential to the accurate construction of all forms of engineering drawings. (See Figs. 1–36, 1–37, and 1–38 for examples of compasses and dividers.) A bow compass has a center

wheel that is used to set and hold the spacing between the center point and the lead. Dividers do not have a center wheel and are used to quickly set off measurements from one view to another, which is extremely useful in the construction of electronic diagrams.

The centering point for the compass is either a tapered point or a short needle point projecting from a wider shaft to create a shoulder. The needle point is better for beginners because it provides a stop to limit the point's penetration into the paper. A small piece of drafting tape can be applied to the drawing at the center of the arc or circle to be constructed, and the centerlines can be drawn over the tape. Using this method will restrict the compass point from penetrating the drawing medium while providing a stable, secure centering point from which to swing an arc or circle. Circles smaller than .50 in. (12.7 mm) are much easier to draw with a template than with a compass. A compass is confined to odd sizes and large circular shapes. Electronic templates contain circle shapes for such components as transistors and meter encasements.

Dividers have two identical tapered metal points, one of which can be replaced with a piece of 4H lead. This lead can be used to set off dimensions instead of the two metal points, which tend to mar the drafting medium.

The compass lead should be a piece of the drafting pencil lead (same grade lead). Then both straight and curved lines will be drawn with the same lead, and it will be easier to maintain uniformity. The lead is secured in the compass with about ⅜ in. (9.52 mm) exposed and is sharpened with a sandpaper block. Use care while sharpening to keep the line through the point and the lead perpendicular to the sandpaper. Make a flat cut that leaves an oval surface, called a *bevel*. The bevel should be about three times as long as the diameter of the lead. The resulting *point* is chisel shaped and should have about the same taper, when viewed from the side, as the cone shape of the drafting pencil. Do not adjust the lead in the compass after it is sharpened because it is almost impossible to properly reposition the chisel shape. Adjust the centering point so that the midpoint of the needle point is even with the end of the lead. The beveled end can be on either side of the lead. Both sides of the lead may be beveled to create a thin, dark curve with a point that is longer lasting. The compass can now be adjusted to the required radius and used to draw a circle or arc. Locate the center of the required circle or arc and draw a horizontal construction line. Set a distance equal to the radius of the circle or arc to be drawn. Then set the compass to this distance and draw a construction circle.

Measure the diameter of the circle. The reading should be twice the given radius. To get an accurate diameter reading, make certain the measurement is taken along a line that passes through the center point of the circle. Any difference between the measured diameter and twice the given radius is twice the error of the compass setting. The width of the line that will be drawn is determined entirely by the width of the lead (thickness of the bevel) at the moment. As a circle is drawn, the point begins to shorten and the line to widen. This tends to make the line thicker, but may also create a fuzzy line. To get a crisp, clean, dark line, keep the bevel very sharp, and draw the line thin and dark by rotating the compass a couple of times. Then resharpen the lead and draw another line touching the first line

but slightly larger or smaller, depending on the required dimension and the size of the first line. This procedure always gives a sharp, clear line. A longer taper on the lead holds a line width longer.

Irregular Curve

Noncircular curves require the use of an irregular curve to make smooth, printable lines. Examples of noncircular curves are the ellipse, an angular view of a circle, and the helix. Irregular curves are manufactured in a great many shapes and sizes, and it is a good idea to have a variety of types from which to choose.

Curves drawn with irregular curves are usually determined from a series of **plotted points,** as in Fig. 3–17. Then a curve is drawn that includes all these points. Good results can be obtained if the following steps are taken:

1. Sketch lightly by hand a smooth line that includes the plotted points. It is easier to set the irregular curve to a line than to a series of points.
2. Set the irregular curve so that it fits to a part of a line, as shown in Fig. 3–17, usually a minimum of four points.
3. Draw the line that fits the curve, but stop one point short, before the end of the fit. In Fig. 3–17, A fits from point 1 to point 5 but is drawn from 1 to 4. B fits from point 3 to point 9 and is drawn from 4 to 8. C fits from point 17 to point 23 and is drawn from 18 to 22.
4. Reset the irregular curve to fit each next part of the curve and include the last portion of the already drawn line. This overlapping will give a smoother curve.

Neater work results if the sketched curve and the first series of fitting the irregular curve are all done on a tracing paper overlay. When using an overlay, mark

FIGURE 3–17
A noncircular curve drawn with the aid of a French, or irregular, curve.

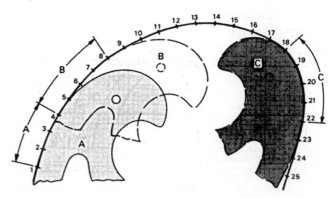

FIGURE 3–18
Inking with an irregular curve and a technical pen.
(Courtesy Koh-I-Noor Rapidograph, Inc.)

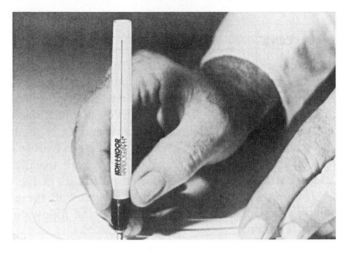

FIGURE 3–19
Noncircular curves plotted on a graph. (Courtesy Simpson Electric Co.)

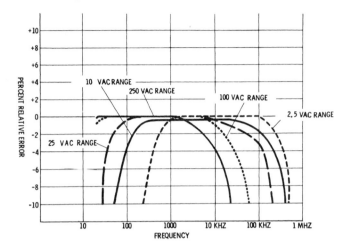

the ends of each segment of the line as it fits the irregular curve so that the same *fits* can be used in the next step. When all fits are made, place the tracing paper overlay under the drawing and carefully align the curve under the plotted points. Trace the curve onto the drawing using the irregular curve fits marked on the overlay. This technique has two advantages. First, all the fitting is made on a throwaway paper so that erasures can easily be made without erasing the plotted points. Second, the accuracy of the fit can be seen before the final drawing of the curve, when the overlay is positioned under the drawing.

The overlay technique is particularly valuable when the curve is symmetrical. For example, an ellipse has four identical curves; two are mirror images of the other two. All are symmetrical about the major and minor axes. It is only necessary to fit one of these curves and then to duplicate this fit on the other three.

The plotting of the points of an irregular curve is particularly important if a smooth curve is to result. A small error in the position of a point can easily cause irregularities in the curve. The plotted points should be spaced farthest apart where the curve is straightest and closest together where the curve is the sharpest.

In Fig. 3–18 the drafter is drawing a curved line with a French curve and technical pen. Note that the pen is held almost vertical to the work. A slight angle away from the irregular curve edge must be maintained in order to prevent the ink from seeping between the plastic curve and the paper.

Graphical data are sometimes represented by noncircular curves, as in Fig. 3–19. **Graph lines** are normally plotted on a given grid pattern on rectangular coordi-

nate paper, representing two variables. The horizontal coordinate is usually plotted as the independent variable and the dependent variable is plotted vertically. The horizontal or X axis is called the abscissa and the vertical or Y axis, the ordinate. The origin of data may be set in a number of different places on the graph, such as the lower-left corner of the graph with 0–0, as in Fig. 3–20, or the center, as in Fig. 3–21. The curved graph lines were plotted with this technique and drawn with an irregular curve.

In Fig. 3–22 a chart has been created with a CAD system. Charts and graphs completed on CAD systems are easily plotted with small multipen plotters, as shown in Fig. 3–23.

FIGURE 3–20
Thermal characteristics of 68-pin leadless package.
(Courtesy Motorola, Inc., Semiconductor Products Sector)

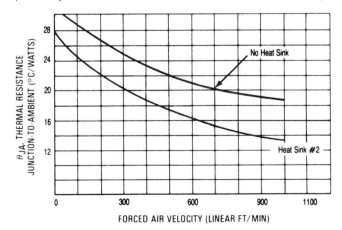

FIGURE 3–21
Voltage versus current
characteristics for an FET.

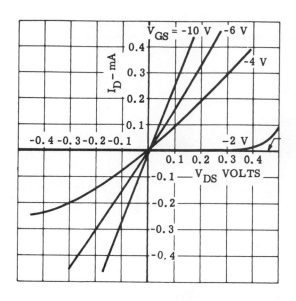

FIGURE 3–22
CAD-generated chart.

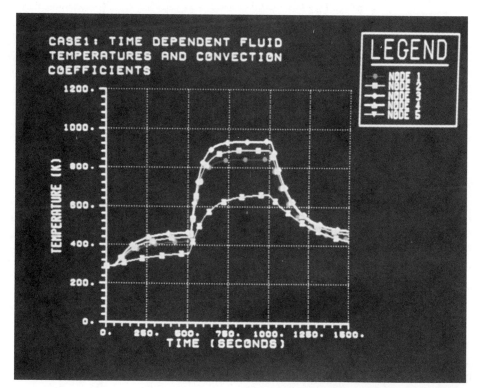

FIGURE 3–23
Houston Instrument's PC Plotters. (Courtesy Houston Instrument)

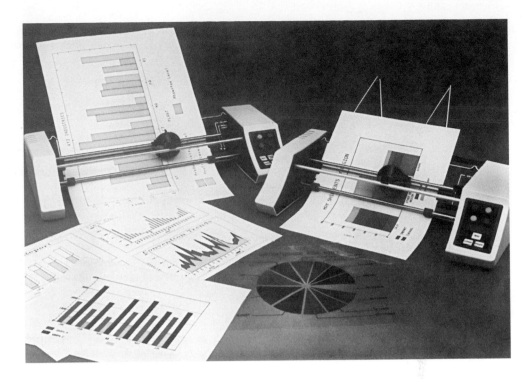

REVIEW QUESTIONS

3–1 *True or false:* A solid, medium-weight line is used to show the hidden portions of an object on a multiview drawing.

3–2 Diagram lines represent ( ) on an electronic diagram.
  a. object lines   c. future circuits
  b. circuit flow lines   d. dimensions

3–3 A cutting-plane line shows where the object is to be ( ).
  a. dimensioned   c. sectioned
  b. projected   d. broken

3–4 *True or false:* CAD-generated lines can be plotted at any desired weight (thickness) depending on the system software capabilities.

3–5 *True or false:* The lineweights for an electronic symbol and circuit flow line should be different thicknesses.

3–6 Which of the following is not considered an electronic diagram drawing?
  a. block   c. pictorial
  b. schematic   d. logic

3–7 List four typical uses for a sketch.

3–8 Which type of inking pen is used throughout industry?
  a. ruling pen   c. inking drop compass
  b. technical pen   d. felt pen

3–9 How are lines used as symbols?

3–10 Technical pens should be held ( ).
  a. at a 60° angle   c. at a 67 ½° angle
  b. almost horizontal   d. almost vertical

3–11 *True or false:* The exact thickness of a line on an electronic diagram defines its use and is covered by an ANSI Standard.

3–12 Define the following terms:
  visible line   break line
  technical pen   cutting-plane line
  diagram line   arrowhead
  section   hidden line
  dimension line   centerline
  leader   component outline
  extension line   emphasis line
  plotted points

PROBLEMS

3–1 Draw the block diagram of the garage-door opener control shown. Use all uppercase lettering. Block lines are to be .7 mm thick and flow lines are to be .5 mm thick.

PROBLEM 3–1
Block diagram of a garage-door opener.

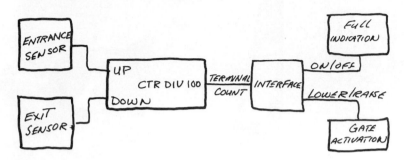

3–2 Draw the flow diagram of the simple computer shown. Use upper- and lowercase lettering. Boxes are to be .7 mm and flow lines .5 mm thick.

PROBLEM 3–2
Flow diagram of a simple computer.

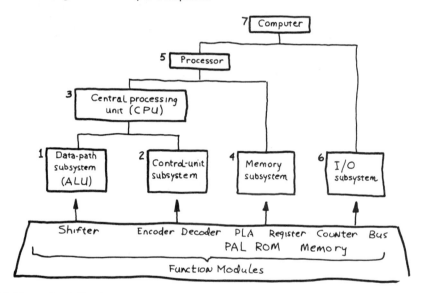

3–3 Draw the following lines 3 in. (75 mm) long.

phantom	diagram	cutting plane
centerline	break	section
construction	circuit	hidden
leader	extension	object
dimension	flow	

3–4 Redraw the diagrams in Figs. 3–3 through 3–8 as assigned. Use proper lineweights and lettering.

3–5 Sketch the diagrams in Figures 3–3 through 3–8 as assigned. Use proper lineweights and lettering.

3–6 Using the graph in Fig. 3–20, plot the following curves:

Curve 1		Curve 2		Curve 3	
X	Y	X	Y	X	Y
0	28	0	20	0	16
100	26	200	14	300	12
400	22	500	6	700	10
800	18	1,100	0	1,000	4
1,100	0			1,200	0

4

PROJECTION AND DIMENSIONING

INTRODUCTION

All forms of engineering and technical work require that a two-dimensional surface (paper) be used to communicate ideas and give the physical description of shapes. Here, projections have been divided into two basic categories: **pictorial** and **multiview.** This simple division separates single-view projections (oblique, perspective, and isometric) from multiview projections (orthographic). Pictorial projections are covered in Chapter 5.

Division of types based on whether the drawing is a one or multiview projection separates projection types into those used for engineering working drawings (orthographic/multiview) and those used for display (technical illustration, manual preparation).

Electronic drawings for production, manufacturing, and assembly use multiview orthographic projection almost exclusively, except where a pictorial projection is needed to explain particular aspects of a design.

In Fig. 4–1, the angle block shows each of four projection types with the same scale. This figure illustrates the difference between the types of projections. It also points out some of their shortcomings.

Pictorial projections are single-view drawings that do not lend themselves to the communication of engineering data except as rough sketches of preliminary ideas. **Perspective** projections are constructed with projecting lines that converge at a point; therefore, they do not show the true dimensions of a part, though this method provides the most *lifelike* appearance. The **oblique** method distorts when a part's depth becomes too great. The **isometric** method uses full-scale dimensions for all lines that are vertical or parallel to the axes (receding at 30°). It is therefore more useful for engineering sketching.

Multiview drawings, because they show the part in more than one view, are not lifelike. This, however, is their only major drawback. Multiview projection presents the part's top, front, and side in related adjacent views. All dimensions are drawn to a predetermined scale, and the three basic views can be used to project any number of needed **auxiliary views** in order to solve for or establish engineering data. An auxiliary view is any projection of a part other than one of the **six principal views:** *front, top, right side, left side, back,* and *bottom.*

MULTIVIEW PROJECTION

Multiview orthographic projection is the primary means of graphic communication used in engineering work. Figure 4–2 shows a multiview drawing that conveys

FIGURE 4–1
Types of projections.

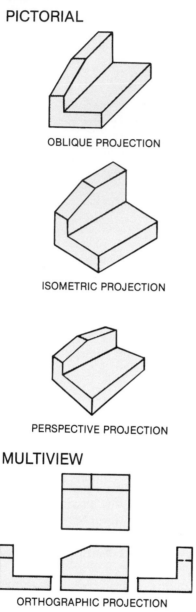

PICTORIAL

OBLIQUE PROJECTION

ISOMETRIC PROJECTION

PERSPECTIVE PROJECTION

MULTIVIEW

ORTHOGRAPHIC PROJECTION

Multiview drawing is the use of orthographic projection to solve for advanced technical data involving the spatial relationship of points, lines, planes, or solid shapes. There are two primary means of making orthographic projections: the **normal method** and the **glass box method.** In the normal or natural method the part is viewed perpendicular to each of its three primary surfaces.

In the glass box method the user must imagine that the part, with its points, lines, planes, and solid shapes, is enclosed in a transparent *box.* A view of the part is established on its corresponding glass box surface or plane by perpendicular **projectors** originating at each point of the part and extending to the box surface, as shown in Fig. 4–3. This box is hinged so that it can be unfolded onto one flat plane (the paper).

In the glass box method, all six sides are revolved outward so that they are in the plane of the paper. With the exception of the back plane, all are *hinged* to the front plane. The back plane is normally revolved from the left side view. Each plane is perpendicular to its adjacent plane and parallel to the plane across from it before it is revolved around its hinge line. A *hinge line* is the line of intersection between any adjacent image plane, including principal and auxiliary views.

The left side, front, right side, and back are all elevation views. Each is a vertical plane. In these views the height dimension, elevation, top, and bottom can be determined and dimensioned.

The top and bottom planes are in the horizontal plane. The depth dimension, width dimension, front, and back can be established in these two horizontal planes.

In the United States the six principal views of an object, or the glass box, are normally drawn in **third angle orthographic projection.** In third angle projection the **line of sight** goes through the image plane to the part, as shown in Figs. 4–3 and 4–4. To obtain each view of the object, the drafter must assume that the part is projected back (along the lines of sight) to the image plane. Projectors are used to illustrate this projection from the part to where they intersect the image.

The lines of sight represent the direction from which the part is viewed, as illustrated in Fig. 4–3. The vertical lines of sight (A) and horizontal lines of sight (B) are assumed to originate at infinity. They are always perpendicular to the image plane, represented by the surfaces of the glass box (top, front, and right side). Projection lines (C) connect the same point on the image plane from view to view, always at right angles. Remember that the part could be any graphical form.

A point is projected on the image plane where its projector or line of sight pierces that image plane. Point

ideas, dimensions, shapes, and procedures for manufacturing a sheet metal part for electronic equipment. Multiview projection completely describes an object's shape and dimensions in two or more views that are normally projected at 90° angles to each other, or at specified angles for auxiliary views.

With the widespread use of computer-aided design (CAD), computers are now being used for many projects. This new medium eliminates hand-drawn linework and lettering but still requires knowledge of multiview projection. Knowledge of engineering projection methods and dimensioning conventions is essential regardless of the method used: CAD or manual.

FIGURE 4–2
Sheet metal enclosure.

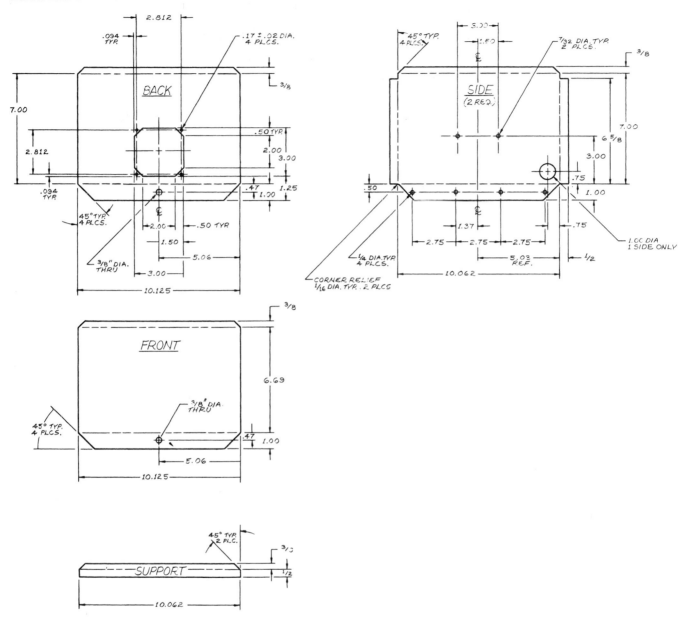

1 in Fig. 4–3, which represents a corner of the part, has been projected onto the three primary image planes.

The Glass Box and Hinge Lines

Each image plane or surface of the glass box is connected at right angles to an adjacent view. The top view is *hinged* to the front view, as is the right side view. These hinge lines are the intersection of the perpendicular image planes. Normally, hinge lines are not shown on technical drawings.

Figure 4–4 (1) shows the part pictorially. In Fig. 4–4 (2), the part is enclosed in a glass box. The top image plane is shown being rotated about the line of intersection and the hinge line, which is between the top image plane and the front image plane. The side image plane is rotated about the hinge line, between the side and the front image planes.

Figure 4–4 (3) shows the glass box opened into the plane of the paper. The front view is assumed to be stationary. Each required view is then rotated until it is in the same plane as the front view.

FIGURE 4–3
Multiview projection.

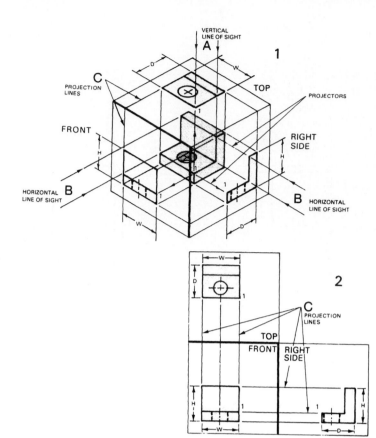

FIGURE 4–4
The glass box and multiview projection.

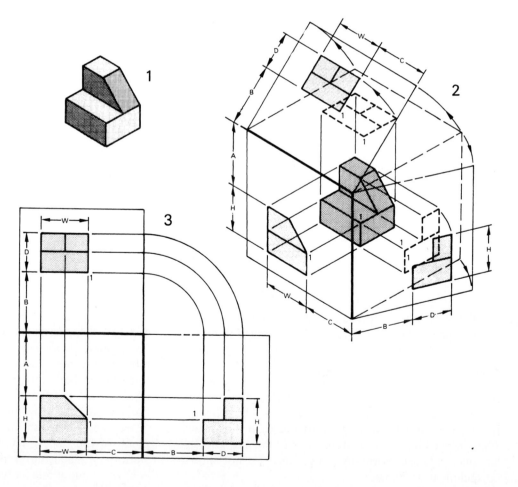

Point 1 is located on the corner of the object and is shown projected onto each of the three image planes.

The line of sight is at right angles to the projection plane. To properly visualize this, you must imagine standing in front of the part with the image plane between you and the part. Your position will change for every view of the object so that your line of sight is at a right angle to each image plane.

Auxiliary Views

Any view that lies in a projection plane other than the horizontal, frontal, or profile plane is considered an **auxiliary view.** This type of projection is essential when the part to be drawn is complex and has a variety of lines or planes that are not parallel to one of the three principal planes.

Primary auxiliary views are projected from one of the six principal views. A primary auxiliary view is perpendicular to one of the three principal planes and is inclined to the other two. *Secondary auxiliary views* are projected from a primary auxiliary view and are inclined to all three principal planes of projection. *Successive auxiliary views* are projected from secondary auxiliary views.

In industry, auxiliary views are used to describe the true configuration of a part and to give its dimensions in views that show inclined lines or planes true size. In most cases only partial auxiliary views are constructed.

Primary auxiliary views can be divided into three types: frontal (front), horizontal (top), and profile (side).

These three types are represented in Fig. 4–5, where auxiliary view A is projected from the top view, aux-

iliary view B is projected from the front view, and auxiliary view C is projected from the side view. Each auxiliary projection in this figure is a partial view, showing only the inclined surface as true shape. This is normal industry practice since the projection of the total part would add little to the understanding of the part's configuration and might actually confuse the view. Hidden lines that fall behind the true shape surface in an auxiliary view are normally not drawn for the same reasons.

Besides being projected from one of the three principal views, each primary auxiliary view has common dimensions with at least one other principal view. The height (H) dimension in the front view is used to establish the H dimension in auxiliary view A. The depth (D) of the part can be found in the top and side views and is used to establish the D dimension in auxiliary view B.

Selection of Views

The proper selection of views and view orientation must take into account the part to be drawn and its natural or assembled position. The choice of additional views, after the top and front views are established, is determined by the configuration of the part and the minimum number of views necessary to describe it graphically and show its dimensions.

For cylindrical parts, only one view may be necessary, since the diameter dimensions describe width and depth but features along the length are dimensioned in the given view. This is called a one-view drawing. Figure 4–6 shows a one-view drawing of a panel. The part is a thin sheet of metal and therefore does not require a second view since its depth or thickness is *called out* (specified).

A three-view drawing of a bracket is shown in Fig. 4–7. In most cases, engineering drawings require three or more views.

Sectioning

An imaginary cut taken through an object on a preestablished plane or planes perpendicular to the line of sight is called a *section*. **Sectioning** exposes the interior shape and dimensions of the part. If hidden lines in the exterior view do not adequately define the part, then a section should be taken for clarification. The exposed, cut portions and surfaces of the section are defined by section lines and cross-hatching, normally drawn at a 45° angle. Section lines are spaced according to the overall size of the sectioned surface and the expected reduction requirements. A distance of .12 in. (3.0 mm) between section lines can be used for most drawings. In some

FIGURE 4–5
Auxiliary views.

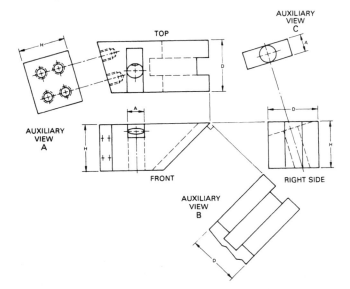

FIGURE 4-6
Panel.

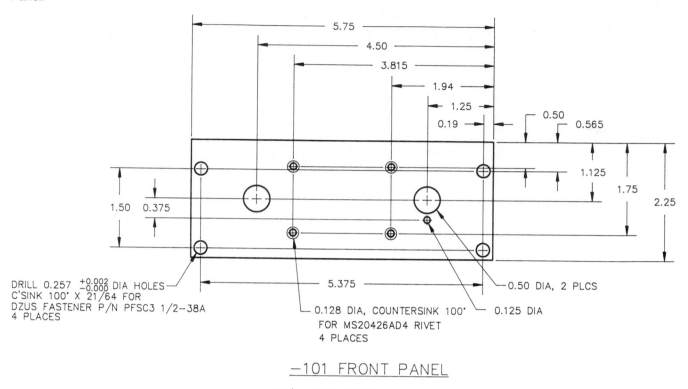

DRILL 0.257 $^{+0.002}_{-0.000}$ DIA HOLES
C'SINK 100° X 21/64 FOR
DZUS FASTENER P/N PFSC3 1/2-38A
4 PLACES

0.128 DIA, COUNTERSINK 100°
FOR MS20426AD4 RIVET
4 PLACES

0.50 DIA, 2 PLCS

0.125 DIA

−101 FRONT PANEL

MAT'L: ALUM SHEET 2024−T3, 0.063 THK

cases the material of the part is identified by a specific section *symbol*. The standard section cross-hatching for all types of materials uses evenly spaced parallel lines.

The *general section* symbol used for all materials is the same as the *cast iron* symbol, or evenly spaced parallel lines (Figs. 4–8 and 4–9).

The **cutting-plane** line is shown on the view where the cutting plane appears as an edge (Fig. 4–8). The ends of the cutting-plane line form a corner of 90°. These lines are terminated by arrowheads to show the direction of sight for viewing. In most cases, the section is defined by capital letters, A–A, B–B, C–C, and so on (Fig. 4–8). Arrows on cutting-plane lines should be larger and of a different configuration than those used for dimension lines.

For simple sections or symmetrical parts, where the location of the section is obvious, the cutting-plane line may be omitted, for instance, where the adjacent view is the section: top view with front section view, front view with side sectioned view. Two or more mating parts must have section line angles at different degrees to differentiate the parts.

Solid round parts such as hardware (nuts, bolts, washers, rivets, shafts, pins) and other solid machine elements that have no internal construction are not shown sectioned even though the cutting plane passes through these features. They are more easily recognized by their exterior surfaces.

Sections that are taken longitudinally through webs, ribs, spokes, gear teeth, or similar *solid* elements should not show section lines even though the cutting plane passes through these elements.

A variety of section types is used on engineering drawings. In Fig. 4–8, sections A–A and B–B are *removed* sections. They are also *partial* sections. Section C–C is a *full* section. The detail of the memory pin in Fig. 4–9 makes use of partial sections E–E and F–F, *detail* sections B, C, and D, and *enlarged* sections taken from A–A.

DIMENSIONING

A wide variety of electronic drawings requires dimensioning. Sheet metal details and developments for elec-

FIGURE 4–7
Bracket.

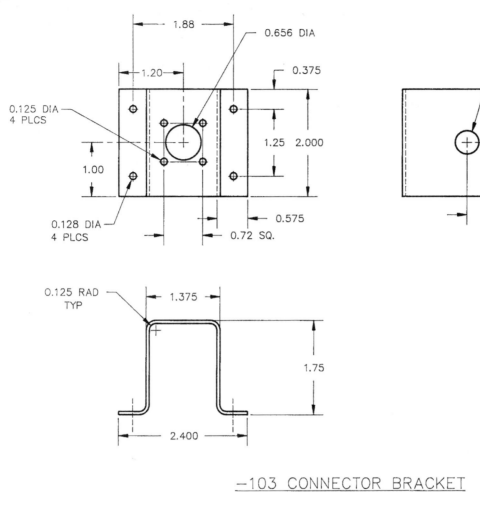

−103 CONNECTOR BRACKET

MAT'L: ALUM SHEET 5052−H32, 0.063 THK

tronic equipment as well as cabinets, enclosures, and panels (Figs. 4–8 and 4–9) need extensive dimensioning. Printed circuit boards also require dimensioning for accurate manufacturing and production.

CAD systems require that the dimensioning be placed and calculated by the drafter. In other words, you still have to know how to dimension the drawing.

Drawings should be dimensioned for the end product with dimensions and notes. Dimensions must be provided between points, lines, or surfaces that are functionally related to each other or control relationships of other parts.

Each dimension on a drawing (Figs. 4–8 and 4–9) has a **tolerance,** either implied or specified. A general tolerance specification is given in the title block, with specific tolerances provided with each appropriate dimension when required. Tolerances are discussed at the end of the chapter.

Units of Measurement, Linear Dimensions

For some drawings in the United States, linear dimensions are expressed in inches. When the metric system (SI units) is used, dimensions are expressed in millimeters or centimeters. In either system, dimensions should be shown only to as many decimal places as accuracy requires. Omit the inch or millimeter symbol unless there is a possibility that the dimension may be misunderstood. When U.S. customary units are used, fractions and decimals should not be mixed. (See Appendix A for U.S. customary and metric equivalents.)

Dual dimensioning (including both U.S. customary and metric units) is sometimes used on drawings. The U.S. customary measurement is normally placed above the metric equivalent. The U.S. customary measurement is in decimal inches and the metric measurement is given in millimeters. The top measurement (or first measure-

FIGURE 4-8
Upper housing fabrication.

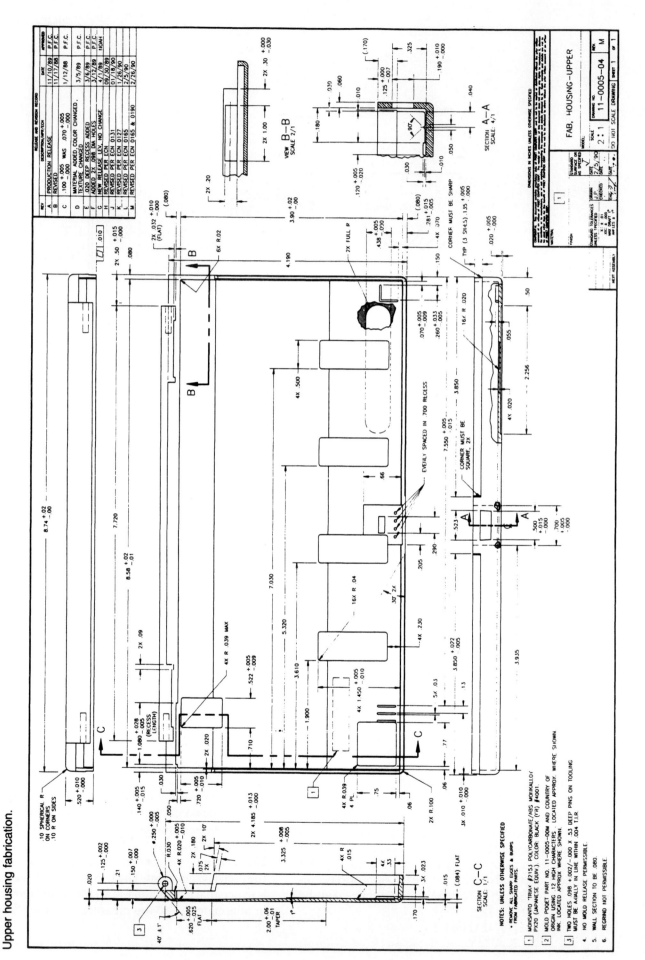

FIGURE 4-9
Memory pin connector.

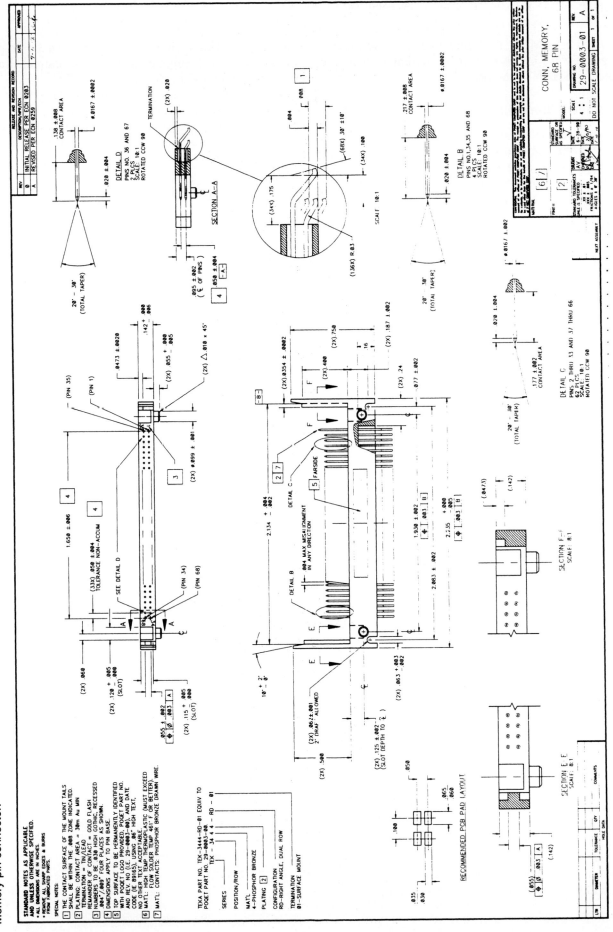

ment when placed on the same line) in dual dimensioning is always in the unit of measurement that was used in the design of the part.

Dimensioning Numerals

Whole numbers in the inch system are normally shown to at least one decimal place (1.0, 2.0, 3.0, etc.). This practice prevents dimensions from being lost on the drawing, a common occurrence when the number 1 is not accompanied by a decimal point and a zero: 1.0.

Common fraction dimensions are seldom used in the electronics industry. Before the adoption of the decimal inch, common fractions were used for subdivisions of the inch to specify nominal sizes and dimensions. Some firms still use this system, especially where the tolerance factor is relatively unimportant. Older company drawings also show this type of dimensioning.

A method of rounding off numbers has been adopted by ANSI (American National Standards Institute). A decimal value may be rounded off to a lesser number of places by the following procedure:

1. Where the digit to be dropped is less than 5, there is no change in the preceding digits.
   Examples:
   .47244 rounds to .4724
   .1562 rounds to .156
   .20312 rounds to .2031
   .35433 rounds to .3543
2. Where the digit to be dropped is greater than 5, the preceding digit is increased by 1.
   Examples:
   .23437 rounds to .2344
   .55118 rounds to .5512
   .03937 rounds to .0394
   .6406 rounds to .641
3. Where the digit to be dropped is 5, round the preceding digit to the nearest even number.
   Examples:
   .98425 rounds to .9842
   .59055 rounds to .5906
   .19685 rounds to .1968
   .4375 rounds to .438

Drawings should be to a scale that allows the object to be easily read and accurately interpreted. Scales should remain constant within a given project where multiple drawings are needed. The choice of scale must take into account the maximum reduction required for the drawing. Scales should be stated in the title block: 1:1 (full scale), ½ (half scale), ¼ (quarter scale), 1:5, 1:10, and so on.

When the detail on a drawing is too small, an enlarged scale can be used: 2:1 (two times size), 5:1, 10:1,

and so on. Printed circuit artwork is normally prepared at a 2:1 or 4:1 enlarged scale when manual drafting and taping methods are used.

In some cases more than one scale is used on a drawing, as when a portion of the drawing is *blown up* (enlarged). The predominate scale is to be shown in the scale area within the title block, and any other scales are placed under the appropriate view.

Dimensioning Elements

Dimension lines were introduced in Chapter 3. Dimensioning elements include **leaders, dimension lines, extension lines,** and **arrowheads.** This chapter concentrates on the actual use and placement of these elements.

Leaders are used to point out a curved feature of a drawing or to reference a portion or surface of an object. In Fig. 4–10 the three most common uses of a leader are shown: to call out a hole diameter, to call out a radius, and to reference a surface or a part with a note. When a leader is used to dimension a circle or radius arc, it must point at or from the center of the circle or radius.

Leaders and their accompanying notes and callouts should be kept to the outside of dimension lines and away from the part being dimensioned. It is poor practice to put a note or dimension on the part itself. Care should be taken not to cross leader lines, although leaders can cross object, dimension, and extension lines.

Spacing Dimensions

Dimension lines should be positioned as shown in Fig. 4–11. The minimum distance from the first dimension to the part outline should be .375 in. (10 mm), and the minimum spacing between parallel dimensions should be .25 in. (6 mm). Note that .50 in. (12 mm) from the part and .375 to .50 in. (10 to 12 mm) between dimensions are suggested for large drawings and those that need to

FIGURE 4–10
Leaders.

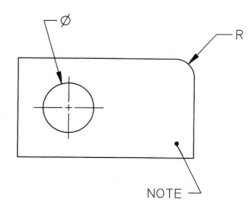

FIGURE 4–11
Spacing dimension lines.

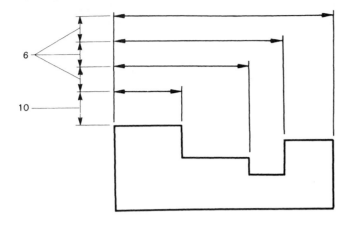

FIGURE 4–13
Unidirectional dimension alignment and grouping.

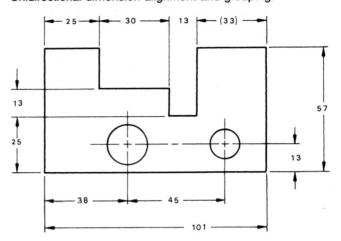

be greatly reduced. The larger sizes should be used whenever there is enough room.

Extension lines should start about .06 in. (1.3 mm) from the part and end approximately .12 in. (2.5 mm) beyond the dimension line and arrowhead.

All dimensions are drawn parallel to the direction of measurement. Numerals should be staggered when there are a number of parallel dimensions, as in Fig. 4–12. For electromechanical drawings the dimension line should be broken for insertion of the measurement (numerals). Centerlines can be used as extension lines, as in Fig. 4–13, but must not be used as dimension lines. In most cases the lines of a part should not be used as extension lines and never as dimension lines.

Grouping Dimensions and Orienting Numerals

Numerals that are placed in a position lined up with the dimension line are called **aligned dimensions.** In aligned dimensions, horizontal dimensions should always be readable from the bottom and vertical dimensions, from the right side of the drawing. **Unidirectional dimen-**

sioning places the numerals parallel with the bottom of the drawing (and therefore readable from the bottom of the drawing), as in Fig. 4–13. This system is preferred since the drawing may be read and lettered without being turned.

It may be desirable to include all the dimensions for reference information or checking purposes. In these cases, a nontoleranced reference dimension is used. In Fig. 4–13, the referenced dimension is placed within parentheses (33); however, this measurement is not necessary for part manufacturing.

Application of Dimension Elements

Size and location dimensions may be given either as linear distances or as angles. **Angular dimensions** should be expressed in degrees, minutes, and seconds or as decimal equivalents of the angle, as in Fig. 4–14:

FIGURE 4–12
Staggered dimensions.

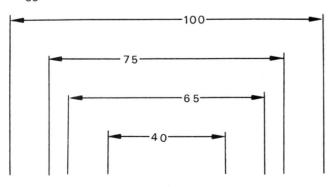

FIGURE 4–14
Angular dimensions.

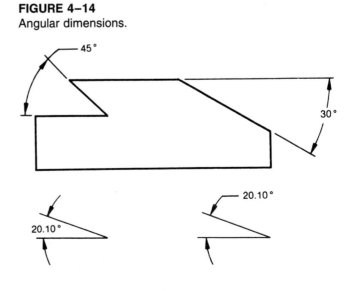

20.10°. When expressing angular dimensions, use symbols for degrees, minutes, and seconds on the drawing. Where angles are less than 1 degree, precede the minute by 0°. Place angular dimensions to read horizontally with no dash between degrees and minutes. The angle may sometimes be given in degrees and decimal parts of a degree. Whenever possible, avoid angle dimensions by locating the endpoints of inclined lines and planes.

The dimension line for an angle is drawn as an arc from a center at the intersection of the sides of the angle. A variety of methods can be used to dimension angles, as shown in Fig. 4–14, depending on the available space.

Dimensioning Arcs

The dimension line for any radius should always be drawn as a radial line at an angle, as shown in Fig. 4–15. Do not use horizontal or vertical dimension lines when dimensioning arcs.

Sometimes the center of an arc is moved on a drawing either because there is a break or because the center lies outside the drawing paper. Then the new position is on a centerline of the arc, and the newly located *false* center leads to a **staggered dimension,** as in Fig. 4–15. The portion of the dimension line touching the arc should be a radial line drawn from the true center, whereas the staggered dimension is drawn parallel to the first radial line. In other words, when the center of an arc lies outside the limits of the drawing, the center is moved closer along a centerline of the arc, and the dimension line is jogged.

The dimension of a radius is preceded or followed by the letter *R* (or in some cases *RAD*) when U.S.

customary units are used. It is always preceded by an *R* when metric units are used, as in Fig. 4–15.

Actual arc lengths are dimensioned as in Fig. 4–15, where the 99-mm measurement is the arc length.

Dimensioning Slots

Slotted holes are found on a variety of electronic sheet metal drawings. Slots may be dimensioned in a number of ways. In Fig. 4–16, three variations are shown. Method 1 shows the slot's centerlines located from an edge of the part and between centers. The slot width is also given as an **R** (radius) pointing to the end of the slot arc. The **R** is normally accompanied by the note **2 PLACES.**

Method 2 uses a leader and a note stating the outside dimensions of the slot, 13 × 50 in. An **R** callout is also included. The slot can also be located from the part's edges (as in the third example), or its centerlines can be located from two controlling edges.

Method 3 shows dimensions of the slot on the view, giving the overall length and width along with its location dimensions. A full **R** callout is provided as in the other methods. (Note that all dimensions in this figure are given as metric units.) Fig. 4–17 is an example of the third method of slot dimensioning. A slot can have arc, radii, and/or rectangular configurations. Use of a *full radius (R)* guarantees a semicircle instead of an arc at the end of the slot.

The choice of methods for dimensioning slots is determined by design factors and the required slot tolerance (fit). If something fits into the slot, accurate tolerance and dimensions must be given for the slot shape

FIGURE 4–16
Dimensions for slots.

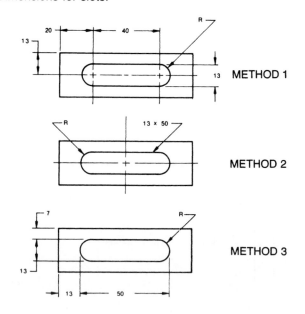

FIGURE 4–15
Dimensions for radii and arcs.

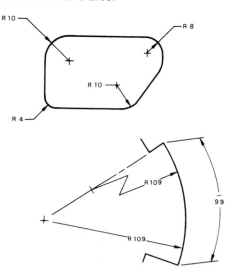

FIGURE 4–17
Mounting panel.

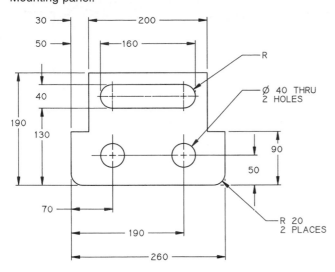

and location on the part. The first and third methods are recommended when accuracy is important. If the slot position and size need not be accurate, the second method can be applied, such as for air-vent slots and loose control-handle travel guides.

Diameters

Diameters can be dimensioned as shown in Fig. 4–18. All diameter dimensions should be preceded by the

FIGURE 4–18
Dimensioning diameters.

international symbol for diameter shown on this figure. The symbol should be a circle drawn the same size as the numerals, and it should have a 60° slanted line passing through its center. On some inch unit drawings the size of the diameter is called out with the abbreviation **DIA** after the numerals, for example, **.375 DIA.**

A number of different methods can be used to dimension diameters, depending on the size of the diameter and whether the diameter represents a hole or a solid shape. In general, **holes** should be called out with a leader and note. The leader must point toward the center of the circle.

A drafter can show the dimensions of a very large hole by drawing the dimension line at an angle through the diameter, as shown in Fig. 4–18 (*lower left*). The area within the shaded 45° section should be avoided when the dimension runs through the diameter.

Solid round shapes should be dimensioned on the side (edge) view, as in Fig. 4–18 (*upper right, Ø27 and Ø50*).

Chamfers

Chamfers may be specified by dimensions or notes. It is not necessary to use the word chamfer when the meaning is obvious. Where the chamfer is other than 45°, dimensions should always be used to show the direction of the slope. In Fig. 4–19 the methods for dimensioning inside and outside chamfers are provided. Note that a drafter can show chamfer dimensions by giving the chamfer angle and one leg, dimensioning both legs, or pointing to the chamfer and giving the angle and one leg as a callout. Inside dimensions can be dimensioned by giving the included angle (here 90°) and the largest diameter, or giving the chamfer angle (here 45°) and the largest diameter. The metric method of dimensioning chamfers is shown on this figure. For inch unit drawings the angle is sometimes given second and the leg first; for example, .25 × 45°. This method, however, is being replaced by the international method, as in ANSI Y14.5M.

Holes

As has been stated, you can show dimensions for holes by pointing to the diameter with a leader and giving a note containing the diameter's size and, when necessary, the type. Where the depth of the hole is not obvious or not dimensioned, the word **THRU,** implying drill through, should follow the size specification.

In Fig. 4–20, several dimensioning methods for holes are provided. The most common method uses a note and leader. The method of drawing a dimension line through the diameter should be used only where the note method may get lost on the drawing, as when the object is complicated and the hole is large,

FIGURE 4–19
Dimensioning chamfers.

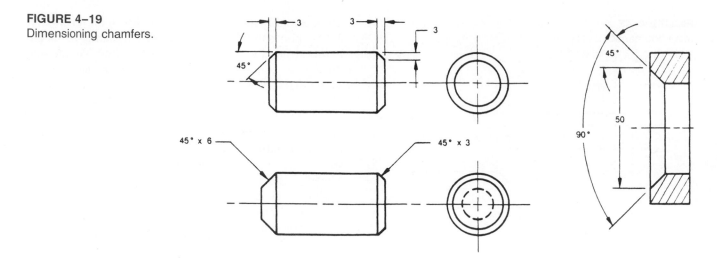

measuring 1.50 in. (38.1 mm) and above. Dimensioning the hole by extending dimension lines from its edges (circumference), as in the mid-sized hole in Fig. 4–20 (far right), should be used for large-diameter holes only.

Dimensioning holes in side views or section views should be used only when the hole cannot be adequately called out where it shows as a circle. The depth of the hole can be dimensioned in this view if it is not included in the note (Fig. 4–20, *bottom*).

Counterbore, Spotface, and Countersink

A counterbore (**CBORE**) is an enlarged hole, normally piloted from a smaller hole to maintain concentricity.

FIGURE 4–20
Dimensioning holes.

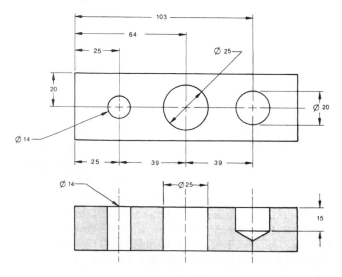

Counterbored holes are machined to a square seat at a specified depth, as shown in Fig. 4–21. The depth is normally established by dimensions showing the thickness of stock remaining under the counterbore rather than the depth of the counterbore, unless the piece is excessively thick. The depth can be called out within the hole note as the distance from the upper surface (beginning surface) to the bottom of the counterbore. Counterbores are used extensively for socket head screws, where the head of the screw must be flush with or below the surface.

A spotface (**SF**), shown in Fig. 4–21, is a method of cleaning up and squaring a surface like a cast metal part so that a screwhead will seat flush against the part.

Countersinking (**CSK**), Fig. 4–21, is a conical seat usually specified by the included angle and the diameter at the large end. Countersinking is used on holes where flathead screws need to be flush with the surface. *Symbology* on *abbreviations* can be used when calling out CBOREs, SFs, and CSKs, as in Fig. 4–21.

THREADS

A simplified form of representing screw **threads** should be used to save drawing time when there is no possibility of confusion with other drawing details. The simplified symbol consists of straight lines, representing the major and minor diameters of screw threads.

The following thread classifications apply to nonmetric threads. When using metric units, consult the appropriate standard for proper specifications. Screw threads are classified in series according to the number of threads applied to a specific diameter. Unified (UN) thread is the standard type of thread for the

FIGURE 4–21
Dimensioning counterbores,
spotfaces, and countersinks.

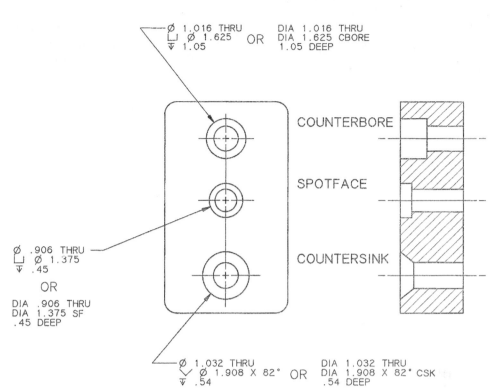

⌀ 1.016 THRU   DIA 1.016 THRU
⌴ ⌀ 1.625   OR   DIA 1.625 CBORE
▽ 1.05   1.05 DEEP

COUNTERBORE

SPOTFACE

COUNTERSINK

⌀ .906 THRU
⌴ ⌀ 1.375
▽ .45

OR

DIA .906 THRU
DIA 1.375 SF
.45 DEEP

⌀ 1.032 THRU   DIA 1.032 THRU
∨ ⌀ 1.908 X 82° OR DIA 1.908 X 82° CSK
▽ .54   .54 DEEP

United States. The UN thread series is designated as follows:

- Coarse Thread   UNC
- Fine Thread   UNF
- Extra Fine Thread   UNEF
- Constant Pitch Thread   UN
- Special Thread   UNS

When specifying screw threads, give the nominal major diameter first, followed by the number of threads per inch and the series designation. Then give the class of fit between positive and negative threads, followed by an A for positive threads and a B for negative threads. For tapped holes, however, the complete note contains the tap drill diameter and depth of hole, followed by the thread specification and the length of the tapped threads. All threads are assumed to be right hand unless left hand is specified by LH following the class. A few examples of screw thread notations are presented next:

- .190–32 UNF–2A or #10–32 UNF–2A
- .250–20 UNC–2A or ¼–20 UNC–2A
- 2.000–16 UN–2A
- 2.500–10 UNS–2A

For specifying the drill and tapping requirements of a hole, give the tap drill size, its depth, the thread specification, and the depth of threads, as in the examples shown here:

- .312 DIA, 1.25 DEEP
  .374–16 UNC–2B or ⅜–16 UNC–2B
  .88 DEEP

- .422 DIA, 1.50 DEEP
  .500–13 UNC–2B L.H. or ½–13 UNC–2B L.H.
  1.12 DEEP

DIMENSIONING METHODS

There are five main methods of dimensioning: rectangular coordinate, datums, hole charts and tabular, centerline, and continuous. This section describes each of these dimensioning methods.

Rectangular Coordinate and Datum Dimensioning

Rectangular coordinate dimensioning locates the features of an object by providing dimensions from two or three perpendicular planes or **baselines.** In general, this type of dimensioning either establishes **datum lines** (X and Y coordinate lines from which all dimensions are taken) (Fig. 4–22) or it establishes the center lines of a symmetrical or circular shape. In Fig. 4–22, the X and Y coordinates are used as datum lines. Here, all dimensions are rectangularly positioned from the datum lines.

Figure 4–23 uses the rectangular coordinate method. Circular cutouts and curved features are dimen-

FIGURE 4–22
Datum line dimensioning.

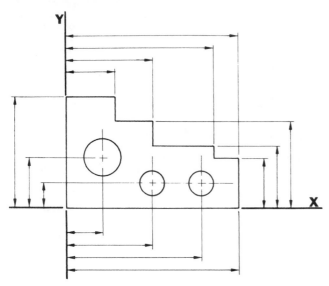

sioned by locating their centers from the left edge of the part. Hole patterns are established in separate details from the cutouts' centers (details A through E).

Datum points, lines, or **surfaces** are features of a part that are assumed to be exact. They act as a baseline or reference for locating other features of the part. A feature selected to serve as a datum must be easily accessible and clearly identified. In most cases, datums are established as the far left and bottom line in a view, as shown in Fig. 4–22.

The vertical datum is designated as the Y axis and the horizontal datum as the X axis when the coordinate system is used. An artificial datum like a construction hole or line edge is sometimes machined in a part for manufacturing and checking purposes only. Printed circuit boards are normally provided with *tooling holes* for the same purpose. A datum surface must be more accurate than any locations measured from it. In some cases, it may be necessary to specify form tolerances for the datum surface to assure that locations can be accurately established.

Figure 4–23 shows a detail of a power distribution box. The top view shows all dimensions taken from the lower left of the part. When dimension lines are eliminated and only measurements and extension lines are shown, this is called *rectangular coordinate dimensioning without dimension lines* (Fig. 4–24).

Where parts must match or mate, the same hole centers must be used as the datum.

Where cumulative tolerances are excessive, dimensioning from a common base or datum will reduce the overall accumulation, but the tolerance on the distance

between any two features, located with respect to a datum and not with respect to one another, will be equal to the sum of their tolerances. Therefore, if it is important functionally to control two features closely, the dimension should be given directly, as in the slot dimensions in Fig. 4–17. Here, the angle bracket was dimensioned from baselines except for the distance between the ends of each slot.

Hole Charts and Tabular Dimensioning

For complicated parts, or where a part has a multitude of holes in one or more surfaces, **hole charts** may be used to simplify the drawing. In hole charts, the surface of hole entry and each hole must be identified on the drawing.

The surface of hole entry must be identified with the names of the principal views. The order of these views for charting purposes is as follows:

1. Front
2. Top
3. Right
4. Left
5. Bottom
6. Rear
7. Auxiliary view (if used)

When using a hole chart, show the surface of entry of each hole, the symbol number that identifies each hole, and the amount each hole is used in this surface. The chart must also show the complete specification for each hole. This information is shown in the logical order of manufacture. Identical holes in a surface may be shown by a single symbol number or letter. Hole charts are commonly used for sheet metal details and drilling drawings for printed circuit boards.

On parts with very complex hole patterns, the locating dimensions for the holes may be shown in the chart as the X and Y positions in each view. This is called *rectangular dimensioning in tabular form*. These dimensions are usually placed on a chart to the right of the hole specifications. Figure 4–25 shows an example of rectangular dimensioning without dimension lines, using a table.

In X and Y coordinate dimensioning, each hole must have a separate identifying symbol. A drafter can group holes by giving diameters the same size and the same letter symbols or by numbering them consecutively. The bottom left or bottom right corner of the front or primary view is usually selected as the zero point for labeling the X and Y axes. The Z axis is also provided for the depth dimension when only one view has holes in it. The direction of dimension measurement is always from a datum/base/coordinate line in the

FIGURE 4-23
Power distribution box.

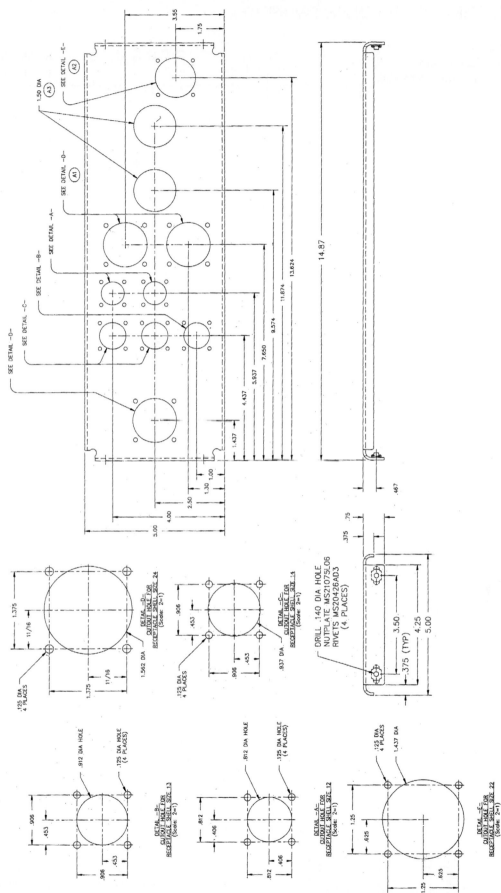

NOTES: Unless Otherwise Specified
1. Remove all burrs and sharp edges
2. Inside bend radii shall be .20
3. Corner relief diameters shall be .270
4. Alodine 1200 (Gold Tint) per MIL-C-5541, Type II, Class 1

FIGURE 4-24
Rectangular coordinate dimensioning without dimension lines.

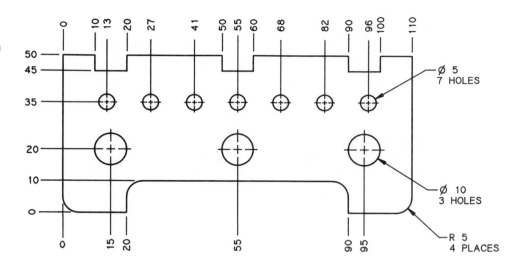

direction of the arrow. All holes are listed in the hole chart. Holes should be listed alphabetically starting from the largest with the letter *A*.

Another method of labeling holes for tabular dimensioning numbers each hole consecutively from number 1 without regard for size. If a numerical controlled (NC) machine is to do the machining operation,

FIGURE 4-25
Rectangular coordinate dimensioning in tabular form.

HOLE	FROM	X	Y	− Z
A 1	X , Y	90	44	10
B 1	"	26	150	30
B 2	"	100	150	30
B 3	"	26	26	30
C 1	"	64	100	40
C 2	"	40	76	40

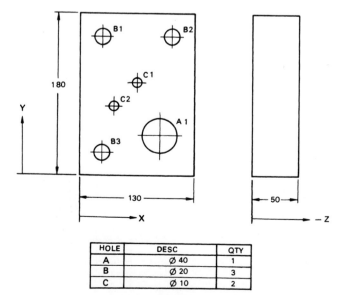

HOLE	DESC	QTY
A	⌀ 40	1
B	⌀ 20	3
C	⌀ 10	2

holes of the same size should be grouped for easy programming.

When the hole is to be completely through the part, **THRU** should be used as the Z dimension. Note that where more than one surface needs to have holes called out, X and Y axes can be established for each surface or view. The depth will need to be specified for each hole. The hole chart will also need to have the *view* noted.

Continuous Dimensioning

Continuous dimensioning is found on a variety of drawings. In this method dimensions are given from each other, or are *progressive,* as in Fig. 4–26, where dimensions along the part's length are *stacked* or in a *chain.* Continuous dimensioning is also used where accurate tolerances are not given for the part features or where automated machining methods are not used, as in the fabrication of many sheet metal cabinets and enclosures for electronic equipment. When continuous dimensioning is used, it is important to remember that any errors in manufacturing will also be progressive, or accumulative. These errors will add up and could produce problems for mating or interchangeable parts. For additional details refer to Chapter 19.

TOLERANCE

Tolerance is the difference between the maximum and minimum limits. The tolerance of any dimension may be specified or implied. Implied tolerances mean that a dimension's maximum and minimum sizes are controlled by a company standard specification noted in the title block.

All dimensions on a drawing are subject to tolerance. When tolerances are not given with the dimension,

FIGURE 4-26
Junction panel hinges.

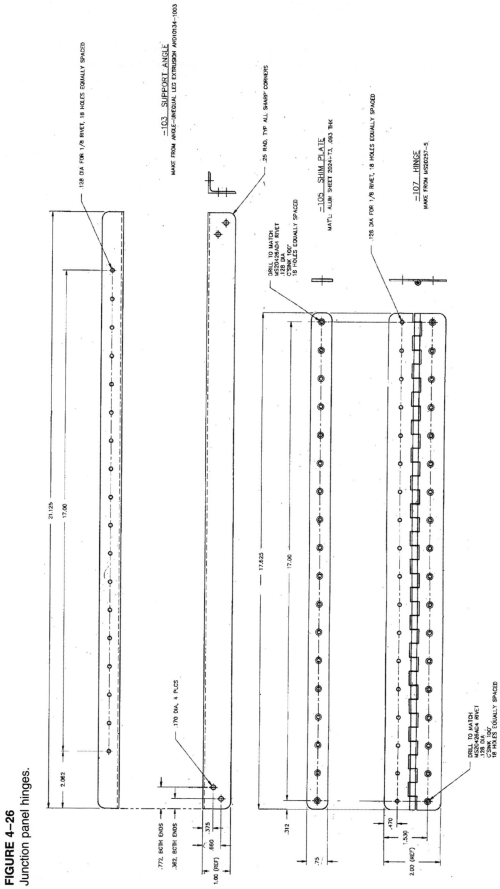

the implied tolerances are specified in the title block. The following is an example of typical company specified tolerances:

General Surfaces

CUSTOMARY (inch) DIMENSIONS:

ONE-PLACE DECIMALS ± .04
TWO-PLACE DECIMALS ± .01
THREE-PLACE DECIMALS ± .005

METRIC (mm) DIMENSIONS:

ALL DIMENSIONS UNLESS
OTHERWISE SPECIFIED ± 0.25

ANGLES:

ANGLES that are machined on both surfaces will have an implied tolerance of ± 0.0° 30′.

Tolerance of size is specified on the drawing when the implied tolerance is not considered satisfactory, for example, tolerance for mating parts or for parts that must fit together when assembled. In Fig. 4–27 a few of the dimensions are given specific tolerances.

Two basic methods of tolerance dimensions are used: **plus and minus tolerances** and **limit dimensioning.** Fig. 4–27 is an example of plus and minus tolerance. Here each tolerance dimension is followed by a plus and a minus expression of tolerance, for example, 9.250 ± .005. This means that the feature should not exceed 9.255 and not be smaller than 9.245. Note that the plus and minus tolerance is also given for the dual metric unit dimension.

Limit dimensioning gives largest and smallest acceptable sizes for a particular feature of a drawing. In general, the high limit is placed above the low limit.

Tolerances and limits are complicated subjects, and this text cannot completely cover their many intricacies. The aspiring drafter or designer should consult ANSI Standard Y14.5M, *Dimensioning and Tolerancing,* for further discussion of tolerances.

Many companies use ANSI Y14.5M as their in-house drafting standard. But there are a considerable number of companies who have developed their own company standards and have not adopted ANSI methods.

FIGURE 4–27
Printed circuit board detail showing dimensions and tolerances. (Courtesy Motorola, Inc., Semiconductor Products Sector)

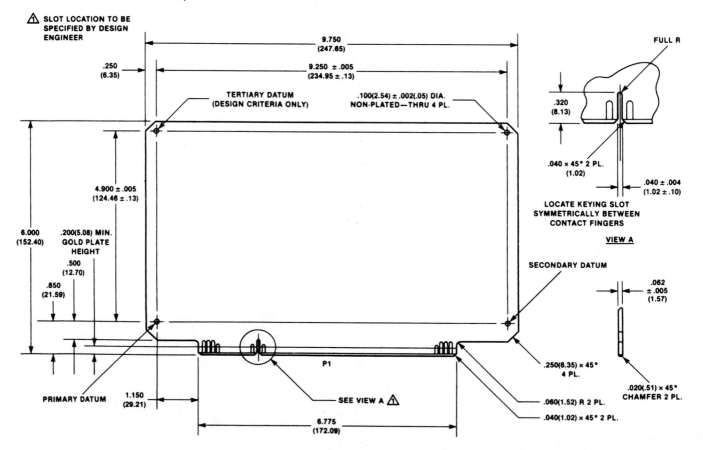

REVIEW QUESTIONS

4–1 How far apart should parallel dimensions be?

4–2 Which type of projection is best for engineering work?
  a. isometric   c. multiview
  b. oblique   d. perspective

4–3 Section lines should be drawn at an angle of (   ) whenever possible.

4–4 Mating parts that are sectioned must have section lines that are at:
  a. the same angles   c. 30 and 45°
  b. different angles

4–5 *True or false:* Screws, bolts, and ribs should not be sectioned.

4–6 When dual dimensioning is used, the (   ) unit is normally on top.
  a. metric   c. U.S.
  b. SI   d. design

4–7 *True or false:* Angles are always given in degrees, minutes, and seconds.

4–8 *True or false:* A chamfer can be dimensioned if the drafter gives its angle and one leg.

4–9 *True or false:* The simplified thread method should be used when screw threads are drawn to simplify the process.

4–10 *True or false:* Continuous dimensioning is the most accurate of the various dimensioning methods.

4–11 Define the following terms:

multiview	CSK
aligned dimensions	glass box method
unidirectional dimensioning	auxiliary view
tolerance	sectioning
plus and minus tolerance	cutting plane
limit dimensioning	leaders
dual dimensioning	threads
baseline	coordinate dimensioning
hole chart	dimension lines
angular dimensions	extension lines
third angle projection	staggered dimension
projector	chamfer
CBORE	datum line
SF	

PROBLEMS

4–1 Draw the dual-in-line package 3:1 and dimension completely. Use dual dimensions. Letter notes in upper left-hand corner of drawing. Use an A sheet size.

PROBLEM 4–1
IC outline.

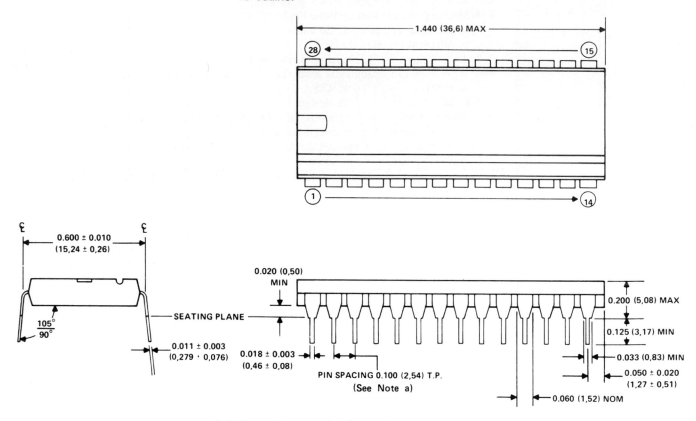

4–2 Draw the mounting frame half-scale on a B size sheet. Dimension the entire project on the three primary views using datum lines. The frame is aluminum.

PROBLEM 4–2
Mounting frame.

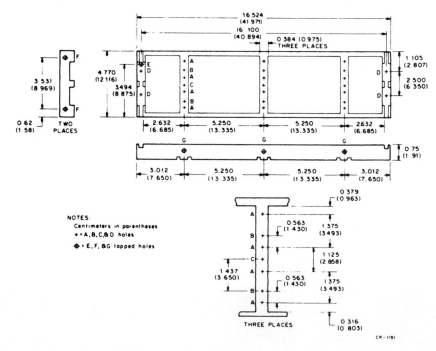

CP-1191

4–3 Draw the mounting panel half-scale on a B size sheet. Dimension using the same system given. The panel is made of aluminum.

PROBLEM 4–3
Mounting panel.

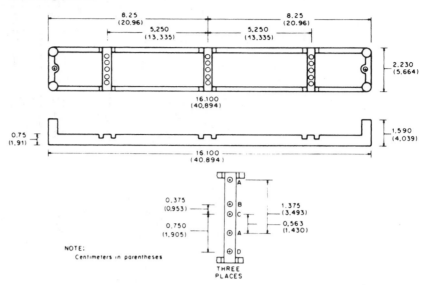

NOTE:
Centimeters in parentheses

THREE
PLACES

4–4 Draw the connector block full scale on an A size sheet. Place the pin side view as a projection on the left. Dimension with the same system as shown.

PROBLEM 4–4
Connector block.

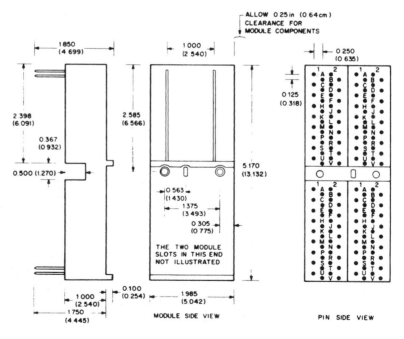

4–5 Draw the component 3:1. Use the same scale for the whole project. Put pin side view as a projection on the left. Show all sections and dimension completely. Use an A size sheet.

PROBLEM 4–5
Component.

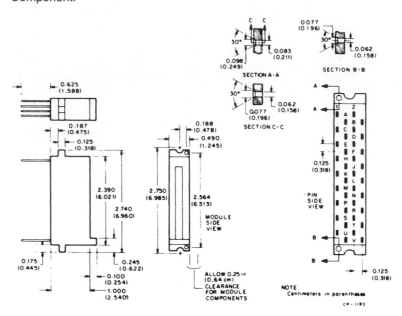

4–6 Using a tabular hole chart, draw and dimension the socket panel. Number the holes sequentially. Draw the project half-scale on a B size sheet.

PROBLEM 4–6
Socket panel.

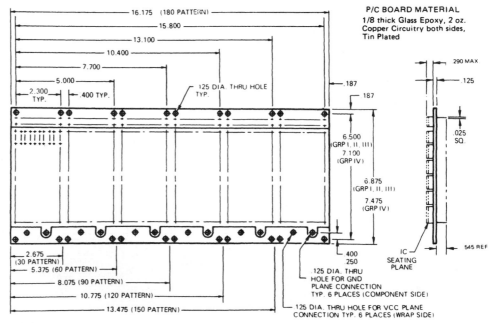

4–7 Draw and dimension the two brackets as separate projects. Draw the brackets three times size. The brackets are steel and will have an inside bend radius of ⅛ in.

PROBLEM 4–7
Brackets for electronic equipment.

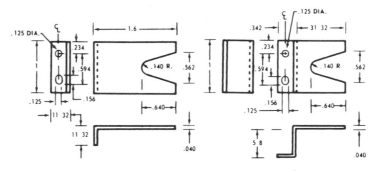

4–8 Sketch the component outlines 3:1 on an A size sheet. Do not dimension.

PROBLEM 4–8
Component outlines.

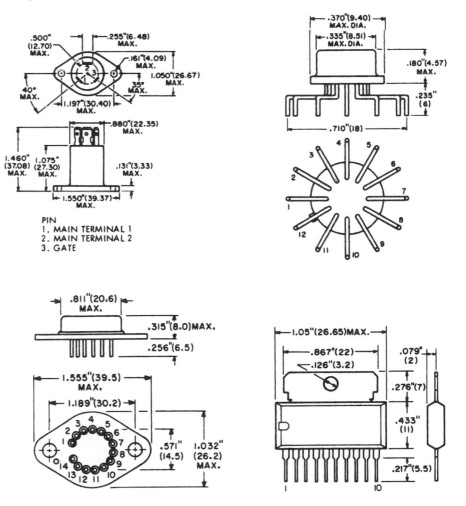

PIN
1. MAIN TERMINAL 1
2. MAIN TERMINAL 2
3. GATE

4–9 Draw and dimension the front panel shown in Fig. 4–6. Use symbology for hole callouts.
4–10 Draw and dimension the three views of the connector bracket shown in Fig. 4–7. Use symbology for hole callouts.

4–11 Draw and dimension the upper housing fabrication shown in Fig. 4–8.

4–12 Draw and dimension the 68-pin memory connector shown in Fig. 4–9.

4–13 Draw and dimension the panel in Fig. 4–17.

4–14 Draw and dimension the power distribution box in Fig. 4–23.

4–15 Draw and dimension the junction panel hinges in Fig. 4–26.

4–16 Draw and dimension the PCB in Fig. 4–27.

5

PICTORIALS

INTRODUCTION

Electronic and electrical drafting and design projects are drawn almost exclusively with multiview orthographic projection for the engineering and production stages of a project. Pictorial sketches are used extensively in the design and engineering stages to clarify design configurations and in working sketches to show optional designs. Pictorials offer a greater freedom in creative communication among engineers, designers, and drafters. Their usefulness, however, is quite limited as a means of communicating production and engineering manufacturing data.

Formal pictorial illustrations with **perspective, axonometric,** and **oblique** projections are used throughout the electronic field to communicate data to nontechnical personnel and to clarify assembly or sales information to equipment users. Technical manuals, sales brochures, user manuals, assembly manuals, service and repair manuals, abstracts, and parts catalogs all make liberal use of pictorial communication. Although a photograph of a particular part or assembly offers the most accurate and realistic representation of an object, the dimensional and assembled characteristics of the object cannot be fully understood without a logical view orientation and technical data associated with line drawings (line

art) and pictorials. Figure 5–1 is a photograph of two wiring structures. Figure 5–2 shows a pictorial drawing of a side entry U-clip terminal used for solderless connections. Each figure has advantages and disadvantages. In many sales catalogs, photographs and pictorials are combined in order to clarify a situation, as in Fig. 5–3, where the Klipwrap Post is depicted in the photograph with a blowup, inset view of an isometric pictorial. The combination of a photograph with a drawing provides a greater degree of information for the user, assembler, and purchaser of the part. Note that the photograph was touched up; the components were outlined directly on the photo.

Wiring and chassis details, as well as connection and assembly instructions, can be best understood by both nontechnical and technical users of electronic equipment if a pictorial view or a photograph accompanies the orthographic drawing. In Fig. 5–4 the wiring diagram of the CRT chassis is shown as an orthographic (top view) pictorial with each of its primary parts called out, or *ballooned*. This pictorial wiring diagram shows the major wiring and cable information needed for a technical manual that accompanies the product.

Normally, an electronic firm will employ a small group of technical illustrators to do pictorial illustrations needed for sales or manual preparation. In some cases

FIGURE 5-1
Wiring structures. (Courtesy Vector Electronic Co.)

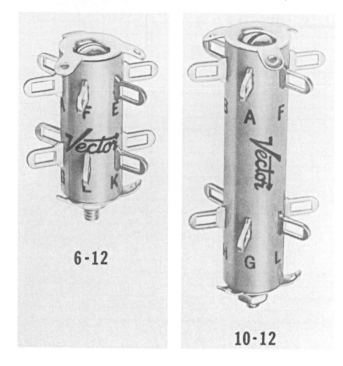

FIGURE 5-3
Wiring post. (Courtesy Vector Electronic Co.)

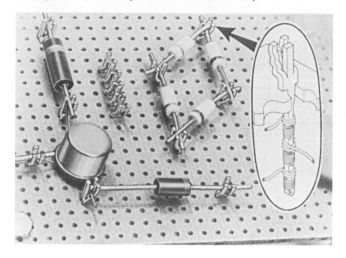

the drafting department is called on to prepare pictorial illustrations. The drafter has extensive knowledge of the object to be illustrated, whereas the technical illustrator is schooled primarily in illustration techniques, with little or no electronic expertise. Drafters usually suffer from limited knowledge of techniques associated with

FIGURE 5-2
Pictorial of a side entry U-clip terminal. (Courtesy Vector Electronic Co.)

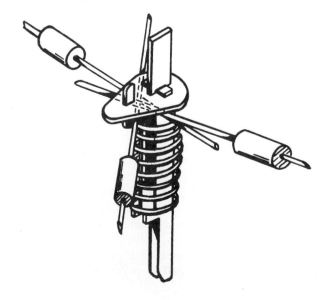

pictorial illustrations: one-plane projections, shading procedures, inking experience, and graphic arts applications for technical illustrations.

One of the greatest drawbacks concerning the use of drafting personnel to do technical illustrations lies in the rigid, exacting nature of engineering drawing. A drafter normally finds it quite difficult to take what is referred to as *artistic license.* A technical illustrator usually shows the object at its best (though perhaps not totally correct) orientation or slightly moves parts of an assembly to better expose them for identification. This ability to adapt and bend the rules is a distinctive skill that must be cultivated for artwork that is not only visually accurate but also artistically pleasing.

The lack of knowledge concerning hand-drawn methods of illustration preparation will no doubt become less important with the increasing use of computer-aided design systems with technical illustration software packages. Many CAD systems are available that will convert multiview design drawings into single-view technical illustrations.

PICTORIAL ILLUSTRATION CONSTRUCTION

In most cases, an illustration is first sketched in two or more possible arrangements. The best possible *view* or orientation is then chosen. Proper arrangement, view orientation, and pictorial projection procedure must be determined before the actual drawing is started in order to eliminate time and cost from false starts. After the illustration technique (orthographic, axonometric, oblique, or perspective) and view orientation have been selected, the drawing must be blocked in with a pencil.

FIGURE 5–4
Wiring diagram of a CRT chassis. (Courtesy Motorola, Inc., Semiconductor Products Sector)

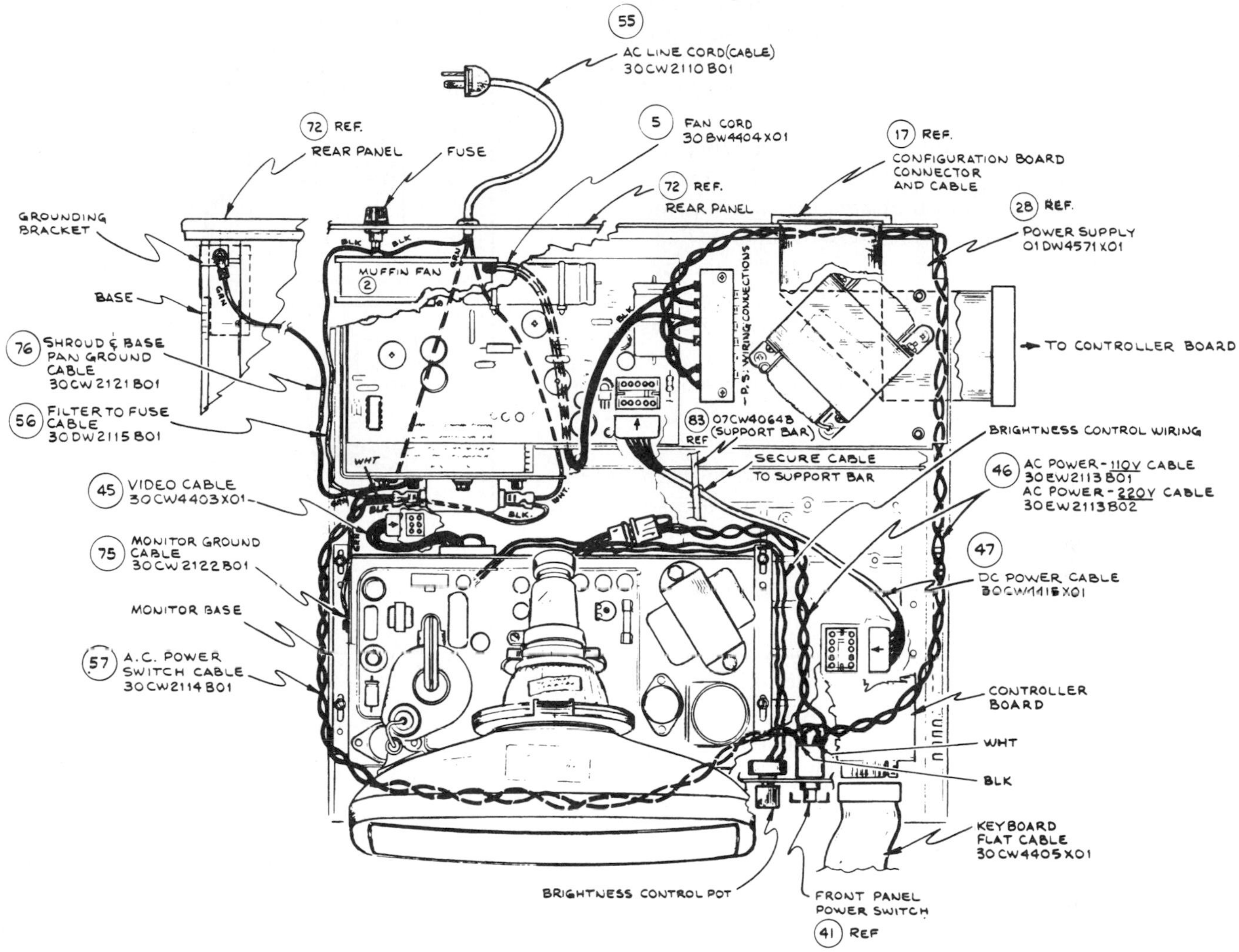

Two separate techniques are used at this stage. The drawing can be completed in 2H or H lead and then traced in ink with an overlay. (A light table is helpful with this technique.) Or the project can be drawn very lightly with a *hard* lead, 3H to 6H, and then inked over the top of the penciled lines. When the penciled illustration is complete, it is checked for accuracy and approved by the designer.

The last step in pictorial illustration involves the inking of the project. Proper line thickness and lettering techniques must be established based on the amount of photographic reduction required. In most cases, pictorials are completed at 2:1 or 4:1 size. Subsequent reduction will *clean up* line and corner discrepancies. In other words, the drawing looks better since any problems are also reduced, if not eliminated altogether, by the reduc-

tion. Proper spacing of lines and forming of letters are essential for the drawing to be acceptable in its reduced stage.

This chapter is meant to be an introduction to the area of pictorial projections. It provides examples of pictorials in all their variations. (Pictorial examples and projects are also provided at the end of Chapter 4 and throughout the text.) To better understand these illustrations, consult the following standards: *Pictorial Drawing:* ANSI Standard Y14.4–1957; *A Guide for Preparing Technical Illustrations For Publication and Projection:* ANSI Standard Y15.1; and *General Drafting Practice:* Military Standard MIL-STD-1.

Pictorial illustrations can be divided into two basic categories: orthographic and 3-plane (simulated 3D). An orthographic illustration normally uses only one of the three common views of the object: front, side, or top. A **3-plane pictorial** is more common and gives a greater lifelike appearance since it exposes three sides (three planes) simultaneously: top, front, and right side, for example. A 3-plane projection is divided into two separate categories: parallel line and converging line. Axonometric (isometric, dimetric, and trimetric) and oblique illustrations are parallel line projections. Perspective illustrations are converging line projections.

The simplest type of pictorial drawing is the orthographic illustration, as shown in Fig. 5–5, which is the front view of a meter measuring DC current. This illustration, communicating the procedure for using the meter, was taken from a user manual. A more complicated perspective, axonometric, or oblique drawing is not needed here.

Axonometric, oblique, and perspective projections are classified as *pictorial* or *3-plane* projections. Axonometric and oblique projections use parallel projection lines, whereas perspective drawings use converging projection lines. Perspective drawings give the most *lifelike* appearance and in many cases are almost photographic, as in Fig. 5–6, which shows a wiring assembly.

Perspective projections are the most visually accurate, but also take a considerable amount of time and money. They cannot be as easily scaled or dimensioned as other 3-plane pictorials. Axonometric and oblique projections, on the other hand, are not theoretically correct and contain their own problems of visual distortion, but are easier to draw since they use simplified projection procedures. These two projections are easy to scale from a multiview drawing or from the object itself. They can be precisely dimensioned, since they

FIGURE 5–5

Pictorial of a meter measuring DC current. (Courtesy Triplett Corp.)

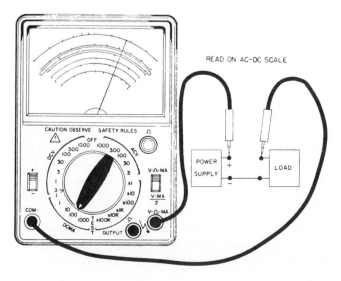

FIGURE 5–6

Perspective pictorial assembly. (Reprinted by permission of HEATH COMPANY)

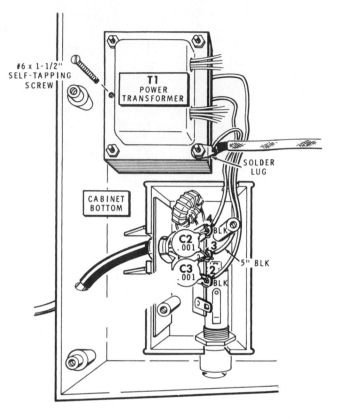

can be drawn to scale from true length measurements. Fig. 5–7 shows a variety of 3-plane pictorial component cases illustrating axonometric projection.

Computer-aided design systems eliminate much of the tedium and difficulty of pictorial projections and technical illustrations (and unfortunately some of the feeling of artistic satisfaction). In Fig. 5–8 the system is used to model and illustrate the cockpit and control panel of an airplane. CAD systems that are not limited to two-dimensional drafting formats normally can design and illustrate in three dimensions, as well as *model* on the screen. The art on the screen can be drawn by the plotter with or without the body elements shown in this figure. Shading and shadows could also be added to create a more lifelike appearance. Technical illustration programs for state-of-the-art CAD equipment, as shown in Fig. 5–8, allow the production of simple to complex line drawings in any desired projection, including perspective. The technical illustration software package normally has the following capabilities:

☐ Unlimited geometric capabilities, including the ability to create drawings from any desired angle
☐ 3D drawing capabilities

FIGURE 5–7
Pictorial drawings of component cases. (Courtesy Motorola, Inc., Semiconductor Products Sector)

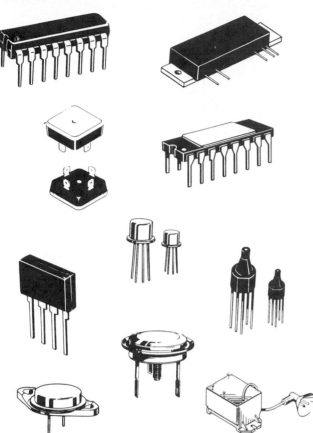

- Shading, overlay, and lineweight selection
- Freehand tablet input
- Axonometric direct-entry drawing functions
- Automated part number insertions and ballooning
- Figure rotation and new view orientation
- Wide variety of screen fonts for possible direct output to a phototypesetter
- High-quality and uniform artwork that can be easily reproduced, stored, and plotted at any scale

ISOMETRIC PROJECTION

Isometric projection is based on the theory that a cube representing the projection axes will be rotated until its front face is 45° to the horizontal plane and then tipped forward or downward at an angle of 35° 16'. In Fig. 5–9, an object has been enclosed in a *glass box* with the enclosed object projected onto each of its corresponding surfaces. The cube has been rotated 45° and tilted forward to show all three faces equally. The viewing plane 1–2–3 is parallel to the projection plane (image plane). The view is an **isometric drawing** in this case since the cube and object were drawn from true length lines and 30° angles for the receding axes.

In **isometric projection** all three axes make equal angles with the projection plane. All three axes are equally foreshortened and make equal angles of 120° among themselves. The three faces of the object are identical in size and shape. The projected lengths of

FIGURE 5–8
Technical illustration using CAD. (Courtesy American Small Business Computers, Design CAD)

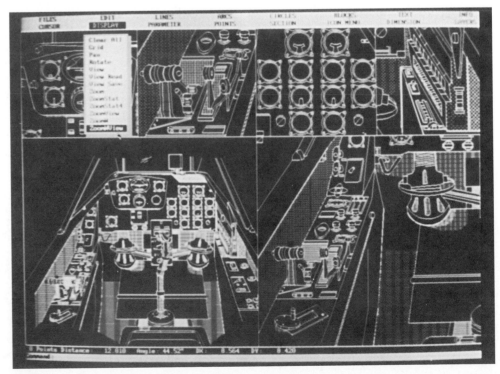

FIGURE 5-9
Isometric projection.

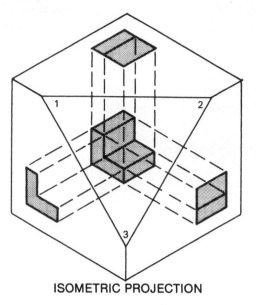

ISOMETRIC PROJECTION

FIGURE 5-10
Isometric pictorial of a PCB showing contact assignments.

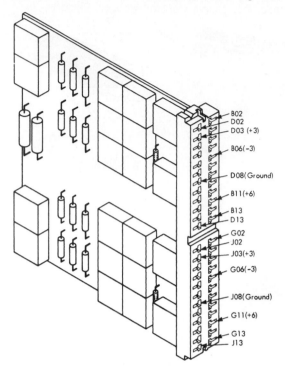

each edge appear foreshortened. This type of projection is a *true isometric projection,* which is seldom used. A true isometric projection is about 81% of the size of an isometric drawing.

In actual industry practice, *isometric drawings* are constructed along the three axes, one vertical and the other two receding at 30° to the right and left. Each dimension is measured true length (not foreshortened). In Fig. 5–10 the printed wiring board showing the component locations and the contact assignments has been pictorially depicted as an isometric drawing. All lines in isometric drawings that are on or parallel to the three axes are drawn true length. Lines not on or parallel to the axes are constructed with offset dimensions.

Isometric drawings are constructed from isometric axes. The Z axis (height) is vertical, the X axis recedes to the right at 30°, and the Y axis recedes to the left at 30° (Fig. 5–11).

Isometric Construction

Isometric construction using the *box method* is illustrated in Fig. 5–12. The procedure for drawing an isometric box is shown in (1) using 30° triangles. Note that the axes are at 120° to one another. The 90° angles of the box transfer to the isometric projection as 60° or 120°, as shown. The three visible faces correspond to the top, front, and side faces in the orthographic (multiview) projection of the same object (3). Different arrangements are possible for the isometric axes, as long as they remain at 120° to one another.

Starting at point A, the three axes are drawn first, one vertical, one 30° receding to the right, and one

at 30° receding to the left. The edges of the box are constructed from the height, width, and depth dimensions transferred from (3). Remember, in an isometric drawing the dimensions are not foreshortened; therefore, each measurement is taken full scale from the multiview projection (or directly from the object), provided that the distance is on or parallel to one of the axes. In isometric drawing, all lines parallel to or on

FIGURE 5-11
Isometric axes.

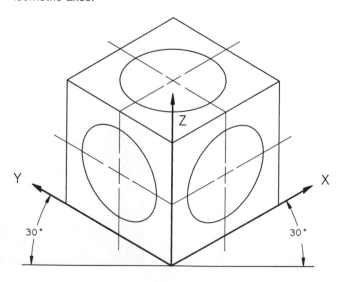

FIGURE 5–12
Isometric construction with the glass box method.

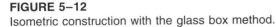

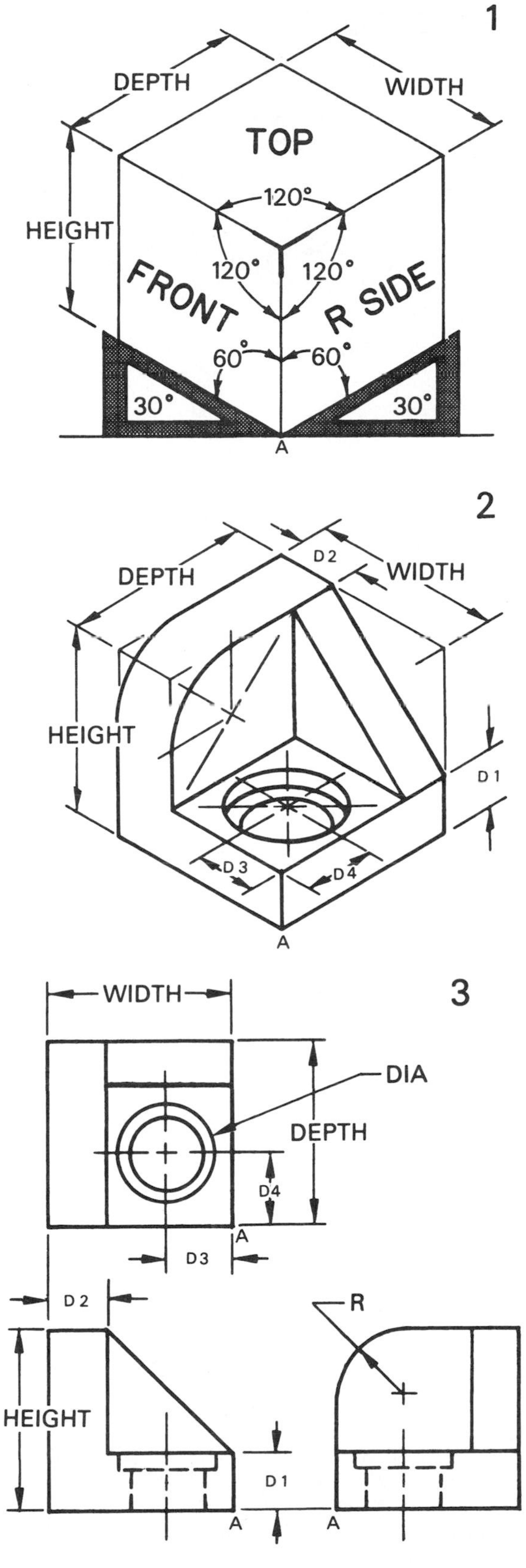

one of the three axes are true length. This type of line is called an **isometric line,** and all lines that are not parallel to or on an axis are called **nonisometric lines** (inclined lines) and will not be true length when shown on the isometric drawing. Nonisometric lines must be established from their endpoints, which are located along isometric lines.

After the object is boxed in, the remainder of the drawing is completed. Dividers (or a scale) are used to transfer each of the dimensions shown in (3) to the isometric view (2). All measurements are taken along lines that are parallel to or on the axes (along isometric lines). Dimension D1 is measured along the vertical axis, and dimensions D2, D3, and D4 are in the horizontal plane and are measured along or parallel to one of the corresponding receding axes as shown.

Hidden lines can be omitted in most pictorial illustrations unless required for clarification. Isometric circle and arc construction are covered later in the chapter.

Isometric Angles

The three major axes along an isometric box (or cube) are at 120° angles to one another. In reality, all lines of a cube are at 90° or are parallel to each other. Because of the distortion created by the isometric view of the box, few angles appear as true angles. Angles must be established by means of *offset dimensions.* Angles appear larger or smaller than true size on isometric drawings depending on their position in the view.

The rectangular plane in Fig. 5–13 has angles of 45° and 30°. These two angles must be constructed from offset dimensions along *isometric* lines using true length measurements from the orthographic view of the plane (1).

In Fig. 5–13 the isometric plane (2) is boxed in with true length dimensions A and B along the isometric axes. The 30° angle was constructed by transferring dimension C. The 45° angle was drawn by transferring offset dimensions E and F to establish the endpoints. The points were then connected, completing the figure. Note that D and H are nonisometric lines and appear distorted, as do both angles. Neither is true size.

The dimensional multiview projection of the integrated circuit shown in Fig. 5–14 was used to construct its isometric pictorial in Fig. 5–15. The part was boxed in and then each feature was *carved out.* (Figure 5–7 shows two examples of similar integrated circuits pictorially represented in a technical manual.)

Isometric Circles

All circles and circular arcs on isometric drawings appear elliptical. A variety of methods is available for **isometric ellipse** construction: *template, trammel, four*

FIGURE 5–13
Angle construction in isometric projection.

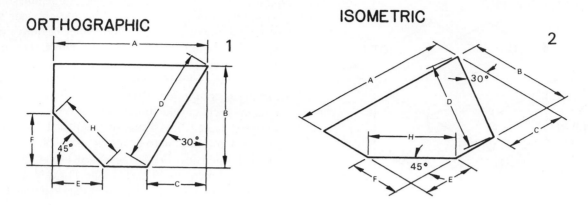

FIGURE 5–14
Multiview-dimensioned drawing of an integrated circuit. (Courtesy Ernie Schweinzger & Assoc.)

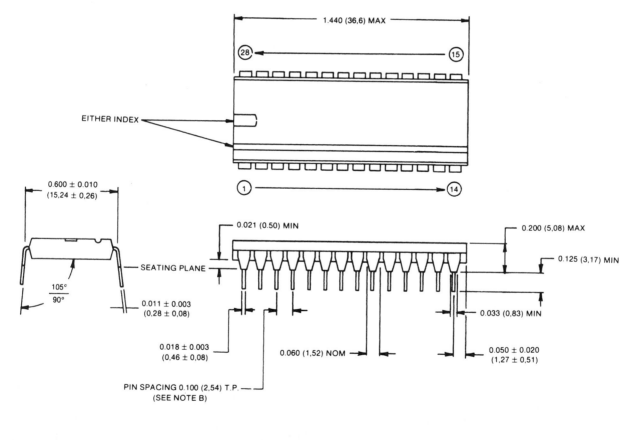

FIGURE 5–15
Construction steps for isometric pictorial of integrated circuit shown in Fig. 5–14. (Courtesy Ernie Schweinzger & Assoc.)

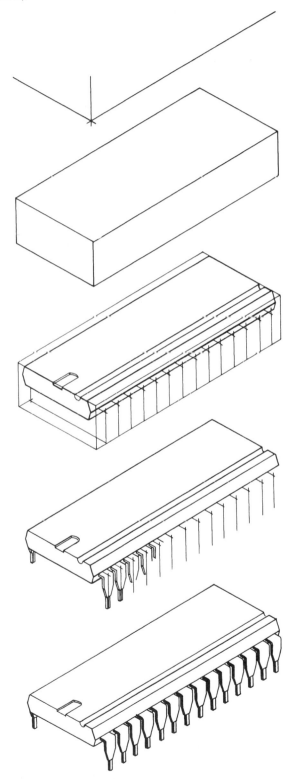

FIGURE 5–16
Isometric ellipse template. (Courtesy Berol RapiDesign)

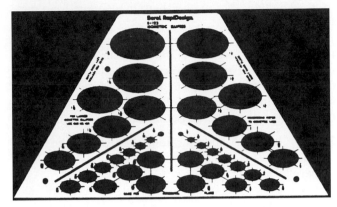

FIGURE 5–17
Ellipse template. (Courtesy Berol RapiDesign)

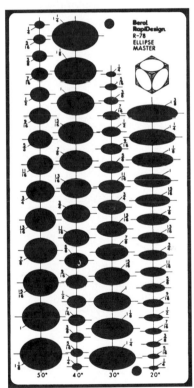

center, and *point plotting,* to name a few. In general, a template should be used whenever the size of the ellipse can be matched with available equipment. Figure 5–16 shows a true isometric projection hole template. Figure 5–17 is an ellipse template with typical angle ellipses: 20°, 30°, 40°, and 50°.

The trammel and point plotting methods are the most accurate, but time-consuming, procedures for constructing circles and arcs in isometric when templates are not available. These two methods should only be used when the other methods are inadequate for a particular project.

The four-center method shown in Fig. 5–18 does not create a perfect ellipse, but is accurate enough for most purposes. This method can be used to draw circles or portions of circles on any isometric face (plane), as shown in Fig. 5–19. The following steps describe the construction of an isometric ellipse (Fig. 5–18):

1. Draw lines DA and DC along the two receding axes (at 30°). Line AB is parallel to DC, and line CB is parallel to AD.
2. Draw construction lines from point D perpendicular to line AB at its midpoint and perpendicular to line CB at its midpoint.
3. Repeat step 2 using point B and lines AD and CD.
4. Use point D and point B to swing arcs R1. Arc R2 originates at the intersection of the construction lines for both sides of the ellipse.

FIGURE 5–18
Isometric circle construction (ellipse), four-center method.

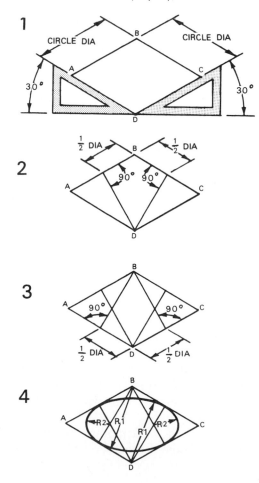

FIGURE 5–19
Isometric circle construction (ellipse) on an isometric cube.

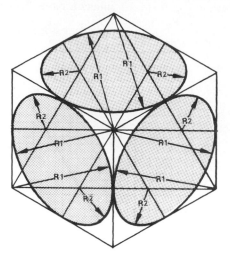

Circles, arcs, or curves that do not lie in isometric planes must be plotted with offset dimensions. This procedure requires that a series of points be established along the curved outline in the orthographic view. Offset dimensions for these points are transferred to the isometric drawing and are laid off along isometric lines. In actual industry practice, templates are used whenever and wherever possible.

OBLIQUE PROJECTION

Oblique drawings are similar to isometric drawings; however, they are produced from parallel projectors that are not perpendicular to the projection plane. The primary difference lies in the use of only one receding axis and the ability to draw one surface as true shape and size in the front plane. In Fig. 5–20, the three axes are vertical Z, horizontal Y, and receding X. In oblique projection the front face of the part is placed parallel to the image plane; therefore, the Z and Y axes are at 90° and are parallel to the projection (image) plane, as shown in Fig. 5–20. The other two faces of the part are on receding (oblique) planes. The front face of the block and the diameter of the hole are drawn true shape and size, and all measurements on this front face are true length. The most commonly used angle for the receding axis is 45°.

There are two basic categories of oblique projection, **cavalier** and **cabinet,** as shown in Fig. 5–21. In a cavalier projection (1), receding lines are not foreshortened, but are drawn full scale. In a cabinet projection (2) the receding lines have been foreshortened one-half their original length (½ scale). The

FIGURE 5–20
Oblique projection, standard position.

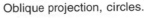
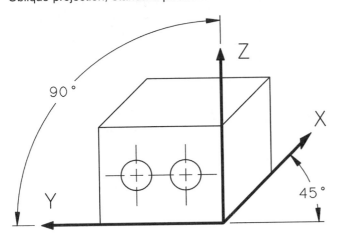

receding axis can be any convenient angle between 1° and 89°. The choice of receding angle is determined by the shape of the object and the most descriptive view orientation.

Objects that are drawn with oblique projection should be oriented so that the surface with curved lines lies in the front plane bounded by the axes that are at 90°. This orientation lessens drawing time since all surfaces that lie on the front plane or that are parallel to it are drawn true shape and size. Circles and arcs are therefore true projections.

Oblique projection is extremely useful for drawings composed of multiple and parallel curved, or irregular,

FIGURE 5–21
Oblique projection: (1) cavalier; (2) cabinet.

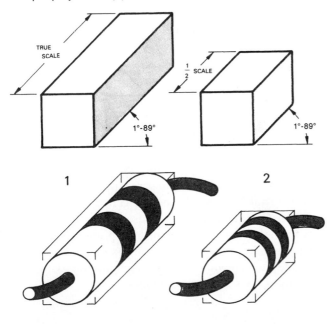

FIGURE 5–22
Oblique projection, circles.

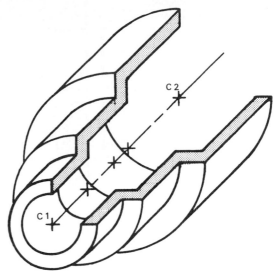

outlines. The longest or most irregular surface should be the front of the projection wherever possible. This is a definite advantage when the surface is composed of curved lines, as in Fig. 5–22. Here, successive circular outlines and surfaces are not distorted since they are all parallel with the front plane.

The face of a part that has a curved outline should be chosen as the front face (the face lying in or parallel to the front plane), as in Fig. 5–23. In this figure a set of steps is provided for pictorially representing the object. This drawing was completed with cavalier projection, so all dimensions (parallel to an axis, even on the receding faces) are true length and can be transferred from the multiview projection in Fig. 5–23:

1. Each dimension of this orthographic projection can be transferred true length to the oblique view.
2. Start by drawing the front face of the object using the height (H) and width (W) dimension. Then determine the receding angle (45° was used here). Transfer the depth (D) dimension along the receding axis and block in the outline of the object. Use dimension A to establish the top of the plane surface.
3. Use dimensions B and C to locate the circular shape. Dimension C is the diameter of the circle.
4. Complete the object by transferring offset dimensions E and F to establish the angled surface. Draw a true circle (half-circle) using dimension C as the diameter. Note that the circular shapes are parallel to the front plane and are therefore true shape. For circles and arcs that appear on oblique faces, the oblique four-center method or template must be used.

FIGURE 5–23
Oblique construction with glass box method and dimensions transferred from multiview drawing.

MULTIVIEW

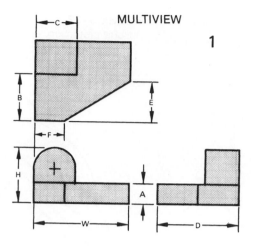

1

OBLIQUE PROJECTION

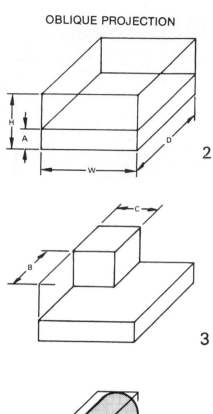

2

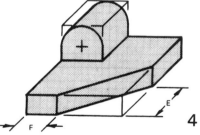

3

4

Note that the construction process for slanted, inclined lines and planes is similar to that of isometric drawings. Locate their endpoints along lines that are parallel to one of the axes. For slanted surfaces, locate both ends of the surface and connect the points.

PERSPECTIVE PROJECTION

Perspective projection is used to provide illustrations that approximate how a particular object looks to the human eye or as a camera would record the object on film. Since a perspective drawing approximates how an object really looks, it is not dimensionally correct. Many lines and planes on perspective drawings cannot be scaled since they are not drawn true length or true shape. Because of this distortion, perspective drawings are seldom found in engineering or design work. Technical illustrations for advertisements, sales catalogs, technical

manuals, and architectural renderings make extensive use of this form of pictorial projection.

All lines in perspective drawings converge at one, two, or three points on the horizon (vanishing points) and therefore are not parallel, as in oblique and axonometric projections.

Three basic categories exist in perspective projection: *parallel, angular,* and *oblique.* Parallel perspectives (*one-point perspectives*) are constructed similar to oblique drawings since the front face of the object is parallel to the projection plane, but all receding lines converge at one point. Angular perspectives are drawn with two vanishing points and are called *two-point perspectives.* Oblique perspectives are *three-point perspectives* since they have three vanishing points (Fig. 5–24). For further information regarding perspective projection refer to one of the many technical drawing and illustration texts that are available.

FIGURE 5–24
Three-point perspective.

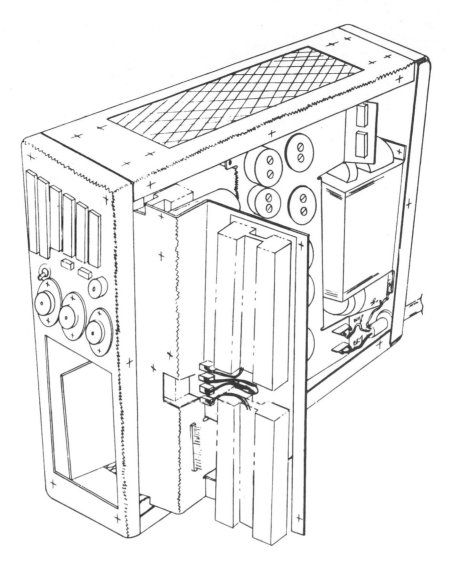

APPLICATIONS FOR PICTORIALS

The most common uses of pictorial illustrations in the electronics field are for assemblies and exploded views. An **assembly drawing** may be needed to show the construction sequence, to visually depict assembly instructions, or to identify parts in technical, user, and service manuals. **Exploded views** are widely used to show an assembly procedure (Fig. 5–25).

A photograph with overlay ballooning may be sufficient for parts identification, but a pictorial drawing is far more descriptive. The Output Scanner Optical System shown in the parts identification pictorial in Fig. 5–26 could be adequately communicated with a photograph and overlay identification if only its exterior configuration and parts were to be presented. Note that the use of a photograph and line drawing in combination is called **photodrafting** or a photographic pictorial. Since

the modulator, mirror 2, photoreceptor, and other important parts need to be identified in this figure, a pictorial drawing with *cutaway* procedure was used so that the interior of the system could be viewed. Notice that hidden lines and centerlines have been left out in this and most of the pictorials shown throughout the text. Hidden lines are shown only where they are absolutely needed for identification. Unneeded hidden lines and centerlines only confuse the user of the drawing. Centerlines are shown where they are needed to define symmetry or for dimensioning.

Exploded views are drawn with any of the previously mentioned pictorial projection techniques. Isometric exploded projections are the easiest to construct. Although a photograph shows the item in its entirety, an exploded pictorial provides a view where each individual part can be called out (ballooned) and shown in its relation to all other parts of the item.

FIGURE 5–25
Exploded assembly.

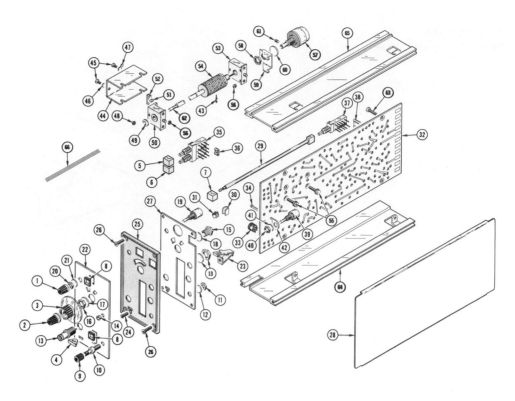

Pictorial rendition allows for the identification of each part and how it is connected or assembled. Figures 5–27 and 5–28 show a photograph and the exploded assembly of the trackman mouse.

Each part of an exploded view is normally drawn separately and then positioned along its axis or centerline, as in Fig. 5–29, where the component and the socket are shown aligned with the centerlines that correspond to the screws and holes. The individual parts must be sufficiently separated to show all the individual items and their relationship to one another in the assembly.

Figure 5–30 shows a variety of transistor and IC socket outlines drawn in pictorial form. Vertical **unidirectional lettering** was used to dimension these drawings.

In Fig. 5–31 the cage side wall plate uses the **aligned method of dimensioning.** The unidirectional lettering is constructed along horizontal guidelines and is the easiest to letter or place (if phototypeset). The aligned method requires that all characters be formed so that they lie on the plane of the isometric face in which they fall. Guidelines for aligned freehand lettering are drawn parallel to the dimension line. Thus, all lettering is aligned with the dimension lines. The vertical lines of all characters are aligned with (parallel to) the extension lines.

FIGURE 5–27
Trackman mouse. (Courtesy Logitec, Inc.)

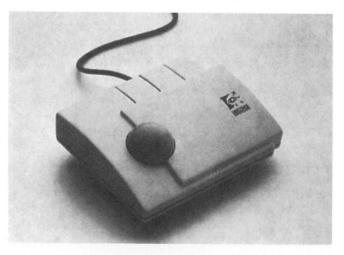

FIGURE 5–26
Output Scanner Optical System. (Courtesy Xerox Corp.)

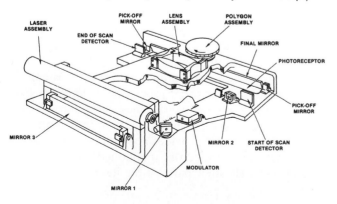

FIGURE 5–28
Exploded assembly pictorial of trackman mouse. (Courtesy Logitec, Inc.)

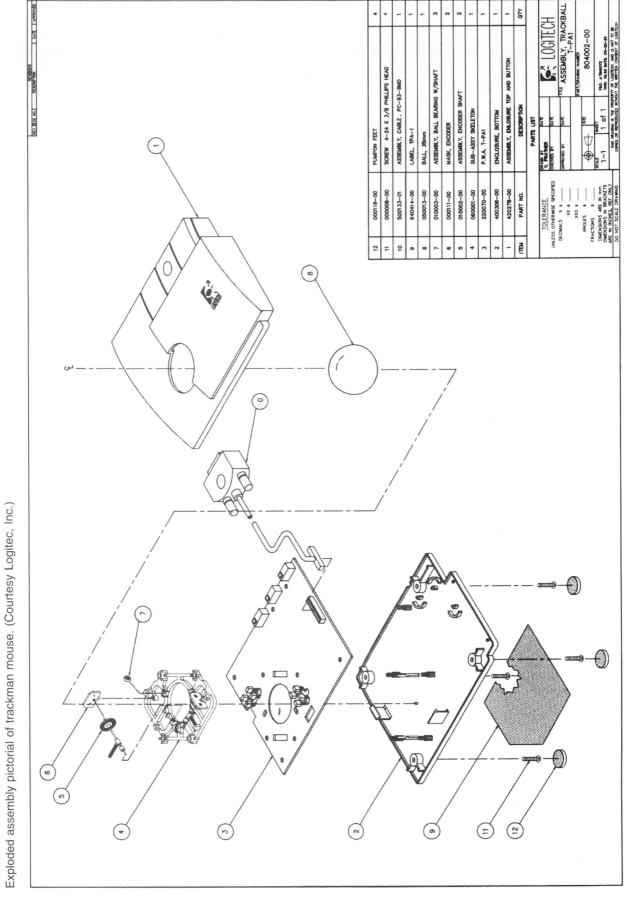

ITEM	PART NO.	DESCRIPTION	QTY
12	000116-00	PUMPON FEET	4
11	000008-00	SCREW 4-24 X 3/8 PHILLIPS HEAD	4
10	500133-01	ASSEMBLY, CABLE , PC-93-9MD	1
9	640414-00	LABEL, TPA-1	1
8	050013-00	BALL, 35mm	1
7	010003-00	ASSEMBLY, BALL BEARING W/SHAFT	3
6	000111-00	MASK, ENCODER	2
5	010002-00	ASSEMBLY, ENCODER SHAFT	2
4	060001-00	SUB-ASSY SKELETON	1
3	220070-00	P.W.A. T-PA1	1
2	400306-00	ENCLOSURE, BOTTOM	1
1	420278-00	ASSEMBLY, ENLOSURE TOP AND BUTTON	1

PARTS LIST

LOGITECH

TITLE ASSEMBLY, TRACKBALL
T-PA1

PART/DRAWING NUMBER
804002-00

DRAWN BY V. PRATHER

1 of 1

SCALE 1-1

103

FIGURE 5–29
Assembly pictorial. (Reprinted by permission of HEATH COMPANY)

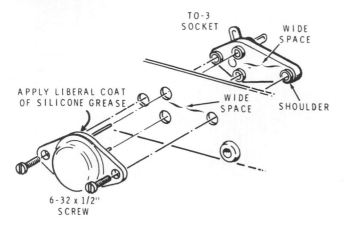

FIGURE 5–31
Pictorial drawing of cage side walls, including aligned dimensions. (Courtesy Vector Electronic Co., Inc.)

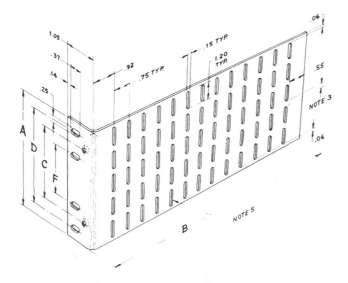

FIGURE 5–30
Pictorial drawings of transistor and integrated socket outlines.

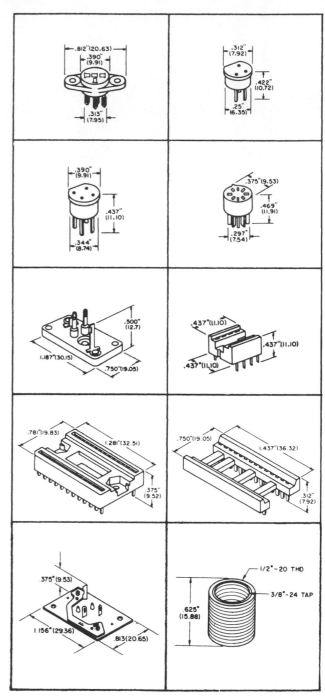

REVIEW QUESTIONS

5-1 A pictorial drawing normally shows (   ) views.
 a. one   c. three
 b. two   d. four

5-2 A cabinet drawing is a form of (   ) drawing.
 a. perspective   c. multiview
 b. oblique   d. orthographic

5-3 The primary difference between cabinet and cavalier drawings is that the depth axis for cabinet drawings is foreshortened:
 a. ½   c. ⅔
 b. ¾   d. ¼

5-4 The angle between axes of an isometric drawing is:
 a. 90°   c. 60°
 b. 180°   d. 120°

5-5 Which face of an object should be placed in the front plane of an oblique drawing?
 a. front   c. longest
 b. curved   d. side

5-6 The advantage of (   ) drawing lies in the ability to vary the degree of axis orientation.
 a. isometric   c. oblique
 b. perspective   d. multiview

5-7 True isometric projections require that the receding dimensions be set off:
 a. full scale   c. ¾ scale
 b. half-scale   d. with an isometric scale

5-8 The main principle of (   ) drawing is that all lines tend to meet at one or two points:
 a. oblique   c. perspective
 b. isometric   d. multiview

5-9 Isometric means:
 a. equal angles   c. 30° angles
 b. three axes   d. equal sides

5-10 Vertical lines are (   ) lines when transferred to the isometric drawing:
 a. slanted   c. true length
 b. inclined   d. vertical

5-11 After the axes are established on an axonometric or oblique drawing, the next step involves:
 a. boxing in   c. scaling
 b. transferring d. inking

5-12 The common receding axis angle for an oblique drawing is:
 a. 30°   c. 60°
 b. 45°   d. 90°

5-13 The right and the left isometric axes of an isometric drawing are drawn 30° in the horizontal. How many degrees are they from the vertical axis?
 a. 30°   c. 45°
 b. 120°   d. 60°

5-14 Which form of 3-plane pictorial projection is best suited for objects with circular faces?
 a. perspective   c. oblique
 b. isometric   d. orthographic

5-15 Nonisometric lines are not:
 a. true length   c. inclined
 b. angled   d. foreshortened

5-16 The primary use for pictorial drawings is for:
 a. engineering sketches   c. manuals
 b. design drawings   d. production

5-17 Which type of pictorial is the easiest to dimension?
 a. multiview   c. perspective
 b. orthographic   d. isometric

5-18 Which type of line is not usually shown on a pictorial drawing?
 a. inclined   c. dimension
 b. hidden   d. centerline

5–19 A(n) (   ) pictorial is used to show how parts of an assembly are related.
  a. blow-up   c. expanded
  b. inset   d. exploded

5–20 Briefly define the following terms:

oblique	isometric drawing
isometric line	perspective
scaled	cabinet
3-plane pictorial	cavalier
axonometric drawings	nonisometric line
exploded view	isometric ellipse
isometric projection	unidirectional lettering

PROBLEMS

5–1 (A–J) Your instructor will assign the item, scale, and projection technique as required. Parts can be scaled from the drawings or dimensions can be used. A size sheets should be used.

PROBLEM 5–1
Heat dissipators and component outlines.

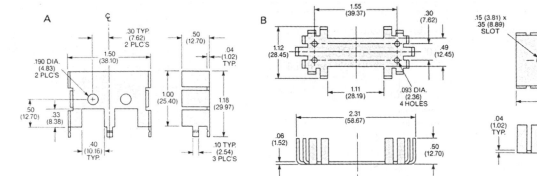

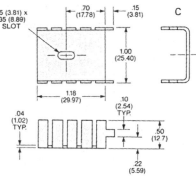

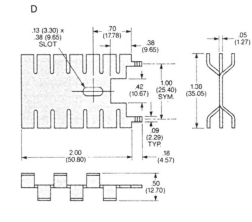

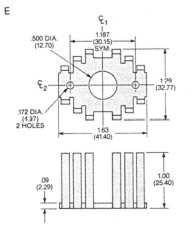

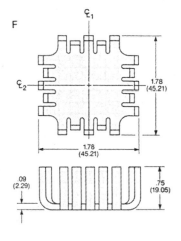

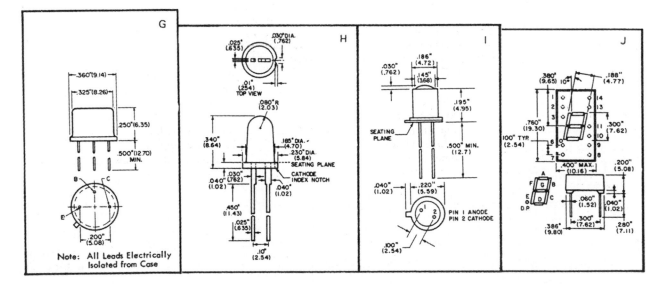

5–2 (A–M) Your instructor will assign the item, scale, and projection technique as required. Parts can be scaled from the drawings or dimensions can be used. A size sheets should be used.

PROBLEM 5–2
Component outlines.

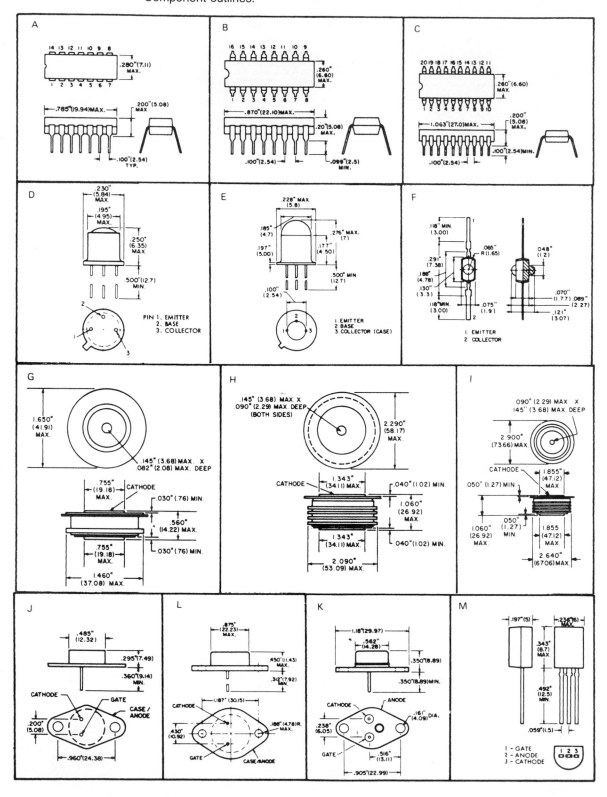

5–3 Draw a pictorial (isometric) of the bracket.

PROBLEM 5–3

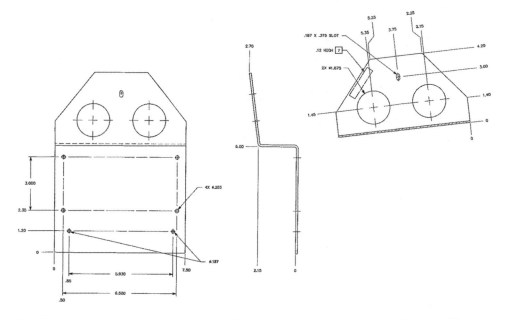

5–4 Your instructor may assign projects to be completed as pictorials from the art in Chapters 4 and 19.

6

COMPUTERS IN THE DESIGN AND MANUFACTURING PROCESS

INTRODUCTION

The production of electronic products involves many distinct but related operations: circuit design and development, printed circuit board (PCB) design (Fig. 6–1), integrated circuit (IC) layout (Fig. 6–2), mechanical design, manufacturing, assembly, and testing. Each of these areas had traditionally been done manually, but now the computer is being used to automate and integrate much of electronic drafting and design.

Anyone wishing to enter industry as an electrical or electronic drafter, designer, or engineer must be familiar with how the computer is profoundly altering the factory floor and the engineering office. **CAE** (computer-aided engineering), **CAM** (computer-aided manufacturing), and **CADD** (computer-aided design and drafting) are all areas where the computer has had an impact. The term **CAD/CAM** refers to the use of computers to integrate the design and production process to improve productivity (Fig. 6–3). **NC** (numerical control) (Fig. 6–4), **CNC** (computer numerical control) machining, and the use of robotics (Fig. 6–5) in manufacturing are examples of CAM. A comprehensive CAD/CAM glossary is provided in Appendix C.

The skill and knowledge of traditional manual draft-

ing and design will be needed and taught for the foreseeable future. But the knowledge of computer-aided systems is also required for the aspiring electronic or electrical engineer, designer, and drafter. A typical **interactive** CAD system is an engineering design and drafting tool. Interactive graphics give you the ability to perform graphics operations directly on the computer with immediate feedback. You can electronically erase selected portions, shrink or enlarge the geometry, copy, and rotate complete parts or selected geometry. To accomplish all this, an integrated combination of hardware and software is required.

An operator of a CAD system (Fig. 6–6) must be able to understand the system's **hardware** configuration and its **software** capabilities. The following are essential for you to understand and use a CAD system effectively in electrical and electronic engineering, design, and drafting:

1. Knowledge of drafting standards and procedures
2. Specific electrical and electronic engineering field conventions
3. Actual electrical and electronic industrial applications
4. Software specifics for a particular CAD system package

FIGURE 6–1
CAD system used for PCB design and CMOS gate arrays.
(Courtesy Prime–Computervision)

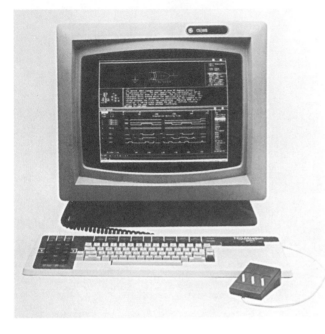

FIGURE 6–3
CAD/CAM system. (Courtesy Prime–Computervision)

FIGURE 6–2
IC design. (Courtesy
Prime–Computervision)

FIGURE 6–4
NC machine using CAD/CAM input. (Courtesy Lockheed–California Co.)

Software programs are available for all areas of engineering and design. Most systems can be mastered with training, but the specifics of the design area (electrical, electronics, piping, architecture, mechanical, structural, and others) must be learned through a combination of education and experience.

It must be stressed that CAD is a drafting and design tool. As a drafter or designer you are still required to create the necessary graphics and, therefore, must have a thorough knowledge of drafting standards and practices. Only the method of creating engineering graphics has changed, not the content.

CAD TECHNOLOGY

The heart of any CAD system is the design **terminal,** or **workstation** (Fig. 6–7). By issuing commands to the system and responding to messages from the system, you create a design, manipulating, modifying, and refining it—all without ever drawing a line on paper or re-creating an existing design element.

As a design is developed, the **computer graphics** system accumulates and stores product-related data, identifying the precise location, dimensions, descriptive text, and other properties of every element that helps

FIGURE 6–5
Robot programming. (Courtesy Cincinnati Milacron, Inc.)

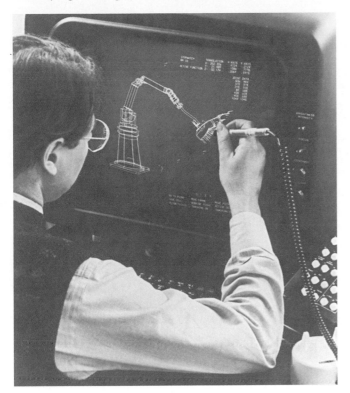

to define the new project. With these design data, the system is used to do complex engineering analyses, generate special lists and reports, and detect and flag (note or indicate) design flaws before the part or system is manufactured.

CAD SYSTEM DESCRIPTION

In addition to a computer, the system includes the workstation (with CRT display, function and alphanumeric keyboard, digitizer, and other input devices), output devices (pen plotter, electrostatic plotter, photoplotter, COM, and others), software (high-level applications and others), and manufacturing applications using CAM, NC, and robotics (Fig. 6–8).

Large-scale host computers are equipped with large amounts of memory as well as tape drives, line printers, card readers, and hard copy devices. A minicomputer (Fig. 6–9) may be used to support, with its own memory and processing capabilities, various graphics programs running on the CAD/CAM system, as well as to do related engineering analyses.

Hardware itself does nothing unless directed to do so by a set of instructions (or software). **Turnkey** means that the system is functional, complete, and ready to

FIGURE 6–6
CAD system. (Courtesy CalComp, Inc.)

FIGURE 6–7
Typical CAD workstation.
(Courtesy
Prime–Computervision)

FIGURE 6–8
Using a CAD system to program
a robot. (Courtesy
Prime–Computervision)

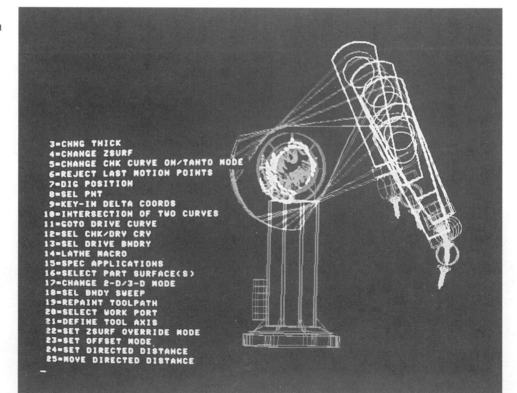

FIGURE 6–9
Minicomputer-based CAD system, the VersaCAD®.
(Courtesy Prime–Computervision)

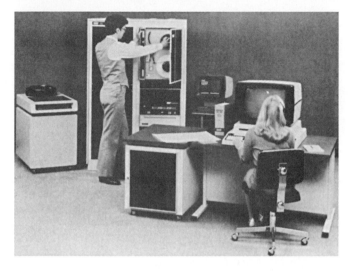

operate upon installation (just turn the key). Software can include programs to perform graphic manipulation functions (such as scale, zoom, and rotate) and to perform graphic drawings functions, such as the creation of lines, arcs, and splines.

A typical PC configuration is shown Fig. 6–10. The computer used in such a configuration is generally a

FIGURE 6–10
PC CAD system. (Courtesy Prime–Computervision)

personal computer and contains its own **central processing unit** (CPU). Therefore, it does not need an interface with a mainframe or large central computer. Personal-computer-based CAD systems can be networked together and interfaced with a central CPU, providing the designer with access to a large database.

HARDWARE AND SOFTWARE

The typical CAD system is modular in both its hardware and software. Engineering firms can select the particular computational processing, ranging from large mainframes or minicomputers to PCs, and graphic output that best serve their needs. Some companies, such as engineering firms, use the system exclusively for drafting and design. Manufacturing companies also use CAD/CAM for analysis, fabrication, and testing.

The most common types of CAD system, which are PC based, using PowerMacs or pentium-based PCs, now compete with workstation-based or terminal systems. AutoCAD®, MicroStation®, CADKEY®, VersaCAD®, P-CAD®, and Personal Designer® software packages are PC-based systems typically found in schools and throughout industry.

On a **networked system,** a number of designers and drafters can work simultaneously at various terminals (Fig. 6–11), each on a different phase of development—such as design, engineering analysis, drafting, or manufacturing—for a single product or for many different products. It is not uncommon to have eight people working on different design tasks, all on the same networked system.

Operation

The workstation of a typical CAD system makes possible a simple yet powerful interaction between you and the computer. You can create, modify, and refine the design interactively, viewing the emerging work on a graphics display. With a single stroke of the pen or keyboard input, you can move, magnify, mirror, rotate, copy, stretch, or otherwise manipulate the entire design or any portion of it (Fig. 6–12).

In addition to the pen, mouse, or puck, you can communicate with the system through a keyboard. Using a combination of numbers and simple phrases, you can type X, Y, or Z coordinates, enter text for drawing annotation, and initiate graphics processing commands.

You can ask the system to automatically retrieve any previously completed drawing needed for reference, as well as the standard design symbols that you expect to use. Symbols and completed designs can all be stored, where they are instantly available. The on-line library

FIGURE 6–11
Networked CAD/CAM system. (Courtesy Prime–Computervision)

FIGURE 6–12
Interactive IC design, the CALMA®. (Courtesy Prime–Computervision)

speeds up the design process by eliminating unnecessary redrafting of commonly used components and subassemblies.

Documentation

As a part is designed on the system, its physical dimensions are defined, along with the attributes of its various components. CAD systems provide the capability to extract the data from the drawing automatically. Information such as part number, material, vendor, and cost are entered into the system when the part is used on a drawing. The information can then be tallied, and a report can be generated by the system. For example, the part-number data of materials can be used to help generate a parts list for the purchasing department. Computer tapes can guide NC machine tools and equipment for quality control or other product-testing uses. Other computer programs can help engineers to check for interferences or tolerances, generate models for engineering analyses, and calculate areas, volumes, and weights of the product under development. All these nongraphic capabilities are automatic by-products of the CAD/CAM design process. For electronic design, data retrieval is one of the most important nongraphic capabilities.

Memory and Storage

Symbol libraries, drawing segments, whole drawings, design models, and submodels complete with text are stored on magnetic tape (if inactive or waiting for scheduled revisions) or on disks (if needed for reference at an adjoining workstation or another CAD system). A completed drawing is placed on a portion of a disk, where it can be found rapidly by the system.

Workstations

The drafting workstation is composed of a large, smooth drafting table called a digitizer (which converts graphics to digits), a display (monitor), and an alphanumeric keyboard. This workstation is the control center for active work input (Fig. 6–13). The designer or drafter enters commands for all the system functions from this station.

A typical interactive CAD workstation provides several functions:

Interfaces to the host computer, either a large mainframe or local minicomputer

Stores digital descriptions of a drawing, possibly locally

Generates a steady image on the display through its own local memory or by other means

FIGURE 6–13
Inputting an electronic diagram sketch at a digitizing station.

FIGURE 6–14
Advanced graphics system with two monitors. (Courtesy Autotrol Technology Corp.)

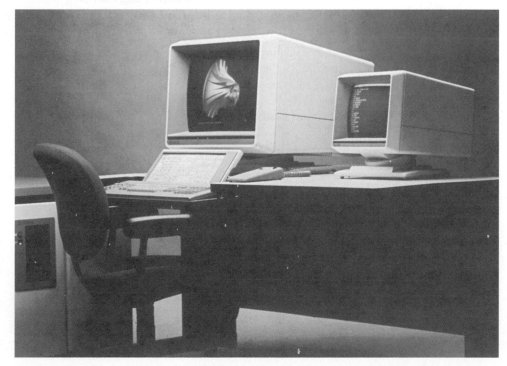

FIGURE 6–15
Light pen tablet and menu.
(Courtesy
Prime–Computervision)

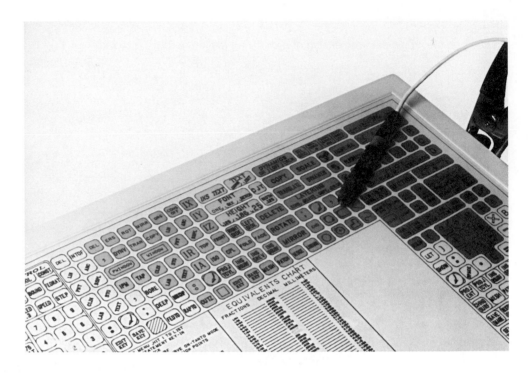

Translates computer instructions into operating functions and routes commands for the various function generators

Contains operator-input devices for communicating with the computer: data tablets, mouse, light pen, digitizer table, cursor controls, or function keys

CRT Devices

All CAD systems use some kind of display device. There may be two screens, one alphanumeric (letters and numbers), the other graphic (pictures) (Fig. 6–14). Two basic picture-generation techniques are used in computer graphics systems: **stroke writing** and **raster scan**. Most CAD systems use the raster-scan CRT.

INPUT DEVICES

A variety of devices allows you to communicate with a computer without the need to learn programming. These devices involve picking a function to enter text and numerical data into the system, to create and modify the graphics on the CRT (inputting, moving, copying, erasing, stretching, rotating, dimensioning), and finally to create the desired finished part.

All CAD systems have at least one operator-input device. Many systems have several such devices, each for a different function. Alphanumeric keyboards, function boxes, light pens (Fig. 6–15), track balls, mice (Fig. 6–16), joysticks, styluses, data/graphics tablets, and digitizing tables are all used on CAD systems.

Most data tablets allow some separation between the stylus and the tablet surface. That is, the stylus need not be in contact with the tablet surface. Therefore, a paper drawing or other sheet can be placed on the data

FIGURE 6–16
Logitech mouse. (Courtesy Logitech, Inc.)

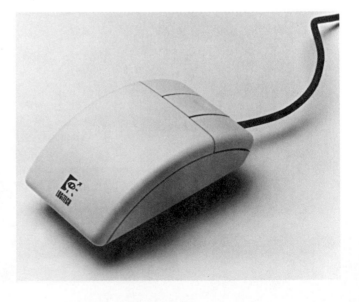

FIGURE 6–17
Data tablet. (Courtesy
Prime–Computervision)

FIGURE 6–18
Optec digitizing table. (Courtesy Auto-trol Technology
Corp.)

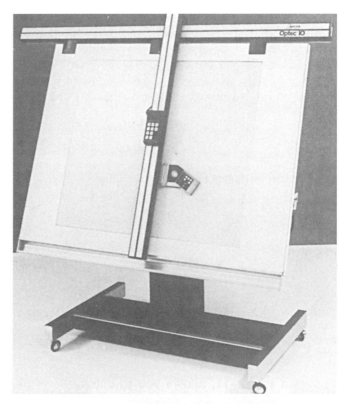

tablet (Fig. 6–17), enabling you to translate drawing
coordinates into digital form. This process is called **digitizing**. This digitizing feature is very important in many
computer-aided design and data-analysis applications.
Digitizers are devices that convert coordinate information into numeric form that is readable by a digital
computer. Digitizer tables are also available as shown
in Fig. 6–18.

Tablets without digitizing options are simply data/
graphics tablets. This graphics tablet has a surface area
that corresponds to the display area of the CRT. By
moving a hand-held puck (a *mouselike* device) with
input buttons (Fig. 6–19), the operator can position the
display **cursor** symbol on the CRT. Instead of a tablet
menu, a **display menu** appears on the CRT. Menu commands are entered by positioning the symbol cursor on
the desired menu function displayed on the screen and
pressing an input button.

Most PC-based CAD systems can be operated with
a mouse as their primary means of input. A mouse is
moved on a flat surface (a pad or table); its movement
controls the position of the screen cursor. Buttons on
the mouse (Fig. 6–20) allow the user to input the screen
menu selections quickly and to pick locations on the
screen.

FIGURE 6–19
Using a puck to pick an option from a tablet menu.

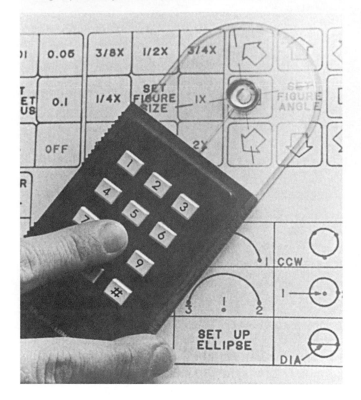

Menus

The menu (Fig. 6–21) shows the commonly used symbols and commands of a particular engineering field. Since not all symbols can be placed on a menu, a typical system has the capability to create and hold a large drafting library. A library contains all the needed symbols, drawings, or figures: nuts, bolts, screws, electronic and electrical symbols, welding symbols, or component outlines. It is basically an unlimited template. This drafting library can be added to or subtracted from as necessary. A drafter typically collects and customizes figures or symbols from the drafting library, creates any special figures that will be used repeatedly, and assembles them into special menus.

The typical menu item is inserted with a minimum of keystrokes. Each symbol can be inserted at any angle or scale. The symbol or figure can be as simple as an electronic diagram symbol (Fig. 6–22) or as complex as a complete printed circuit board. Once the menu symbol or figure is created, it can be stored and used any number of times in other drawings.

Electronic Symbols

Symbols are graphical representations of standard parts or items used repeatedly by the designer or drafter. In traditional drafting methods, a template is used to draw common features, such as a resistor. With CAD, these symbols are drawn once, stored, and then recalled and used when needed. When symbols are created, information (**attributes** and properties) about the part can be included. This information can then be used to create a parts list or to do calculations.

Creating and Defining Symbols

Forming a symbol involves creating the graphics in the form of a block or subpart and storing the symbol for future use. Symbols are created using standard graphic commands. A symbol can be altered by recalling the symbol, making the desired changes, and re-storing the symbol. A modified symbol can be stored with its old name, thereby replacing the old symbol, or stored under a new symbol name, creating a new symbol. A new version of a symbol can only be replaced in any existing drawings by issuing an update command. All other drawings on which the old version appears are left unaltered.

OUTPUT OR HARD-COPY DEVICES

Output from CAD systems can be in many forms. The most common is a drawing, just like the one created on a drafting board. This drawing is created by a **plotter**. Output can also be a copy of what is on the screen, sometimes called a hard copy. A hard copy normally comes from a printer or plotter attached directly to the workstation. A drawing cannot only be obtained from a plotter, but it can also be drawn on microfilm. This is called **computer-output microfilm** (COM). Output is not

FIGURE 6–20
Three-button mouse. (Courtesy Logitech, Inc.)

FIGURE 6–21
Electronic design menu. (Courtesy Intergraph Corp.)

ELECTRONIC DESIGN SYSTEM   INTERGRAPH

TEXT — PLACE, PLACE FITTED, EDIT, COPY & INCREMENT; .XX XX XX, -XX XX XX, XX XX XX.

GRAPHIC ELEMENT MANIPULATION — COPY, SCALE, ROTATE, PARTIAL DELETE, CHANGE SYMB; MOVE, DELETE, MODIFY, MIRROR (V H L); EXTEND LINE, COPY PARALLEL

GRAPHIC ELEMENT PLACEMENT — LINE, LINE STRING, ARC, ELLIPSE, CIRCLE, SHAPE, TERM

ENTER DATA — FILL-IN, COPY; COPY & INCREMENT; JUSTIFY (L C R)

AUTO DIMEN — PLACEMENT (AUTO SEMI MAN); WITNESS LINES (ON OFF); INT/EXT; STRING RADIAL; ANGULAR; AXIS

FENCE — MOVE, COPY, SCALE, ROTATE, DELETE, CHANGE SYMBOLOGY; MIRROR (V H L); DROP COMPLEX SET, DROP WORKING SET

CELLS — PLACE, DEFINE ORIGIN, SELECT, IDENTIFY

FAST DISPLAY ON/OFF — TEXT, FONT, CELL

ANGLE INPUT — 0, 90, 180, 270

SCHEMATIC

SCHEMATIC CREATE — PAGE, SYMBOL, LABEL, REF. FIELD, PIN, TYPE FIELD

REVIEW — LEVEL, ID

PAGE — PLACE, SELECT

SYMBOLS — PLACE, ASSIGN, COPY, DELETE, FREE PIN, MOVE UNATTACHED, ASSIGN DEFAULT PIN, EDIT REF., EDIT TYPE, EDIT PIN

LABELS — PLACE BLANK, LABEL POINT, EDIT, DELETE, COPY/CLEAR LABEL, MOVE, FREE LABEL, DUPLICATE LABEL

INTERCONNECT — ORTHO ROUTE, FREE ROUTE, DELETE

CIRCUITS — CREATE CIRCUIT CELL, PLACE CIRCUIT CELL, COPY CIRCUIT

DISPLAY — WEIGHT, LEVEL SYMBOLOGY, ENTER DATA FIELD, GRID, NET, TRACE WIDTH ON/OFF, CONST LINES, BOARD

SCREENS — L R UPDATE BOTH, VIEW; L R DELAY, SCROLL SCREEN; STOP DRAWING; VIEW ON, VIEW OFF; ZOOM IN, ZOOM OUT; WINDOW (FIT, AREA, CENTER); COPY, SWAP, OVER VIEW ON/OFF

PRINTED CIRCUIT

COMPONENT CREATE — PC PART, VIA, REF. FIELD, PIN, TYPE FIELD

REVIEW — LEVEL, ID, LAYER

LEVELS — DISPLAY ON, DISPLAY OFF, ACTIVE, CHANGE

ELEMENT DISPLAY SELECT — PACKAGE OUTLINE, PINS, MOUNTING HOLES, REF. DESIG., ROUTE BORDER, REGIST. MARKS, SOLDER 1, FORCE VECTOR, NET NAMES, BOARD OUTLINE, PADS, VIAS, TYPE DESIG., BOARD ORIGIN, LAYER 'N', SOLDER 2, SLI'S

RASTER CONTROL — ZOOM IN, ZOOM OUT, PAN; EXIT, EXIT & SAVE, DRAG ON/OFF

PARAMETERS — SET PINCODE, SET PIN TABLE, DEFINE BOARD AREA, SELECT COMP. OBSTRUCT, SET PADCODE, SET PAD TABLE, DEFINE BOARD ORIGIN, SELECT SCHEMATIC FILE, SET VIA NAME, SET TRACE WIDTH, SET ROUTE LAYER SYM, GET COLOR TABLE, SET POINT OBSTR, SET ROUTE LAYER, DEACTIVATE LAYER

ROUTE OBSTRUCT — PLACE LINE OBSTRUCT, PLACE POINT OBSTRUCT, PLACE NO ROUTE AREA

PACKAGE PLACEMENT — INTERACTIVE DATA PREP, PLACE; CIP DATA PREP, PLACE; TYPF, PACK; MOVE PACKAGE ATTACHED, UNATTACHED

SLI — PLACE, DELETE, TWO POINT, ORTHO, FREE

TRACES — ADD VERTEX, MODIFY VERTEX, DELETE VERTEX, FREE TRACE, CHANGE WIDTH

MULTIWIRE — SET WIRE LAYER, PRE WIRE, ROUTE WIRE, REROUTE WIRE

MEASURE — X,Y,Z, ΔX,ΔY; DISTANCE (CUMULATIVE, MINIMUM)

LOCKS — GRID, UNIT, SNAP; ANGLE, SCALE, LEVEL; REF FILE SNAP (R1, R2, R3)

PAGE — SELECT R1, SELECT R2, SELECT R3

DRC — SETUP (PARAMETERS, DATA PREP); INTERACTIVE (BEGIN, SKIP, NEXT, EXIT, RECORD)

PLOTTING — PC ETCH, DRILL MAP, SILK SCREEN, SOLDER MASK, ALL; P1, P2, P3, P4, P5

PADS — ADD, MOVE, DELETE; EDIT FREE PAD, EDIT PAD CODE, DELETE

VIA — PLACE, DELETE

FILES — COMPRESS DESIGN FILE, COMPRESS CELL FILE, DESIGN OPTIONS; FILE DESIGN, CREATE INDEX, DISPLAY HEADER

DIGITIZER — DEFINE ORIGIN, ROTATION ANGLE, FIXED SCALE

always a drawing. Sometimes the output is a magnetic tape containing instructions for a particular machine to make a part that has been designed. This form of output is called **numerical control** (NC). The output can also be a report produced on a printer. Artwork for printed circuit or integrated circuit design can be produced on a photoplotter (Fig. 6–23).

Hard-copy devices include printers, plotters, and photocopy equipment. Printers provide you with alphanumeric readouts and material lists. The plotter allows you to produce ink drawings on paper, vellum, or drafting film in a multitude of colors. Some plotters are limited by the size of the plotting surface, like those shown in Fig. 6–24. Although they are limited to standard paper widths, others (Fig. 6–25) can plot drawings of any length. Pen plotters can use ball-point, felt-tip, and liquid-ink pens or pencils (Fig. 6–26). Check copies are normally run with inexpensive ball-point pens. Original, high-quality drawings are plotted with India ink and liquid-ink pens. When plotting a drawing, you have a number of options: to scale the drawing, rotate it, select the colors to plot, and even substitute different line widths. Not all pen plotters have these options. A variety of plotters is available, including drum plotters (Fig. 6–27), flat-bed plotters (Fig. 6–28), electrostatic plotters (Fig. 6–29), digitizer plotters, laser plotters, and recently inkjet plotters capable of producing standard D and E size plots.

FIGURE 6–22
CRT menu.

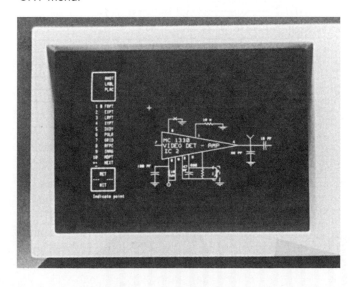

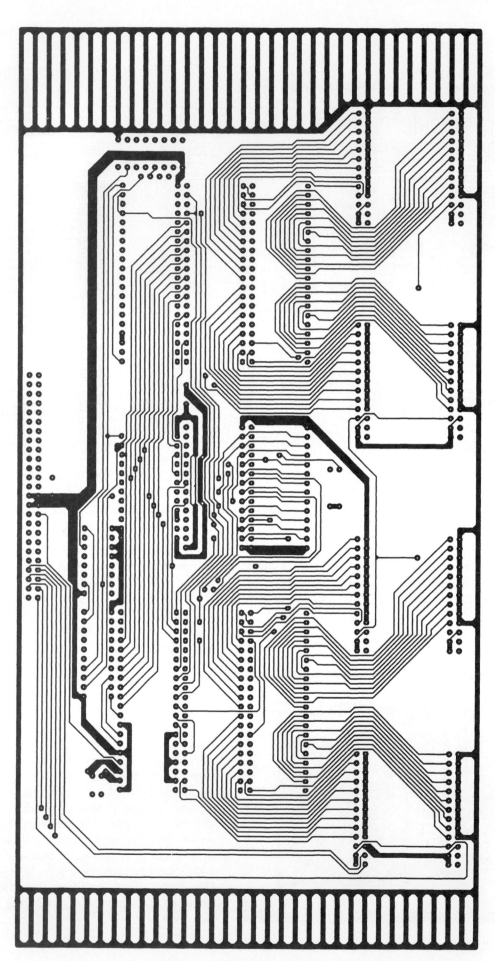

FIGURE 6–23
PCB artwork plot.

FIGURE 6–24
Small desktop plotters.
(Courtesy Hewlett–Packard Co.)

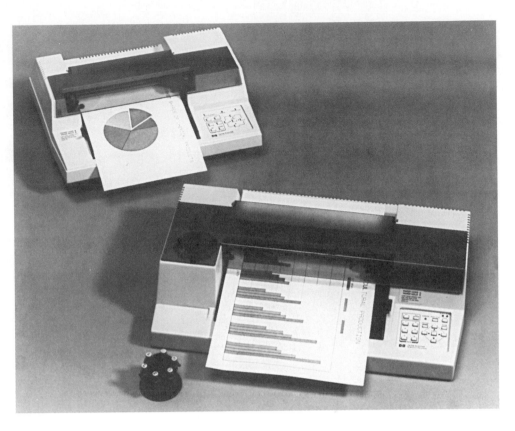

FIGURE 6–25
Thirty-six-inch plotter.

FIGURE 6–26
Pen plotting. (Courtesy Lockheed–California Co.)

FIGURE 6–27
Drum plotter.

FIGURE 6–28
Precision flat-bed plotting. (Courtesy Lockheed–California Co.)

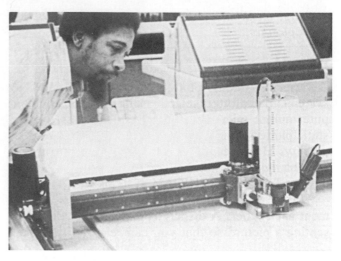

FIGURE 6–29
Electrostatic plotter. (Courtesy Hewlett–Packard Co.)

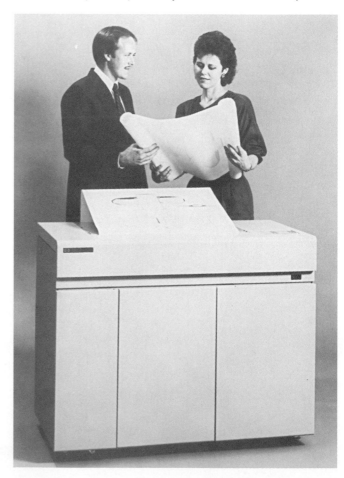

FIGURE 6–30
Printed circuit board design.
(Courtesy
Prime–Computervision)

ELECTRONIC DESIGN CAD SOFTWARE

One of the most important applications for CAD is integrated circuit and *printed circuit board design and documentation* (Fig. 6–30). The complexity of these circuits makes them very good candidates for automated drawing tools. CAD increases productivity by automating and integrating the key steps in the design and production of printed circuit boards (Fig. 6–31). The typical PCB program uses automatic and manual editing modes to design the entire board, from the drawing of the schematic to the final manufacturing and testing stages. Schematics, text, and board geometry are entered interactively into the system. *Automatic assignment, placement,* and *routing* routines are used to complete the design of the board. A variety of PC board sizes and types can be designed. Manual input can be used to override the automatic routines.

FIGURE 6–31
Automated NC drilling and printed circuit board component placement. (Courtesy Prime–Computervision)

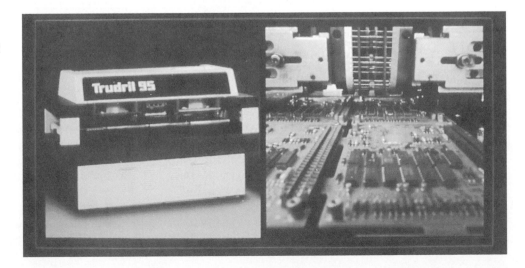

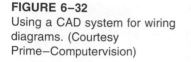

FIGURE 6–32
Using a CAD system for wiring diagrams. (Courtesy Prime–Computervision)

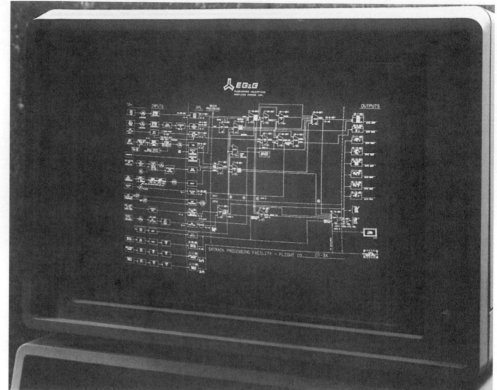

A CAD system can also automate and integrate key steps into the design, documentation, and checking of design rules in *wiring diagrams* (Fig. 6–32). It reduces the time and expense in capturing, checking, updating, and extracting design information. This capability is applied to many types of diagrams: logic (Fig. 6–33), elementary (ladder) (Fig. 6–34), wiring (Fig. 6–35), schematic, wire harness, and interconnections. Designs are developed faster, with fewer errors and higher quality.

USING CAD SOFTWARE

The remainder of this chapter contains an introduction to some of the capabilities of AutoCAD, one of the popular CAD programs in use today. However, even if you have a different software package available, you should be able to complete the same activities. All CAD software should contain the same functions, though menus and steps may be a little different. You should be able to customize most of the different screen elements using commands like preferences, options, or settings (Fig. 6–36). These elements would be the system variables, environment variables, location of associated files, and other miscellaneous features. Any software that

runs on a PC (both Mac and Windows) will have a common interface. There will always be an active window with a title bar and a menu bar, with pull-down menus generally including File and Edit and other options (Fig. 6–37). These similarities will help you to move from one software to another. The main window in a CAD program is the drawing window (Fig. 6–38), which replaces the sheet of drawing film. The other tools you will find in CAD programs are standard toolbars, command line windows, and perhaps additional floating toolbars or on-screen menus. Because of improvements in graphics interfaces and floating toolbars, typically, on-screen menus are not as useful or necessary as they once were.

Objects are placed on the page using a combination of mouse clicks and keyboard entries. The keyboard entries are typed in the command line window. Enlarging this window will display a command history for this drawing session, as seen in Fig. 6–39. There may be a status bar at the bottom of the screen. In windows, as you move the mouse over a tool, a short description of the tool or button will be displayed in the help area of this status bar. There will be other information, including items like X, Y coordinates of the current mouse location.

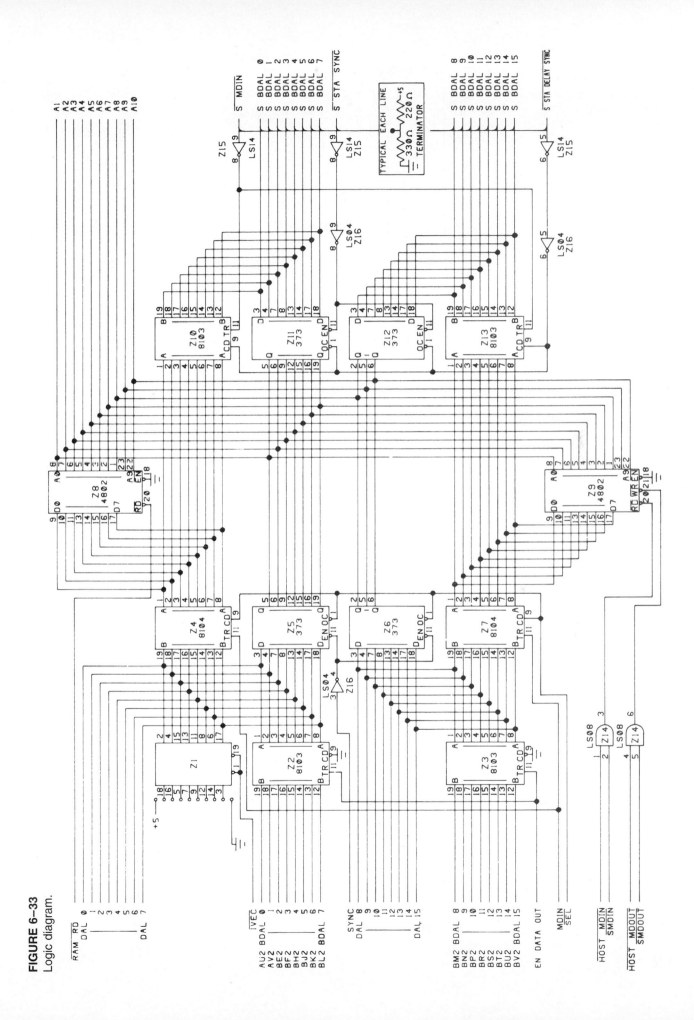

FIGURE 6-33
Logic diagram.

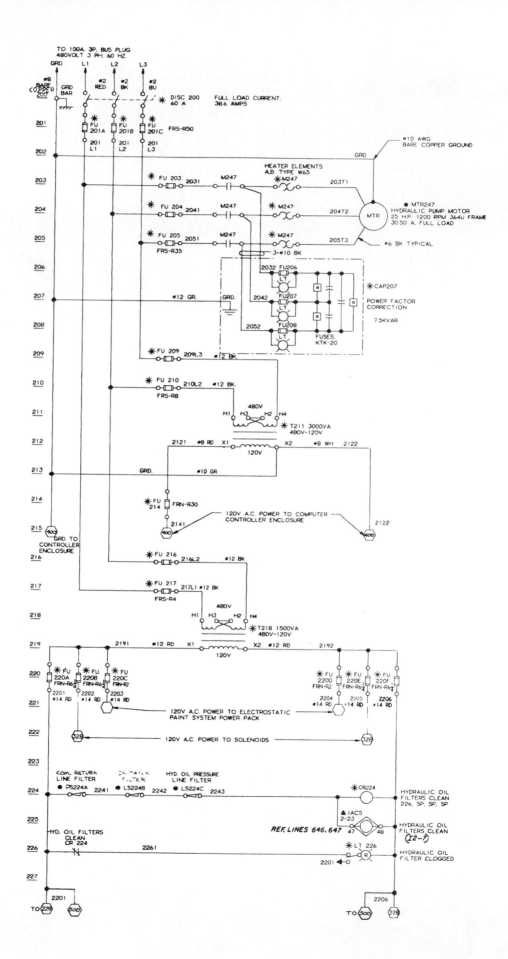

FIGURE 6–34
Ladder diagram for NC production robot. (Courtesy GMF Robotics)

FIGURE 6–35
Wiring diagram.

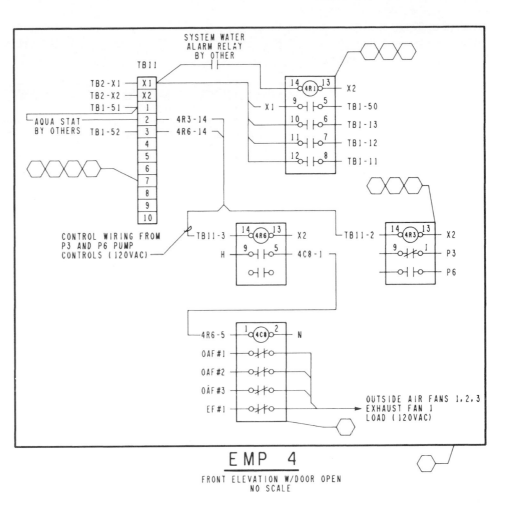

FIGURE 6–36
Customizing screen elements.

FIGURE 6–37
Parts of AutoCAD active
window.

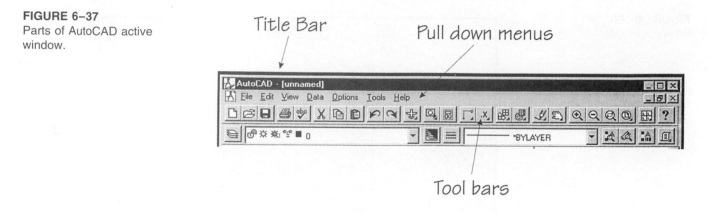

Drawing an Object

When you start AutoCAD, a new untitled document will be displayed. You are then ready to either start a new drawing or choose to *open* or retrieve an existing drawing to modify or plot. You will always begin a new drawing by selecting a tool or issuing a command. Again,

even if you are not using AutoCAD, there will always be a way to select tools or draw objects.

Correcting Mistakes

If you are using the command line to type commands and you notice that you have typed incorrectly, you can

FIGURE 6–38
Main drawing window.

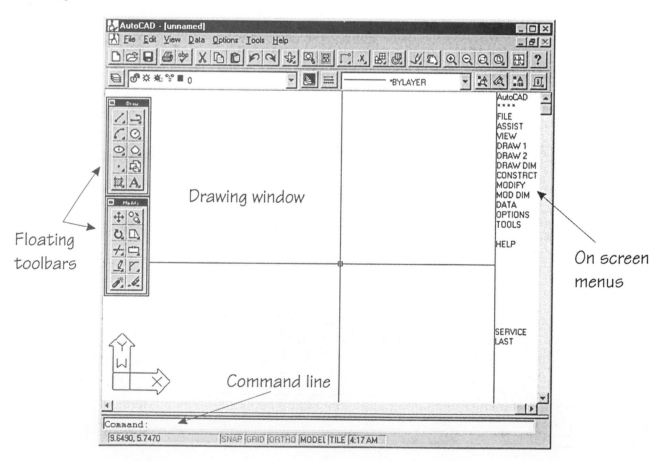

FIGURE 6–39
Command history window.

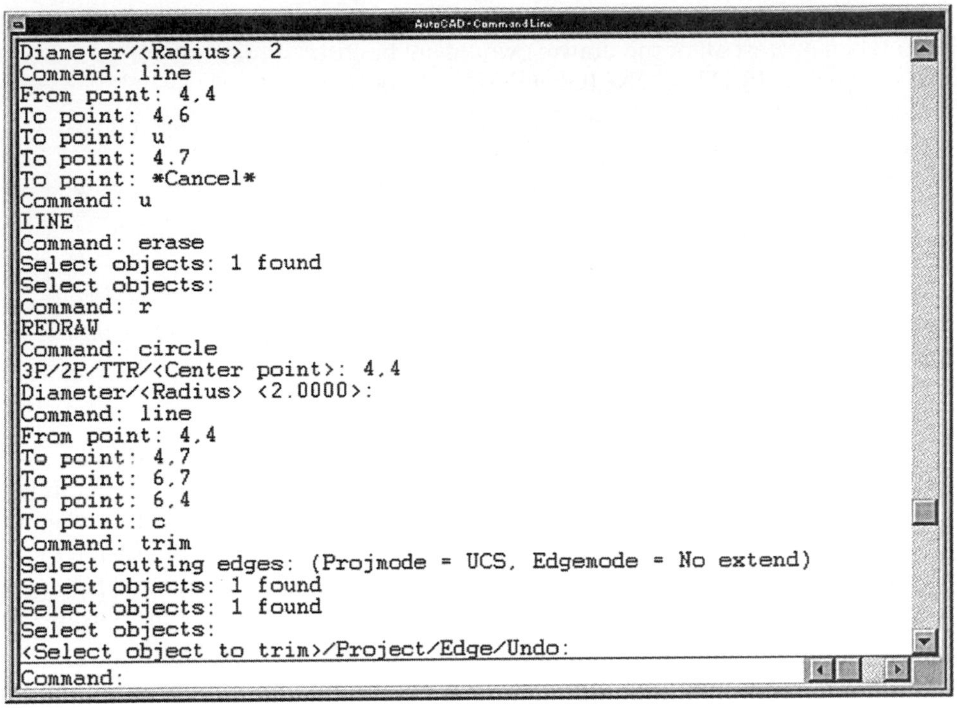

```
                          AutoCAD - Command Line
Diameter/<Radius>: 2
Command: line
From point: 4,4
To point: 4,6
To point: u
To point: 4,7
To point: *Cancel*
Command: u
LINE
Command: erase
Select objects: 1 found
Select objects:
Command: r
REDRAW
Command: circle
3P/2P/TTR/<Center point>: 4,4
Diameter/<Radius> <2.0000>:
Command: line
From point: 4,4
To point: 4,7
To point: 6,7
To point: 6,4
To point: c
Command: trim
Select cutting edges: (Projmode = UCS, Edgemode = No extend)
Select objects: 1 found
Select objects: 1 found
Select objects:
<Select object to trim>/Project/Edge/Undo:
Command:
```

simply backspace and retype the correct command. If you have completed typing and pressed the enter key, typically you will get a response indicating that an "unknown command" was entered (Fig. 6–40), and you will then have the opportunity to type the correct command. Or, if you issue a command for the wrong type of object, you can usually push the escape key to cancel that command. Another very helpful command in most software programs is the *undo* command. AutoCAD allows the undo feature with the ability to go back through a history of commands and undo only the one command that you want to change, without having to change all the things you did after that. Another way to correct mistakes or to modify drawings is to use an *erase* or *delete* object command. This can be used to make changes to past drawings, as well as to the current or new drawing.

The tools used to produce drawings will be the same no matter what kind of drawing is produced: electrical, mechanical, piping, or other. These tools will need to draw lines, circles, arcs, ellipses, boxes, etc. Drawing

libraries may be available, with commonly used parts or symbols. And there should also be a way to create additional symbols to add to custom libraries.

Creating a Drawing Using AutoCAD

Remember, when you start AutoCAD, it always opens with a new blank page. As you start drawing, your work will be *stored* in the computer memory and will only remain there as long as there is power to the computer and the application is running. To save *often* is perhaps one of the most important lessons to learn when using computer tools to produce drawings. Unfortunately, it is not all that uncommon for computer failures to occur caused by system problems or even power failures. Anything that has not been written to disk, or saved, will be lost when you restart the computer. Also, if the power failed or the computer froze during a save operation, the drawing or file previously saved is lost. The best way to avoid this type of disaster for critical drawings

FIGURE 6–40
Entering an "unknown command."

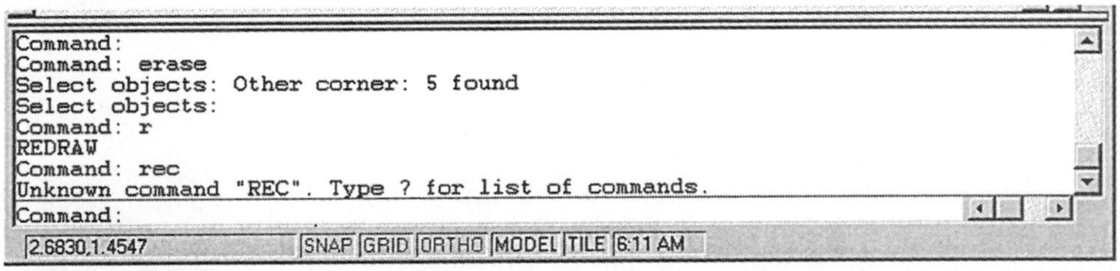

```
Command:
Command: erase
Select objects: Other corner: 5 found
Select objects:
Command: r
REDRAW
Command: rec
Unknown command "REC". Type ? for list of commands.
Command:
2.6830,1.4547          SNAP GRID ORTHO MODEL TILE 6:11 AM
```

is to save the drawing with a different name at least once. Then at least all of the drawing would not be lost. The commands PLOT, CONFIG, and SHELL seem to cause the most system failures, and it will only be a matter of time before this happens to you. Therefore, it is always a good idea to SAVE before issuing any of these commands.

In a classroom activity or even in a work environment, if files are stored to a common directory area of a hard drive, you must be aware of common file names that might be used: LAB1.DWG, for example. The next person to save with the same name would overwrite your file and all your work would be lost. If you need to save your files to a floppy disk, never save them directly to floppy. During the save process, if the disk becomes full, AutoCAD could freeze and your drawing would be lost. It is always a better idea to save on a hard drive and then transfer it to a floppy when you are finished working. Also, it is never a good idea to *load* or retrieve a file directly from a floppy. AutoCAD creates and writes a lot of temporary files to the disk while it is running, and if this causes the disk to become full, it might also freeze the system, resulting in lost work or maybe even destroying the original file.

When you finish one drawing and want to begin a new drawing, do not use QUIT. This will actually quit the entire application. Though this will not cause problems, it simply takes more time, because the application would have to be restarted to begin the next drawing. If you simply select New or Open from the File Menu, you will be forced to exit the existing drawing, and a new blank drawing or previously created drawing will then open. Also, earlier versions of AutoCAD used the command END to quit the drawing and return to a *main menu,* which no longer exists. So END will save the drawing and exit the drawing editor or completely quit AutoCAD. Again, no problems are created; it is just not as efficient if you want to start a new drawing because the loading time of the program itself can be very slow.

AutoCAD drawings will open with default settings. As you begin to become more comfortable with the software, you may wish to change some of these settings. For example, the UCS icon (Fig. 6–41), which represents the coordinate system used for the drawing, is not as critical when producing two-dimensional drawings, and in many cases it may take up valuable screen space and interfere with drawing space. Using the options menu, the setting for this could be set to off (Fig. 6–42) so that it would not show on the drawing.

All drawings, no matter how simple or how complex, are made up of objects (lines, circles, or arcs) and text. Identifying a starting point or location is the beginning of drawing any of these objects. Most electri-

FIGURE 6–41
UCS icon.

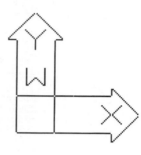

cal drawings do not have a scale. The lines that are drawn are often functional representations and are not based on true length. This means that the placement of these objects on the page has more to do with readability, and not true size or scale. Placement and spacing on the page using AutoCAD to produce drawings are just as critical as for manual drawings. You must have trial sketches and trial layouts before you begin the final drawing so that you can produce a readable drawing that is also esthetically pleasing.

The coordinate position of points or the current location of the cursor is displayed on the status line as shown in Fig. 6–43. You can enter all the points of a drawing using absolute coordinates. This would require keying in every point based on the absolute distance

FIGURE 6–42
Options menu.

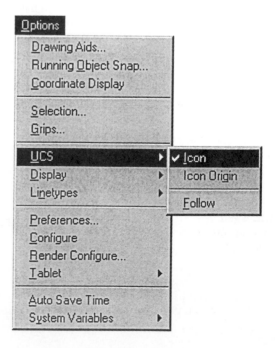

FIGURE 6–43
Status line.

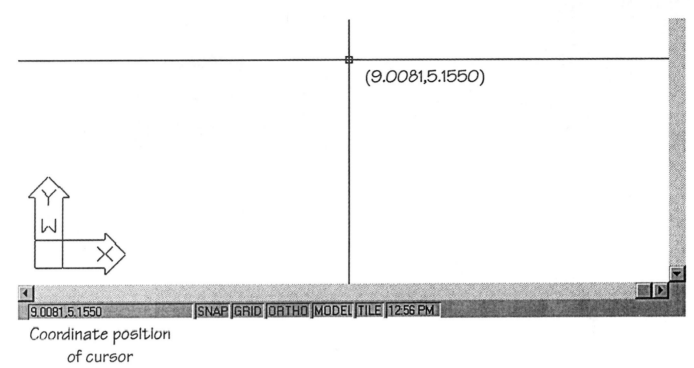

(9.0081,5.1550)

9.0081,5.1550   SNAP GRID ORTHO MODEL TILE 12:56 PM

Coordinate position
of cursor

that the point was from the origin. This would be an overwhelming and very tedious task. Fortunately, drawings are more often created using relative coordinates or objects placed relative to other objects.

Using Layers

For the majority of electrical drawings, there will not be much need to use different layers to complete the drawing. In production drawings or PCB drawings, it might be necessary to have a series of drawings to produce a two-sided printed circuit board and component layout drawing (Fig. 6–44). Using multiple layers makes it easier to maintain registration, since the different layers are all part of one drawing. A simple schematic may be divided by text or annotations on one layer, with another layer for symbols and other schematic elements. You can then assign different colors or line types to different layers to help to distinguish them from each other.

A separate layer could be used to draw construction lines. Though construction lines are very useful in the preparation of a drawing, they need to be removed or erased before the final drawing is plotted. The use of a separate layer containing only construction lines would very effectively solve this problem. The layer could be visible during creation of the drawing and then invisible

when plotted or simply removed or erased when the drawing is finished. When using the LAYER command, multiple layers may be created at the same time by entering multiple names and separating them with commas.

Using Grids

Most electrical drawings are done using primarily horizontal or vertical layouts. One AutoCAD tool that is very helpful in producing drawings is the grid function. When you enable the grid, a series of dots will appear on the page (Fig. 6–45) to help locate parts and points. The spacing on this grid is a user-defined option. Be careful not to choose a grid that is too dense or too many points will be placed on the drawing, which could make the drawing more difficult to see.

Editing Existing Drawings

Once objects have been placed on a drawing, it is very easy to modify them. This may be one of the biggest advantages of using an electronic medium for producing drawings. Drawings that are completed manually are very difficult to edit or change. The three basic kinds of editing you are most likely to do include changing, copying, or erasing. The location of an object could be

FIGURE 6–44
PCB artwork. (Courtesy Motorola, Inc., Semiconductor
Products Sector)

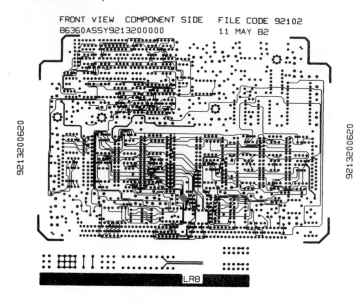

changed using MOVE or ROTATE commands. The length of lines or arcs could be changed using TRIM, EXTEND, SCALE, STRETCH, or LENGTHEN commands. A portion of a drawing, like a power supply, might be copied to place on another drawing, or a portion, object, or lines of a drawing may be erased or deleted. Objects that are erased by accident may be unerased using the command OOPS. However, this only works for the most recently erased selection.

The command SCALE is used to permanently make individual objects larger or smaller. To simply plot the drawing to a different size or scale, it is much better to use the scale option within the PLOT command.

Using Blocks

Blocks are drawing elements that are grouped together so that they may be easily used in current or future

FIGURE 6–45
Turning on GRID display.

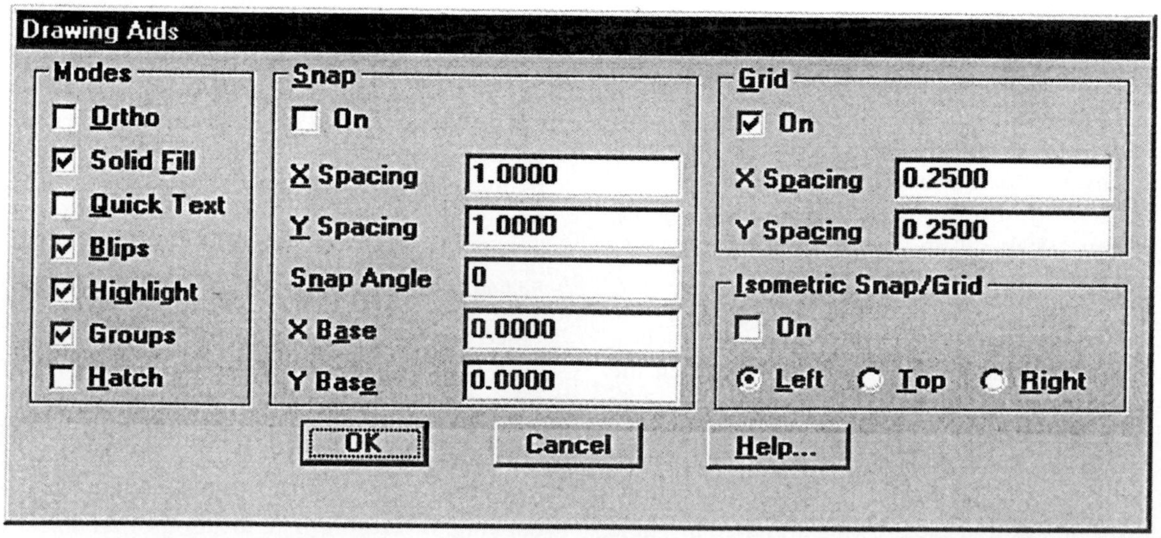

FIGURE 6-46
Use of blocks.

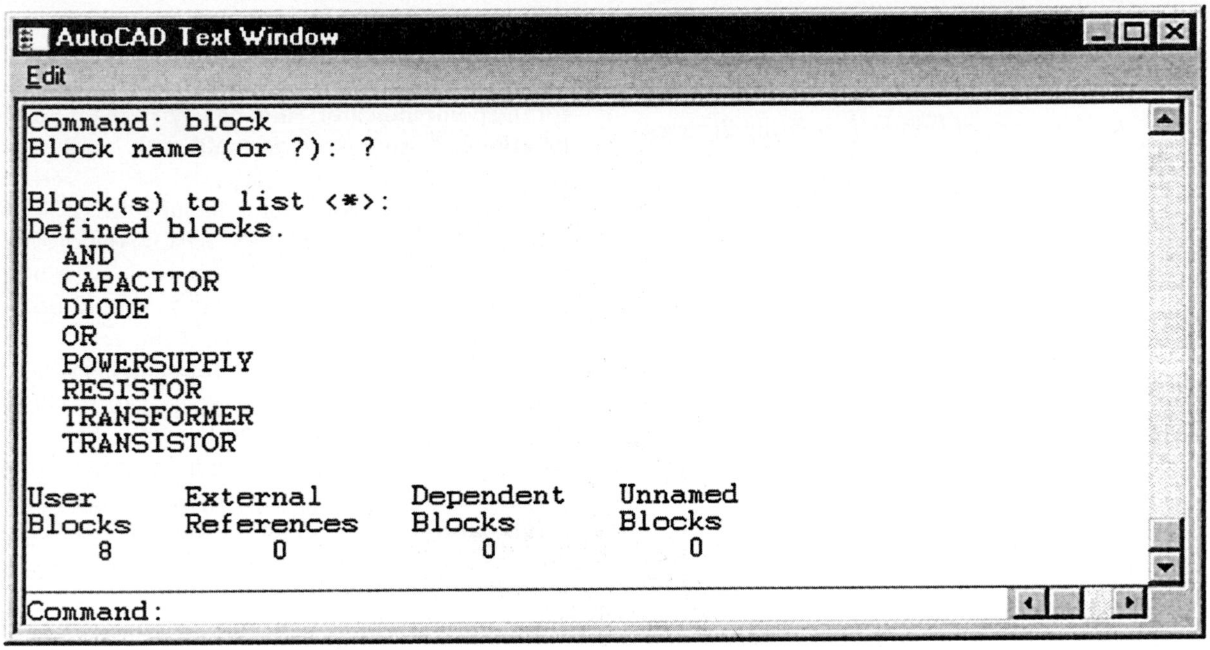

```
┌─────────────────────────────────────────────────────────────────┐
│ ▤ AutoCAD Text Window                              [_][□][✕]      │
├─────────────────────────────────────────────────────────────────┤
│ Edit                                                              │
├─────────────────────────────────────────────────────────────────┤
│ Command: block                                              [▲]   │
│ Block name (or ?): ?                                              │
│                                                                   │
│ Block(s) to list <*>:                                             │
│ Defined blocks.                                                   │
│   AND                                                             │
│   CAPACITOR                                                       │
│   DIODE                                                           │
│   OR                                                              │
│   POWERSUPPLY                                                     │
│   RESISTOR                                                        │
│   TRANSFORMER                                                     │
│   TRANSISTOR                                                      │
│                                                                   │
│ User        External       Dependent     Unnamed                 │
│ Blocks      References      Blocks        Blocks                  │
│    8           0              0             0              [▼]    │
│                                                    [◄][ ][ ][►]   │
│ Command:                                                          │
└─────────────────────────────────────────────────────────────────┘
```

drawings. This would be similar to using a template to draw symbols or objects in the manual completion of a drawing. The use of blocks eliminates the need to redraw common symbols that are used often and also ensures that each time the symbol is used it will be a standard size (Fig. 6–46). The use of blocks in an individual drawing is not the same as the use of symbol libraries. You may create blocks on any drawing that requires repetition of the same object. You can create a library of symbols by completing a drawing that is a collection of symbols or blocks of symbols.

Prototype Drawings

If you are doing lots of similar drawings you will most likely be using the same units, scale factors, etc. One drawing may be created with all the standard settings that would normally be used and saved as a prototype

FIGURE 6-47
Creating a new drawing using a prototype.

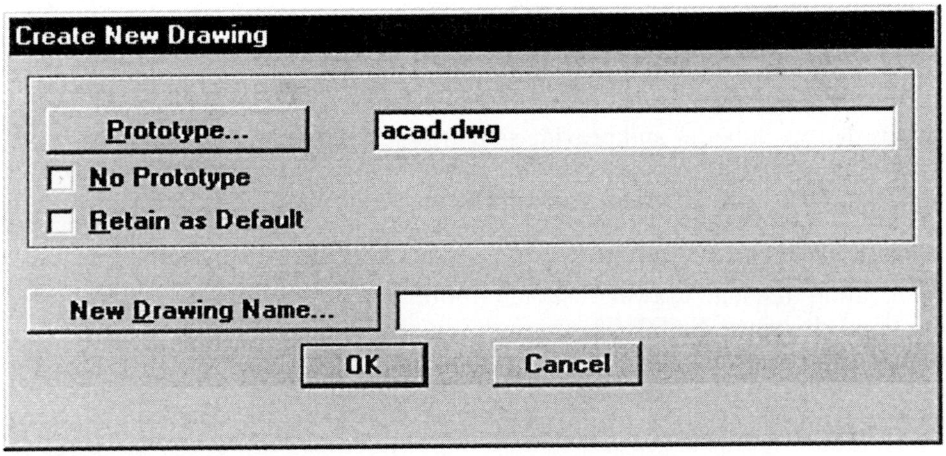

```
┌─────────────────────────────────────────────────────────────────┐
│ Create New Drawing                                                │
├─────────────────────────────────────────────────────────────────┤
│                                                                   │
│   ┌─────────────────┐   ┌─────────────────────────────────────┐  │
│   │   Prototype...  │   │ acad.dwg                            │  │
│   └─────────────────┘   └─────────────────────────────────────┘  │
│   □ No Prototype                                                  │
│   □ Retain as Default                                             │
│                                                                   │
│   ┌─────────────────┐   ┌─────────────────────────────────────┐  │
│   │ New Drawing Name...│ │                                     │  │
│   └─────────────────┘   └─────────────────────────────────────┘  │
│                                                                   │
│           ┌──────────┐       ┌──────────┐                        │
│           │    OK    │       │  Cancel  │                        │
│           └──────────┘       └──────────┘                        │
└─────────────────────────────────────────────────────────────────┘
```

drawing. Then anytime you start a new drawing by selecting this prototype (Fig. 6-47) all of your typical settings will already be selected.

Drawing Aids

One of the most important things in creating drawings is accuracy. In mechanical drawings, this accuracy is very critical in parts manufacturing. Placement accuracy is not quite as important for schematic or some other electrical drawings. The distance between parts is primarily based on esthetics or readability and has very little to do with actual location of parts or lengths of wires. However, the same aids that are used to improve true lengths and line placement will also be useful tools for any layout. In addition to the grid feature already discussed, there is a snap feature. The grid itself will not help in the accuracy of the drawing; it is only a pattern, visible on the screen and very useful as a visual aid. Snap, on the other hand, will allow you to place points or other objects, but only in precise increments. If you have snap turned on and set to .5, you can only place points on exact .5-unit increments. In addition to point snap, there is a special form called object snap that enables you to accurately locate geometric objects in your drawing.

If you are only going to draw horizontal and vertical lines, you can use a special snap called *ortho* mode. This limits all lines to right angles. There are many different ways to use snap to align objects and points with other objects or points. As you become more proficient in the use of AutoCAD, you will have projects that may need these features.

There is a special *isometric mode* that could be used to create isometric drawings, but it is generally not the preferred way to produce them. Since Release 10 of AutoCAD, the preferred method is to construct the model using 3D functions and then use the commands DVIEW or VPOINTS to view the design in isometric mode.

A number of commands may be found in AutoCAD that are *leftovers* from earlier versions. They are still available for use, but often have been replaced by commands that work better and provide similar functionality. The TRACE command is one of these and, for all practical purposes, has been replaced by the PLINE command.

The SKETCH command, though somewhat useful for creating freehand drawings, should not be used for electrical drawings. SKETCH produces thousands of objects for a drawing, which overloads the drawing file and causes screen regenerations to be unacceptably slow.

Drawing Basic Objects

An object may start with the simplest item—the point. A point object in AutoCAD is created using the POINT command. The command locates a point object with X, Y, and Z coordinates. A number of styles is available for the point indicator, ranging from a plain dot to combinations of squares, circles, and hash marks, as shown in Fig. 6–48.

Other object commands include LINE, CIRCLE, ARC, DONUT, POLYGON, RECTANG (for rectangle), and ELLIPSE. Using these tools with other features, like grid, snap, and ortho, it should be possible to create most of the drawings in this text.

To draw a series of lines connected to each other, you would start by using the mouse to select the line tool or the keyboard to type the line command. You will be prompted to enter a beginning point. You either use the mouse to position this point or enter the X, Y coordinates. Pressing enter on the keyboard completes the selection of the starting point. Then you will be prompted to enter the next point. And, again, you would either select a point by using the mouse or by typing in the coordinates of the second point. You will continue to be prompted for additional points until you *CLOSE* the command by typing a C and pressing enter (Fig. 6–49).

FIGURE 6–48
Variety of different point indicators.

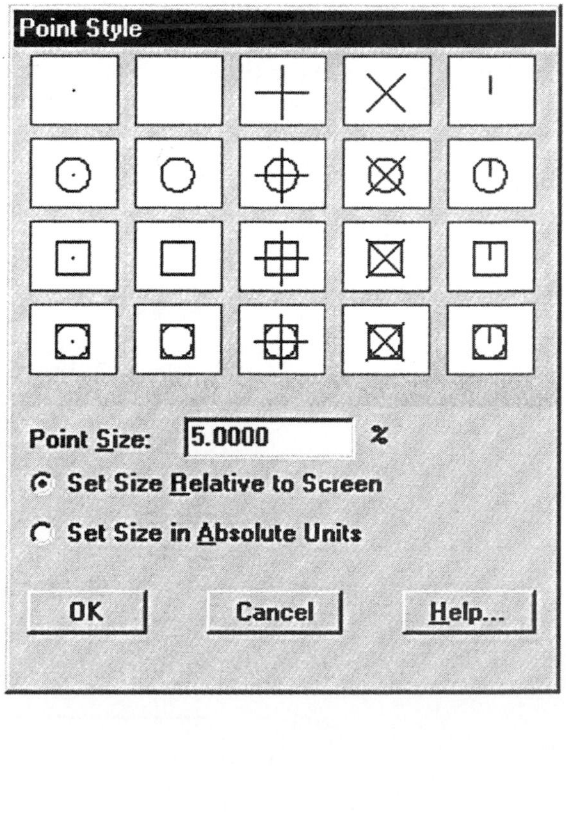

FIGURE 6–49
Drawing a series of lines.

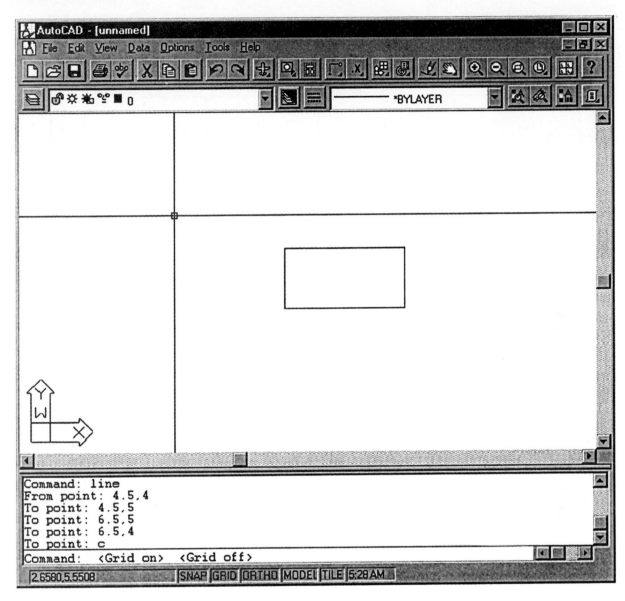

FIGURE 6–50
Using U to erase the last line drawn.

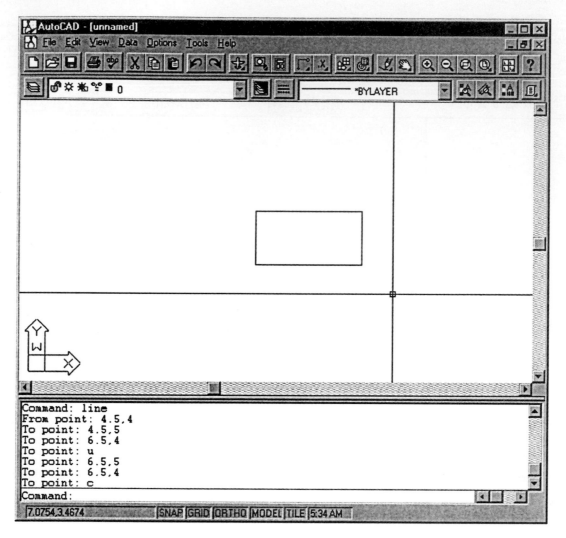

You will learn to use many shortcuts as you become more proficient in the use of the software. One such shortcut allows you to quickly erase the line just drawn from within LINE command. You simply type U when the computer prompts for a beginning point for a line (Fig. 6–50). The line just created is erased, leaving you in the LINE command where you may continue drawing new lines or erasing the previously placed lines.

Editing Drawings

Once objects have been drawn, a number of tools may be used to edit the drawing. Objects can be erased, rotated, moved, copied, and even resized or scaled. The TRIM command can be used to cut away a portion of an object by selecting the cutting edge and indicating which lines to remove (Fig. 6–51). The FILLET command will automatically produce a rounded corner by creating an arc between two lines, or a line and a circle, and so on (Fig. 6–52). The EXPLODE command allows you to take an object like a rectangle and explode it to four individual lines so that any of these individual lines may be modified or edited.

Scaling the Drawing

When you are creating the drawing in AutoCAD, you don't have to worry about the scale of the drawing; you simply draw it as if it were full size, picking a unit of measure that is appropriate (inches, feet, or other). The only time that scale is a concern is when the drawing needs to be plotted; then you need to determine a scale

FIGURE 6–51
The TRIM command.

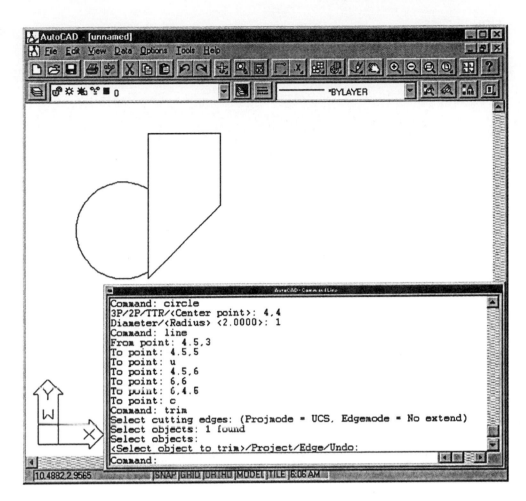

factor. If the finished plot is to be done on an 11″ × 17″ piece of drawing paper with a scale of ¼″ = 1′, how big could the original or full-sized item be? If ¼″ = 1′ (or 12″), the scale factor would be 1:48. So, on this size of paper, the original item could be 11″ × 48 = 528″ (or 44′) by 17″ × 48″ = 816″ (or 68′). Using this scale would allow a fairly large original item to be drawn full size and then scaled and plotted on this size of paper.

Adding Text to Your Drawing

AutoCAD includes text tools that allow you to place notes and other annotations on your drawing. There is a simple text tool that allows you to type one line of text. The command is TEXT (Fig. 6–53), though it is not used all that often since there are two other, more versatile methods of adding text to drawings. The command DTEXT (Fig. 6-54) allows multiple lines of text to be dynamically added to a drawing by continuing to click the cursor where each line of text needs to appear. Thus, once you initiate the command you can type words

on different parts of the drawing until you close the command. This could be very useful for identifying components and symbols on a schematic diagram.

MTEXT is probably the most versatile text tool. It allows you to enter text as a paragraph, with word wraps as in word processing, and the whole paragraph can be resized or reshaped. All the text entered can be formatted with different fonts or styles as discussed in Chapter 2. Once the text has been entered, it may still be edited. This method of entering text is very useful for adding a list of notes or a parts list, for example (Fig. 6–55).

Text can also be aligned to an angle; right, left, or center justified; and rotated about an axis. It is also possible to underline and overscore (or bold) text by using embedded commands when entering the text. To turn underline on or off, use %%u. To turn overscore (bold) on or off, use %%o. Entering "This %%uword%%u is underlined" will produce

This <u>word</u> is underlined

FIGURE 6–52
The FILLET command.

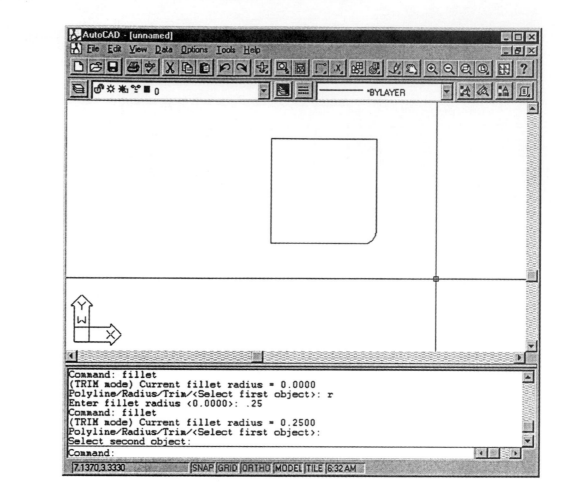

Note that there are no spaces between the embedded commands and the individual word. If there were spaces, those spaces would also be underlined. For the computer, a space is a character, and the embedded command says turn on underlining before the next character and turn it off after this last character, with no concern for what that character might be. Some additional special symbols may be used in the text as follows:

%%d   produces a degrees symbol
%%p   produces a plus or minus symbol
%%c   produces a circle diameter symbol
%%%   must be used to actually insert a % symbol in the text

When adding text to a drawing, not only is the location of the text important but also, the style or basic shape of the text. The standard or default text in AutoCAD is a simple, straight line text called TXT. Because a number of font types are included with Auto-

CAD, you are not limited to this simple text. Fig. 6-56 shows examples of some of these different font faces. Text can also be inserted justified or aligned with an angle of rotation as seen in Fig. 6-57.

Text is also affected by the scale factor for plotting. If the final text is supposed to be ⅛″ high on the drawing and the scale is ¼″ = 1′, the original text would have to be 6″ tall full-size on the screen.

Line Types

There is a default group of line types in AutoCAD, some of which are shown in Fig. 6–58. For each line type shown, there are two additional line types for each named type included in this default group. For the line type CENTER, there is also CENTER2 and CENTERX2, and so on. Each line type has similar lines ending with a 2 or an X2. Line types ending in 2 are made with spaces and dashes that are half as long as the original line type, whereas line types ending in X2

FIGURE 6–53
The TEXT command.

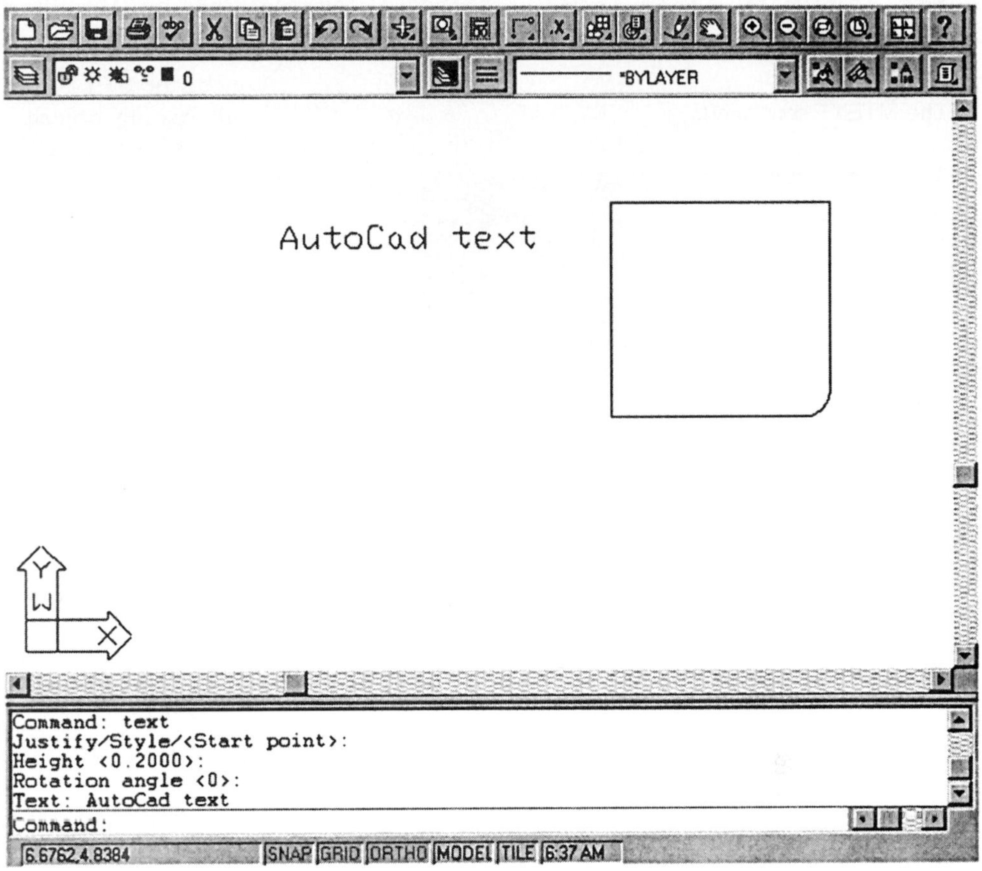

```
Command: text
Justify/Style/<Start point>:
Height <0.2000>:
Rotation angle <0>:
Text: AutoCad text
Command:
```

FIGURE 6–54
The DTEXT command.

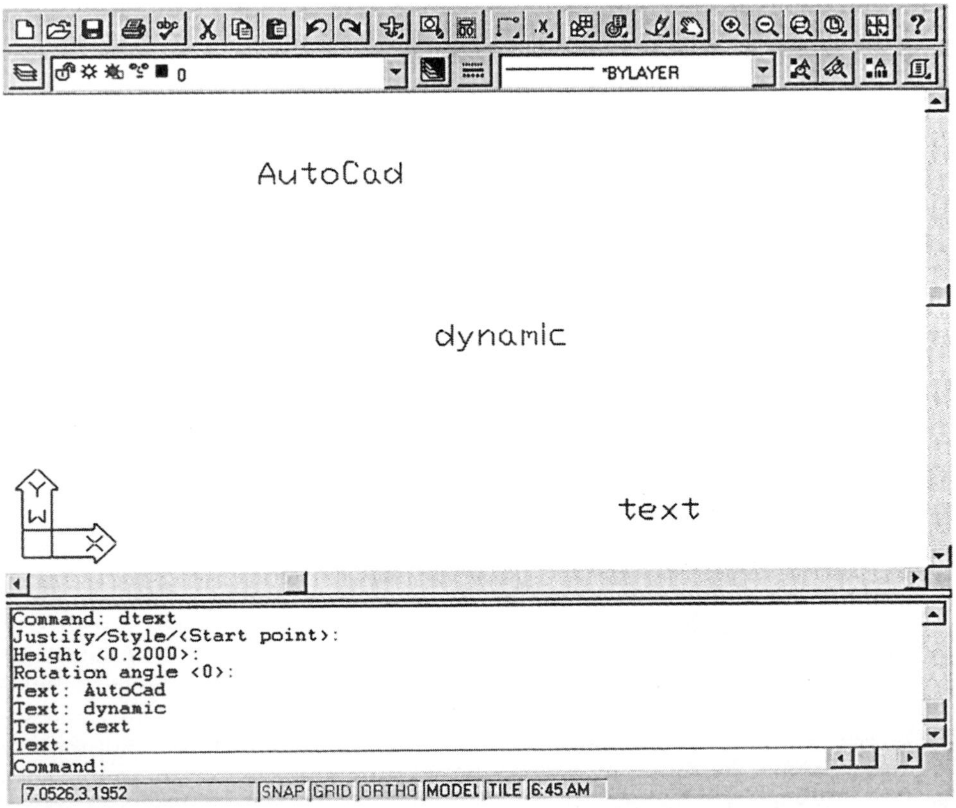

```
Command: dtext
Justify/Style/<Start point>:
Height <0.2000>:
Rotation angle <0>:
Text: AutoCad
Text: dynamic
Text: text
Text:
Command:
```

FIGURE 6–55
The MTEXT command.

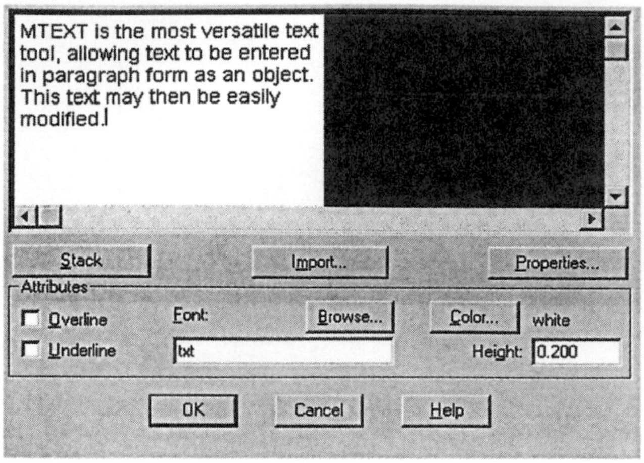

have spaces and dashes that are twice as long as the original line type. All objects will be drawn with the current line type. Use Select Line Type command to set this line type to the desired style. If all construction lines were going to be drawn on a separate layer, the line type for that layer could be set to a construction line and then the current line type could use the setting BYLAYER. The majority of electrical and electronic drawings use a continuous line type since the objects on these drawings are mostly lines and functional symbols. The completion of mechanical drawings would require the use of most of the default line types.

Some line types, like dash lines, can be affected by scaling a drawing. For example, if a hidden line was completed with .50″ dashes and .25″ spaces and plotted at a 1 = 1 scale, it would be fine. However, plotted at a 1 = 4 scale, would also scale the hidden line, resulting in dashes of .125″ and spaces of .0625″. The lengths of the dashes and spaces could be changed to compensate for a plot scale by the LTSCALE com-

mand or CELTSCALE command. LTSCALE affects all existing objects and any objects drawn after the change of scale. CELTSCALE, on the other hand, only affects the objects drawn after the change.

Changing LTSCALE from the default setting of 1 to a value equal to the plot scale will result in the plotted hidden lines (or other dashed lines) maintaining their correct lengths. LTSCALE is set to the inverse of the plot scale. A plot scale of ¼″ = 1′ gives a ratio of 1 : 48, resulting in LTSCALE set to 48. Typing LTSCALE on the command line (Fig. 6–59) allows you to change this setting. Remember, changing LTSCALE will affect all the objects in a drawing: those that already exist and all new objects. To change the scale of only new objects, or the objects drawn after the change of scale, use the command CELTSCALE.

Setting both LTSCALE and CELTSCALE to values other than 1 can be very confusing since both of these factors interact to determine the final scale of an object. If CELTSCALE has been set to .5 and LTSCALE is set to 2, the final result would be .5 × 2, or a net result of line-type scale of 1.

Dimensioning

Dimensioning is critical in mechanical drawings. Even though many electrical drawing are schematics and other drawings representing function rather than size, you may need to draw a number of production items that fall into the category of mechanical. Packaging or enclosures for electronics (Fig. 6–60) are examples of an associated drawing for the electronic drafter. Therefore, it is important to know how to add accurate dimensions to the drawings. To produce accurately dimensioned drawings, you should be able to set up the drawing correctly, as well as know how to place the dimensions on the drawing. Two commands are used most often in dimensioning: DIMLINEAR, which creates both horizontal and vertical dimensions, and DIMRADIUS, which dimensions circles and arcs. To prepare a drawing for dimensioning, it is helpful to understand the use of layers. Good CAD practice includes creating different layers and then including different types of information on different layers. Five basic steps for dimensioning include the following:

1. Create a separate layer for dimensions.
2. Create a text style to use only for dimensions.
3. Set your dimension units to the type of measurement that you are using (decimal, architectural, or other).
4. Set the overall scale factor (DIMSCALE variable) to the plot scale factor of the drawing.
5. Set the running object snap to speed up picking points on objects.

FIGURE 6–56
Sample font faces in AutoCAD.

Architect Hand Lettered
Bank Gothic
Commercial Script
Dutch
monospaced font
Swiss
θνιωερσαλ ματη (universal math)
Vineta (shadow)

FIGURE 6–57
Rotating text.

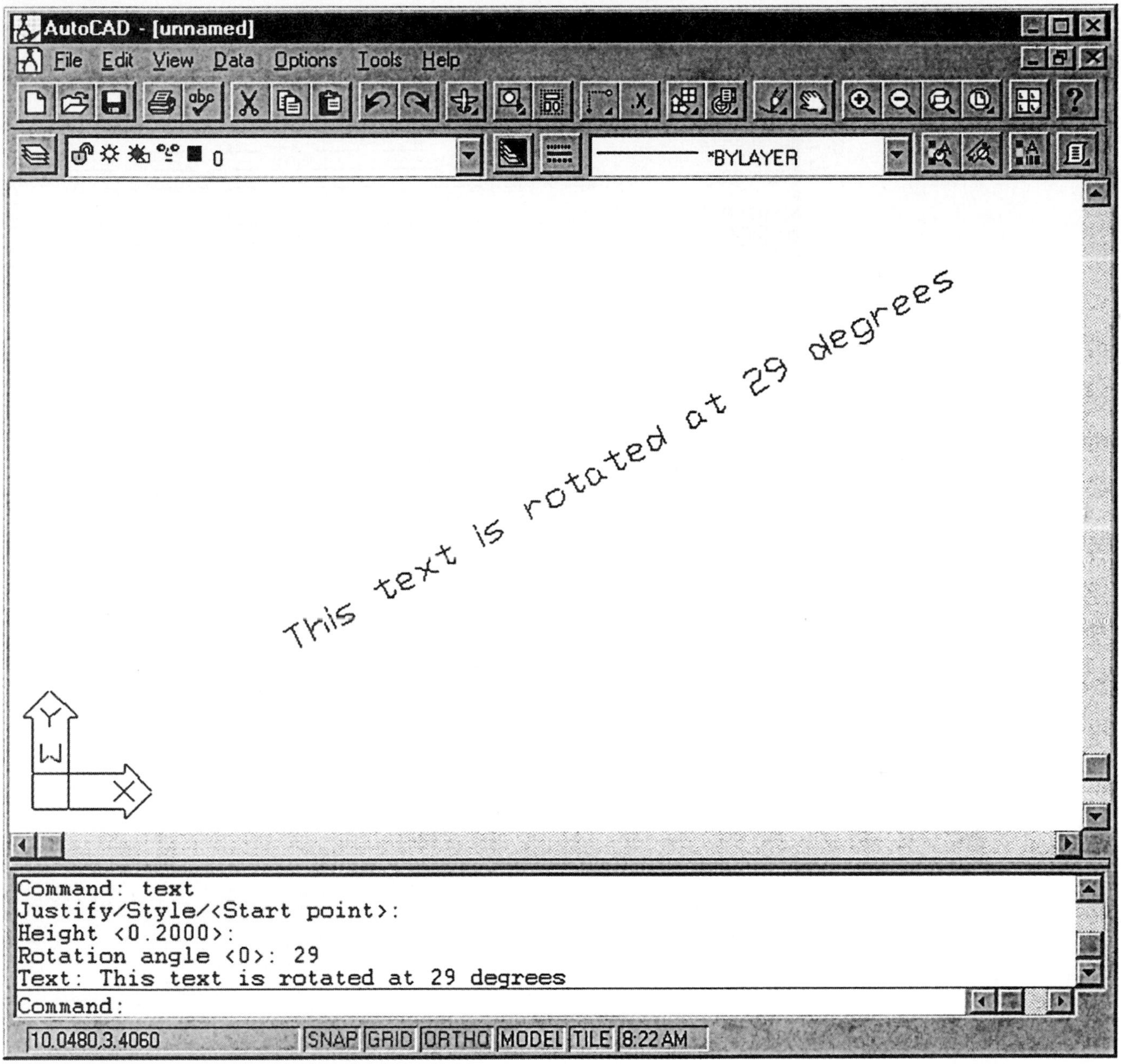

FIGURE 6–58
Variety of standard line types.

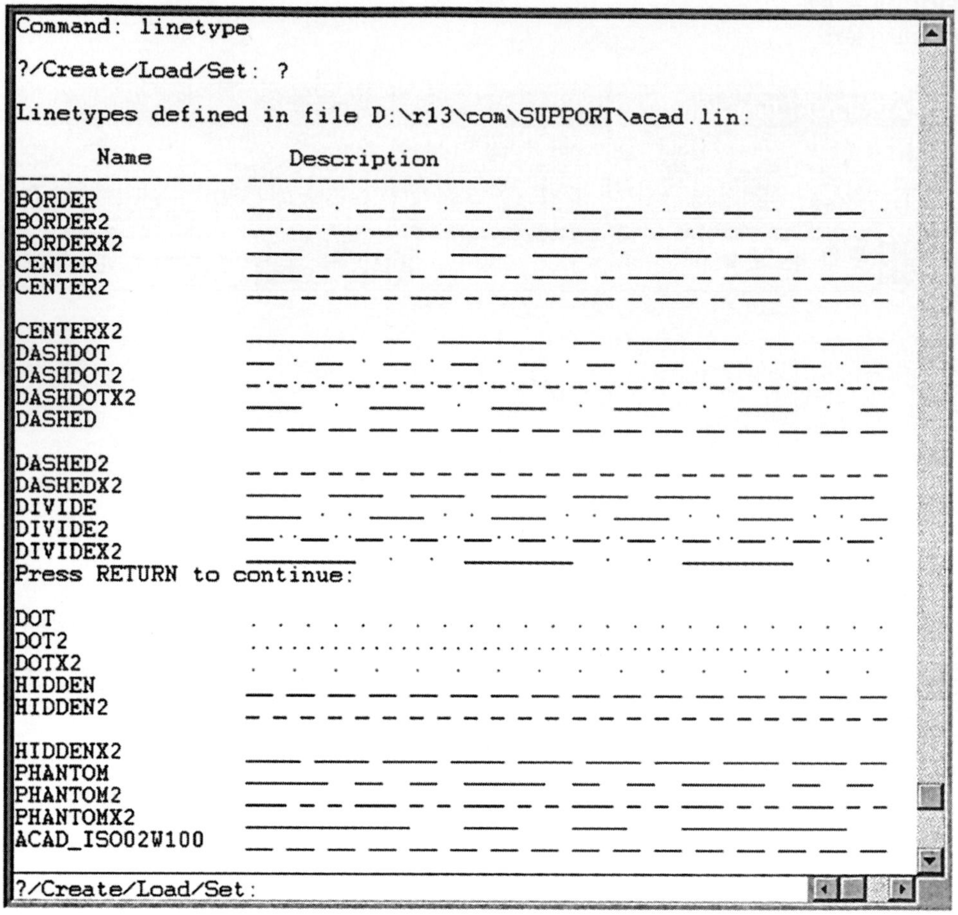

```
Command: linetype

?/Create/Load/Set: ?

Linetypes defined in file D:\r13\com\SUPPORT\acad.lin:

        Name                Description
_____    _____
BORDER
BORDER2
BORDERX2
CENTER
CENTER2

CENTERX2
DASHDOT
DASHDOT2
DASHDOTX2
DASHED

DASHED2
DASHEDX2
DIVIDE
DIVIDE2
DIVIDEX2
Press RETURN to continue:

DOT
DOT2
DOTX2
HIDDEN
HIDDEN2

HIDDENX2
PHANTOM
PHANTOM2
PHANTOMX2
ACAD_ISO02W100

?/Create/Load/Set:
```

FIGURE 6–59
Using LTSCALE.

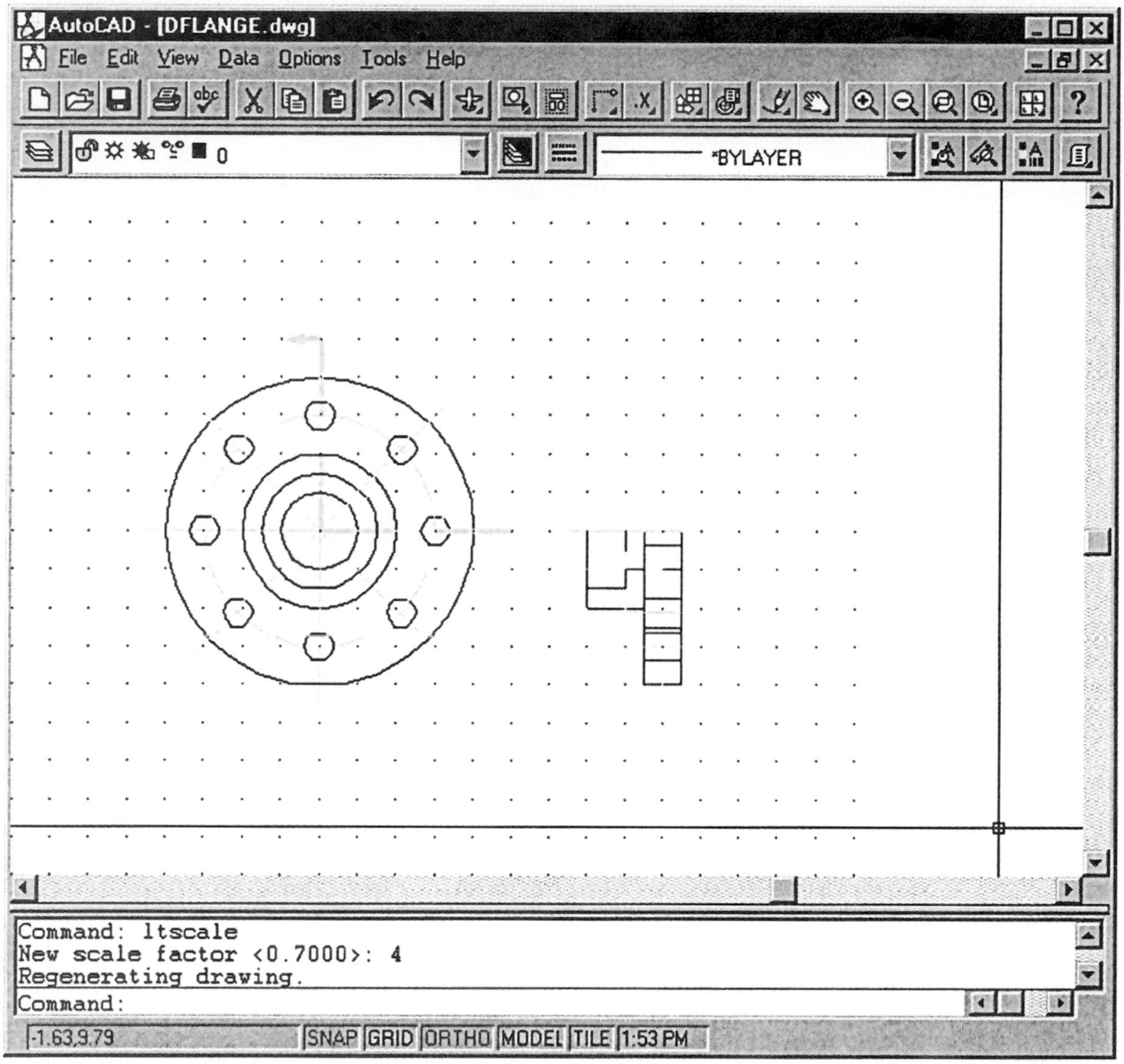

FIGURE 6-60
Electronic packaging.

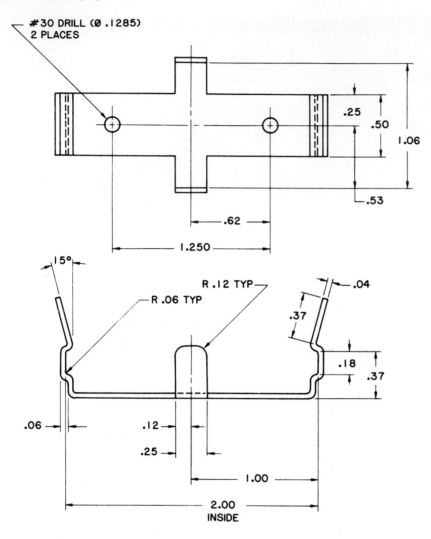
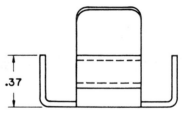

NOTES:

1. HEAT TREAT TO CONDITION R.H. 950
   (REF. JORGENSEN STEEL BOOK)
   HEAT CONDITION "A" MATERIAL
   TO 1750° F FOR 10 MIN. COOL TO 100° F AND HOLD FOR 8 HOURS
   HEAT TO 950° F AND HOLD FOR 1 HOUR. COOL IN AIR TO ROOM TEMP.

FIGURE 6-61
Preparations for dimensioning.

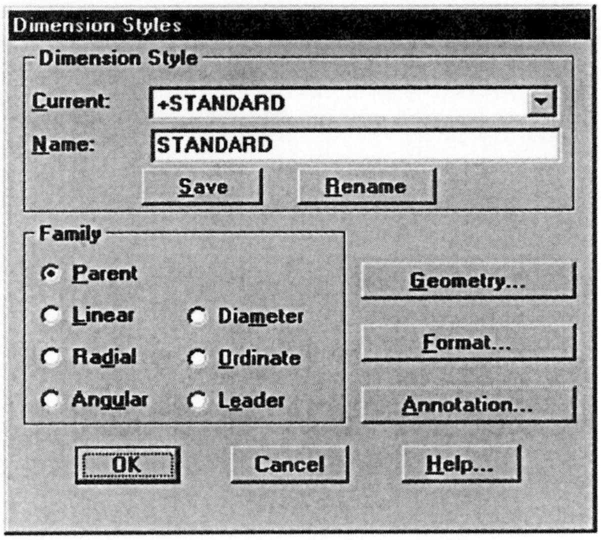

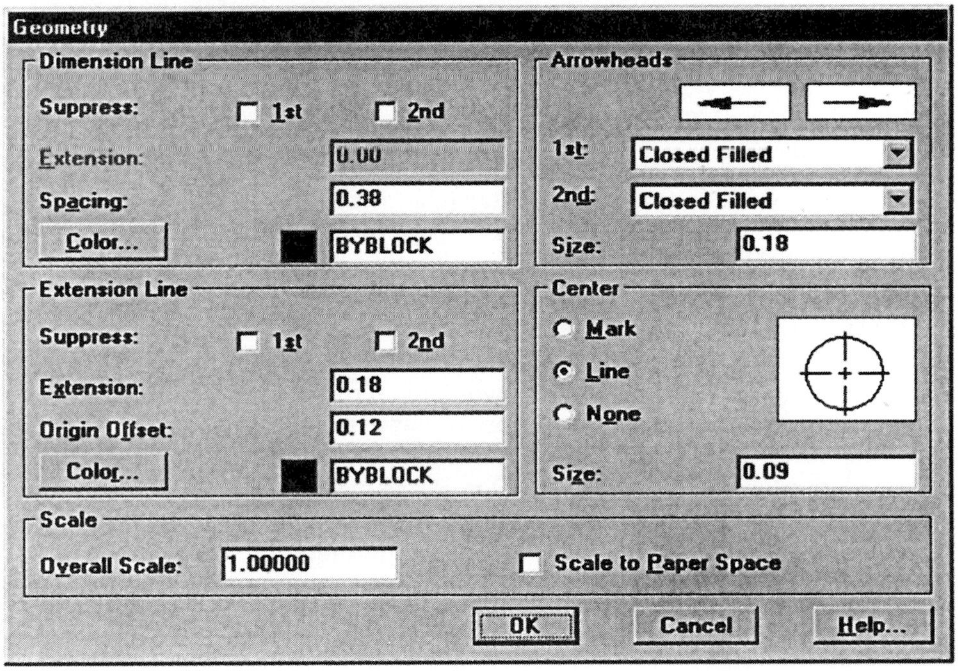

All dimensions include some standard components, extension lines, definition points (where the extension lines originate), dimension lines, dimension text, and arrowheads or other termination marks (Fig. 6-61).

Placing dimensions automatically on a drawing may be a little frustrating at times. Using manual tools, the tight areas can be manipulated a little to come out correctly, but AutoCAD may position dimensions in a different way, and it may seem like there is no way of controlling this placement. This is one reason it is very important to make sure that the drawing is set up in the first place to include dimensions.

FIGURE 6-62
Continuous dimensioning.

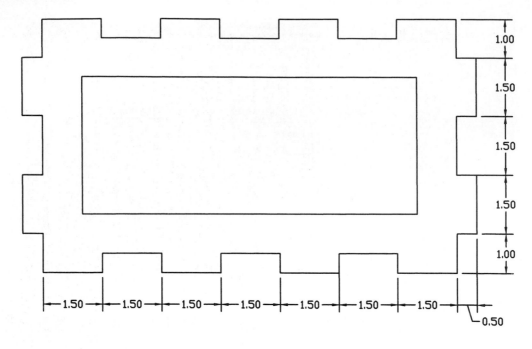

Automated Linear Dimensions

AutoCAD allows dimensioning in different forms or methods (as covered in Chapter 4): rectangular coordinate, datum (or baseline), hole charts, centerlines, and continuous.

Continuous dimensioning (Fig. 6–62) involves dimensions that are stacked or progressive. Each measurement is made to the last object measured. The biggest problem with this type of dimension is that it allows small errors to be multiplied as the part is drawn and later manufactured. This happens because the errors are cumulative for each measurement. Dimensions from a datum or baseline (Fig. 6–63) have the potential to be more accurate, since all points are dimensioned from one baseline, thus eliminating accumulated errors. Hole chart and coordinate dimensioning (Fig. 6–64) is most useful when there are many holes or other circles on the drawing.

In addition to horizontal and vertical placement of dimensions, some dimensions may be aligned at other angles. If the angle of the dimension line is parallel to the extension line origin points, the DIMALIGNED command (Fig. 6–65) is used. Otherwise, a rotated dimension using the rotate option of DIMLINEAR is used. Rotating a dimension is not the same as rotated text. Dimension lines will be vertical or horizontal unless they are rotated or placed using the DIMALIGN command. The dimension annotation or text could be rotated independently of the dimension lines. Fig. 6-66

FIGURE 6-63
Datum or baseline dimensioning.

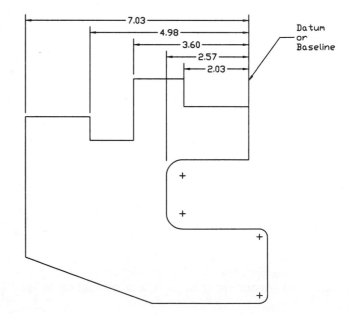

FIGURE 6–64
Hole chart and coordinate
dimensioning.

SIZE SYMBOL	A	B	C
HOLE DIAMETER	1	.5	.25

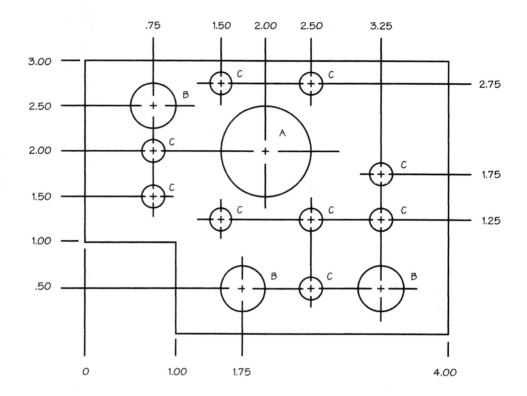

FIGURE 6–66
Using rotate with DIMLINEAR.

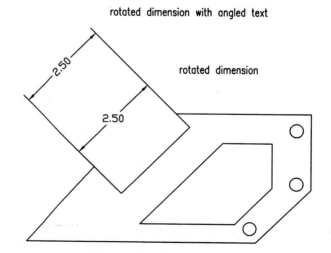

FIGURE 6–65
DIMALIGNED command.

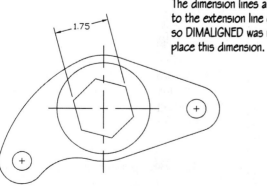

The dimension lines are parallel
to the extension line origins,
so DIMALIGNED was used to
place this dimension.

FIGURE 6–67
Dimensioning curved objects.

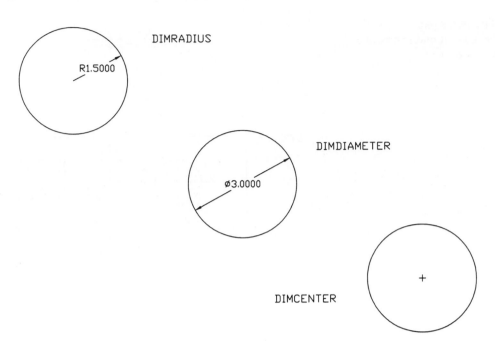

was dimensioned twice using DIMLINEAR. Both times the dimension was rotated 45 degrees, but the second time the angle of the text was also rotated 45 degrees.

To dimension curved objects, use DIMRADIUS, DIMDIAMETER, or DIMCENTER (Fig. 6–67). DIMCENTER also allows you to add center marks for arcs and circles. To dimension angles on the drawing, use DIMANGULAR (Fig. 6–68).

Using Leaders

In addition to dimensions, drawings often need some additional text or information. The STANDARD di-

FIGURE 6–68
Dimensioning angles with DIMANGULAR.

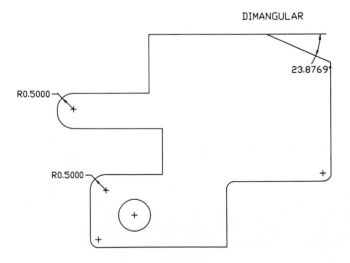

mension command places standard dimension information on your drawing. The LEADER command allows you to point at an object and add text or other information about that object (Fig. 6–69).

Formatting Dimensions

Like everything else, the way dimensions appear can be controlled somewhat. The number of decimal points displayed, the size and style of the text, the length of extension lines, the size and shape of the arrowheads, and the rotation of the text can all be controlled (Fig. 6–70).

One of the biggest challenges of drafting, both manual and CAD, is completing a drawing with dimensions that are both esthetically pleasing and technically correct. To do this using AutoCAD or other CAD software, you must have a thorough knowledge of basic drafting practices, a good understanding of the dimensioning commands in the software, and a fairly good understanding of the technical requirements of the drawing.

Plotting

To produce a hard copy of your drawing, use the PLOT command. The output device does not have to be a plotter to issue this command. It could also be a laser printer or other hard-copy device. The basic procedure to produce an output is as follows:

1. Make sure the plotter or printer is attached and ready.

FIGURE 6-69
Using LEADERS.

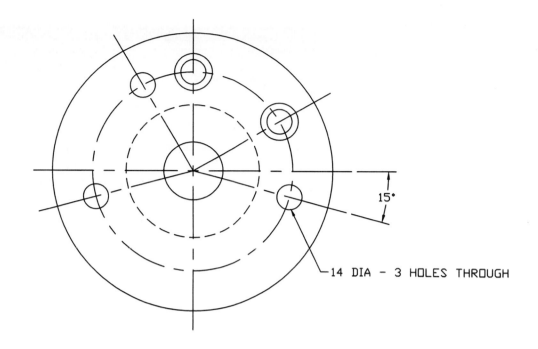

15°

14 DIA - 3 HOLES THROUGH

2. Issue the PLOT command.
3. Select the plotter or printer that you want.
4. Set the pen assignment if the device is a plotter.
5. Specify the area of the drawing.
6. Select a sheet size.
7. Set a plot scale.
8. Preview the settings and start the plot.

Many options are available when you select PLOT (Fig. 6–71). Zooming in on only the part of the drawing desired and then selecting DISPLAY as an option will plot only what is displayed. If different VIEWS were saved earlier, plot by VIEW could be the option. Layers can be turned OFF and not plotted. Different pen widths could be selected by assigning colors to these different

FIGURE 6-70
Formatting dimensions.

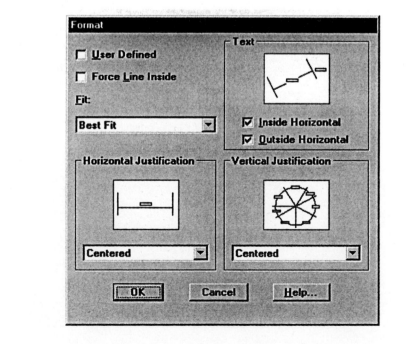

FIGURE 6–70 (cont.)

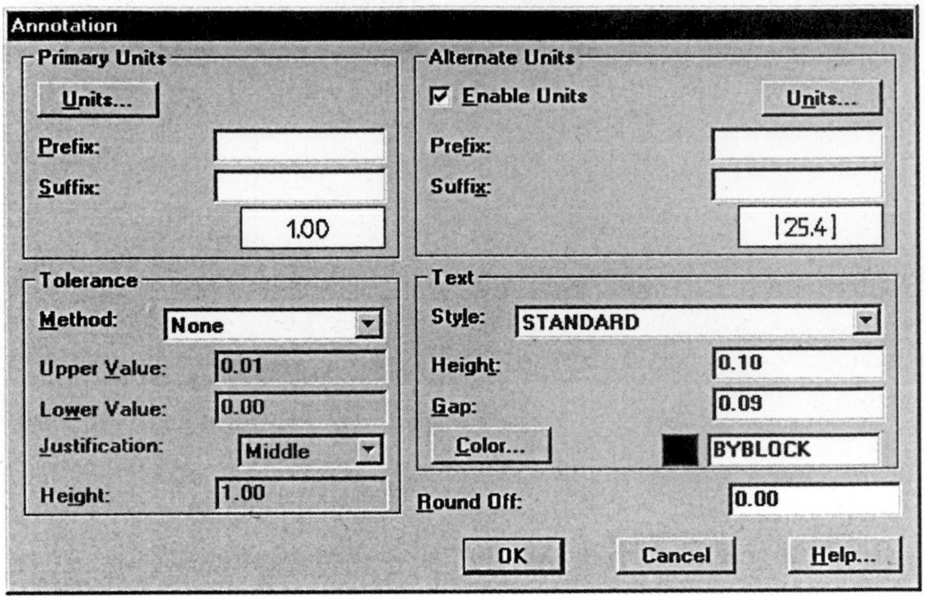

Annotation

Primary Units

Units...

Prefix:

Suffix:

1.00

Alternate Units

☑ Enable Units   Units...

Prefix:

Suffix:

|25.4|

Tolerance

Method: None

Upper Value: 0.01

Lower Value: 0.00

Justification: Middle

Height: 1.00

Text

Style: STANDARD

Height: 0.10

Gap: 0.09

Color...   BYBLOCK

Round Off: 0.00

OK   Cancel   Help...

pens, and then lines could be made different widths by making them the appropriate color. All lines of the specified color would be plotted with the same pen.

All the features found in Release 13, including the

graphics and true type fonts, may not be adequately reproduced with a pen plotter. Drawings of this type may require a raster plotter for high-quality output. Large-media laser printers and the new large-bed color inkjet plotters may give acceptable prints.

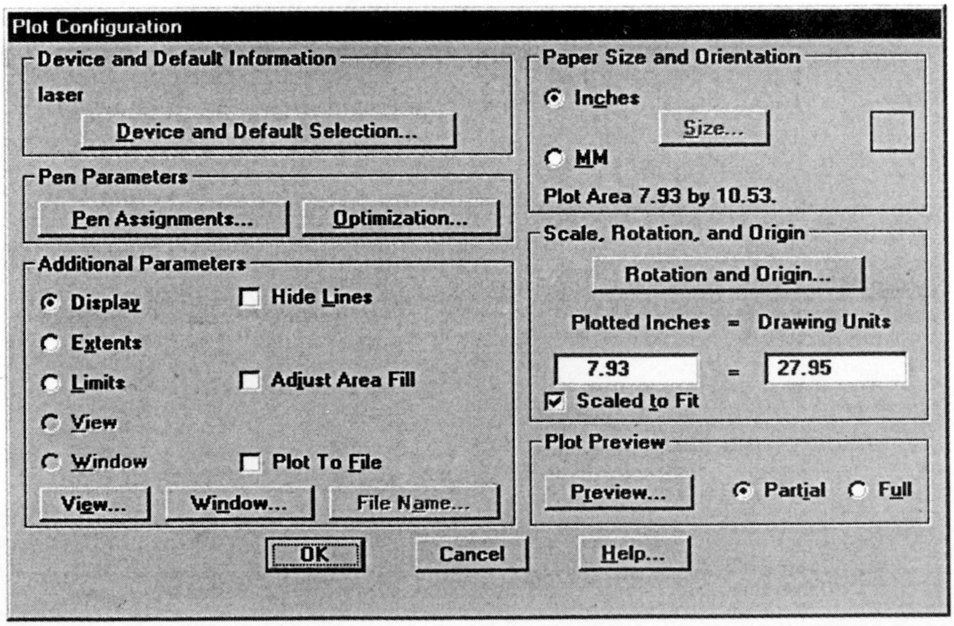

FIGURE 6–71
PLOT command.

Plot Configuration

Device and Default Information

laser

Device and Default Selection...

Pen Parameters

Pen Assignments...   Optimization...

Additional Parameters

◉ Display   ☐ Hide Lines

○ Extents

○ Limits   ☐ Adjust Area Fill

○ View

○ Window   ☐ Plot To File

View...   Window...   File Name...

Paper Size and Orientation

◉ Inches   Size...

○ MM

Plot Area 7.93 by 10.53.

Scale, Rotation, and Origin

Rotation and Origin...

Plotted Inches = Drawing Units

7.93 = 27.95

☑ Scaled to Fit

Plot Preview

Preview...   ◉ Partial ○ Full

OK   Cancel   Help...

SUMMARY

AutoCAD and other CAD software packages can be used to produce standard 2D drawings. In addition, many packages can produce 2D isometric and 3D drawings. None of these software packages can be a substitute for learning the basic requirements and practices of drafting. The operator must still control how the drawing will appear and make certain that it adheres to company standards. This chapter could only introduce some of the common commands of AutoCAD and should not be used in place of a reference manual.

REVIEW QUESTIONS

6–1 What does **interactive** mean in a CAD system?

6–2 What is the *heart* of any CAD system?

6–3 Name five typical hardware components of a CAD system.

6–4 Define **turnkey**.

6–5 What is the advantage of a **networked system**?

6–6 What is the difference between a data tablet and a digitizing table?

6–7 Name five types of input devices.

6–8 Describe the difference between a display menu and a tablet menu.

6–9 Name five types of output devices.

6–10 What is a hard-copy device? Describe its function.

6–11 Name three types of plotters and describe their differences and uses.

6–12 What is application software?

6–13 Explain how CNC and CAD work together.

6–14 How does CAD/CAM affect the design sequence of a PC board?

6–15 Define the following:

CAD	minicomputer
CPU	cross hairs
CAM	robotics
CNC	CAD/CAM
hardware	CRT
computer graphics	menu
display menu	data tablet
printer	graphics tablet
joystick	track ball
terminal	alphanumeric keyboard
digitizer	photoplotter
drum plotter	flat-bed plotter
digitizer table	symbol
display	cursor
light pen	hard copy
mainframe	software

PROBLEMS

Complete the following exercises using the CAD system software that is available.

6-1 Using the LINE command, produce the object shown in Fig. P6-1. The dimensions shown are to be used to construct the object and should not be in the final drawing.

PROBLEM 6-1

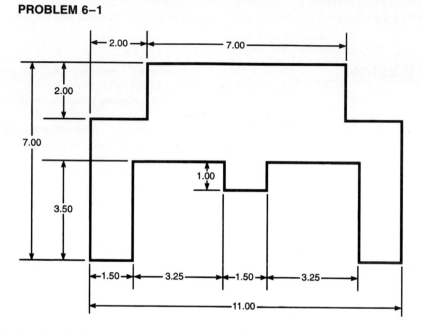

6-2 For Fig. P6-2, calculate the absolute coordinates for each point.

PROBLEM 6-2

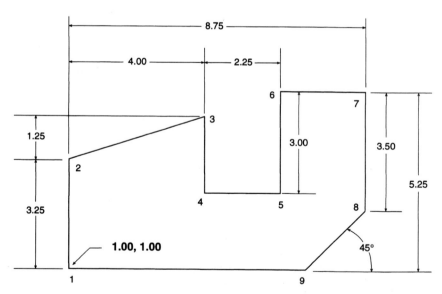

6–3 Draw the sleeve in Fig. P6–3 full scale on an A size sheet.

PROBLEM 6–3

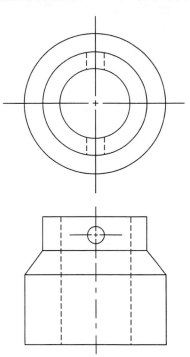

6–4 Construct Fig. P6–4 using the RECTANGLE, CIRCLE, and EXPLODE commands. Do not include dimensions in final drawing.

PROBLEM 6–4

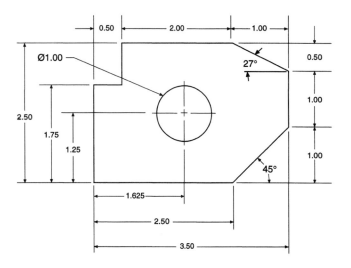

6–5 Use ARC, CIRCLE and LINE commands to complete the drawing in Fig. P6–5.

PROBLEM 6–5

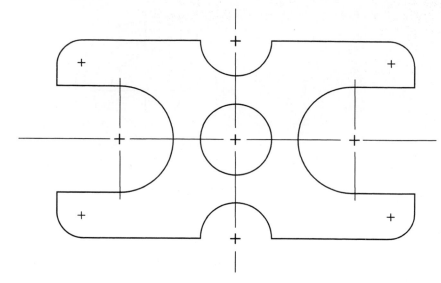

6–6 Produce top, front, and side multiview drawing for Figure P6–6. Assume that the hole goes through the object.

PROBLEM 6–6

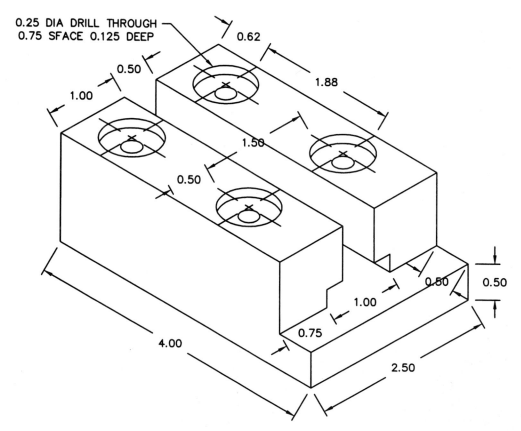

0.25 DIA DRILL THROUGH
0.75 SFACE 0.125 DEEP

6–7 Produce a multiview drawing of Figure P6–7. Assume that the hole and notch go through the object.

PROBLEM 6–7

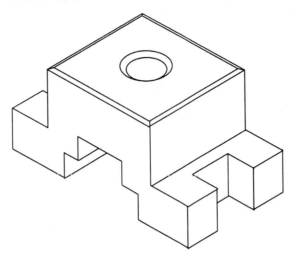

6–8 Dimension and counterbore in Fig. P6–8. The diameter of the hole is .50″ and the counterbore is .50″ deep. Use the DIMLEADER command for the diameter.

PROBLEM 6–8

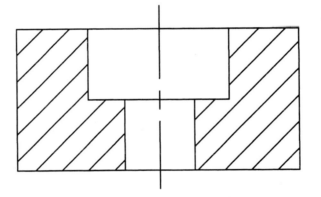

6–9 (Advanced Problem) Draw the base plate in Fig. P6–9 full scale on an A size sheet. Dimension using the same system given. Plot using a plotting scale of 1 = 1 to produce a full-sized plot.

PROBLEM 6–9

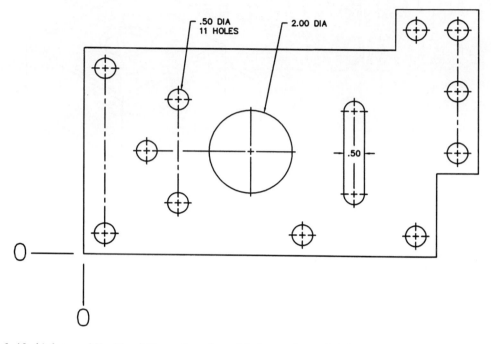

6–10 (Advanced Problem) Draw the adjustable brace shown in Fig. P6–10. Include dimensions. Plot using a plotting scale of 2 = 1 on an A size sheet. Make sure to use the correct line type.

PROBLEM 6–10

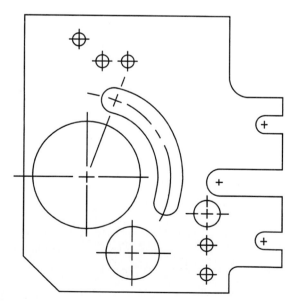

7

FUNDAMENTALS OF ELECTRONICS

INTRODUCTION

This chapter is an overview of electronic and electrical circuits, usually called simply the study of electronics. A basic understanding of the fundamentals of electronic and electrical circuits is necessary for understanding material in later chapters. In the study of electronics, perhaps more than many other fields, your success rests largely on your ability to learn the *language*. The language of electronics is quite extensive. Every quantity, every device, and all characteristics have names (with symbols for the names) and units (with symbols for the units). Some typical examples are shown in Fig. 7–1. Familiarity with all the symbols and letter abbreviations and quantities can only come with time and practice. However, without mastery of this language, you cannot successfully communicate with engineers and other designers and drafters.

Electronic and electrical circuits are so closely related that it is sometimes difficult to define the differences between the two. A simplistic definition is that electronic circuits contain some type of semiconductor device and electrical circuits do not. There are very few circuits that are described as either electrical or electronic since today there are very few circuits that do not contain some type of semiconductor device. There are still a few circuits, however, that are thought to fall into either the electrical or the electronic category. For example, circuits such as the control circuits (other than solid state), as shown in Fig. 7–2, and power distribution circuits, in Fig. 7–3, are generally considered to be electrical circuits. Logic circuits, amplifiers, and computers, or the recording device shown in Fig. 7–4, must be described as electronic circuits.

HISTORY

The study of electronics is based on physical science and atomic theory. This study is often defined as the science of a specific part of the atom, the electron. Knowing the history of electronics should make it easier to understand where all the names of devices, components, and units come from.

Electricity and electrical circuits are thought to trace back to ancient civilizations, where the Greeks and Romans gave credit to their gods for fantastic displays of electricity in the form of lightning. Around 640 B.C. was the first recorded history of electrical science and technology: a magic rock was discovered. This magic rock became known as the magnet or the lodestone. After only mild interest, the science of electricity fell into the background for many centuries.

Between A.D. 1200 and 1600, the subject of electric-

FIGURE 7–1

Device and characteristic symbols and notation. (Reprinted by permission of HEATH COMPANY)

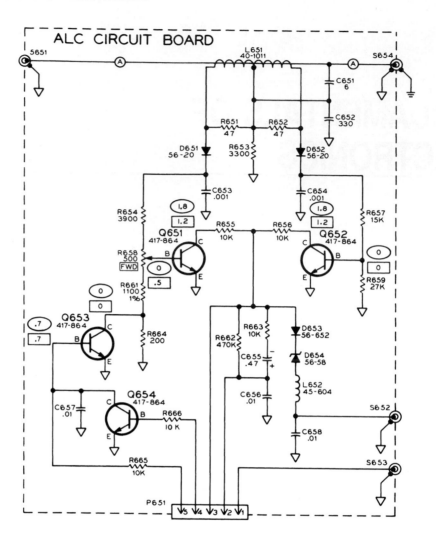

ity sprang to life again. Much of this early experimentation was done by Queen Elizabeth's personal physician, William Gilbert. He was the first to propose that the earth itself was a huge magnet. As early as 1729, experimenters were attempting to find some applications for these new and wondrous phenomena. In England, Steven Gray caused electricity to flow along a string. It was this experiment that led to the belief that electricity was a fluid and therefore came to be called **current.**

In the mid-1700s, scientists began experiments that led to the discovery of many important electronic components and theories. Names such as Luigi Galvani, Count Allesandro Volta, Humphrey Davy, and Michael Faraday are associated with the development of the battery cell. In 1820, Hans Christian Oersted of Denmark discovered the correlation between electricity and magnetism. Oersted's discovery led to the beginning of practical applications of electricity and magnetism, with the development of motors and the telegraph in the 1830s.

In 1831, John Henry in America and Michael Faraday in England observed that electricity flowing in one wire could cause a similar flow in a second wire located close to the first wire. This led to later development of the transformer. In 1865, James Maxwell developed an equation for electromagnetic waves, and in 1887, Heinrich Hertz proved the validity of Maxwell's equations. He produced evidence of the existence of electromagnetic radiation even though it could not be seen. The primary breakthrough in the study of electronics came in 1897 when Joseph Thompson discovered the electron.

All matter is made up of atoms, the smallest piece of a material that is still recognizable as that material. Atoms contain electrons, neutrons, and protons. They have a nucleus that contains the protons and neutrons, and orbiting around the nucleus are the electrons, as shown in Fig. 7–5. The only thing that makes one atom different from another is the number of electrons, protons, and neutrons it contains, as shown in Fig. 7–6. These protons and electrons have *electrical* charges. The

FIGURE 7–2
Motor control circuit. (Courtesy Ruth Ann Paynter)

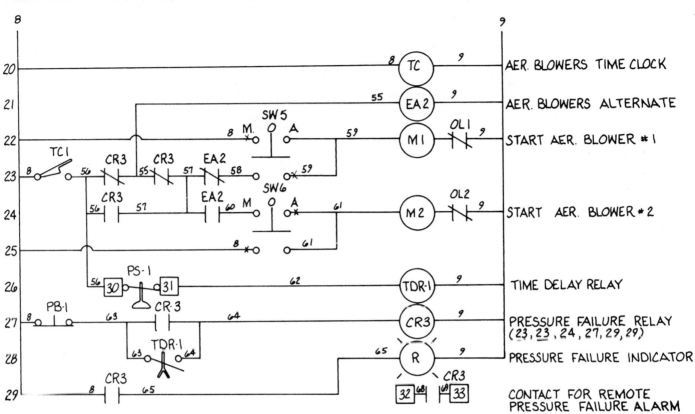

FIGURE 7–3
Typical power distribution
system. (Courtesy Interactive
Computer Systems, Inc.)

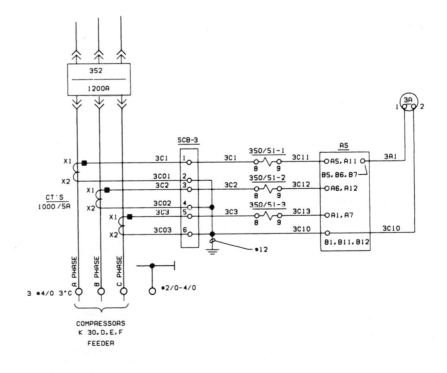

FIGURE 7–4
Typical electronic circuit. (Courtesy International Business Machines Corporation)

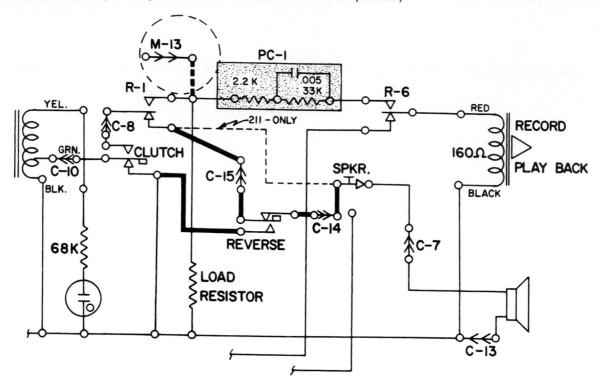

proton has a charge that has been called *positive,* and the electron, thought to have an opposite charge, has been called *negative.* The naming of these charges was initially arbitrary, and early experimentation simply showed that these two parts of the atom repelled each other. To make certain that these two parts appeared opposite, they were given such opposite identification.

FIGURE 7–6
Atoms containing different numbers of protons and electrons. (A) Hydrogen has one proton and one shell with a single orbiting electron. (B) Carbon has six protons and two shells containing six orbiting electrons—two in the first and four in the second. (C) Copper has 29 protons and four shells containing 29 orbiting electrons—two in the first, eight in the second, eighteen in the third, and one in the fourth.

FIGURE 7–5
Bohr model for a hydrogen atom.

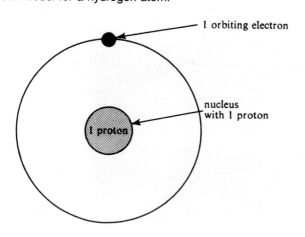

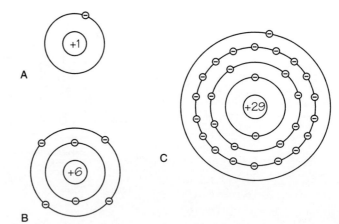

FIGURE 7–7
Hydrogen and helium atoms. Hydrogen contains a single electron and a single proton. Helium contains two electrons and two protons and in addition contains two neutrons.

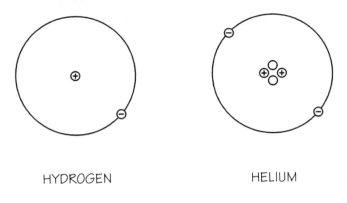

HYDROGEN   HELIUM

In the beginning, all matter contains a balanced number of electrons and protons so that the *charge* associated with the atoms must be zero. If one charge is considered positive and one negative and there are equal numbers of positives and negatives, then there will be a net result or charge of zero. Figure 7–7 demonstrates atomic structure by comparing the simple atoms of hydrogen and helium. Hydrogen contains only one electron and one proton. Helium, on the other hand, has two protons and two electrons; in addition, it has two neutrons. Neutrons are neutral and have neither a positive nor negative charge. The primary purpose of neutrons appears to be simply to add mass to the atom.

The orbits that contain the electrons are called *shells*. Shells are considered full when an appropriate number of electrons is orbiting in each shell. This is very important to the study of electronics, because only atoms that contain partially full shells can be easily acted upon. If the atoms are left alone, the electrons in each atom will be content to merely orbit indefinitely and will produce no energy because of their net charge of zero. When a force causes the electrons to vacate their existing orbits, energy is made to do useful work.

ELECTRICAL QUANTITIES

The force that causes electrons to move is called the **electromotive force,** or **EMF.** This EMF, like any other physical force, provides the potential for energy to occur. Therefore, the greater the force is, the more energy that is produced. When electrons are caused to vacate their orbits, energy is transferred from the force, causing the electrons to move. The higher the applied force is, the more electrons that are moved.

All electrical quantities have names and units, with symbols for each. The EMF is commonly called potential or **voltage,** and the symbol for voltage is V. The unit for measuring voltage is the **volt,** and the symbol for the unit volt is V. If the force applied can cause the electrons to move in an orderly fashion in the same direction, this movement is called **current** (symbol I) and has the unit **ampere** or **amp** (symbol A). The unit amp is a measure of the rate of flow of electrons past a given point in a given period of time. One amp is equivalent to one coulomb (6.25×10^{18}) of electrons in one second.

Since all matter is made up of atoms and all atoms have electrons, it seems that current can be caused to flow through any matter so long as the force or applied voltage is large enough. Theoretically, this is true. In reality the atoms that have unfilled outer shells or orbits are the atoms that will allow current flow or movement of electrons through the material most easily.

Materials that allow easy flow are called **conductors** and are normally used for permitting the current to flow in a desired path. Metals like aluminum, copper, gold, and silver allow current to flow fairly easily (with a small applied voltage) and are therefore good conductors. The conductors most often used in electrical circuits are copper and aluminum because they are much less expensive than either gold or silver. In integrated circuits, or microelectronics, many circuit paths are microscopic, as shown in Fig. 7–8. Gold is used in these circuits. It is a better conductor, and because such a tiny amount is used the cost is not prohibitive.

FIGURE 7–8
Integrated circuit, showing conductors of very fine strands of gold wire. (Courtesy Motorola, Inc., Semiconductor Products Sector)

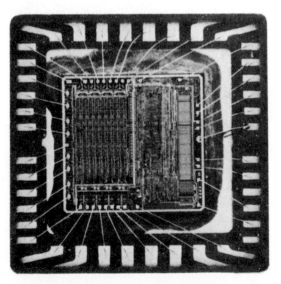

Certain materials that contain full shells restrict electrons from flowing. This type of material is called an **insulator.** Some common materials used as insulators are glass, plastic, rubber, porcelain, and dry paper.

When a voltage is applied to a circuit path made up of conductors, a certain amount of current can be caused to flow. The only limit to this current flow is any opposition to current in the circuit. The opposition to current flow is called **resistance** and has the symbol R. The unit of measure of resistance is the **ohm,** with the Greek symbol omega (Ω). A good conductor must have very little resistance, and a good insulator must have very high resistance. When electrons are forced through a resistance, they use a lot of energy. This energy is converted into heat. If the heat cannot be properly dissipated, then the wire could become hot to the touch, indicating that the wire diameter was not large enough to handle that many electrons and the heat energy given off.

The ultimate goal in electronics is to cause work. The measure used is **power,** which is the rate of doing work. The symbol for this quantity is P, and the units are **watts** (symbol W).

Table 7–1 summarizes electrical quantities and their symbols.

Measuring Devices

Since electrical quantities themselves cannot be seen, there must be devices capable of measuring these quantities. One of the most common instruments is the volt-ohm-ammeter (VOM). The **VOM** is called a multitest meter or simply multimeter because it has the capability of measuring different electrical quantities.

TABLE 7–1
Summary of Electrical Quantities

Quantity	Symbol	Unit	Symbol
Voltage	V	volts	V
Current	I	amp	A
Resistance	R	ohm	Ω
Power	P	watts	W

There are many models, manufacturers, and specifications for instruments that measure electricity. Primarily, all instruments fall into two categories: analog or digital. The analog meter uses some form of needle movement to indicate the presence of an electrical quantity, as shown in Fig. 7–9. The digital meter displays the same information in the form of digits, as shown in Fig. 7–10.

Most analog meters use a d'Arsonval meter movement, like the one in Fig. 7–11, which works on the principle of a DC motor. When current goes through a coil of wire, a magnetic field is developed. This field is translated into mechanical movement, causing a needle to move. The larger the current is, the more the needle moves. This type of meter movement always indicates the amount of current in the meter. It does not matter if the instrument is being used to measure current, voltage, or resistance, it still indicates the amount of current going through the meter. Figure 7–12 illustrates the parts of an analog meter. When the face cover is placed on the meter, each scale is calibrated so that the desired quantity may be read directly. When the meter is used to measure voltage, the scale is marked in volts.

FIGURE 7–9
Analog meters. (A) Model 60-NA (Courtesy Triplett Corp.); (B) Model 260-7M (Courtesy Simpson Electric Co.); (C) "The Grabber™" Model 30. (Courtesy Triplett Corp.)

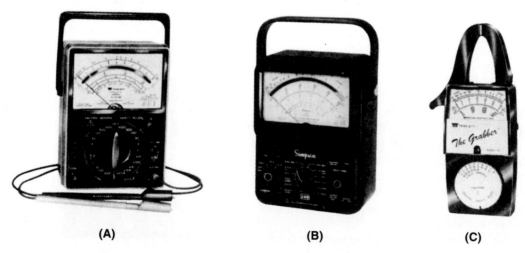

(A)   (B)   (C)

FIGURE 7–10
Digital meters: (A) Model 4200 (Courtesy Triplett Corp.); (B) Model 467 (Courtesy
Simpson Electric Co.); (C) Model 360-2. (Courtesy Simpson Electric Co.)

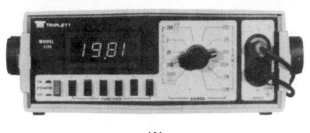

(A)

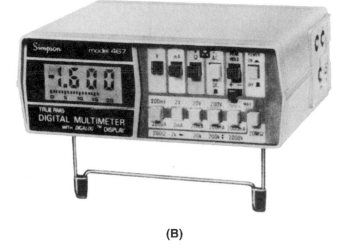

(B)

(C)

FIGURE 7–11
D'Arsonval meter movement.

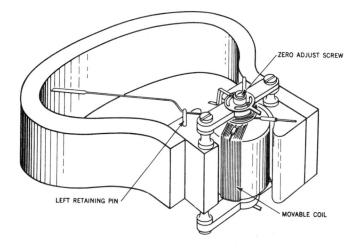

Digital meters use a number of different electronic and logic circuits to produce their indications. Unlike analog meters, digital meters do not always measure the amount of current.

When measuring devices are connected to existing circuits, it is important to connect them correctly. A meter incorrectly connected not only may give the wrong reading, but it may also be damaged or destroyed. Since voltage is the difference in potential between two points, to measure voltage, the meter is placed either across the two points as in Fig. 7–13 or in parallel with the device being measured. Current, on the other hand, is said to flow through a circuit, so to measure current the meter must be connected so that the current flows through both the circuit and the meter. The only way this can be done directly is to make a break in the

FIGURE 7–12
Pictorial representation of meter movement and scale (Courtesy Simpson Electric Co.) and photo showing ranges and scales. (Courtesy Triplett Corp.)

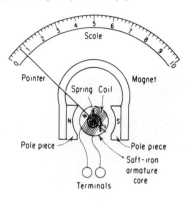

FIGURE 7–13
Circuit when a voltmeter takes measurements.

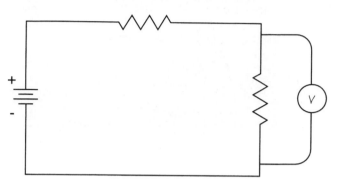

age supply that causes current to flow through the resistor under test. The current then is translated into ohms on the meter scale, as shown in Fig. 7–15.

Electrical Circuits

An electrical circuit must represent a complete path for the flow of electrons. It must contain a voltage, conductors, and some resistance. A simple electrical circuit is represented pictorially by Fig. 7–16. This circuit contains a voltage source or battery, conductors (the case and the terminals of the battery), an on-off switch, and a resistance. The resistance of the circuit is the light bulb, which represents an opposition to current flow. As the energy is transformed into heat energy, the tiny wire in the light bulb gets hot and radiates light.

One source of voltage is a battery. Cells are constructed of two metal plates surrounded by a chemical solution. The metal plates are normally two different types of metals that react when surrounded by the chemical, causing the transformation of chemical energy directly into electrical energy. Batteries are arrangements

FIGURE 7–14
Circuit when an ammeter takes current readings directly.

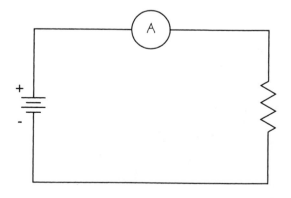

circuit and insert the ammeter in series with the circuit components, as in Fig. 7–14.

Shown in Fig. 7–9 (C) is an ammeter, which works on a different principle and therefore does not have to be inserted into the circuit. This meter uses the principles used in transformers. Electric current passing through the wire causes a current to be *induced* in the meter. This current is displayed and represents the current through the wire. Ohmmeters must be used to measure only components that are not in circuits or to measure portions of circuits where there is no applied voltage or current. The ohmmeter has an internal volt-

FIGURE 7–15
An ohmmeter measures the resistance of a component.

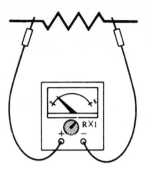

FIGURE 7–17
Symbol representation of the simple flashlight circuit.

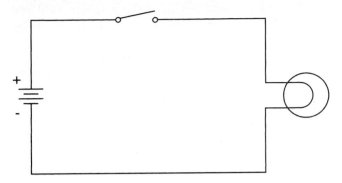

of cells in a particular configuration to render the desired result. In the manufacturing process of the battery, the ends (or terminals) of the two types of metal are made to be connected into a circuit.

Figure 7–17 shows a diagram representing this electrical circuit. This diagram is called a *schematic* and represents the total operation of the circuit. (Schematic diagrams are covered extensively in Chapter 11). Not all circuits are this simple, but all electrical circuits must contain the basic quantities of a voltage source, conductor paths, and resistance. And these must be connected in such a way as to allow the electrons to move in a complete closed path so that they end at the place that they start.

Electricity and Magnetism

As was indicated in the history of electricity, magnetism is a physical science property essential to the study of electronics. A complete discussion of magnetism is not presented in this text; however, it is important to under-

stand that all conductors allowing current to flow through them have a magnetic field surrounding them. Inversely, when a conductor is moved through a magnetic field, a current is caused to flow through the conductor. These two major principles are the basis of generator operation.

With the proper connections to a conductor rotating in a magnetic field, a voltage is generated. A simple single-conductor generator is shown in Fig. 7–18. As the conductor is rotated through the magnetic field, it cuts the magnetic lines of force, generating a voltage

FIGURE 7–18
A single-loop generator connected to a load resistor.

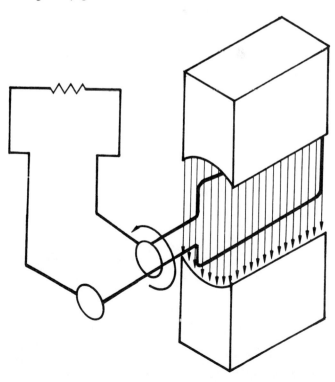

FIGURE 7–16
Pictorial representation of a very simple electrical circuit.

and causing current to flow through the slip rings and through the load resistor. This voltage is directly proportional to the size or strength of the magnetic field and the speed of the conductor moving through the magnetic field.

The amount of voltage generated can be represented on a diagram, as shown in Fig. 7–19. This diagram demonstrates that the size or strength of the field is different in different areas of the magnet. When describing this voltage, it can be stated as peak, peak-to-peak, or RMS (root mean square) voltage. The peak value is the maximum voltage obtained; it is reached twice in each cycle. The peak-to-peak voltage is obviously the value between the maximum positive and maximum negative peaks. The RMS voltage value is 70.7% of peak voltage. The time duration of each cycle is used to calculate the frequency. Also, the direction that current flows depends on the direction of the movement through the field. If the voltage obtained from the generator looks like this diagram, it is called alternating voltage, or **alternating current** (AC). The voltage and current alternate in both magnitude and direction. A battery produces a steady or constant current, which is called **direct current** (DC). Many circuits, both electrical and electronic, contain both AC and DC.

Ohm's Law

It is not enough to have a special *language* for electronics; there are also a number of laws. These laws are nothing more than the relationships between various electrical quantities, generally expressed as equations. **Ohm's law** shows the relationship between voltage, current, and resistance. Current is directly proportional to the amount of applied voltage and inversely proportional to the amount of resistance:

$$I = \frac{V}{R}$$

From this equation, it is apparent that one volt of DC voltage would cause one amp of DC current to flow through one ohm of resistance.

Closely related to Ohm's law is the power formula. Power is directly proportional to the current strength and to the voltage level at which the circuit operates:

$$P = IV$$

If other quantities are substituted following Ohm's law, the following equations can be used:

$$P = I^2 R$$
$$P = \frac{V^2}{R}$$

Ohm's law is used in *all* analysis of electrical and electronic circuits. If it is combined with two of Kirchhoff's laws, all circuits can be mathematically analyzed. Kirchhoff's laws state that all voltages in a closed circuit path add algebraically and all currents that join at a point add algebraically.

Conventional Current Versus Electron Flow

It is important to understand the difference between conventional current and electron flow. These are notations describing the direction of current in a circuit. There has been much discussion as to which method is correct, and yet both notations are describing the exact same physical quantity. So far our presentation has been limited to electron flow or the movement of electrons, which assumes a direction of *current* flow from the negative terminal of the voltage source to the positive terminal, as illustrated in Fig. 7–20. This notation is preferred by some who feel that it most closely represents the physical reality of electron motion.

On the other hand, the conventional flow notation uses a positive to negative direction of current flow

FIGURE 7–19
Waveform from a single-loop generator in two complete revolutions.

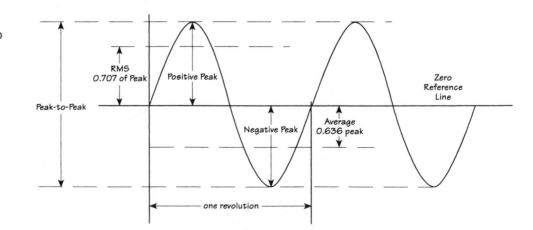

FIGURE 7–20
Electron flow versus conventional current.

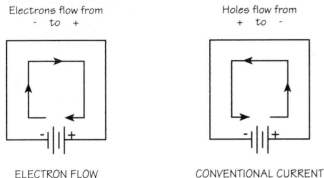

ELECTRON FLOW   CONVENTIONAL CURRENT

FIGURE 7–21
Crystal structure of germanium atoms with covalent bonding of valence electrons. Shared electrons are indicated.

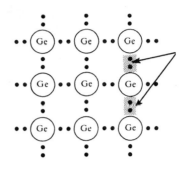

because of the correlation of positive to push. It seems easy to think of the positive terminal pushing current through the circuit. In addition, many standard device symbols are oriented to conventional current direction. For these reasons, many prefer the conventional current notation, or current traveling from the positive terminal to the negative of a source. This has sometimes been called *hole flow* since the absence of electrons would indicate a vacancy or a hole. If the electrons moved in a clockwise direction, the *holes* that they vacated would appear to move in a counterclockwise direction.

Because of this difference in notation, there will be circuits demonstrating both notations in this text. Many of the illustrations are provided by industry, and both notations are used in industrial practice today.

Semiconductor Material

Somewhere between insulators and conductors are materials called *semiconductors.* They were so named because they do not allow current flow as easily as conductors but do not block current flow as well as insulators.

The two most common types of semiconductor material are silicon and germanium. These materials have four electrons in the outer shell. When silicon or germanium atoms combine to form a solid material, they arrange themselves in a crystal structure, and these atoms are held together by a covalent bond, as shown in Fig. 7–21, that is, a bond formed by the sharing of electrons between adjoining atoms. In the crystal structure, each atom seems to have eight electrons in the outer shell, rather than the four each atom actually contains. This structure is not very useful as a conductor because atoms that have full outer shells do not allow current to flow very easily, and this crystal structure forms full outer shells for each atom.

It is not quite as difficult to cause electrons to move in the crystal structure as in a true insulator. As a matter

of fact, the semiconductor material is simply a high resistance. If a material with either five or three electrons in the outer shells is added to the silicon or germanium, an imperfect bond or crystal is formed. If a high enough voltage is then applied, a current is caused to flow, making the material act like a conductor.

The adding of other materials to the silicon or germanium is called *doping.* If material is added that has five electrons, then the semiconductor material becomes n-type material since it appears to have extra electrons. If the added material has only three electrons, the semiconductor material is called p-type. If a section of p-type and a section of n-type material are formed together, a solid-state device called a *diode* is produced.

Many semiconductor devices contain at least one p–n junction. The p–n junction can be considered a building block of solid-state devices, so an understanding of the basic operation of the single junction helps you understand many different semiconductor devices. The primary characteristic of the p–n junction is its ability to conduct current easily in only one direction.

Like electrical quantities, all semiconductor devices have names, symbols, and abbreviations. These are described fully in Chapters 8 and 9.

GRAPHICAL PRESENTATIONS

There are many opportunities in electronics where data showing the relationship between two variables must be presented. In almost every one of these cases, the data could be presented as a list of numerical data. However, in almost all cases this would be cumbersome, lengthy, and not visually clear. Presenting information so that a quick glimpse presents a quick overview is always desirable. There are many different situations where this would be useful and many different types of

FIGURE 7–22
Typical frequency response curve.

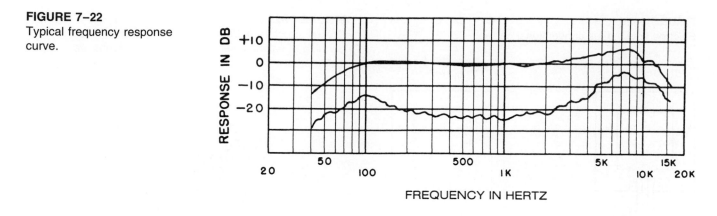

graphs that can be produced. One common graph is the frequency response graph (Fig. 7–22). On any blank audio tape, a frequency response curve indicating the output value expected over a range of frequencies is shown. Again, these data could be presented in tabular form, but a graph is much less cumbersome and occupies far less space.

Throughout this text, much information concerning the fundamentals of electronics is displayed by graphs, charts, and diagrams. Most technical graphs are produced on preprinted grid paper. Many types of graph paper are available. The different types are described by the way the grid lines are arranged on the paper. If the lines in both directions are all evenly spaced, it is normally referred to as a square graph paper. These squares can be any size, but commonly they are found with either five squares or ten squares to the inch.

Paper with equally spaced lines in each direction is

also called linear. To plot data on this type of paper, a linear numbering system must be used. Figure 7–23 shows two examples of electrical data presented on linear graph paper. In Figure 7–23A, note the Y axis is marked in increments of 10 mA and the X axis in increments of 1 V. The physical space between the major values is exactly the same. The same thing is demonstrated in Fig. 7–23B, which shows current versus temperature. Remember, however, that even though these examples are straight-line curves, this does not imply that all lines or curves plotted on linear graph paper will represent data points that are straight lines.

All lines or curves on graphs represent an infinite number of points. It is possible to identify the ratio demonstrated between the variables represented at any point or location along this curve. This ratio may also be referred to as the *coordinate* of the point. To locate these points on a graph, values must be located on each

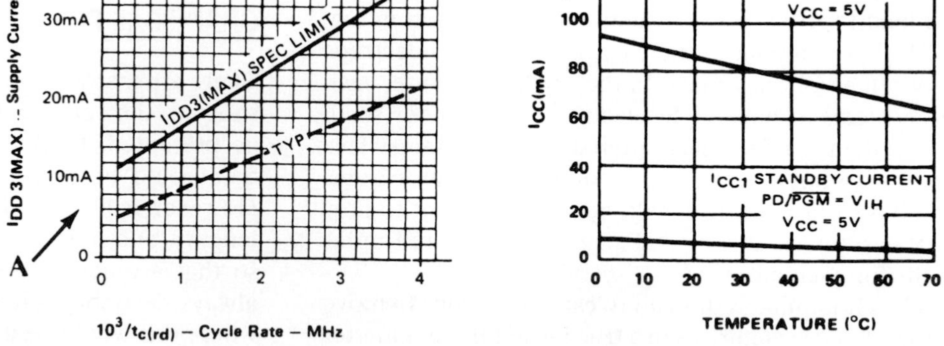

FIGURE 7–23
(A) Rulings and scales. Heavier rulings are used for scale divisions. (B) Scale designators in multiples of two. (Courtesy Texas Instruments, Inc.)

FIGURE 7–24
Linear graph paper with heavy scale lines.

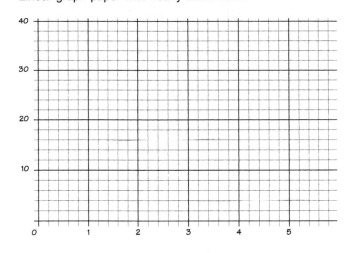

Logarithmic Graphs

When the range of values to plot is small or the values are close together, a linear graph can be used. When the range of points is considerable, a linear graph could misrepresent the data visually. In cases such as this a logarithmic scale can be used. This scale allows a large range of data to be represented in a smaller space. Often one variable has a large range and one does not. In a case like this, semilogarithmic graph paper can be used. This paper has one equally spaced, or linear scale and one logarithmic scale. Paper in which both scales are logarithmic is sometimes called log–log paper. Both types of paper are shown in Fig. 7–25.

When locating the number scale on the graph, the baseline is often the reference line. This works well if all the values are positive. However, if both positive and negative values are needed, the reference line may be located either in the middle of the paper or at any location along the axis. There is no "correct" place for the reference line. It should be placed wherever it is most logical for the data you need to present.

axis. These values are called *scales*. Before you can choose a scale, you must know the entire range of values to be displayed. For example, if you want to show currents from 10 to 40 mA, your scale must include this range. Often the maximum value on the axis is one increment or a portion one larger than the maximum values. For clarity, the values used for scales are often multiples of 1, 2, 5, or 10. Other values can be used, however, if they more closely represent the units being displayed.

When using graph paper with accent or heavier rulings, the scale designators are usually placed on the heavier lines. These lines are not required, but are often used to highlight major divisions on the graph. The use of these lines can be seen in Fig. 7–24.

Multigraphs

Sometimes a number of curves is shown on the same graph. For example, power gain, voltage gain, and current gain versus frequency may all be shown on one graph, as in Fig. 7–26. Such a graph shows the relation of each individual value to frequency and also the relationship between the three concepts. The number scales for the vertical axis are placed on each side of the graph. An extra line is used to indicate the third value. Note that this graph has been drawn on semilogarithmic graph paper. Also, it is often desirable to show a characteristic

FIGURE 7–25
Samples of semilogarithmic and logarithmic grid paper.

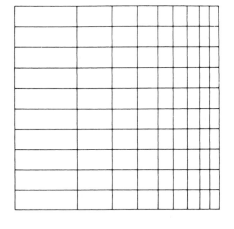

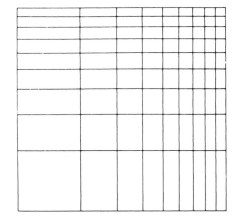

Semilogarithmic Grid   Logarithmic Grid

FIGURE 7–26
Multigraph showing power gain, voltage gain, and current gain.

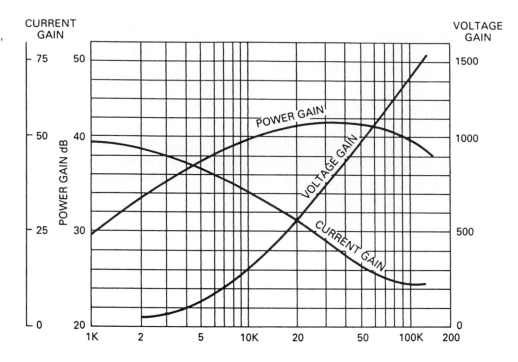

family of curves for an individual component. For example, transistor specifications show a family of curves for voltage versus current for various base currents. These are helpful in choosing appropriate transistors for a particular application. A typical transistor family of curves is shown in Fig. 7–27.

There are many other types of graphs and charts, such as bar charts, pie charts, and scatterplots; many of these are not used in electrical applications.

All the previous examples demonstrated values that are all positive and fall in the same quadrant of the Cartesian coordinate system. For many of the graphs this is true, but occasionally a graph shows both positive and negative values. In such a case, the reference lines can divide the graph into four quadrants. A zener diode

has both forward and reverse characteristics; to plot these, a four-quadrant graph is needed, as shown in Fig. 7–28. Even though the curve extends to only two quadrants, all four are shown, and both positive and negative voltage and current values are shown.

Miscellaneous Graphs

There are a number of special-purpose graphs that lend themselves well to specialized applications. For exam-

FIGURE 7–27
Typical characteristic family of curves for a transistor.

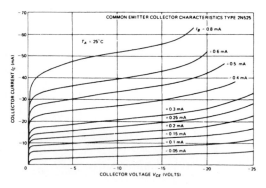

FIGURE 7–28
Characteristic curve of a zener diode.

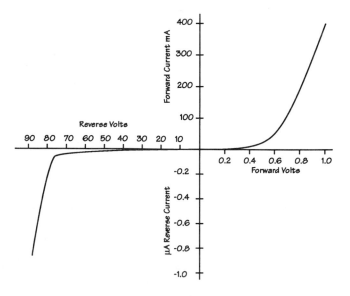

FIGURE 7–29
Sample of polar graph paper. (Courtesy of Keuffel & Esser/ Kratos)

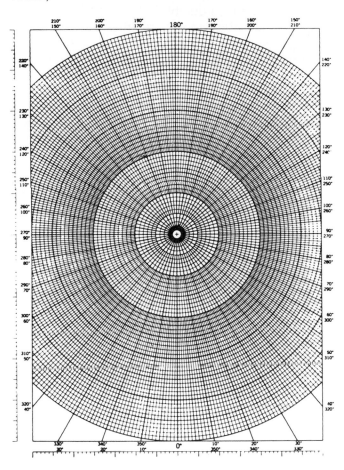

ple, the *nomograph* is used when one variable depends on another variable, such as degrees Fahrenheit and degrees Celsius. By placing two different scales immediately adjacent to one another, it is very easy to convert back and forth between the two values. The primary advantage to this type of graph is that an infinite number of values can be obtained simply by locating points on this graph; no calculations are required. *Footprints* are graphs used to illustrate satellite TV signal strength. Such a graph gives a visual picture of the signal strength across the country based on the midpoint. The graph is drawn as ellipses on top of a map. Polar diagrams are specialized graphs consisting of concentric circles. A sample sheet of polar graph paper is shown in Fig. 7–29. Most often the polar graph is used to represent the directional response curves for microphones (Fig. 7–30) or the field pattern of an antenna.

Preparation of Graphs

1. Choose an appropriate grid paper.
2. Mark scale lines and divisions.
3. Determine scale designators. Designators must be centered around the scale division lines. They should be oriented to be read from the bottom. They should be lettered $\frac{1}{8}$ in. high and located approximately $\frac{1}{8}$ in. from the scale line.
4. Locate scale captions. They should be centered and located parallel to each axis. Captions usually include the name of the variable and the unit of measure. Captions should be written out in full; if space is limited, only standard abbreviations should be used. These are demonstrated in Fig. 7–31 (A).

FIGURE 7–30
Microphone response curves.

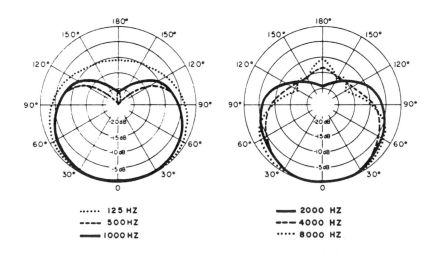

TYPICAL POLAR PATTERNS

FIGURE 7–31
Labels. (A) Scale captions are centered and parallel to each axis. (B) Title is located in clear area centered at top. (Used with permission from Radio Shack, a division of Tandy Corporation, Fort Worth, Tex. 76102)

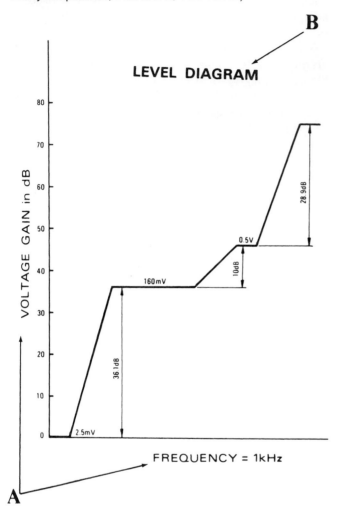

interfere with the displayed data, as shown in Fig. 7–32. A good place to locate much of this information is outside the border, off the main scale on the grid, or in a preprinted title block.

7. Plot the data. Represent the position of all data points by placing a small pencil dot or circle at the appropriate position on the grid paper, as illustrated in Fig. 7–33. With more small squares between major divisions, more accurate representation can be made.

8. Draw the curve. If the data points represent a straight line, simply use a straightedge and connect the dots, as shown in Fig. 7–34. Typically, the curve is neither a straight line nor a perfect arc. In this case the curve

FIGURE 7–32
Notes on the graph but away from the curve.

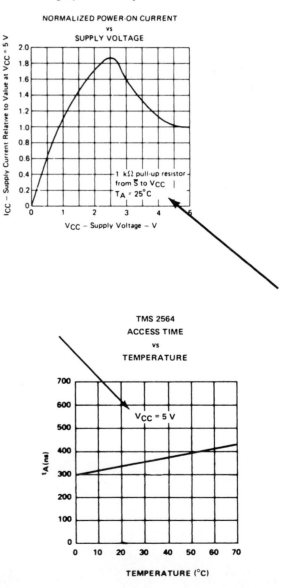

5. Locate the title of the graph. The title must be in an open area well separated from any plotted data. Ideally, the title should be centered near the top. Wherever it is located, it must be evident that it is the title and not related notes or captions for the plotted material. Some prefer to reserve this step for later after the location of the curve is established to ensure that it does not interfere with the presented data [Fig. 7–31 (B)].

6. Add all additional material. Such additional material includes signatures, dates, drawing numbers, sometimes equations, curve labels, key or legend, and written notes. This type of information should be kept to a minimum and must be placed so that it will not

FIGURE 7–33
Small dots or circles for data points.

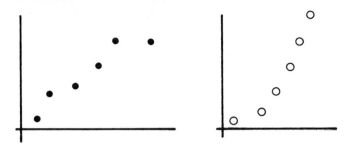

FIGURE 7–34
Data points in a straight line, connected with a straightedge.

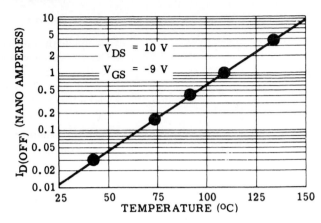

is drawn that best represents the general shape of the plotted data points. Graphs are never completed as a *connect-the-dots* project. Sometimes a very light construction line is used to produce an idea of the finished curve. To produce the final graph, an irregular or French curve can be used. Frequently, one curve cannot be used to finish the entire graph, but small sections of the graph are drawn and then the irregular curve is moved (see Fig. 3–17). This method was described fully in Chapter 3, and you are encouraged to return to that chapter and practice the techniques discussed.

The importance of drawing clear, easily understood graphs should not be overlooked. Much of the information about devices is demonstrated in graph form. These include characteristic curves of devices, response curves of devices and entire systems, relationships of one quantity to another, waveforms in AC circuits, and timing diagrams in logic circuits, to name just a few.

REVIEW QUESTIONS

7–1 Why is it important to understand the *language of electronics?*
7–2 What could be considered the primary difference between electrical and electronic circuits?
7–3 List several types of circuits and identify them as either electrical or electronic.
7–4 How early were discoveries, experiments, and applications of electronics made?
7–5 What are the three basic parts of an atom?
7–6 Why were charges identified for the parts of an atom?
7–7 What is another word for voltage?
7–8 What would cause heat to be generated in a wire?
7–9 What are the symbols for voltage, resistance, current, and power?
7–10 What are the units for voltage, resistance, current, and power? What are the symbols for these units?
7–11 Define conductor and insulator.
7–12 A basic relationship between voltage and current can be represented by what law?
7–13 How does conventional current differ from electron flow?
7–14 What semiconductor materials are used most frequently in electronic circuits?
7–15 What type of structure is formed when semiconductor material forms a solid?

PROBLEMS

Prepare graphs to represent the following data.
7–1 Plot the following points first; then complete a smooth curve starting at zero. (Use small triangles to mark data points.)

X = 4	6	8	10	12	16	20
Y = 2	3.5	6	8.3	10	12.3	14

7–2 Plot the following points first; then complete a smooth curve starting at the first X–Y coordinate. (Use small circles to mark data points.)

X =	−8	−7	−6	−5	−4	−3	−2	−1	0
Y =	10	7	5	3.5	2.5	1.5	1	0.5	0.4

7–3 Complete the other side of the curve in problem 7–2 by making all the X values positive.

7–4 Current versus voltage for a resistance value of 1000 Ω. Use a range of current from 0 to 10 mA and voltage of 0 to 10 V. Use 10 × 10 grid paper with each major division a multiple of 2. Plot no fewer than 10 points.

7–5 Plot a *family of curves* (or individual curves for each set of data points) for the following data. Each value of voltage (X axis) will have four values of current.

Collector–Emitter Voltage	Collector Current in Milliamperes			
1	10	20	30	40
1.5	12	22	33	43
2	12	23	33	44
2.5	13	23	34	44
3	13	24	34	45
3.5	14	24	35	46
4	14	25	35	46
4.5	15	26	36	47
5	15	27	37	48

7–6 Prepare a graph for a diode with the following characteristics with volts on the X axis and current on the Y axis.

Forward Voltage in Volts	Forward Current in Milliamperes
0.1	0
0.2	0.08
0.3	0.2
0.4	0.5
0.5	0.6
0.6	0.7
0.7	0.8
0.8	1.0
0.9	7.7
1.0	7.9

7–7 When a graph is made of frequency versus voltage gain, semilog paper is normally used. For this problem, plot the following data on a graph using 10 × 10 square divisions.

Frequency	1 Hz	5 Hz	10 Hz	50 Hz	100 Hz	500 Hz	10 kHz	100 kHz
Gain	0	100	5000	5000	5000	5500	5300	3500

7–8 Plot the data from problem 7–7 on semilog paper with frequency on the log or X axis and gain on the linear or Y axis. Compare this graph to the one prepared in problem 7–7.

8

COMPONENTS AND SYMBOLS

INTRODUCTION

Each different component used in electrical equipment has a unique function, so it is necessary to have a method to illustrate the components. You could simply use rectangles for every component and letter the function that the component performs. This method is not very practical, however, because it would take some time to analyze every drawing and the lettering requirement would be overwhelming. It would be impossible to depict pictorially all circuits used in electronic equipment. However, by representing each of these components with a unique symbol and/or reference designation, all the relationships and interconnections can be shown. These drawings are used to illustrate circuit functions. The unique symbols are a combination of various circles, rectangles, arcs, and squares that may or may not have anything to do with physical size or characteristics.

Many different manufacturers supply components, which makes it important for a set of standards to be established for the physical characteristics and the functional properties of each component. These standards are referred to as *specifications*, and manufacturers adhere to these specifications when they produce components. The same organizations publishing electronics standards have published standards for components and for symbols representing those components.

STANDARDS

It is very difficult to trace circuit functions by actually looking at the existing equipment. For this reason, diagrams are used to describe the circuit functions. The unique shapes used to represent the individual components are used in the development of these diagrams.

When symbols were first used, drafters attempted to draw the physical characteristics of the components. As the development of integrated circuits progressed, many components became integrated into one package. To make certain the symbols stayed simple enough to provide quick visual representation, it became more important to reflect the basic circuit function rather than the physical characteristics of a component.

The symbols that are used today have been adopted and published in *USA Standard Graphic Symbols for Electrical and Electronics Diagrams* (ANSI Standard Y32.2d–1970) from the American National Standards Institute, Inc., and by such organizations as the Institute of Electrical and Electronic Engineers (IEEE) and the American Society of Mechanical Engineers (ASME), in addition to all those detailed in Chapter 9. Use of these symbols, though not mandatory, is highly recommended in most industrial practices. Most drafting room policy includes use of one of these standards. The Department of Defense has made use of these standards mandatory. A number of these symbols taken directly from the published standards can be found in Appendix F.

Many companies set their own standards, which are often modifications of these published standards. One problem with published standards is that they can be obsolete almost as soon as they are printed. New components and equipment are developed continually, and some older components have been discarded.

Tolerances

When many components are produced, it is prohibitively expensive to make them too precise. Thus, each component is given a desired value with a possible variation on either side of that desired value. This variation is called a *tolerance* and is expressed as a plus or minus deviation from the desired value. Typical tolerances used by manufacturers are ±5%, ±10%, or ±20%. To keep component cost lower and to be certain the desired components are available, many circuits are designed around a ±10% tolerance. A list of the standard numbers used to cover each tolerance range is shown in Table 8–1. For each value listed in the table, all multiples of tens are available, for example, 10 Ω, 100 Ω, 1000 Ω, 10,000 Ω, and so on.

TABLE 8–1
Preferred Values for Standard Tolerances

±5%	±10%	±20%
10	10	10
11		
12	12	
13		
15	15	15
16		
18	18	
20		
22	22	22
24		
27	27	
30		
33	33	33
36		
39	39	
43		
47	47	47
51		
56	56	
62		
68	68	68
75		
82	82	
91		

It is not necessary to include every number between 10 and 100 because a resistor with a value of 100 Ω and a tolerance of ±10% in reality can fall within the range of values from 90 Ω to 110. Because of this overlap of ranges, manufacturers feel it is too expensive to attempt to manufacture each and every value of resistance. In some critical circuit applications, variations as large as 10% are totally unacceptable, and more precise values are required. These precision components are available, but not as readily as ones with standard tolerances. This increase in precision also increases the cost of the component. When precision components are to be used, they must be clearly marked on the drawings.

Reference Designations

Every component that appears on electronic or electrical diagrams or drawings is shown as a unique shape or symbol reflecting the circuit and function. It includes a note indicating code number, value, tolerance, part number, or other important information. This information is covered extensively in Chapter 9, and a list of abbreviations or designations can be found in Appendix E.

Color Code

Many components are very small, and the lettering on them would be too small to be easily read. Also, assembly time would be increased to ensure that the part was mounted lettered side up. Because of these problems, a system of colored stripes has been developed for such components as resistors. Since the stripes go completely around the component, the mounting problem is eliminated. It is also easier to recognize a color stripe than to attempt to read lettering on the component.

A color code with numbers from 0 to 9 has been established. Each number is represented by a different color. As Fig. 8–1 shows, the code for each component includes multiple stripes that are used to identify the numerical value to two significant figures, as well as a multiplier and a tolerance. Typically, four or five stripes are used. The first two stripes determine two significant figures in the value. The third stripe is a multiplier or some multiple of 10 to establish the level of resistance.

FIGURE 8–1
Colored stripes showing a number code.

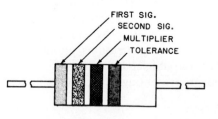

TABLE 8–2
Standard Color Code Used for Resistors

Color	Number	Multiplier	Tolerance
Black	0	1	
Brown	1	10	
Red	2	100	
Orange	3	1000	
Yellow	4	10,000	
Green	5	100,000	
Blue	6	1,000,000	
Violet	7	10,000,000	
Gray	8	100,000,000	
White	9	—	
Gold	—	.1	5%
Silver	—	.01	10%
body	—	—	20%

The fourth stripe indicates the tolerance in terms of a percent. A fifth stripe is used to indicate the military failure rate and is not essential for commercial applications. The standard color code is shown in Table 8–2, along with the number that each color represents.

As seen in Fig. 8–2, if the first stripe is brown, which is 1, and the second stripe is black, which is 0, the two significant figures are 10. The third stripe, or multiplier, tells how many zeros to add to the two significant figures. In this case yellow is 4, which means you should add 4 zeros, or multiply by 10,000. The fourth band is silver, which indicates a tolerance of 10%. The value of this resistor is 100,000 Ω, plus or minus 10,000 Ω.

RESISTORS

A **resistor** is an electrical component that opposes or resists the flow of current. The resistor has a value determined by the amount of *resistance* it contains within a tolerance range. The basic unit of measure for resistance is the ohm. Recall that the Greek letter omega (Ω) is used as an abbreviation for ohms. Resistance is used to describe opposition to current flow, and it can refer to

any electronic part or circuit, not just a resistor. One ohm is the resistance of a conductor when a constant current of one amp through it produces a voltage of one volt between its ends. The maximum value of resistance is infinity, which means no current could flow. This can be represented by an open circuit. The minimum amount of resistance is zero, which can be represented by a short circuit.

An ohm is a rather small amount of resistance, and it is common for values to be several hundred, several thousand, or even millions. The numbers representing the amount of resistance must be lettered on the drawing. For clarity and ease of lettering, all typical prefixes are used to signify very large or very small values. For example 10,000 Ω can be changed to 10 kΩ, where k (an abbreviation for kilo) means to multiply by 1000, or 10^3. The standard color code is used to identify this value of resistance. For 10 kΩ, the first stripe is brown for 1, the second black, for 0. These two numbers establish the significant numbers. The third stripe or the multiplier is to multiply by 1000, so an orange stripe is required. The fourth stripe indicates the tolerance and is gold for 5%, silver for 10%, or no fourth stripe for 20%.

Resistors are manufactured in a variety of different methods out of several types of material. A very common material is a *carbon composition*. This type of resistor is basically a slug of carbon mixed with some filler. It is fairly easy to control the amount of resistance, and the material can be molded into a fairly rugged cylindrical shape. The overall physical size of the resistor does not reflect the overall resistance. Instead, it determines the amount of heat that can be safely dissipated through the resistor, as shown in Fig. 8–3. This is generally referred to as the *power rating* of the device.

Resistors can also be made from a *carbon film*. Here, the resistance is determined by the amount of carbon or carbon film. The film is molded within an epoxy coating or surrounded by a nonconductive core of glass or ceramic. These resistors are similar in size to carbon composition. They are also marked with the standard color code.

Another resistor is a *metal film* type. Metal film resistors are constructed with a thin layer of metal oxide deposited around an inert core. The metal film forms

FIGURE 8–2
Brown, black, yellow, and silver, showing resistance of 100 kΩ ±10%.

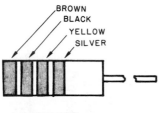

FIGURE 8–3
Carbon composition resistor; the amount of power safely dissipated is determined by size.

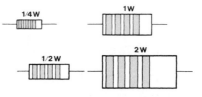

FIGURE 8–4
Wire wound resistor, resistance wire wrapped around a ceramic core.

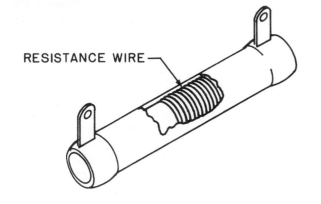

RESISTANCE WIRE

the resistor element, which is covered with a ceramic rod and then coated with a fire-retardant conformal multilayer coating. These resistors are also marked with the standard color code. The primary advantage of this material is in the use of the nonflammable coating, which keeps the resistor from flaming even if subjected to overload conditions. The use of a metal film or a metal oxide resistance element allows for a resistor with a higher power rating to be produced and still maintain a relatively small overall size.

Another popular type is a *wire wound* resistor. The name is derived from the method in which the resistor is manufactured. For this resistor, one or more layers of a resistive-type wire are wound around an insulating material like ceramic, as shown in Fig. 8–4. The wire is usually designed to handle heavier currents, and the core of ceramic, which is frequently hollow, can dissipate more heat than the smaller carbon composition resistor. Typically, these resistors do not use the color code to identify their resistance values, but instead are marked on the body with all appropriate specifications. Because of their shape, wire wound resistors are normally only mounted in one direction, ensuring that the lettering is visible after it is mounted.

Resistors are also classified as axial or radial. This designation refers to the method by which the leads are connected to the body of the resistor. An axial resistor has the leads exiting from both ends of the resistor or, as the name implies, on the same axis as the cylindrical body. A radial resistor has both leads coming out of one end or the other. The type of resistor used is determined by the physical space available for mounting.

FIGURE 8–5
Standard symbol for fixed value of resistance.

FIGURE 8–6
Template used to draw symbols. (Courtesy Berol RapiDesign)

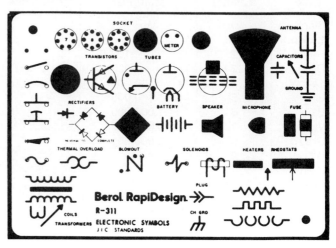

Resistor Symbol

The symbol for the resistor is shown in Fig. 8–5. This symbol, like others, can be drawn with the aid of a template, such as the one shown in Fig. 8–6. It can also be constructed with a straightedge and the appropriate angles.

The resistor symbol is usually constructed with 60° angles between adjacent points, with only three points needed on the top and bottom. The width of the resistor symbol from the top of a peak to the bottom of a peak should be from 1/8 to 3/16 in., as shown in Fig. 8–7. Since this is only a symbol representing a circuit function, there is really no true size or scale used. If a 1/4-in. grid is used in the preparation of the drawings, one square from point to point is a convenient measure, with an overall length equal to about three squares. The most important practice is to keep the size the same for all resistors on any one drawing. This is obviously much easier to do if a template is used. Also, a convenient proportional relationship between the resistor and other symbols is determined when a template is used.

FIGURE 8–7
Resistor formed with a zigzag with three points on the top and bottom.
A = 60°
B = 1/8 to 3/16 in.

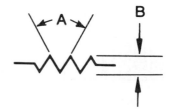

FIGURE 8–8
Wire wound resistor with multiple terminal lugs allows user to select resistor value.

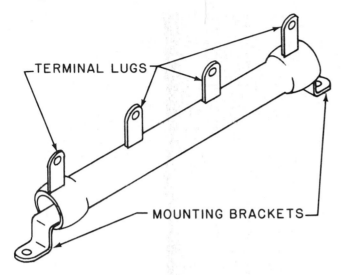

FIGURE 8–10
Potentiometers.

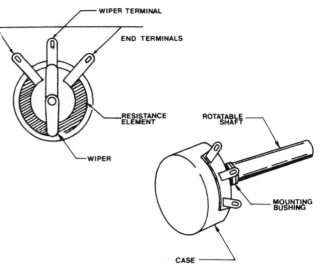

Another method of producing symbols is a dry transfer method, either rub-on or stick-on symbols. These aids produce very consistent, exactly repeatable symbols and make very neat and accurate drawings.

The method of producing the symbols is determined by the use of the drawing. Drawings that will be used *in house* most likely can be completed with the use of templates. On the other hand, drawings that will be used as technical illustrations for publication should be done with stick-ons or other means to ensure clean, crisp, and clear drawings suitable for photocopying.

Variable Resistors

Resistors come in two primary forms, either a fixed value or a variable value. The method that makes the resistor variable can be accomplished in a number of

ways. The wire wound resistor can come with additional points of connection other than the two endpoints, as shown in Fig. 8–8. This resistor is variable in the sense that different values can be selected. A true variable resistance allows the user to select an infinite number of values from the minimum to the maximum on a continuous scale. This type of resistor can consist of a resistance wire connected between two end terminals and a sliding terminal or wiper that makes contact with the wire resistor, as shown in Fig. 8–9. If one end terminal and the sliding wiper are connected in a circuit, a variable resistance, called a *rheostat,* can be obtained. If this resistance is formed in a circle about a shaft and the wiper is connected to the shaft, it is called a *potentiometer,* illustrated in Fig. 8–10.

The unique symbol for variable resistors starts with the same basic resistor symbol. A line with an arrowhead is then drawn at 45° to the basic symbol, or additional connections are shown at locations between the two endpoints. These symbols are shown in Fig. 8–11. The potentiometer includes a short line with an arrowhead drawn perpendicular to and touch-

FIGURE 8–9
Movable metal band, allowing a continuous range of resistances.

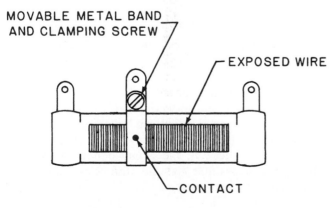

FIGURE 8–11
Methods used to indicate a variable resistance.

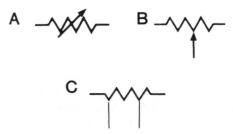

FIGURE 8–12
Rectangles used to represent resistors. (Courtesy Joint Industrial Council)

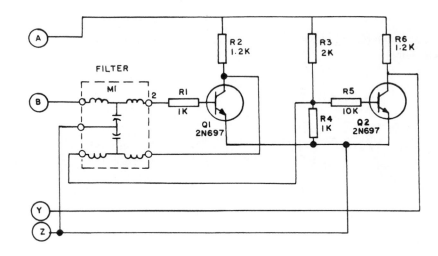

ing the apex of the zigzag symbol. This is also shown in Fig. 8–11.

In addition to the zigzag symbol often found in electronics, a rectangle has been used primarily in electrical and control applications. The use of the rectangle, illustrated in Fig. 8–12, has been adopted by some companies, and it can be found in some maintenance books. It is the responsibility of the drafter to check company policy and use whichever symbol has been approved.

CAPACITORS

A **capacitor** is a component that stores electrical energy between conductive plates separated by a **dielectric.** Place two pieces of aluminum foil separated by a larger sheet of wax paper in a manila folder and you've got a crude capacitor. Apply a DC voltage to the leads of a capacitor and one plate will charge to a point equal to the voltage of the battery. If the battery leads were reversed, the capacitor would charge in the other direction. Thus, a capacitor *stores* electricity.

Like resistors, capacitors are either fixed or variable. And, like resistors, capacitors can be axial or radial. You will be dealing with fixed capacitors about 95% of the time. Generally, variable capacitors are used in circuits that must be frequency dependent. Capacitors are classified according to the dielectric material that they employ. The most common types include

- ceramic
- mica (also glass and enamel)
- film
- electrolytic

The dielectric for a *film capacitor* is a plastic resin or metallized plastic resin film (polycarbonate, polyethylene, polystyrene, and polypropylene). In an *electrolytic capacitor* the dielectric is an oxide of either aluminum

or tantalum. Subcategories of tantalum are based on the material used for the conductive components of the capacitor. They include foil, *wet slug* (a gel of sulfuric acid), and solid (manganese nitrate). Electrolytic capacitors are polarized devices, meaning they can only work in a circuit if they are oriented in the right direction.

Other types of capacitors include paper and oil filled for high-voltage applications and special-purpose, custom-built units for power storage.

The use of a curved line or two straight lines for the capacitor has continually changed through the years. One method of drawing the symbol for a capacitor is to use two parallel straight lines. A point of confusion arises because in electrical systems two parallel lines are used to represent open contacts of an electromechanical relay. Although the contact is normally drawn with a larger gap between the lines, finding contacts and capacitors on the same drawing can still be confusing. Generally, the designation indicates whether it is a capacitor or a set of contacts. Again, it is the responsibility of the drafter to conform to the procedures or practices of the company. If the company drafting manual does not stipulate which symbol is preferred, then consistency should be the key.

Capacitors can be drawn with templates, or dry transfer aids are available. If it is necessary to construct the symbol for the capacitor, the following proportions can be used. The space between the straight line and the curved line should be about one-half the length of the straight line. The curved line should be an arc with a radius equal to the length of the straight line, as shown in Fig. 8–13.

Symbols for the capacitors shown in Fig. 8–14 include both polarized and nonpolarized. The primary difference between the two symbols is the small plus sign found on the polarized capacitor to ensure that it is oriented correctly in the circuit configuration. The plus sign indicates that that end of the capacitor should be connected to the positive side of the circuit.

FIGURE 8–13
Proportions for constructing a capacitor.
A = Radius of curved line
B = ½ A

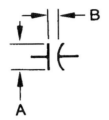

The two main specifications of the capacitors are its storage ability, or *capacitance,* and its operating voltage. The unit of capacitance is the farad. (*Farad* comes from Michael Faraday, a 19th-century British scientist.) However, capacitance values are often very much smaller than 1, and to avoid a great number of decimal places, standard abbreviations for micro (μ, 10^{-6}) and pico (p, 10^{-12}) are used to label capacitors. There are no set rules governing whether a capacitor is marked in micro- or picofarads. So one manufacturer's 0.015-μF device is another's 15,000 pF, just as a 1.2-MΩ and a 1200-kΩ resistor have the same values.

A capacitor color code does exist, although capacitors rarely use this color code. The same numbers are used from 0 to 9 as are found in the standard color code for resistors, and they indicate two significant figures. The difference is in the multiplier band. Capacitors generally do not have very large values, but instead have values in the range of 10^{-6} to 10^{-12}. There are different codes based on the types of material found in the capacitor.

It is often necessary to use variable capacitors in tuning circuits, such as radio receivers or transmitters. The symbol for variable capacitors is the standard capacitor symbol with an arrow drawn through it at an angle of 45°, as shown in Fig. 8–15.

INDUCTORS

The **inductor** or induction coil can best be described as wire wrapped around a core made of either air (or other insulating material) or iron (or other magnetic material).

FIGURE 8–14
Symbols for capacitors.

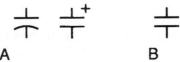

A   B

FIGURE 8–15
Symbol for capacitor, including arrow to indicate variable.

An inductor is a component that stores electrical energy in the form of a magnetic field. The *henry* is the unit that induces a counter electromotive force of one volt when the inducing current is changing at a rate of one amp per second. The terms *choke* and *coil* are often used to identify inductors. Inductors are frequently used in tuning circuits with capacitors, so it is often necessary to make them adjustable. Unlike capacitors or resistors, which are called *variable,* inductors are normally referred to as *adjustable.*

Typical symbols are shown for inductors in Fig. 8–16. The symbol used to represent the inductor reflects the physical characteristics of the inductor. The inductor is coils of wire wrapped around a cylindrical shape. Typical inductors are shown in Fig. 8–17. The symbol looks like loops of wire. The symbol shown in Fig. 8–16 (B) shows loops and does appear to be a coil of wire. Generally, the symbol is completed with only three or four semicircles, as shown in Fig. 8–16 (A). On a standard template the symbol for inductor is usually included. If it is necessary to construct the inductor symbol, simply draw four semicircles with diameters of ⅛ to 3/16 in. on the same centerline tangent to each other. This should produce a symbol with approximately the same overall size as the resistor. If a ¼-in. grid is used, the diameter of the semicircle could be ¼ in. for ease in preparation.

The standard symbol for the inductor is modified as shown in Fig. 8–18 to include additional information, such as the type core that is used. As shown, two solid lines parallel to the basic symbol indicate a magnetic core, and broken lines indicate a ceramic core. A single broken line with an arrowhead is used to indicate an adjustable inductor. An arrow drawn at 45° is sometimes used to indicate an adjustable inductor, although the use of the 45° line generally implies variable, and the inductor is treated slightly differently. The single broken line with the arrowhead more closely represents the physical characteristics of the adjustable inductor be-

FIGURE 8–16
Symbols for inductors.

PREFERRED

AIR CORE

FIGURE 8–17
Photos of typical inductors.

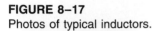

cause the core material is moved in or out of the coil of wire to produce differences in inductance. When a current flows through this coil, a magnetic field is created. When inductors are used in circuits, this characteristic must be considered. Often a metallic shield of some sort is used around the inductor. If this is done, it may be represented by a dashed-line box around the coil.

A **transformer** is two inductors placed close together with no electrical connection made between them. They work principally because of electromagnetic action and interaction between the two coils or inductors. They are called transformers because they *transform* electrical energy from one side of the device to the other. Because they are electrically isolated, they are used to isolate one

FIGURE 8–18
Standard symbol for inductor, modified to include the type of core material and whether it is adjustable.

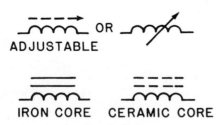

FIGURE 8–19
Various transformers. (A) & (B) Air core. (C) Magnetic core. (D) Multiple windings.

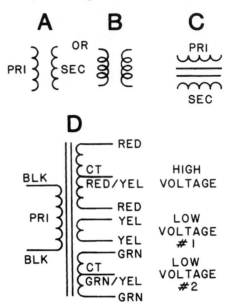

stage from another in audio circuits or radio frequency circuits. They are also used to provide different voltages than can be obtained from a primary source, such as a wall plug.

Transformers may consist of two single windings or multiple windings. A single winding may also contain multiple taps or connections at various points along the main winding. As can be seen in Fig. 8–19, the symbols used for transformers are very similar to those for inductors.

ADDITIONAL SYMBOLS

In addition to the basic electrical components already discussed, there are symbols to represent all the different switches, plugs, jacks, batteries, meters, speakers, or other portions of an electrical circuit. For a complete list of all approved symbols, you can send for a copy of the standards from ANSI, IEEE, or other standards organizations. A quick reference to symbols is contained in Appendix F.

ELECTRONIC DEVICES

A distinction is usually made between electrical components and electronic devices. When the term electronic device is used, it frequently refers to semiconductor devices. These solid-state devices have generally replaced vacuum tubes. Symbols for solid-state devices

FIGURE 8–20
Symbol for a diode.

have been created from some of the circuit functions. Two-terminal devices consist mainly of types of diodes. A **diode** is a two-terminal device made of semiconductor material that allows current to flow easily in only one direction. This physical characteristic is indicated in the symbol shown in Fig. 8–20, which resembles an arrow pointing in the direction of *conventional* current. Although this is a physical characteristic, it is very much the circuit function of the diode. The diode has been called a rectifier because it is used frequently for changing alternating current into current that flows in only one direction.

Diodes can be found in the sections of the circuits identified as the power supply and are used to change AC voltages to DC voltages. Diodes come in many different sizes and shapes. The diode itself is really very small and is contained in different case styles. These case styles vary from small cylindrical shapes made of glass or plastic to much larger metal types with a threaded end, as shown in Fig. 8–21. The case allows the diode to be connected to a heat sink or simply to the chassis so that more heat can be drawn away from the actual diode. The heat sink, shown in Fig. 8–22, is simply a piece of metal that can be attached to a component to help radiate the heat away from the component. The differences in the physical size have to do with their ability to dissipate the heat. Diodes that can more easily dissipate heat can handle larger values of current and power without damaging the device.

Diodes are marked with type numbers and some indication of which end is which. Like the polarized

FIGURE 8–21
Typical diode shapes, including LED and photodiode.

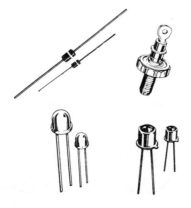

FIGURE 8–22
Heat sink, used to help dissipate heat away from the device.

capacitor, the diode can be destroyed if connected improperly, but, equally important, the diode may not function properly. The mark on a diode can be either a black band or a stripe around one end, as in Fig. 8–23, to indicate which end should be connected to the more negative of the voltage supply. It is not uncommon to find a white or colored stripe. This band is the same as the solid line drawn in the symbol and is called the cathode end. For some of the larger case styles, the actual diode symbol is stamped or inked on the case. Every diode has a number of specifications concerning its operation and use. There is not one individual value, as with the resistors or capacitors, but a number of values that are important. All manufacturers who produce diodes accept these specifications and, regardless of which manufacturer produces it, any diode with the

FIGURE 8–23
Case style of diode showing stripe in comparison to diode symbol.

same reference number will have in general the same specifications. The reference designation used on drawings need contain only the reference number of the diode.

The basic diode allows current to flow in only one direction and is meant to block current in the reverse direction. The *zener diode* is very similar to the basic diode except it is designed to be operated in the reverse-biased mode and, when operated at the reverse breakdown voltage, provides a constant voltage when connected in this manner. Other types of diodes include the *tunnel diode* and the *varactor* or *varicap diode*. Symbols for these diodes are shown in Fig. 8–24. Case styles for these diodes are similar to those for basic diodes.

Two other types of diodes include one that is light sensitive and one that is light emitting. The light-sensitive diode is called a *photodiode* and will allow current to flow in one direction only when light is directed into it. The *light-emitting diode,* or LED, will glow when conducting or allowing current flow. It is encapsulated in colored plastic that forms a lens to intensify the light. The symbol used must include the standard symbol for light, which is two small arrows. The direction of the arrows, shown in Fig. 8–25, indicates whether the light must fall on the diode or is given off by the diode.

Transistors

Most transistors have three leads that extend from the case. It is extremely important that these three leads be labeled and identified correctly. They must be connected to the circuit correctly so that it can function properly. There are a variety of transistors as there are a variety of diodes. It is not the purpose of this text to cover all the aspects of electronics but to simply make you aware of the differences in the devices to emphasize the importance of producing a correct symbol on the drawings and diagrams. The transistor is made up of two pn junctions. To form the two junctions, p material, then n

FIGURE 8–24
Symbols for (A) zener; (B) tunnel; and (C) varactor diodes.

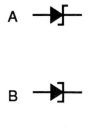

FIGURE 8–25
Small arrows indicating the direction of light, either toward the diode (A) for a photodiode or given off by the diode (B), such as an LED.

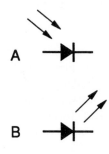

material, and then p material are fused together, creating a PNP transistor. The opposite combination is used for NPN transistors. Although these transistors are biased with opposite voltages (positive and negative), they each work in precisely the same way. It is important to note that the arrow points in the direction of positive, or conventional, current flow. Typical symbols for transistors are shown in Fig. 8–26.

Most transistors perform some type of amplification. One of the most common transistors is the *bipolar transistor,* which functions as a current amplifier. The leads of the bipolar transistor are the base (B), collector (C), and emitter (E). These names came from the original method of constructing the transistor and the functions they perform. For example, the base received its name because it was in fact the base material on which the transistor was made. The emitter and collector sections were then diffused onto the base. The emitter was so named because it emitted major current carriers, and the collector collected them. The circle should be drawn around the basic symbol of the transistor to complete the symbol. Frequently, the engineer or the designer will leave this off, but it should be included in the final drawing.

Other types of transistors include the FET, or *field effect transistor,* the MOSFET, or *metal oxide FET,* and the UJT, or *unijunction transistor.* These are just a few of the more common types of transistors, and their symbols are shown in Fig. 8–27.

It is unlikely that you will ever be called on to

FIGURE 8–26
NPN and PNP bipolar transistor symbols.

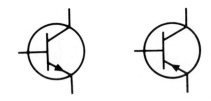

FIGURE 8–27
FET, MOSFET, and UJT.

actually construct these symbols. All symbols should be produced from a standard template or from any of the many dry transfer aids that are available. If you do need to construct symbols from scratch, the following guidelines can be used. For bipolar transistors, start with a circle of about ⅝-in. diameter. Draw the baseline ⅛ in. from the outer edge and parallel to a center line, usually vertical or horizontal. Add the collector and emitter lines by drawing 60° lines from near the end of the base line. The arrowhead is added to the emitter lead, and it should not touch the baseline.

There are subtle differences in the symbols used to represent the different types of transistors. For example, the field effect transistor (FET) shown in Fig. 8–27 has three leads similar to the base, emitter, and collector of the bipolar transistor: the gate (G), source (S), and drain (D). This type of transistor starts with the gate; the drain and source are added and drawn at 90° perpendicular to the gate. The arrowhead is added to this symbol on the gate lead but does not touch the bar. This symbol and other variations of transistors should also be enclosed in a ⅝-in. circle.

Integrated Circuits

Most integrated circuits are usually drawn in rectangles, squares, or triangles representing their function. A few examples of these circuits are shown in Fig. 8–28. The operational amplifier, or OP AMP, is represented by the triangle. Lettered information generally includes pin numbers and identification of input or output leads,

FIGURE 8–28
Squares, rectangles, and triangles used to represent circuits.

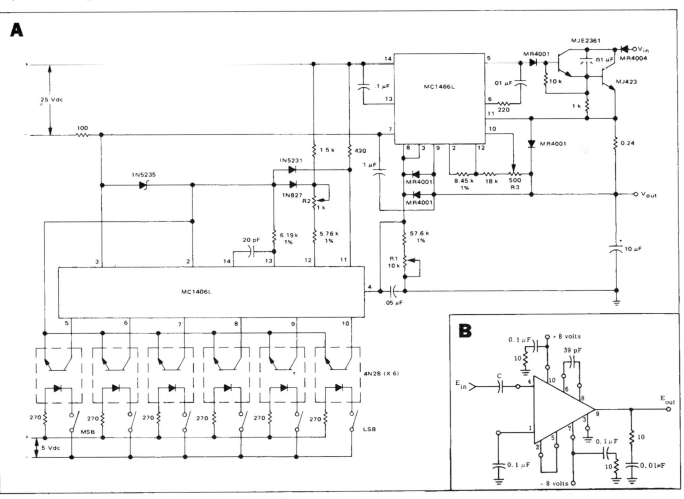

ground connections, and voltage supplies. Instead of having the symbol that represents a dry cell as seen earlier, the voltage connection is simply noted as 8 volts, as shown in Fig. 8–28B. Often the pin numbers are not sequential around the op amp or other integrated circuit. The circuit functions are included and displayed in the most easily understood configuration, with inputs on the left and outputs on the right. Biasing voltages and grounds are placed on the top and bottom of the triangle.

For drawings, an attempt is made to minimize the number of corners and crossovers. Frequently, this requires that the pin numbers be represented out of their true physical locations. Manufacturers produce the integrated circuit in the most economic fashion, which may not give the most convenient physical location. It is also common for the integrated circuit to have pins that do not need to be connected in the circuit. These pins need not be shown on the drawing. Sometimes they are included and labeled NC, which indicates *no connection.* This label helps the technician in assembly or during maintenance and repair.

Component Outlines

In addition to using unique symbols, it often is necessary to actually outline the physical characteristics of a device or component. This is especially true for assembly drawings or printed circuit board design. The overall physical characteristic is called the *case style.* Case styles are standardized by the Joint Electronic Devices Engineering Council (JEDEC), and all manufacturers conform to these standard sizes. Devices like transistors are very critical in terms of lead spacing, as are larger-power rating resistors. Some typical case styles are shown in Fig. 8–29. All these case styles have dimensions governing their physical size. Resistors, capacitors, and inductors also have physical dimensions that are standardized, as are integrated circuits or semiconductor devices.

For additional outlines and dimensions, see any number of data books supplied by the manufacturers. Most outlines for semiconductor devices are identified as DO for diode outline or TO for transistor outline. Many integrated circuits also fall into the TO categories. The dimensions for integrated circuits are also standardized and are generally determined by the number of

FIGURE 8–29
Typical case styles for diodes and transistors. (Courtesy Texas Instruments, Inc.)

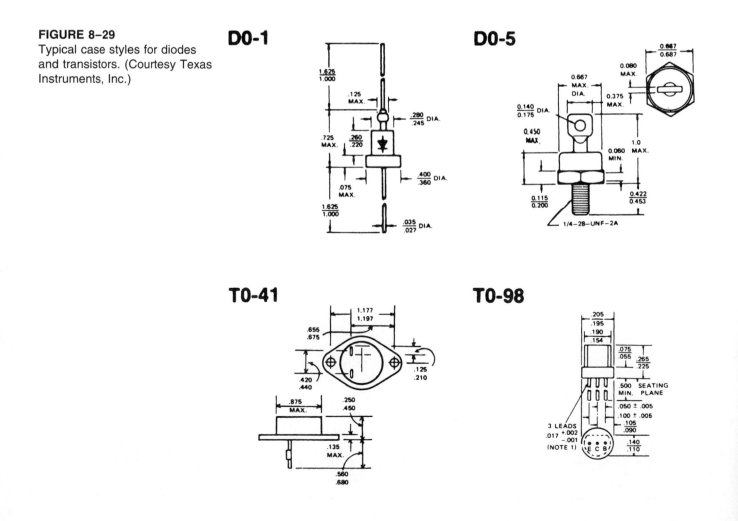

connecting leads or pins. The types of integrated circuits are discussed more fully in Chapter 10.

USE OF CAD

As the use of CAD systems to prepare drawings continues to gain popularity, it will be essential for you to understand the function of the system. The primary function is to eliminate some of the repetition necessary in preparing many drawings. The use of CAD does not eliminate the need for you to know and fully understand good drafting principles and practice. Specifically, in the preparation of component symbols, the storage capability of a CAD system can be very beneficial. It becomes necessary to construct a desired symbol only once and then store that symbol in the library of parts. Placement of that symbol in drawings will still be the ultimate responsibility of the drafter. Figure 8–30 shows how a user locates symbols in the library of parts.

FIGURE 8–30
A CAD system and library of parts to form circuit drawings.

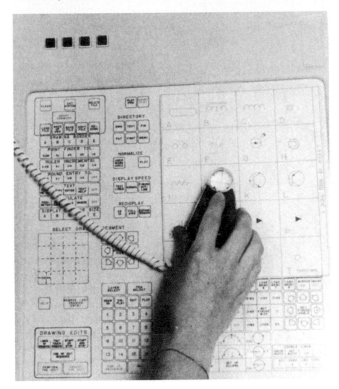

REVIEW QUESTIONS

8–1 What standards are used to ensure that all drawings will be easily interpreted?

8–2 Why are sizes for component symbols much more flexible than for mechanical drawings?

8–3 When would a polarity indication be needed for a capacitor?

8–4 Does the symbol for an inductor reflect its physical characteristics or functional properties?

8–5 The symbol for a diode looks like an arrow. What functional property does this represent?

8–6 The use of two parallel lines for capacitors has a major drawback. What is it?

8–7 Why are rectangles used to represent resistors?

8–8 Using the color code chart for resistors, determine the value of resistance, including tolerance, for the following:
a. yellow, violet, red, silver   c. red, red, red
b. brown, black, green, gold   d. orange, white, yellow, gold

8–9 Using the color code table, determine the color of the stripes for the following resistance values:
a. 100 Ω ± 5%   c. 5600 Ω ± 20%
b. 4.7 kΩ ± 10%   d. 27 kΩ ± 5%

8–10 What does the physical size of the resistor indicate?

8–11 Why do wire wound resistors have the ability to dissipate more heat?

8–12 Why are symbols frequently drawn much larger or smaller than the actual components?

8–13 Why are basic symbols modified, and what do the new symbols represent?

8–14 Why are inductors called adjustable when other components are called variable?

8–15 When using resistors from the standard tolerance ranges (5%, 10%, or 20%), why are some values not included in the list of preferred values?

PROBLEMS

8–1 Practice drawing symbols for the following. Templates should be used and at least two of each should be completed. Repeat as needed any component not drawn very well.
   a. resistor   c. diode
   b. capacitor   d. variable resistor (two methods)

8–2 Make freehand drawings for the following symbols using a grid paper and an estimation for the recommended overall sizes.
   a. resistor   c. transistor
   b. inductor   d. LED

8–3 Obtain a manufacturer's data book, choose 10 symbols, and produce two of each. One should be constructed, and the other should be produced using a template.

8–4 Do a pictorial sketch of the wire wound resistor in Fig. 8–4. Use approximately 8× size in book. Finish this drawing in pencil and follow the guidelines for isometric drawings given in Chapter 6.

8–5 Follow the same general instructions as for problem 8–4; however, complete the drawing as an oblique drawing.

8–6 Enlarge the wire wound resistor found in Fig. 8–4 to 10× size. This project should be completed as a finish drawing in ink.

8–7 Using a template, produce drawings of inductors similar to those shown in Fig. 8–16.

8–8 Using a circle template and straightedge, produce drawings of inductors similar to those shown in Fig. 8–18. Be sure to include examples of air core, iron core, and ceramic core and the preferred method of adjustable types of inductors.

8–9 Draw the typical component outlines shown in Fig. 8–29. This should be completed by transferring dimensions from the illustration. Make this drawing large enough to fill an A size sheet.

8–10 (Advanced problem) From the photographs shown in Fig. 8–22, draw a multiview of both heat sinks. Take approximate measurements directly from the photograph and enlarge 4×. This should be completed in pencil or ink as instructed.

8–11 through 8–17 Draw or sketch the diagrams using B or C size paper with a grid. Place each of the symbols as shown and provide the proper designations. Draw the symbols correctly, using a template when possible.

PROBLEM 8–11
Schematic diagram.

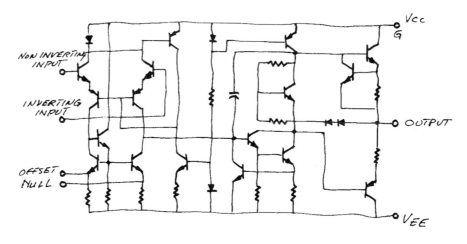

PROBLEM 8–12
Schematic diagram.

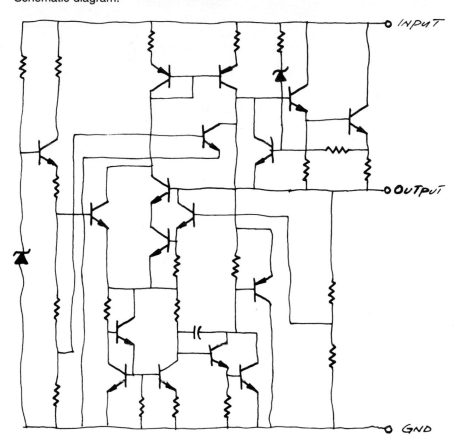

PROBLEM 8–13
Schematic diagram.

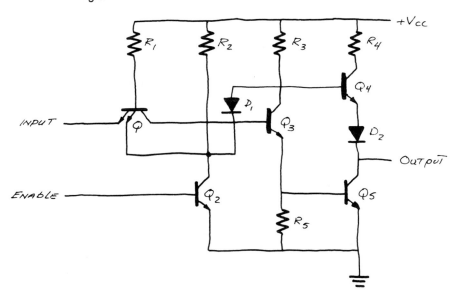

PROBLEM 8–14
Schematic diagram.

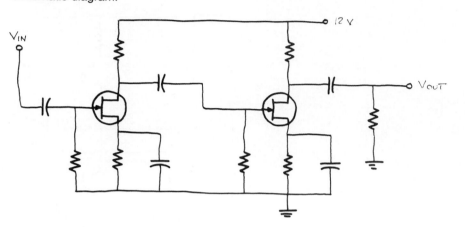

PROBLEM 8–15
Schematic diagram.

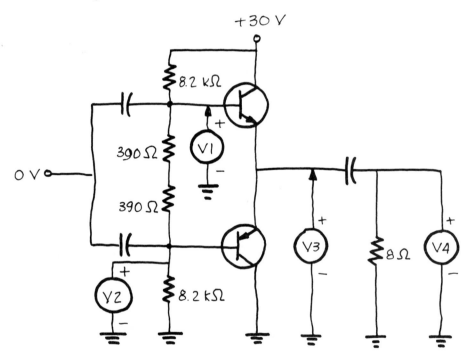

PROBLEM 8–16
Schematic diagram.

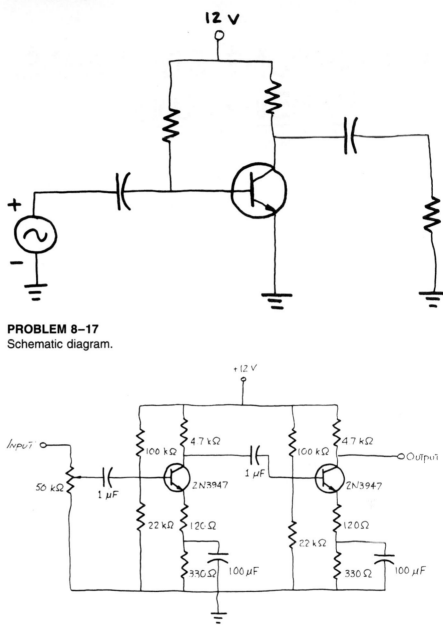

PROBLEM 8–17
Schematic diagram.

9

DESIGNATIONS, STANDARDS, AND ABBREVIATIONS

INTRODUCTION

This chapter is a continuation of the *language of electronics.* It covers **designations, standards,** and **abbreviations** used to describe components, parts, and functions in electronics. Designations, standards, and abbreviations are used in the preparation of all diagrams and schematics to communicate desired designs. It is extremely important for you to become acquainted with these terms and their uses.

REFERENCE DESIGNATIONS

On a typical electrical or electronic drawing, various symbols are used to represent the different components of the circuit. In any one drawing, there is seldom only one of any particular component or part. A method of identification is essential so that these parts can be labeled for parts lists, production drawings, and service manuals.

A code called a *reference designation* is used for this purpose. A reference designation is a *letter symbol* of the name of the component and is not to be confused with an abbreviation of the name. For some components, the reference designation does, in fact,

look like an abbreviation. However, reference designations were developed primarily as one- or two-letter symbols. Abbreviations are usually a shortened version of words, but generally are longer than reference designations.

An alphanumeric code enables us not only to identify the type of part but also to identify which particular part is being discussed. Each designation is composed of two aspects—a component class designation letter and a sequence number. For example, the letter **R** is used for resistors. Therefore, the first resistor is labeled R1, the second R2, and so on, as shown in Fig. 9–1. Capital letters are always used for designations, and when numbers are added, they are the same size as the designation letters. Table 9–1 shows typical reference designators.

Since diagrams, symbols, and designations are the language of electronics, there must be some form of standardization so that engineers, designers, drafters, and technicians can communicate with each other. A large group of industrial advisors has formed the American National Standards Institute (ANSI). This institute acts as a clearinghouse to distribute the acceptable standards. ANSI is only one of a number of organizations responsible for the dissemination of information and acceptable standards. For example, IEEE has become

FIGURE 9–1
Numbered reference designations. (Courtesy Jerry Rye, Evergreen Valley College, San Jose, CA.)

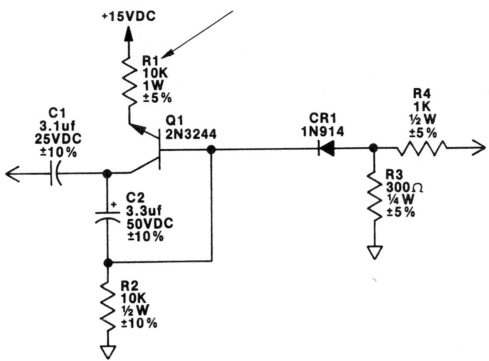

very involved in the publication and passing of standards. Component class designation letters are specified by ANSI in publications Y32.2 and Y32.16. (A partial list of common components and their class designations is included in Appendix E.)

Placement

The location of the reference designation is critical because there must be absolutely no confusion about which component it is identifying. It is permissible to locate designations above, below, or beside the particular components, depending on the space available.

When the diagrams are prepared, it is important to leave space available around each component and between neighboring components. Figure 9–2 shows different designations and their placement. Notice that some of the designations are on the left side of the device, whereas others are on the right side or above the component. Placement on the side or top of a device is frequently determined by the orientation of the device itself. If the device is horizontal, then the designation is usually found at the top or bottom. If it is vertical, the designation is found on either side. Additional information is given on the drawing as well as in the parts list. This information includes the value of the device or

TABLE 9–1
Reference Designations

A	assembly	MK	microphone
B	fan	PS	power supply
C	capacitor	Q	transistor
CR	diode	R	resistor
DS	annunciator, lamp, LED	S	switch
E	misc. electrical part	T	transformer
F	fuse	U	integrated circuit, microcircuit
G	generator	V	electron tube
J	jack	VR	voltage regulator
K	relay	X	socket
L	coil, inductor	Y	crystal unit
LS	meter	Z	tuned circuit

FIGURE 9–2
Placement of reference
designations.

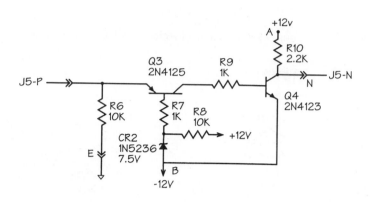

component, often a number indicating the type, terminal numbering, tolerance, and rating.

It is acceptable to *split* the identifying material, as shown in Fig. 9–3. The resistor on the right shows the reference designation R27 on the top and the value of the resistor, 2.7 K, below. In this particular example, there would probably be plenty of room on the top or on the bottom of R27 for the entire designator. It is very important to consider the space required around each component before the drawing is actually started. There are very few instances when these designations need to appear cramped or crowded or when splitting of designations is required. Crowding or splitting of designations can be avoided by considering space needs before laying out the drawing.

When a drawing is going to be used for the general public, as in a magazine, sometimes designations are not used. Frequently, not all the people that might be using the information would be familiar with the language of electronics. To ensure that everyone is able to understand the presented matter, all the components and devices are simply identified by their names and their functions, as in Fig. 9–4. Such identification, however, is for illustration and information and is not common practice for industrial applications. To aid in the preparation of diagrams, drawings, or illustrations, it is essential to become accustomed to designations and their meanings as soon as possible.

Component Numbering

Even though the numbering of components may seem arbitrary, there is a recommended and often-used method for numbering. In general, two methods are used to number components; one method is to assign numbers starting at the upper left and continuing down the left side as if it were a column, numbering all of the same type of component. After the first column is finished, return to the top and form another column, proceeding in this manner until all components of the same type are numbered. Then start again at the upper left for the second type of component and follow the same pattern. Continue until all components have been identified and numbered. The second method of numbering follows the direction of reading, from left to right. The components of one type are numbered starting at the upper left-hand corner and going across the diagram and then across again, forming rows of components.

The first system was developed to follow the basic circuit flow of input to output. It is the recommended method for the numbering of devices. Slight variations in this basic pattern can be found, but the essential rule to follow is that sequential numbers should be used on parts located close together. Note the general flow from input to output in Fig. 9–5. It is very easy to locate sequential numbers. All different types of devices or components must follow the same numbering proce-

FIGURE 9–3
Split reference designations.

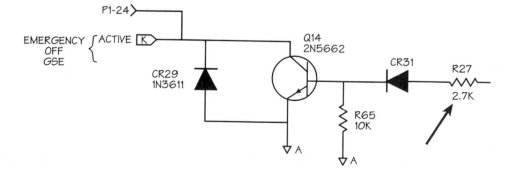

FIGURE 9–4
Labels for illustrations in journals. (Courtesy American Radio Relay League)

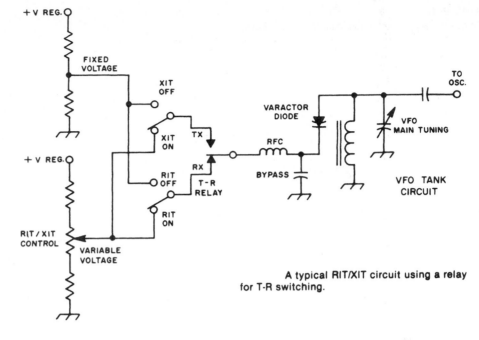

A typical RIT/XIT circuit using a relay for T-R switching.

dure. Each new category starts with 1, and sequential numbers are assigned. A more complex drawing is shown in Fig. 9–6. This is only a portion of a much larger drawing, but it is possible to identify the pattern used in assigning the numbers.

Reference Designation Tables

After a circuit has been tested, some changes may have to be made. When a device or component is added or deleted, a number must also be added or deleted. When a component is added, the new number is a continuation of the sequence already used. For example, if the last resistor was labeled R223 and another resistor is added anywhere in the drawing, it is labeled R224. Or if, for example, a resistor labeled R108 were removed, that number would simply be deleted. If R108 were simply changed to a new value and not removed totally, it still would be handled as an added or new resistor. Therefore, it would have a new label, and the label R108 would be deleted.

All these changes will be confusing unless a list is developed explaining the last numbers used and which numbers were not used at all. This can be done in tabular

FIGURE 9–5
Flow of numbering sequence.

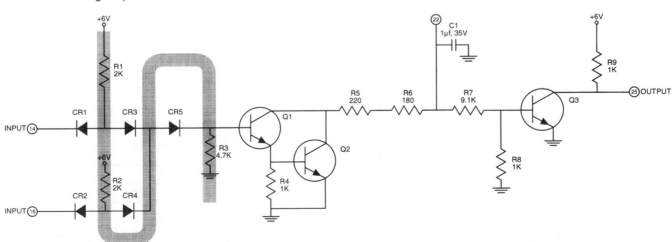

FIGURE 9–6
Numbering sequence found in a complex circuit.

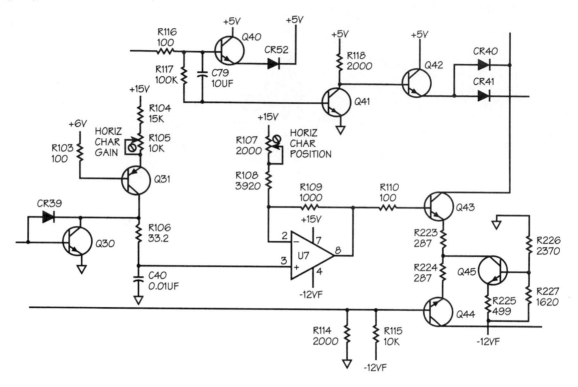

form, as shown in Fig. 9–7. The type of component or device is identified in one column, another column shows the last number used, and still a third column shows numbers not used. Tables are frequently included on the drawing or with other supporting documentation.

Sometimes more information is needed. When necessary, a table like the one in Fig. 9–8 can be used. This table identifies the individual part by reference designation and number and, in addition, shows the component value and the stockroom part number. Another example of a more extensive table is shown in

FIGURE 9–7
Designation table, prepared to indicate which designations are used, the last number found, and numbers omitted.

REFERENCE DESIGNATION	LAST USED	NOT USED
R	R228	R116, R208, R220
Q	Q14	NONE
C	C146	C66, C101

FIGURE 9–8
Reference designations included in parts list.

REF NO.	COMPONENT	PART NO.
R501	75K	216488
R502	240K	216499
R503	10K	216468
R504	200K	216498
R505	11K	216469
R506	47K	216484
CR531	GY	2414891
C511	5μf. 12v	1207557
C512	.01μf. FILM	492411
C513	.033 μf. FILM	217049
Q521	SI NPN. BLK	1207601
Q522	SI NPN. BLK	1207601

FIGURE 9–9
Parts list with additional information.

Electrical Parts List

Ref. Desig.	Description	3C Part No.
C1	CAPACITOR, FIXED, TANTALUM: 1.0 μf ±10%, 35 vdc	930 217 054
C2	CAPACITOR, FIXED, PLASTIC DIELECTRIC: 0.033 μf ±20%, 50 vdc	930 313 016
CR1-CR5	DIODE	943 083 001
Q1, Q2	TRANSISTOR: Replacement Type 2N2369	943 720 001
Q3	TRANSISTOR: Replacement Type 2N3011	943 722 002
R1, R2	RESISTOR, FIXED, COMPOSITION: 2 K ±5%, 1/4w	932 007 056
R3	RESISTOR, FIXED, COMPOSITION: 4.7 K ±5%, 1/4w	932 007 065
R4, R8, R9	RESISTOR, FIXED, COMPOSITION: 1 K ±5%, 1/4w	932 007 049
R5	RESISTOR, FIXED, COMPOSITION: 220 ohms ±5%, 2w	932 006 033
R6	RESISTOR, FIXED, COMPOSITION: 180 ohms ±5%, 2w	932 006 031
R7	RESISTOR, FIXED, COMPOSITION: 9.1 K ±5%, 1/4w	932 007 072

Fig. 9–9. The parts list shown accompanies the drawing in Fig. 9–5. The method used in the preparation of a parts list and other supporting material is ultimately determined by company policy. In spite of the variations, however, all parts lists provide the same basic information.

Control Device Designations

In addition to the designations discussed earlier, which are primarily classified as **electronic designations,** there is a class of **electrical designations** used in control devices and other industrial equipment. Some of the letters are the same as those used earlier, but they have entirely different meanings. Table 9–2 shows a few of the common designators.

Control device designations follow the same procedures stated earlier for electronic device designators. Numbering starts in the upper left-hand corner and follows the column or row rule. These designations must also be made standard, and these standards must be made available to all who might use them. Similar to the American National Standards Institute is the Joint Industrial Council (JIC). This group is primarily concerned with the area of controls and is responsible for publishing standards related to control circuits.

Applications

The use of reference designations is common in identifying actual parts or the location of parts that are mounted on circuit boards, as shown in Fig. 9–10. The same designations that are used on the layout illustrations were used on the circuit drawing. For example, C136 on the illustration is the same as C136 on the circuit drawing. The layout illustration shows the actual component or device shape and location rather than a

TABLE 9–2
Control Device Designations

CR	control relay	R	reverse
ET	electron tube	S	switch
LS	limit switch	T	transformer
M	motor starter	X	reactor
PB	push button		

FIGURE 9–10
Reference designations on a
pictorial layout illustration.

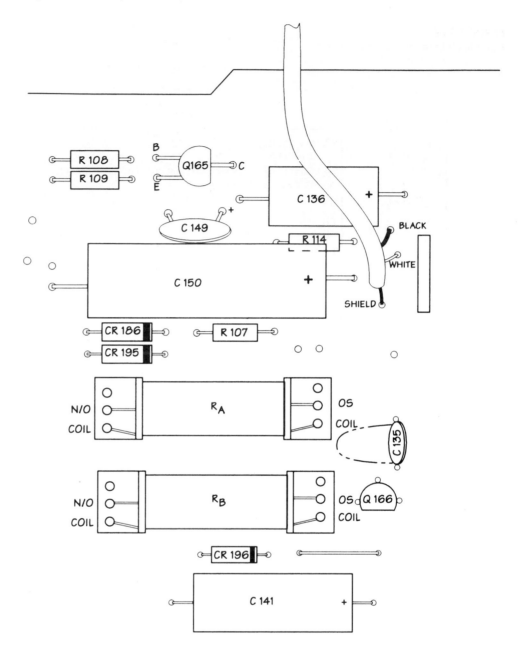

symbol of that device, so the physical size is usually much different for the pictorial layout than for the schematic drawing.

It may be very difficult to try to find a numbering pattern for the actual circuit board. Frequently, parts that were close together on the circuit drawing end up rather far apart in the actual circuit board because of the size of the different parts. For the board layout in Fig. 9–11, the column method of sequentially numbering parts was probably used on the circuit drawing. This pattern is not at all apparent for the component layout shown.

Service manuals contain all types of pertinent information for equipment. In Fig. 9–12 a parts list and diagrams page is shown for an oscilloscope. This page contains the following information: symbols and reference designations used in the preparation of supporting documents listing ANSI standards; a complete list of reference designations, special symbols, or abbreviations appearing on diagrams; and notes depicting values of capacitors or resistors.

It is indeed possible to find variations of all methods listed. In general, it is the best policy to use the reference designations set by ANSI or other similar standards. If these standards and policies are adhered to, then the circuit drawings can be interpreted by any other drafter–designer or engineer. Often a company will establish its own set of designations. Whenever drawings are pre-

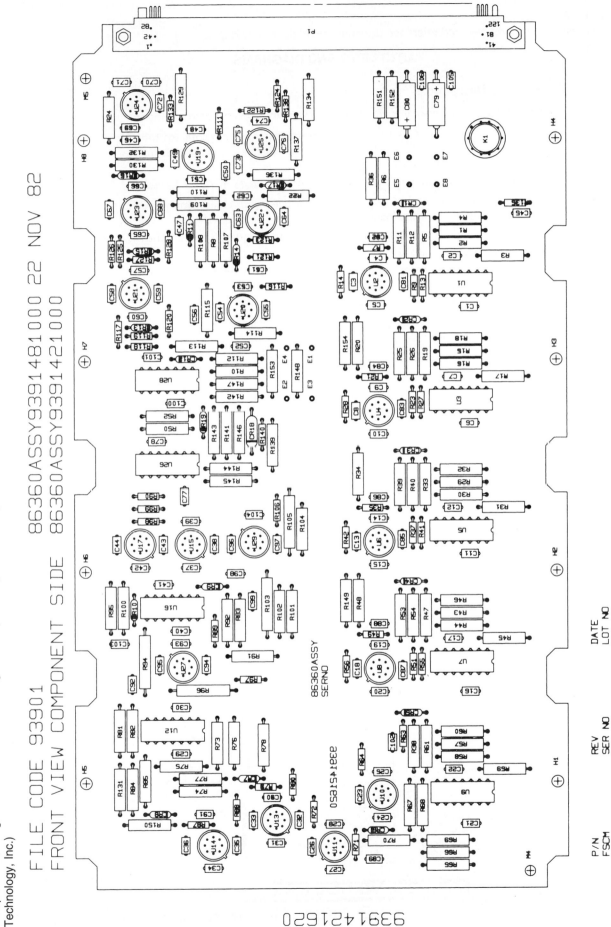

FILE CODE 93901

86360ASSY93914B1000 22 NOV 82

FRONT VIEW COMPONENT SIDE 86360ASSY93914Z1000

FIGURE 9–12

Operations manual with symbols and reference designations. (Courtesy Tektronix, Inc.)

PARTS LISTS AND DIAGRAMS

Symbols and Reference Designators

Electrical components shown on the diagrams are in the following units unless noted otherwise:

Capacitors = Values one or greater are in picofarads (pF).
Values less than one are in microfarads (μF).

Resistors = Ohms (Ω).

Graphic symbols and class designation letters are based on ANSI Standard Y32.2-1975.

Logic symbology is based on ANSI Y32.14-1973 in terms of positive logic. Logic symbols depict the logic function performed and may differ from the manufacturer's data.

The overline on a signal name indicates that the signal performs its intended function when it goes to the low state.

Abbreviations are based on ANSI Y1.1-1972.

Other ANSI standards that are used in the preparation of diagrams by Tektronix, Inc. are:

Y14.15, 1966	Drafting Practices.
Y14.2, 1973	Line Conventions and Lettering.
Y10.5, 1968	Letter Symbols for Quantities Used in Electrical Science and Electrical Engineering.

The following prefix letters are used as reference designators to identify components or assemblies on the diagrams.

A	Assembly, separable or repairable (circuit board, etc)	H	Heat dissipating device (heat sink, heat radiator, etc)	S	Switch or contactor
AT	Attenuator, fixed or variable	HR	Heater	T	Transformer
B	Motor	HY	Hybrid circuit	TC	Thermocouple
BT	Battery	J	Connector, stationary portion	TP	Test point
C	Capacitor, fixed or variable	K	Relay	U	Assembly, inseparable or non-repairable (integrated circuit, etc.)
CB	Circuit breaker	L	Inductor, fixed or variable	V	Electron tube
CR	Diode, signal or rectifier	M	Meter	VR	Voltage regulator (zener diode, etc.)
DL	Delay line	P	Connector, movable portion	W	Wirestrap or cable
DS	Indicating device (lamp)	Q	Transistor or silicon-controlled rectifier	Y	Crystal
E	Spark Gap, Ferrite bead			Z	Phase shifter
F	Fuse	R	Resistor, fixed or variable		
FL	Filter	RT	Thermistor		

The following special symbols may appear on the diagrams:

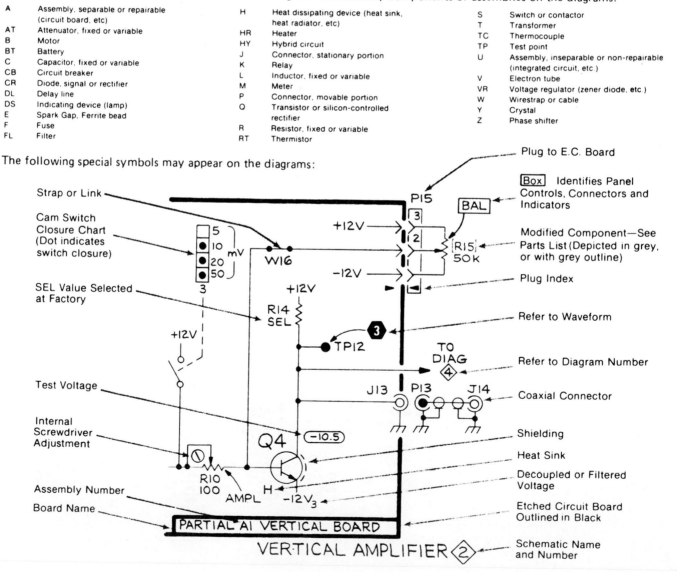

FIGURE 9-13
Company policy establishes standards used. (Courtesy Interactive Computer Systems, Inc.)

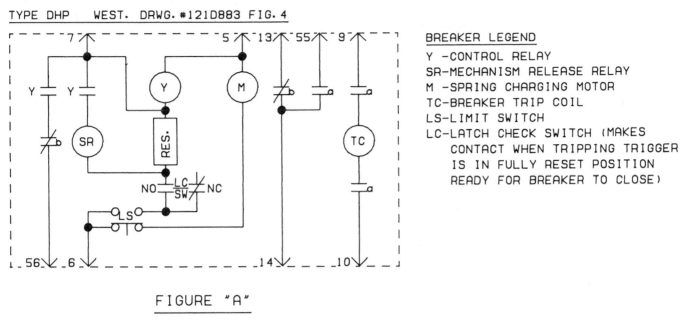

TYPE DHP   WEST. DRWG. #121D883 FIG. 4

BREAKER LEGEND
Y -CONTROL RELAY
SR-MECHANISM RELEASE RELAY
M -SPRING CHARGING MOTOR
TC-BREAKER TRIP COIL
LS-LIMIT SWITCH
LC-LATCH CHECK SWITCH (MAKES
    CONTACT WHEN TRIPPING TRIGGER
    IS IN FULLY RESET POSITION
    READY FOR BREAKER TO CLOSE)

FIGURE "A"

pared, it is essential that company policy be adhered to. If there are no established company policies, following the published standards will always produce acceptable drawings. Anytime the set standards are not used, it is essential to include a list of the designations used on the drawing and the meaning of the designations, as shown in Fig. 9–13. As long as the circuit drawing includes the *breaker legend,* as shown on the figure, it is possible for others to interpret the symbols and designations used in this control circuit.

STANDARDS

As more and more companies begin producing electronics parts and equipment, standardization has become essential. Two kinds of standards dominate the electronics industry: ANSI Standards and Military Standards. Other organizations, however, also develop and issue standards.

ANSI Standards

Many different organizations have compiled standards covering the various aspects of construction, identification, size, overall performance, testing, and other requirements for electronic components and equipment. One such organization is the American National Standards Institute (ANSI). This institute is composed of more than a hundred engineering, technical, and trade

associations as well as government representatives. ANSI does not write the standards; it acts as a distributor of them. Although several technical societies recommend these standards, their use is strictly voluntary. Drafting practice, however, requires the use of—and therefore the familiarity with—present ANSI standards. Table 9–3 lists common ANSI standards with which you should be particularly familiar.

Military Standards

Much of the electronic equipment is produced or purchased by the Department of Defense. This equipment

TABLE 9-3
ANSI Standards

Y14.15–1966	Electrical and Electronics Diagrams
Y14.15a–1970 (R1973)	Supplement
Y14.15b–1973	Supplement
Y32.2–1975	*Graphic Symbols for Electrical And Electronics Diagrams (Including Reference Designation Letters)
Y32.14–1973	*Graphic Symbols for Logic Diagrams
Y32.16–1975	*Reference Designations for Electrical and Electronic Parts and Equipments
Y32E–1976	A collection adopted by the Department of Defense for mandatory use; contains items marked with an *.

must be built to military specifications and standards. In addition, all supporting documentation must be completed according to the same stringent requirements. Military Standards (MS) have been developed to establish standardization for codes, designations, drawing practices, electronic circuits, form factors, procedures, revision methods, symbols, and tests. Adherence to these standards is much more critical than in commercial applications. You should be aware of Military Standards and understand the necessity for compliance with these standards. The standards are developed and issued through the Department of Defense and other governmental agencies, such as

☐ U.S. Army Armament Research and Development Command (ARDC)
☐ Federal Communications Commission (FCC)
☐ General Services Administration (GSA)
☐ National Aeronautics and Space Administration (NASA)

Some of the standards that are distributed by ANSI for electronics and electrical application have been adapted for military use, and these standards have replaced the various similar military standards. Table 9–4 shows common Military Standards.

Other Standards

In addition to ANSI and the government, other organizations have developed standards. Engineering standards are available from the Institute of Electrical and Electronics Engineers (IEEE) and the Joint Industrial Council (JIC):

☐ IEEE Standard 200–1975 Reference Designations for Electrical and Electronics Parts and Equipments
☐ IEEE Standard 315–1975 Graphic Symbols for Electrical and Electronics Diagrams

TABLE 9–4
Military Standards

MIL-STD-8C	Superseded by ANSI Y14.5–1973
MIL-STD-12	Abbreviations for Use on Drawings and in Technical-Type Publications
MIL-STD-27	Designations for Electric Switchgear and Control Devices
MIL-STD-108	Definitions of and Basic Requirements for Enclosure for Electric and Electronic Equipment
MIL-STD-196	Joint Electronics Type Designation System
MIL-STD-242	Electronic Equipment Parts, Selected Standards
MIL-STD-275	Printed Wiring for Electronic Equipment

TABLE 9–5
Sources of Standards

Aeronautical Radio Inc. (ARI)
American Society of Mechanical Engineers (ASME)
Institute for Interconnecting and Packaging Electronic Circuits (IIPEC)
Joint Industrial Council (JIC), a part of ANSI
National Electrical Manufacturers Association (NEMA)
Radio Technical Commission for Aeronautics (RTCA)

☐ JIC Standard EMP–1/EGP–1–1967 Electrical Standard
☐ JIC Standard EL–1–1971 Electronic Standard

(These standards are the same as ANSI standards Y32.16–1975 and Y32.2–1975 listed in Table 9–3.) Other sources of available standards are shown in Table 9–5.

It is very important for you to become familiar with the specifications and standards available from the various organizations. As a drafter, you have the ultimate responsibility in assuring that the drawings, designs, and documentation follow the required standards. (Additional references to the standards and portions of the standards can be found in Appendixes B and E.)

ABBREVIATIONS

The list of abbreviations produced by ANSI was developed not to encourage abbreviations but to provide a standard for their use. Abbreviations should be used only where it is necessary to save time and space. Sometimes abbreviations can lead to confusion because they may have to be interpreted by people with very different backgrounds. In some cases a single abbreviation can refer to a number of different words. If a drawing will be used only within a particular organization, the use of abbreviations may be desirable. If the abbreviations do not follow published standards, a legend should be included on the drawing, as shown in Fig. 9–14.

An abbreviation should always be clear and unambiguous. When using abbreviations on drawings, always follow accepted standards, such as ANSI Y1.1–1972, MIL–STD–12, or Z10.1 from American Standards Association (ASA). As a rule, capital letters are used for all abbreviations on drawings. When abbreviations are used in text or documentation material, it is permissible to use lowercase letters; however, it is still acceptable to use capitals for emphasis and clarity. Subscripts are never used with abbreviations. Periods are never used with abbreviations except in instances where an abbrevi-

FIGURE 9–14

A legend must be included for clarity when using company-defined designations. (Courtesy Interactive Computer Systems, Inc.)

```
LEGEND

A    -AMMETER
AS   -AMMETER SWITCH
BKR  -BREAKER
CLFU -CURRENT LIMITING FUSE
CPT  -CONTROL POWER TRANSFORMER
CS   -CONTROL SWITCH
CT   -CURRENT TRANSFORMER
CTD  -CAPACITOR TRIC DEVICE
FU   -FUSE
SCB  -SHORT CIRCUITING TERMINAL BLOCK
STA  -STATIONARY AUXILIARY SWITCH
V    -VOLTMETER
VS   -VOLTMETER SWITCH

50/51 -TIME/OVERCURRENT RELAY
51    -OVERCURRENT RELAY
52    -1200A CIRCUIT BREAKER
64    -GROUND FAULT RELAY
```

ation might form a word and a period is necessary to avoid confusion, as in atomic (at.).

Although most abbreviations are for single terms, there are combinations of words that through common use have become acceptable abbreviations: ANSI for American National Standards Institute, AFT for automatic fine tuning, and PNP for positive–negative–positive (transistors) are examples.

In general, an abbreviation for multiple words is used with no space between the letters when the abbreviation is formed from the first letters of each word. When the abbreviation for two words is made from two abbreviations, there should be a space between the two, as in REF DES for reference designation or MEM ADRS for memory address. A third variation might include an abbreviation made up of first letters combined with another abbreviation, as in SPDT SW for single-pole, double-throw switch. An abbreviation should not be used when a reference designation is more appropriate.

As a general rule, reference designations are used on schematic diagrams to identify components or graphical symbols. Abbreviations are always used to identify the electrical quantities and units. Farad is not spelled out; instead the abbreviation F is used. On drawings like the one in Fig. 9–15, it is necessary to include both abbreviations and reference designations. The documentation along the left-hand side includes abbreviations such as GND for ground and DC for direct current. At the same time it shows the reference designation C for the capacitors. Interestingly, TP, commonly used for test point, is not approved for either an abbreviation or a reference designation.

We cannot stress enough the necessity for standardization in the use of designations and abbreviations. Drawings are sometimes the only means of communication between different departments or people in an organization. It is not possible to verbally explain what a drawing, symbol, or designation means. It must be readily apparent to all who might need to interpret this information. The only way you can become familiar with these expressions is with much practice and constant attention to the particular standards used.

FIGURE 9–15

Abbreviations and designations on the same drawing. (Courtesy Interactive Computer Systems, Inc.)

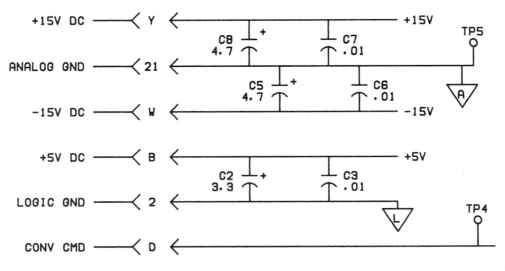

REVIEW QUESTIONS

9–1 Why are reference designations used?

9–2 What method is used to number components in both simple and more complex circuits?

9–3 Who is responsible for maintaining current standards in the drafting room?

9–4 What types of organizations distribute the standards?

9–5 List additional information besides the designator and component number that might be found near a circuit component.

9–6 Explain why abbreviations and designations are not used in technical manuals or publications.

9–7 Explain why abbreviations and designations are said to be a part of the *language of electronics*.

9–8 Discuss the primary difference between reference designations and abbreviations.

9–9 Discuss the location of reference designations on a circuit drawing.

9–10 Are the standards ever changed and who changes them?

9–11 When a part is changed or replaced, what are the steps needed to update the drawing and numbering?

PROBLEMS

These problems will provide practice in recognizing and remembering designations and abbreviations. They will also provide practice in lettering skills. Drafting paper should be used, and proper lettering techniques should be followed.

9–1 List the designations for the following: amplifier, antenna, battery, capacitor, circuit breaker, diode, lamp, light-emitting diode, inductor, jack, meter, transistor, resistor, microcircuit, test point, connector (fixed), and connector (cable end).

9–2 List the abbreviations for the following: abbreviate, alternator, ammeter, ampere, amplifier, antenna, battery, capacitor, circuit breaker, clock, commutator, computer, conductor, diode, double-pole double-throw switch, electronics, electronic voltmeter, engineer, generator, jack, light-emitting diode, inductor, meter, microfunctional circuit, resistor, and transistor.

9–3 List the names of five organizations that supply standards.

10

Microcircuits

INTRODUCTION

The study of microelectronics started as early as 1948 with the development of the transistor. As time progressed, more and more elements were added in a production process that allowed many elements to exist in an area of approximately .25 × .25 in. It is not uncommon for this area to include as many as 68,000 transistors and associated passive components, as shown in Fig. 10–1.

Frequently, a microcircuit is more than a single device. It is, in fact, typically an entire circuit or portion of a circuit capable of performing a multitude of functions. A *discrete* circuit has separate components connected to one another. An *integrated circuit* (IC) is fabricated as an inseparable assembly of components in a single structure. Fig. 10–2 shows a number of discrete components and integrated circuits.

There are three basic types of integrated circuits, named primarily by the process by which they are manufactured: **monolithic, film** (thick film and thin film), and **hybrid.** In addition, ICs can be categorized as *small-scale* integration (generally less than 100 semiconductor components), *medium-scale* integration (between 100 and 1000 semiconductor components), and *large-scale*

integration (more than 1000 elements). The scale size refers to the number of components integrated into the same package and not to the size of that package. An additional category of *very large scale* integration (more than 10,000) has emerged.

ICs can also be classified according to their general function. The two most common categories are analog and digital. Digital ICs contain circuits whose input and output are limited to two states—a low voltage or a high voltage. This can also be described as *on* or *off.* For circuit operation, digital logic is perhaps the easiest because it is limited to only two states. The output is obtained with a logic function. Analog input and output are mostly linear and can be any number of values on a continuum. The only limitation in analog circuits is the limitation of the individual device. Output in an analog circuit is normally proportional to the input.

Digital circuits include logic gates, flip-flops, counters, digital-clock chips, calculator chips, memory chips, and microprocessors. Analog circuits include amplifiers, voltage regulators, and voltage comparators.

Integrated circuits play a major role in the electronics industry today. Practically every modern electronic device, from clock radios to sophisticated computer systems, uses integrated circuits. The astonishing growth

FIGURE 10–1
Microprocessor chip. (Courtesy Motorola, Inc., Semiconductor Products Sector)

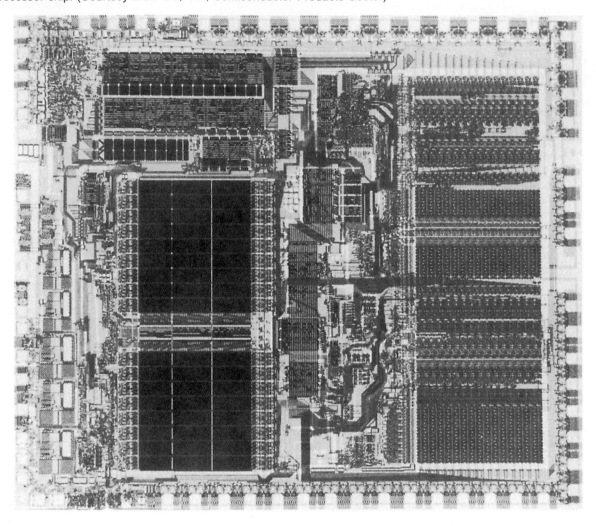

of computerized toys, pocket calculators, and wrist watches is an indication of how solid-state technology affects our lives.

Integrated circuits were first used almost exclusively in military and avionics applications. They offered improved reliability, reduced size, and lower power consumption. But they were also very expensive because it was difficult to design the ICs and to control the processes used to manufacture them. These early circuits were quite simple by today's standards, often consisting of no more than 20 or 30 transistors.

Use of integrated circuits is no longer confined to military and avionics applications. Cost of the circuits has been dramatically reduced by advances in design and manufacturing technologies. The complexity and usefulness of ICs have increased considerably. They are no longer a small collection of simple gates, but are very complex systems.

MONOLITHIC INTEGRATED CIRCUITS

Industries involved in the manufacture of ICs have changed considerably during the past few years. In the beginning, the manufacturer of the IC also manufactured the equipment used in its production. Today, however, there are industries who primarily produce this processing equipment. The equipment is quite sophisticated and costly. Automation has become increasingly more important in the manufacture of ICs. To significantly reduce the possibility of error and increase reliability, microprocessors control many of the production steps.

One critical problem in the production of ICs is the yield level. Output is frequently limited to 20% to 30% good individual circuits from each IC wafer. One method used to attempt to increase the yield rate is the *clean room* environment. The rooms are Class 100

FIGURE 10-2
Typical discrete and integrated component packages. (Courtesy Motorola, Inc., Semiconductor Products Sector)

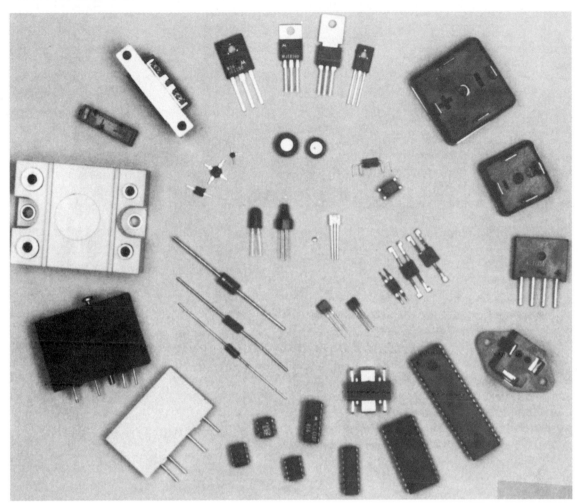

rooms, which are *cleaner* than hospital rooms. Workers in clean rooms are required to wear robes, boots, gloves and hats, as shown in Fig. 10-3, in an attempt to prohibit any foreign substances in the production process. Unfortunately, clean room environments, in addition to larger wafers and automation, have produced an increase in yield of no more than a few percentage points.

The term **monolithic** is derived from Greek words meaning *single* and *stone,* or **single solid** structure. This means that in an IC impurities are diffused into regions of the underlying structure, generally a silicon crystal. Long tubular **ingots** of semiconductor material like silicon are *grown* when seed material is inserted into silicon in a liquid state (under high heat) and then withdrawn. As the seed is withdrawn the liquid around it cools, forming the crystal structure. Ingots can be relatively small, less than an inch as Fig. 10-4 shows, or they can be as large as several inches.

The product finally delivered to the IC fabrication facility is a very thin, circular slice of this crystal called a **wafer,** shown in Fig. 10-5. On each wafer as many as 100 ICs are produced at the same time, as shown in Fig. 10-6. Wafers must be highly polished and extremely flat in order to produce functional circuits or, as they are often called, **chips.** Once the ICs have been produced on the wafer, they must be broken into individual circuits or chips of the whole wafer, as shown in Fig. 10-7. It is very easy to see the pattern in those wafers. Each small square can be broken off to become an individual IC or chip. Impurity or dopant atoms are introduced into the wafer through diffusion or ion implantation. By precisely defining and controlling the regions of diffusion (or ion implantation), their relationship to other regions of the wafer, and the concentration of the dopant atoms in those regions, the manufacturer can construct

FIGURE 10–3
Workers in a clean room environment. (Courtesy Motorola, Inc., Semiconductor Products Sector)

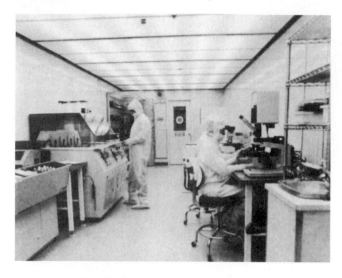

transistors, resistors, and diodes. These are the components of which integrated circuits are made.

Once the many thousands of transistors, resistors, and diodes are built into the wafer, very fine metal lines are deposited on the wafer surface to provide the electrical interconnection between the components. The designer must pay particular attention to the regions where dopant atoms are to be introduced. Precise definition of the location and size of these regions is of critical importance, as well as their relationship to other regions in as many as 12 layers. A region misplaced by as little as .00005 in. can cause a critical failure in the integrated

FIGURE 10–4
Small semiconductor ingots and thin wafer slices, showing size as compared to a quarter. (Courtesy Motorola, Inc., Semiconductor Products Sector)

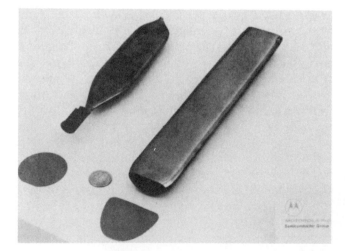

FIGURE 10–5
Large semiconductor ingot and thin wafer slice. (Courtesy Motorola, Inc., Semiconductor Products Sector)

circuit. Let us briefly outline the typical steps in the fabrication process. This will give you a better understanding of the complexities of IC design.

Fabrication Steps

The fabrication of an integrated circuit is an extremely detailed process. The steps in this process are shown in Fig. 10–8. They begin with a wafer that has been meticulously inspected for flatness and surface irregularities.

FIGURE 10–6
Multiple individual integrated circuits on wafers. (Courtesy Motorola, Inc., Semiconductor Products Sector)

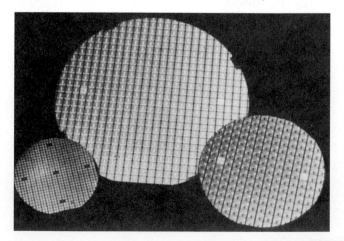

FIGURE 10–7
Individual chips shown on a wafer. (Courtesy Intel
Corporation)

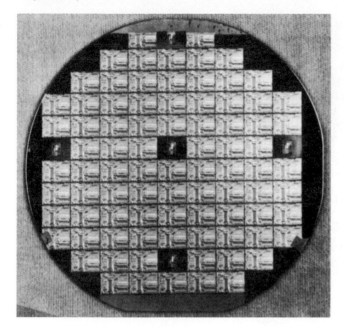

FIGURE 10–7
Individual chips shown on a wafer. (Courtesy Intel
Corporation)

Substrate

The wafer is the **substrate**, or support surface for the
IC. It is essentially a supporting structure that adds
thickness and increased strength to the chip. If it is not
flat or has localized surface irregularities, the chances
of using that wafer to produce working ICs are severely
compromised. A thin layer of glass is *grown* over the
entire surface of the wafer. Next the wafer is coated

with **photoresist,** a substance that is very sensitive to
ultraviolet radiation. When exposed, the photoresist un-
dergoes a change in molecular structure. This change
causes it to react to certain chemical solvents in a manner quite different from the way it reacts if not exposed.

Mask

Next comes a step that uses a **mask** made of glass covered with an opaque substance, such as chrome. The
chrome has been removed from the mask in a pattern
that corresponds to regions on the wafer where dopant
atoms or metal for component interconnections is to be
deposited. Since hundreds of ICs can be fabricated on
a single wafer, the mask will typically contain hundreds
of repetitions of the same circuit.

At one time masks were prepared entirely by hand.
A very large scale stabilene drawing of all layers was
created. This artwork was then transferred to a clear
film coated with a red plastic called **Rubylith**®, shown
in Fig. 10–9. All the sections that needed to be exposed
were very precisely cut from the Rubylith® pattern.
These patterns allowed diffusion precisely in these
areas. When complete, the pattern was photographed
and reduced 500 times.

Today the same stabilene drawing is produced on
a CAD system, shown in Fig. 10–10. All layers of the
mask are digitized directly into the computer memory.
When the mask was completed by hand, each individual
chip had to be redrawn a number of times to form
the entire mask. With the computer memory, only one
circuit needs to be drawn. All layers are stored in the
computer as a cell that can be repeated as many times

FIGURE 10–8
Fabrication sequence for
manufacturing monolithic ICs.
(Courtesy Texas Instruments,
Inc.)

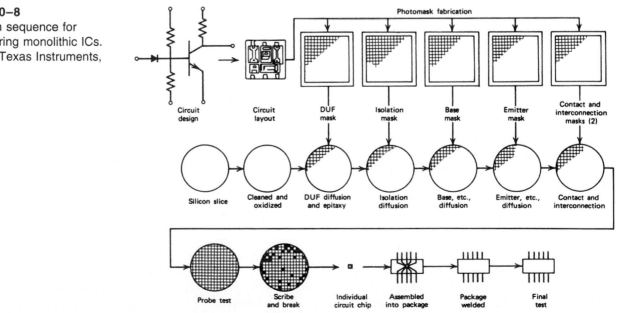

FIGURE 10–9
Rubylith® cutting table used in the preparation of masks for IC manufacture. (Courtesy Motorola, Inc., Semiconductor Products Sector)

as necessary, literally at the push of a button. The master mask (or reticle) can be produced in much less time with much greater reliability, which leads to increased yield. When changes are needed in the circuitry, they can be made to the existing database, which was previously digitized.

Etching

Once completed, the mask is positioned against the coated wafer. The wafer is then dipped into a chemical solvent where the photoresist is **etched** or dissolved in regions that have been exposed (or perhaps left unexposed, depending on the nature of the photoresist and the solvent used). The wafer now has some regions with a photoresist and other regions without a photoresist. In regions without a photoresist, the glass is etched away, and all remaining photoresist is stripped from the wafer.

Diffusion

The wafer is placed in a furnace, shown in Fig. 10–11, that maintains a constant high temperature and that contains a gas with an extremely high concentration of the desired dopant atoms. The very high temperature imparts thermal energy to the dopant atoms and causes them to **diffuse** into the wafer where there is no glass. After the wafer has been through the diffusion process, it is removed from the furnace and the glass remaining on the surface of the wafer is removed.

The wafer is now ready for the next processing steps, which are similar to those just described except that the mask patterns are different. Other processes may also be different: the dopant ionization process, shown in Fig. 10–12; and other processing details, such as temperatures and the oxide coating process in Fig. 10–13. Each processing step is preceded by a masking step, because masks define the regions where the wafer actually experiences the effects of a given processing step. A finished wafer is shown in Fig. 10–14.

In recent years, circuit density has multiplied to nearly a million devices per chip. It is a major challenge to develop a *megatransistor* chip that will be functional the first time around, without exceeding competitive time and cost constraints. Few companies today can afford to debug such a very large scale integrated circuit in production or to redo months of costly product development.

FIGURE 10–10
An intelligent color terminal used in the design stages of multilevel IC manufacture.
(Courtesy Prime-Computervision.)

FIGURE 10–11
Furnace used in the manufacturing process of IC wafers. (Courtesy Motorola, Inc., Semiconductor Products Sector)

FIGURE 10–13
Oxide coating process. (Courtesy Motorola, Inc., Semiconductor Products Sector)

FIGURE 10–12
Ionization chamber used in the diffusion steps of IC manufacture. (Courtesy Motorola, Inc., Semiconductor Products Sector)

FIGURE 10–14
A completed wafer with ICs before separation into individual chips. (Courtesy Motorola, Inc., Semiconductor Products Sector)

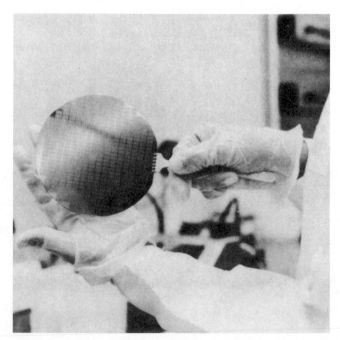

FILM AND HYBRID INTEGRATED CIRCUITS

The monolithic type of integrated circuit is generally preferred to other methods of manufacture. However, **thick-film, hybrid,** and **thin-film** methods are still being used.

Film Circuits

Thick-film and thin-film processes are formed directly on the surface of glass or ceramic, rather than on a wafer of semiconductor material. Components are formed either from evaporation techniques (for thin film) or from silkscreen processes (for thick film). Discrete semiconductor components such as diodes or transistors must be processed in much the same method as integrated circuits are manufactured. The semiconductor components are assembled on the circuit. For NPN transistors the process starts again, as shown in

Fig. 10–15, with the wafer of semiconductor material, in this case a P-type substrate [Fig. 10–15 (a)]. A thin coat of **oxide** is grown on the substrate, which is then coated with photoresist. This is masked and the collector region is etched. To form the collector, phosphorus ions must be **implanted** [Fig. 10–15 (b)] into the substrate. A second oxide coat [Fig. 10–15 (c)] is grown and the collector region is diffused. This second oxide coat is covered with photoresist; then the base region [Fig. 10–15 (d)] is masked and etched. To form the base region, boron ions are implanted [Fig. 10–15 (e)] into the substrate. A third oxide coat [Fig. 10–15(f)] is grown, and the base region is diffused. The third oxide coat is also covered with photoresist so that the emitter region [Fig. 10–15 (g)] can be masked and etched. Phosphorus ions can then be thermally diffused [Fig. 10–15 (h)] to produce the emitter region [Fig. 10–15 (i)]. The phosphorus ions for the collector and boron ions for

FIGURE 10–15
Three separate diffusions used to form an NPN transistor. (Courtesy TRW LSI Products)

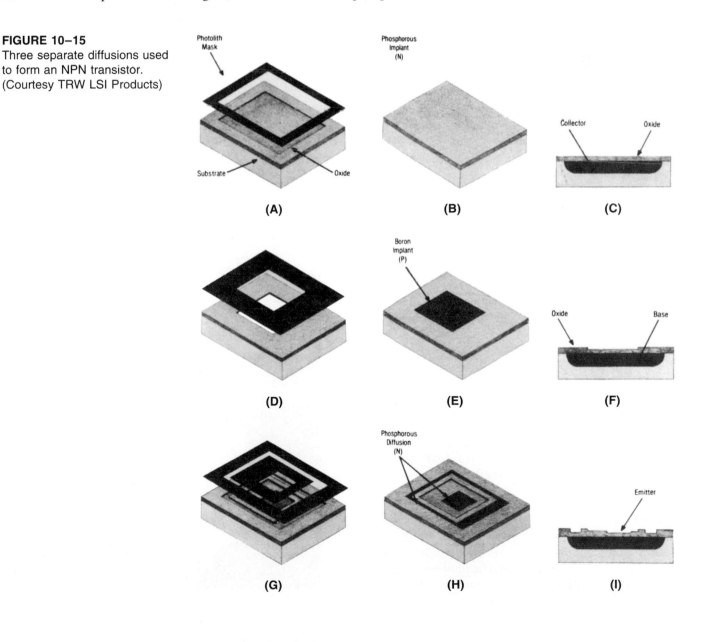

FIGURE 10–16
Process steps to form SLT module. (Courtesy International Business Machine Corporation)

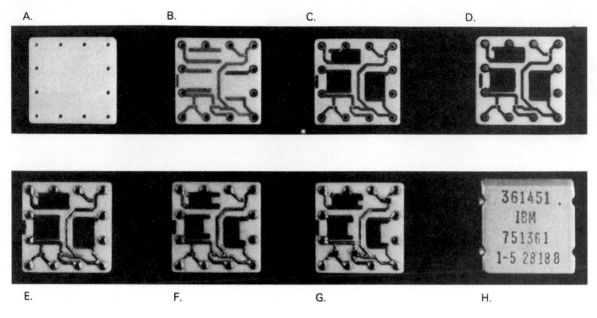

the base are implanted into the substrate to provide better control of the resistivity of these sections of the transistor.

Thick-Film Circuits

Thick-film ICs are used to make capacitors, resistors and conductors and other passive circuit elements. When active devices such as transistors or diodes are assembled to the circuit, it becomes a hybrid circuit. To produce either a thick-film or a hybrid circuit, a number of steps must be followed. IBM used a method of producing modules called **solid logic technology** (SLT). The production steps for an SLT module are shown in Fig. 10–16. Start with a bare substrate or half-inch square ceramic wafer approximately ⅟₁₆ in. thick (Fig. 10–16 (A)]. The holes around the outer edges are for connecting pins added in a later step. The conducting paths or circuit pattern [Fig. 10–16 (B)] is printed on the bare substrate. For the *printing* process, a plastic paste made of conductive or nonconductive material is deposited on the substrate. Paste is deposited by silk-screening techniques in several thin layers to produce the thick film. Resistors are *printed* next [Fig. 10–16 (C)] in a similar process. Connecting pins are inserted [Fig. 10–16 (D)], and the module is dipped into solder [Fig. 10–16 (E)] to provide connections from the circuit pattern to the connecting pins. The resistors are then trimmed to the correct value [Fig. 10–16 (F)] within tolerance ranges with abrasive machines or with highly accurate laser trimmers. The final step is to attach the chips [Fig. 10–16 (G)] or semiconductor material

FIGURE 10–17
Vacuum pencil positioning the transistor chip on the SLT module. (Courtesy International Business Machine Corporation)

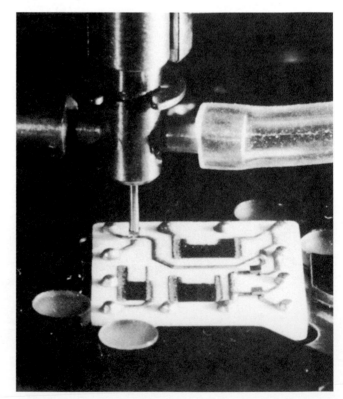

that was produced separately to complete the thick-film or hybrid IC. Fig. 10–17 shows a vacuum pencil positioning the transistor chip on the hybrid module. To ensure that this circuit cannot be physically harmed, it is encased in a number of different forms. One method is to seal the circuit in a metal shell [Fig. 10–16 (H)].

Hybrid Circuits

A hybrid circuit is a combination of film (for passive) components and monolithic (for active) components. Some hybrid circuits are sealed in an epoxy conformal coating as shown in Fig. 10–18. Other hybrid circuits include thick-film resistors and conductors, as well as discrete semiconductor components already encapsulated or packaged, instead of the small wafer chip. Some examples are shown in Fig. 10–19. Occasionally the hybrid circuit created with thick-film components and discrete components is used as a portion of a larger circuit. Figure 10–20 shows the coated circuit mounted on a printed circuit board with a transformer and additional discrete components.

Thin-Film Circuits

Thin-film circuits are produced in much the same way as thick-film circuits. The primary difference is in the method of *printing* the circuit components on the substrate. Figure 10–21 shows a thin-film IC.

For thin-film circuits, the conductive or nonconductive materials are *sputtered* or vacuum-deposited on the

FIGURE 10–18
Hybrid ICs sealed in epoxy-type conformal coating. (Courtesy Centralab, Inc., A North American Philips Company)

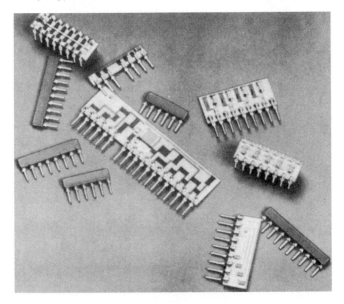

FIGURE 10–19
Hybrid circuits containing discrete components and ICs. (Courtesy Centralab, Inc., A North American Philips Company)

substrate. They can be deposited as either a thin film or as a pattern, which is later trimmed to the required values with abrasive trimming. In general, thin-film methods provide better stability and higher resolution than thick-film methods. However, they are more expensive, and large resistance values cannot be produced.

STANDARD PACKAGE CONFIGURATIONS

Three standard packages are used generally for integrated circuits today. The TO package is a cylindrical metal can popular for discrete transistors. When this type of package is used for integrated circuits, it is usually larger than the individual transistor and contains more leads accessible for circuit connections, as shown in Fig. 10–22. *Flat packs,* shown in Fig. 10–23, and *dual-in-line packages* (or DIP), Fig. 10–24, are very similar in body construction. The basic difference is in the method of lead termination. Both these packages are usually plastic and have a long, rectangular shape. Both will contain as few as 6 and as many as 64 or more leads.

Figure 10–24 shows a pictorial representation of a 64-pin DIP. The size of the IC itself has not increased significantly; however, the overall package size has increased for access to additional leads and increased heat

FIGURE 10–20
Hybrid circuit covered with a conformal coating mounted as a device on a printed circuit board. (Courtesy Centralab, Inc., A North American Philips Company)

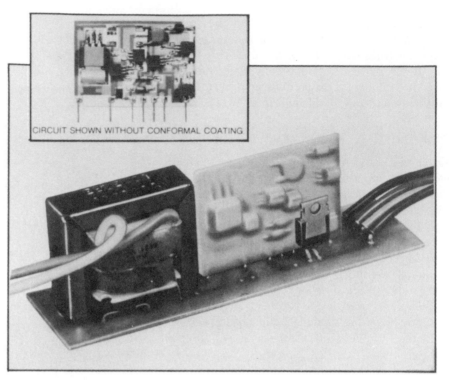

CIRCUIT SHOWN WITHOUT CONFORMAL COATING

FIGURE 10–22
Typical TO packages for ICs. (Courtesy Motorola, Inc., Semiconductor Products Sector)

FIGURE 10–21
A thin-film IC after destructive physical analysis. (Photo is 50X magnification.) (Courtesy Fairchild Industries, Inc.)

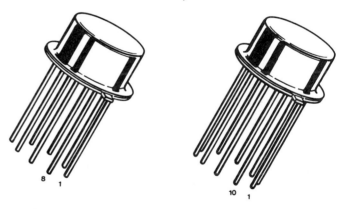

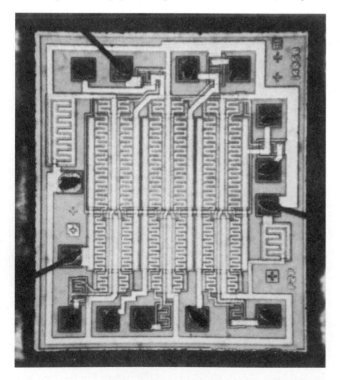

FIGURE 10–23
Typical IC flat-pack packages. (Courtesy Motorola, Inc., Semiconductor Products Sector).

FIGURE 10–24
Typical dual-in-line (DIP) packages. (Courtesy Motorola, Inc., Semiconductor Products Sector)

music, games, calculators, and even the time. It would be a little awkward to say the least if this wristwatch had to accommodate a 3-in.-long integrated circuit. To overcome this inconvenience, a square flat pack with a number of pins on all sides was developed, shown in Fig. 10–25.

Component sizes become extremely important for the design of printed circuit boards. Since boards contain preprinted circuit connections, it is important for the circuit parts to fit into the areas reserved. Some components have flexible leads and therefore some flexibility in final mounting. On the other hand, a DIP package containing 40 pins must be inserted straight into 40 holes, correctly spaced and with circuit connections made to the correct leads. There are aids made specifically for the placement of integrated circuits in the production of printed circuit boards. These aids are discussed fully in Chapter 18.

dissipation requirements. Since the functions of integrated circuits have increased, typically the number of leads necessary for connection have also increased. TO packages do not offer the size needed for these increased numbers of leads.

A DIP integrated circuit containing 64 leads or pins is approximately 3 in. long. This package provides reliable service, but uses a considerable amount of valuable space. An integrated circuit used in a digital wristwatch might contain 64 pins if that watch contained some of the many functions found today, including timer, alarm,

THE ROLE OF CAD/CAM

One may think of IC design as having three distinct procedures: **circuit design, process design,** and **mask design.** Each of these procedures is unique to the integrated circuit industry, vitally important to that industry, and greatly enhanced by the use of CAD. No one of these procedures is independent of the others, since each, to some degree, depends on data derived from the other two.

FIGURE 10–25
Multiple-row square pin connections, allowing more pins in less space. (Courtesy Motorola, Inc., Semiconductor Products Sector)

Circuit Design

The first procedure to examine is circuit design. The circuit designer or design team prepares schematics and logic diagrams to describe how the circuit function is to be implemented. At one time this was strictly a manual job entailing extensive use of drafters. Changes were often so extensive that many drawings could not be modified but had to be entirely redrawn. This is no longer the case.

Today's circuit design team can use CAD systems that partly automate and totally streamline the circuit design process. Fig. 10–26 shows a designer at a CAD terminal using a design system that enables him to create all masks simultaneously. The designer sits at an interactive graphics terminal that provides a very comfortable interface between the designer and a computer. The designer uses the terminal with its CRT display as a *scratchpad* to manipulate circuit elements, develop the design, and view the results. When satisfied with the design, the designer stores the data in the computer, where it may be readily retrieved for examination or modification, or it may be processed for further analysis.

The completed design stored in the computer is called a *circuit design database.* This database serves as the input to the design analysis programs, which ensure that a design both fulfills the desired function and performs according to specifications, such as speed and power consumption. The design analysis programs are of two major types: circuit and logic simulation. Circuit simulation verifies the performance specifications of the circuit. Logic simulation verifies that the logic of the design is what the designer intended. If either of these analysis programs yields undesirable results, the circuit design database must be modified.

When modification is necessary, the CAD tools provided to a designer at the interactive graphics terminal simplify the job. Powerful programs allow for quick retrieval of the circuit design database and easy manipulation of the design elements. High-speed plotters provide drawings in minutes. The masks are plotted and the circuit layout can be checked for flaws, as shown in Fig. 10–27.

Process Design

The processes used in manufacturing ICs are exceedingly complex. Circuits will not function properly—if at all—unless certain dopant atoms are located precisely at strategic points in the crystalline structure of the wafer. To achieve the necessary precision, the designer must take into account the temperature of the wafer, the concentration of dopant atoms, and the effect of previous and subsequent processing steps. With the help of a computer, the designer can extract values for all the parameters applicable to these processes.

The process design engineer of today can use very efficient *process simulation* programs that provide user interaction as well as graphical and statistical output. The designer can *see* what is taking place, both in the wafer and on the wafer. Process design must rely heavily on CAD. Calculations involved in studying the interactions of atoms during some period of time would be very difficult to undertake manually. The only other technique that could be used is *trial-and-error experimentation.* This is usually much too expensive and time consuming to be cost effective, although it is how many of the advances in processing technology have been achieved. The processes that will be used in the future will undoubtedly reflect a strong CAD influence.

Mask Design

The mask designer must preserve the electrical and logical functions of the circuit, observe all process-related constraints, and minimize the area the circuit will occupy. As many as 12 masks may be used to define a circuit for current process technologies. These masks must register (align) exactly with one another as shown in Fig. 10–28. They must also contain all the regions of dopant atoms, precisely sized and located. Finally, they must be designed and manufactured with a minimum of delay.

Like the circuit designer, the mask designer has CAD systems that partly automate and totally streamline the mask design process. The mask designer also has an interactive graphics terminal and can create designs from the signals of an electronic pen. System commands permit the display, editing, retrieval, and storage of data. A well-defined *stack* of regions (that is, all regions on the various masks required to define a transistor) may be stored as a single component and then used over and over again throughout a circuit design. The designer does not have to redefine the same stack every time a certain type of transistor is wanted. Furthermore, the stack will not be missing any elements.

The mask design database contains a massive amount of data. The ability of a CAD system to manipulate these data in a very short time allows a designer to manipulate the design for the most effective placement of all regions. Because of the sophisticated design rules for checking the software, a CAD system provides the tools to ensure that the desired circuit does not violate any process-imposed constraints. It always guards against error.

CAD systems should enable a designer to check the mask design database for any topographical errors. A wide variety of topographical errors may be intro-

FIGURE 10-26
CAD workstation for the layout of very large scale integrated circuits and printed circuit boards. (Courtesy Prime-Computervision)

FIGURE 10–27
A designer examining a Versatec plot of a circuit produced using the Symbolic Interactive Design System program. (Courtesy American Microsystems, Inc.)

duced, for example, regions that have improper internal dimensions or that are located too close to adjacent regions. Such errors on the large, dense ICs of today are impossible to discover by eye. Even one error on a circuit may render it (and others like it on the wafer) useless.

When CAD tools are used, an error can be found well before the design is committed to fabrication, thus minimizing correction costs. The designer can rectify errors at an interactive CAD terminal, using powerful

software aids to manipulate data and reposition offending regions in the mask design. After rectification, the designer checks the entire database. This ensures that existing errors have been corrected and no new ones introduced.

The final step in IC manufacture is to translate the design from data in the CAD computer to integrated circuits on the shelf. This involves all three components of design just discussed. The circuit design component provides input to the program that tests the circuits at the completion of fabrication, as shown in Fig. 10–29. The process design furnishes the specifications to be

FIGURE 10–28
A computer used to align the mask for each level in the IC manufacturing process. (Courtesy Motorola, Inc., Semiconductor Products Sector)

FIGURE 10–29
Computers used to test individual ICs during manufacturing process. (Courtesy Motorola, Inc., Semiconductor Products Sector)

used at all steps of fabrication. The mask designs furnish data for mask fabrication. These data will actually drive the machines (pattern generators) that *paint* the regions in the mask design onto a mask.

APPLICATIONS

Probably one of the most significant applications developed from large-scale integration is the **microprocessor,** sometimes referred to as a computer on a chip. This reference is not altogether correct, but it is the center of a digital computer, the *CPU*. The CPU is the central processing unit, which can be considered the *traffic cop* of a computer. The microprocessor can be as extensive as one with an arithmetic logic unit (ALU), three registers, an accumulator, control logic, address capabilities, and minimal memory. With added memory and input and output capabilities, the microprocessor becomes the CPU of a microcomputer.

Large-scale integration has been the dominant factor in the growth and major developments in electronics. Without large-scale integration we would not have been able to package so much into so little space. Developers began to create the small or personal computer concept with the idea that many small, inexpensive computers could perform just as efficiently as one major, more expensive computer. Small computers started with industrial applications and have expanded to include all commercial products available. The computer market continues to grow daily, and more features are continually being added. Functions that used to be performed by hardware are now being done with software. Microprocessors are being used to control just about everything. If the function needs to be changed, it is a simple matter of changing a program instead of changing a wiring configuration.

The applications discussed here have been primarily restricted to the digital electronics area, where the components used in the integrated circuits are called gates. (Gates and logic functions are discussed in Chapter 15.)

In addition to the advances made in digital applications, there are also advances in analog applications. An integrated circuit is a number of components contained in the same operational device. In analog electronics, perhaps the best-known example of integrated circuits is the **operational amplifier**. One of the most popular *op amps,* the 741, is shown in Fig. 10–30. The 741 contains 20 transistors, 12 resistors, and a capacitor. Although the op amp is a rather complex circuit in itself, it is usually treated as a single device capable of performing many different functions with very specific input and output characteristics. It is, however, a good example of integrated circuitry. The op amp has become

FIGURE 10–30

Actual op-amp schematic contained in integrated circuit. (Courtesy Ken Floering)

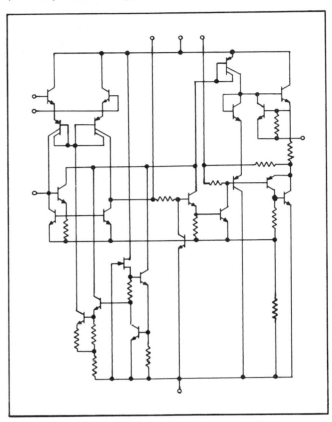

the building block of all kinds of applications and is the most widely used amplifier.

The term operational amplifier normally refers to a high-gain, directly coupled voltage amplifier with a differential input. The operational amplifier with the proper connections is used as a comparator, an amplifier with no phase inversion, an amplifier with phase inversion (what was positive becomes negative) and a voltage follower, which does not provide amplifying capabilities but is used to match two different circuit outputs and inputs.

A number of different packages for op amps are shown in Fig. 10–31: the round metal can or TO style, a MINI-DIP or 8-pin DIP, and 14-pin DIPs (containing a single op amp or two op amps in the same package with each accessed individually). The primary advantages of two op amps in the same package are the single voltage source required and the lower power consumption.

Another application of an integrated circuit is AMI's S3610 speech synthesizer, shown in Fig. 10–32. This chip generates high-quality male or female speech from linear coded data. It contains enough memory to produce 17 seconds of speech with a maximum vocabu-

FIGURE 10–31
Different configurations used to
package operational amplifiers.

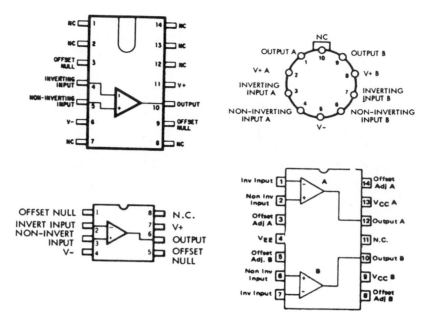

lary of 32 words. It is ideal for applications with automobiles, appliances, industrial controllers, toys, and games.

In addition to the bipolar methods of manufacture discussed, there are many other methods of manufacture and types of circuits. One such technique is referred to as a **complementary metal oxide semiconductor** (CMOS). The primary difference between CMOS and bipolar ICs is in the power consumption of the device itself. Figure 10–33 shows a TLC271 op-amp chip manufactured with Texas Instrument's silicon gate LinCMOS® technology.

FIGURE 10–32
A partitioned photomicrograph of AMI's S3610 speech synthesizer with an on-chip 20K
speech data ROM. (Courtesy American Microsystems, Inc.)

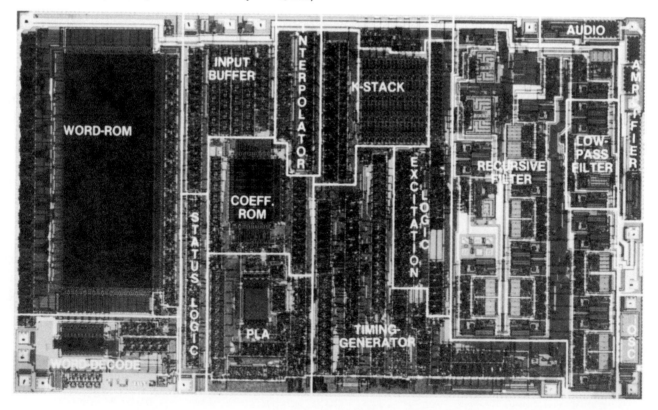

FIGURE 10–33
TLC271 op amp fabricated with silicon gate LinCMOS® technology. (Courtesy Texas Instruments, Inc.)

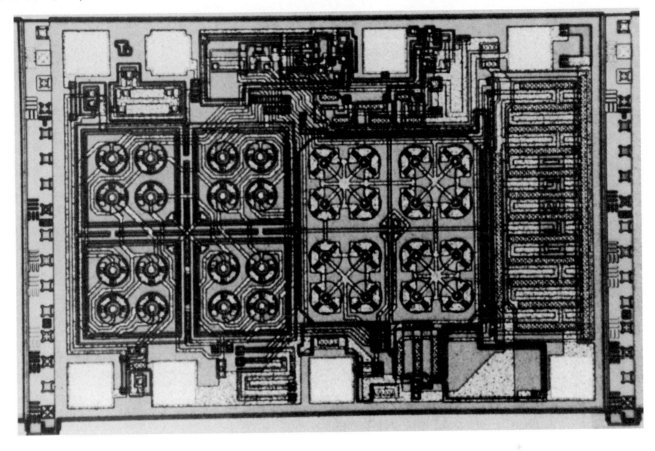

REVIEW QUESTIONS

10–1 Why is the study of microcircuits important for the drafter and designer?

10–2 Discuss reasons why IC fabrication is normally limited to large production quantities.

10–3 What is being done to increase the yield of ICs from the wafer slices?

10–4 Describe the process steps required in the fabrication of monolithic ICs.

10–5 What are the three basic package configurations used for ICs?

10–6 What is a hybrid IC and how does it differ from a monolithic IC?

10–7 What types of drafting problems can you anticipate in the preparation of masks for IC fabrication steps?

10–8 Why are the masks drawn to as much as 500× scales?

10–9 What do you consider the primary advantage of using CAD systems in the IC fabrication process steps?

10–10 List the basic production steps in the process of thick-film hybrid circuits.

11

SCHEMATIC DIAGRAMS

INTRODUCTION

It is important to communicate ideas effectively through **schematic diagrams.** A schematic diagram shows the function of a circuit with little emphasis on its physical characteristics. This diagram has been called the most important drawing for use in equipment production, testing, and analysis. The schematic drawing describes all circuit functions and values of components, using standard symbols in most cases. (See Chapter 8 for a review of standard symbols.)

Although the schematic diagram shows the interconnections between electrical components, it should not be confused with the wiring diagram. In the wiring diagram (fully discussed in Chapter 13), components are arranged showing their physical relationship rather than their functional relationship. Often, in a wiring diagram, the parts are drawn as pictorials rather than as graphical symbols. The wiring diagram is used, as its name implies, in the actual wiring stage. It can be very difficult to analyze because the physical relationships are so complex.

The primary goal of the schematic diagram is to show the functional relationships between parts, allowing the circuit to be easily analyzed. Generally, the schematic, like block and logic diagrams, is laid out to

be read from left to right with the input coming into the upper-left and the output going out the lower-right sides of the drawing, although a "top to bottom" flow can be used when a circuit does not flow naturally from left to right.

The very first conception of an idea drawn by an engineer is usually in the form of a schematic diagram, although it is often a very rough sketch. Once the engineer has completed the ideas for the design, it becomes the drafter's responsibility to prepare a finished schematic diagram. A drafter must be aware of all accepted drafting standards used in industry and in individual company practice. Frequently, the engineer's sketch will not be prepared to meet these standards. Schematic diagrams, like many other electronic diagrams, should be drawn to standards set up by such groups as ANSI, JIC, and IEEE.

The schematic diagram is the master drawing and is used in the preparation of wiring diagrams, printed circuit board layouts, mechanical layouts, parts lists, and other associated drawings. Basically, three elements are included in all schematic diagrams: graphic symbols representing the components; reference designations detailing component values, tolerances, or other ratings; and the interconnections between all components. These elements make up the original layout of the draw-

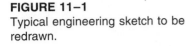

FIGURE 11–1
Typical engineering sketch to be redrawn.

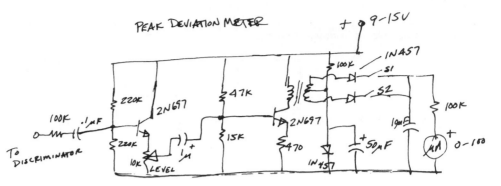

ing and are clarified in the final drawing. The final drawing must be checked to ensure that these elements are clear and accurate.

LAYOUT

Most of the time it is impossible to simply *redraw* a sketch provided by an engineer. It would be a simple task if all that was needed was to take the layout and put it in standard symbols with straight vertical and horizontal lines. Unfortunately, many sketches have to be completely rearranged to provide a finished diagram that is not only functionally correct but also easily interpreted and, just as important, esthetically pleasing. Drawings that are not pleasing to the eye will be difficult to work with. Circuit components must be well spaced around primary or active components. Lines should be horizontal or vertical, with crossover lines minimized.

The drafter may have to make one or more trial sketches to be sure that this balance is achieved. The use of grid paper is strongly recommended for trial sketches and for the final layout. Whenever major rearrangement is necessary, keep one thing in mind: any

changes or alterations that are done must not affect the technical correctness of the drawing. You must be careful not to alter the electrical function of the circuit.

Figure 11–1 shows a freehand sketch that could have come from an engineer's pad. The sketch was prepared with little regard for drafting practice because the engineer was concerned primarily with the electrical function of the circuit. The correct drafting techniques were left for the drafter.

If you received this sketch, many things would be required for you to prepare the finished drawing. No reference designations are used for any components. The transistor symbols are incomplete. It might be a good idea to prepare another sketch to see where some crossovers could be eliminated, how vertical and horizontal alignment could be achieved, and how better spacing could be maintained. Figure 11–2 shows how some of these problems were corrected, and Fig. 11–3 shows a final drawing.

A drawing can be laid out in a number of different ways. For a simple circuit, visually divide the number of components in half. Components associated with input functions should be arranged on the left half of the page, and those associated with the output should be on the

FIGURE 11–2
Trial sketch and possible alignment from original sketch.

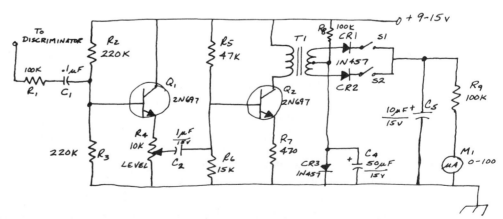

FIGURE 11–3
Final drawing completed from trial sketch.

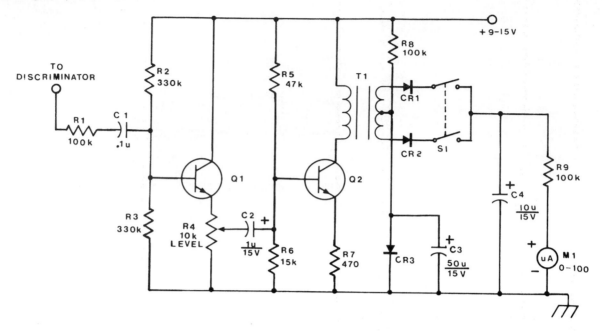

right half. Input signals flow through the circuit in various stages. Components that complete a stage should be aligned vertically. Horizontal alignment is used for the relationship of one stage to another.

A more complicated circuit is normally made up of sections or functional units. These units are arranged in an order, allowing for left-to-right signal flow. Each small group of components representing a functional unit follows the same spacing and alignment procedures used for a simple circuit. Keeping this in mind will help prevent the more complex circuits from becoming overwhelming tasks. A complex circuit is nothing more than the sum of the individual parts.

Do's and Don'ts

A schematic diagram must always show clearly and precisely the circuit design and function, as well as the relationships among the various components. The following drafting practices will help you prepare final drawings:

1. Use medium-weight lines for all symbols and interconnection lines.
2. Ensure adequate space between parallel lines for notes, reference designations, and so on.
3. If two wires are to be connected, use a heavy dot at the junction to indicate a connection. Using the dot for a connection will leave no question or cause for confusion.

4. Avoid crossovers whenever possible. If it is not possible to avoid a crossover, indicate clearly that the line is a crossover and not a connection. The older method of *jumping over* or forming a half-circle is considered obsolete and time consuming. The method of *dot/no dot,* as shown in Fig. 11–4, will help eliminate confusion. (For a detailed discussion of the handling of connections and crossovers, see Chapter 12.)
5. Avoid diagonal lines and curves. Besides representing the electrical function of the circuit, the schematic represents the capabilities of the drafter. Long

FIGURE 11–4
(A) Connections; (B) crossovers.

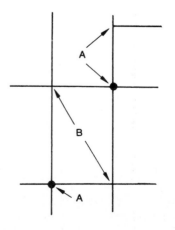

diagonal lines or curves upset the balance and symmetry of the drawing and detract from the overall pleasing quality.

6. Avoid long lines; use interrupted lines instead (covered later in this chapter). The use of long lines requiring a number of changes of direction can be very confusing and require much time to draw and interpret.

7. Ensure a balance of lines and blank spaces in the final drawing. If large open or blank areas can be found on the drawing or if portions of the circuit or groups of components appear to be very crowded, the best layout probably has not yet been found.

8. Show signal flow from left to right and top to bottom. For more complex circuits you may have to use more than one row or column. The general flow in this case should start in the upper left-hand corner and finish in the lower right-hand corner of the diagram.

9. Try to ensure that all technical data are correct in order to present the most accurate and highest-quality drawing possible, even if you are not responsible for checking the accuracy of the schematic.

10. Check and recheck the drawing for completeness and accuracy at all steps of the preparation. This includes all intermediate drawings that are completed to make the final layout easier.

The Final Drawing

After all spacing and layout have been determined, the final drawing can be prepared. All graphical symbols should be constructed as described in Chapter 8. The use of templates like the one shown in Fig. 11–5 is highly recommended to ensure that all symbols are uniform and consistent. This will also save a lot of construction

FIGURE 11–5
Template used to draw graphic symbols on schematic diagrams. (Courtesy Berol RapiDesign)

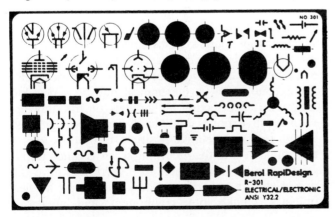

time. If dry transfer symbols are available, as illustrated in Fig. 11–6, they can also be used to save time and produce a uniform symbol. When dry transfer symbols are used, the interconnecting lines and additional symbols should be done in ink to avoid the drastic difference in appearance that would occur if pencil were used to draw the connecting lines.

Since the majority of the lines are horizontal and vertical, a grid underlay is very helpful in the preparation of the final drawing. Grid paper is used to draw the original or additional trial sketches. The finished drawing, however, is prepared on either plain vellum or *fade-out grid* paper. Using a grid underlay gives the advantage of having a grid without having the final drawing on a grid paper. If the same size grid is used for the sketch and the underlay, it is easy to transfer the spacing and placement of components and connecting lines from the trial sketch to the finished drawing.

Interrupted Lines

When the simple circuit was drawn in Fig. 11–7, it contained a single voltage value and was depicted as the symbol for a battery or voltage source. Frequently, in the more complex circuits, a number of different voltages are supplied at different points in the circuit. All these voltages might be supplied by the same functional unit, called a *power supply circuit*. The use of **interrupted lines** helps reduce any confusion that might arise if long connection lines were used to make all voltage connections to the power supply circuit.

A typical power supply is shown in Fig. 11–8. This supply is drawn separate from the rest of the circuit, and the termination points are given labels. Labeled points, wherever they appear in the circuit, are in reality connected to the power supply. The power supply circuit is generally found at the lower left of the drawing because it must be kept away from the signal path part of the circuit.

When long connection lines extended from the power supply throughout the drawing would cause clutter and confusion, interrupted lines should be used. Make certain to indicate clearly the destination of all interrupted lines by using letters, numbers, or unique symbols. The unique symbols in Fig. 11–9 include ground, plus or minus signs indicating voltage polarities, terminal bars, jacks, and plugs. If a number of lines will go to the same place, they are frequently bracketed and noted as in Fig. 11–10.

The labels used for interrupted lines may be enclosed in triangles, circles, or other shapes. This helps call attention to the fact that an additional connection is made and may help to locate that connection elsewhere on the drawing.

FIGURE 11–6
Dry transfer aids: stick-ons or PUPPETS™. (Provided courtesy of Bishop Graphics, Inc.)

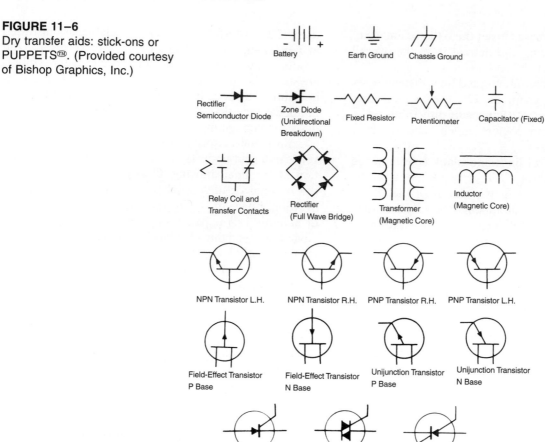

Switches

Some switches are quite easy to represent in a circuit. The single-pole, single-throw (also called SPST), single-pole, double-throw (SPDT), and the double-pole, double-throw (DPDT) are a few of these simple switches. They are used frequently in control circuits and are discussed in Chapter 14. Other types of switches can be very difficult to represent on a drawing. Rotary switches can be especially confusing. A rotary switch rotates a movable contact connected to a center shaft, completing a circuit by touching the stationary contacts around an insulating material. A pictorial representation of a rotary switch is shown in Fig. 11–11. The rotary switch is found frequently in test equipment and is used to select meter function, scale, or range.

To add to the confusion, rotary switches can contain multiple layers, sometimes called **decks** or **wafers.** In general, when multiple layers are used, the movable contacts or wiper parts are all connected to the common center shaft, and all the contacts are changed simultaneously. Since the different layers have no electrical relationship, but merely a mechanical one, it is possible to represent each section individually on the drawing. Because each section is individually represented, it must maintain the same relative position. In other words, if the terminal next to the mounting screw is called terminal 1 for the first layer, then all other layers have the terminal closest to the same mounting screw labeled terminal 1. Unlike some components and integrated

FIGURE 11–7
Simple circuit showing all connections and standard graphic symbols for all components and voltages.

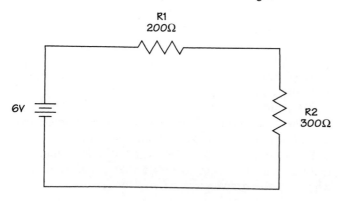

FIGURE 11–8
Typical power supply.

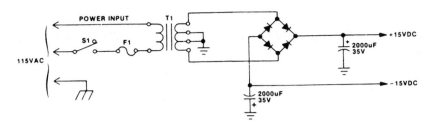

FIGURE 11–9
Unique symbols used for interrupted lines: (A) earth ground; (B) chassis ground; (C) plug; (D) jack; (E) plus or minus for voltages; (F) terminal connectors.

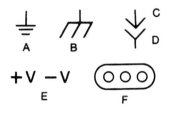

FIGURE 11–10
Grouped lines going to a common destination.

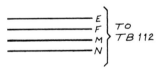

FIGURE 11–11
Pictorial presentation of two-layer rotary switch. (Reprinted by permission of HEATH COMPANY)

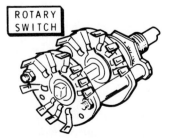

circuits, rotary switches often do not come with numbers on the contacts. They can be numbered arbitrarily. Therefore, whenever a switch is drawn, the relative position to mounting screws and other parts must be made clear. Additionally, some terminals may not be used or even physically there. In this case a space is left where the terminal would be. To avoid confusion, all terminals should be numbered in sequence, whether used or not. The switch in Fig. 11–12 shows space for lug 8 even though it is not in use and in this case is not even shown in the diagram. The direction of rotation (clockwise or counterclockwise) should also be included and noted on the switch detail, as well as how the switch is to be viewed—from front or rear side.

Switch Tables

To help users understand the changes that occur when the switch is rotated, data are frequently presented in a table near the switch. As illustrated in Fig. 11–13, the table indicates which position the switch is in, what function is being done, and what connection is being made. Normally, the switch is drawn in the *de-energized*, or OFF, position. In the case of rotary switches, the *No. 1* position is frequently shown.

FIGURE 11–12
Numbered switch lugs, whether used or not. (Reprinted by permission of HEATH COMPANY)

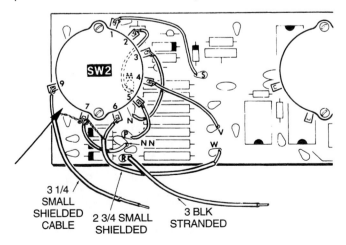

FIGURE 11–13

Switch table showing relationship between functions and positions.

S2B		
POS	FUNCTION	TERM
1	OFF	1-2, 6-8
2	TEST	1-3, 6-9
3	OPERATE	1-4, 6-10

Switch Layout

The decision to show the switch as separate layers or as an integrated version can only be made after a trial layout is prepared. If the switch is shown all in one area and the functions that it controls are scattered throughout the circuit, then long connection lines result. Even though interrupted lines help reduce the clutter and confusion, they are still more confusing than the short connection lines. Figure 11–14 shows the best way to illustrate scattered functions, which is to split up the layers and locate them in the portion of the circuit that they control.

Templates are available to help in the layout and correct spacing of the switch contacts. A typical switch template is shown in Fig. 11–15.

Whenever a mechanical connection occurs between moving parts, dashed lines are used to show the mechanical linkage. For multiple-layer rotary switches, how-

FIGURE 11–14

Switches shown in layers for representing connections to each wafer or deck.

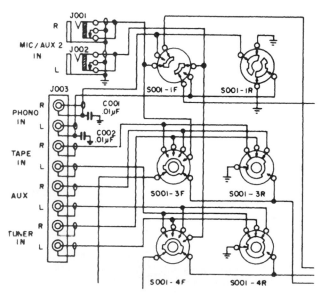

FIGURE 11–15

Template used in drawing of switches for schematic diagrams. (Courtesy Berol RapiDesign)

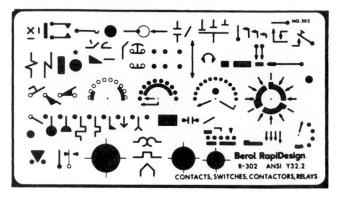

ever, this mechanical linkage is hardly ever shown, because it is understood that all layers rotate with a common shaft. Every layer has a label identifying what view it has (front or rear) and what its relationship is to the other decks.

Figure 11–16 shows the schematic of a VOM or multitest meter. The individual layers of the switch have been separated and are located in the portions of the circuits where their electrical connections are being made. They are all identified as parts of SW1 and include additional letters A, B, and C for different wafers or layers. In addition, each wafer is identified as a front (F) or rear (R) view. Notice that each section of the switch is shown with the same orientation, and the mounting screws are identified to ensure proper location of terminals. Instead of a switch table, inset diagrams and notes are included on the drawing showing rotation, position, common connections, and identification of wafer locations on the actual switch.

Figures 11–17 and 11–18 show additional representations of switch connections. The switch in Fig. 11–17 is similar to the one in Fig. 11–16. The major difference is that this meter includes five wafers, showing both front and rear views of each wafer, in the rotary switch. Inset diagrams identify the knob position, deck locations, and the relationship of position No. 1 to a *strut* or mounting screw. Figure 11–18 displays the switches in a straight-line method even though the switch is a rotating switch. The circuit for this meter (Fig. 11–18) is not terribly complex, and this layout is easier to construct.

Scale and Grouping

Schematic diagrams are not drawn to any particular scale. The size of the final drawing is determined by

FIGURE 11–16
Schematic of meter with switch layers located in the functional areas. (Courtesy Triplett Corp.)

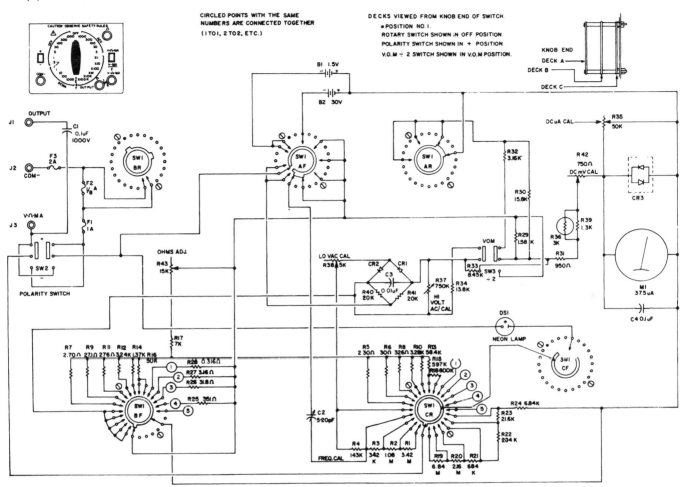

the number of components, connections, and notes that must appear on the drawing. The size of the symbols used to represent particular components does not reflect the physical size of the component itself. Therefore, a larger diagram does not necessarily reflect the need for a larger physical circuit, only a larger number of components.

For a drafter, the preparation of the schematic diagram is the opportunity to fit together all previously learned skills. Drawing a schematic includes good drafting skills, lineweights, lettering, and symbols, as well as use of reference designations.

The schematic diagram layout is very similar to the block diagram layout. The circuit can usually be divided into functional units or blocks. These blocks are divided into areas of the page. Figure 11–19 shows a schematic with each of the functional units identified with a heavy outline surrounding it. For manufacturing or production purposes, these identifying blocks can add clutter and confusion. For the preparation of a printed circuit board (PCB) or multiple PCBs, the identification of the functional units can be an aid. (For a more complete presentation of PCB design and layout refer to Chapter 18.) The identification of functional units is a definite aid in the analysis of the total circuit operations.

Space Estimations

Starting with primary components such as transistors, you can make space estimations for the final drawing. After you are satisfied that the general placement has been achieved, you can determine final placement. Typically, the transistor is enclosed in a circle $5/8$ in. in diameter when the grid size used is eight squares per inch. Using this size as a reference, you can determine spacing for all components, lettering, and notes.

FIGURE 11–17
Meter with five switch decks. (Courtesy Triplett Corp.)

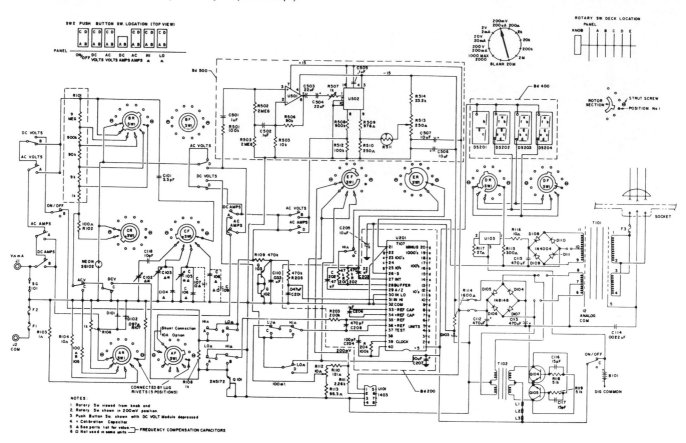

Figure 11–20 shows good spacing for notes, reference designations, and other marks around the components. Lettering should be no less than ⅛ in. when transistor enclosures are ⅝ in. if the grid used is eight squares to the inch. Sometimes a smaller-sized grid is used for less complex circuits. A common size of smaller grid is ten squares to the inch. When this size of grid is used, lettering can be as small as .100 in. When a reduction of 2 to 1 is desired, .100 in. is the minimum lettering size that should be used.

In general, the overall size of the drawing is not limited. Using the larger squares (eight squares to the inch) will present a drawing that can be reduced and yet will not be too small to read. When using the ⅛-in. squares, you can make resistors as large as 1 in., or eight squares, whereas capacitors will probably be four to six squares, as shown in Fig. 11–21. A space of at least five squares should be left between components for reference designations, as shown in Fig. 11–22. If your trial layout makes the lettering appear crowded or confusing, try another layout with better spacing before you finish the drawing.

Diagonals

In most cases, avoid using diagonals. It is impossible, however, to eliminate them totally, and in some applications the use of diagonals is appropriate. For example, a circuit called a **bridge circuit** is normally constructed with diagonals so that it looks like a diamond. This configuration promotes quick recognition for analyzing the circuit function.

Figure 11–23 shows a typical *full-wave rectifier bridge*. The corners are constructed at 90°, and the square that is formed is rotated to appear as a diamond shape. Since this particular type of circuit is a popular configuration, the shape is as important as the components within the drawing.

There may be other times when diagonals or crossovers simply cannot be avoided. When diagonals must be used, an angle of between 45° and 90° and consistency are most important. If you must use a crossover, be sure that the viewer can quickly recognize it as a crossover and not a connection. The use of saddles or lines that appear to *jump* over other lines is not recommended

FIGURE 11-18
Rotary switch represented with straight lines. (Used with permission from Radio Shack, a division of Tandy Corporation, Fort Worth, TX 76102)

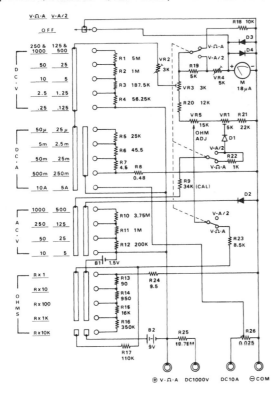

practice. The dot method discussed earlier should be used.

Inset Diagrams

Keep in mind that one of the more important qualities of a schematic is clarity. When inset diagrams are to be included, these small illustrations will be set apart from the main part of the diagram. They must be placed so they do not confuse the main portion of the drawing. Frequently, the inset drawing may be a pictorial representation of the component or switch assembly, as shown in Fig. 11-24. For assemblies of multiple-layer switches, this pictorial can be quite helpful for later preparations of wiring diagrams or assembly diagrams. On some schematics the lead connections for transistors or arrangements of pin connections are included as inset diagrams. Since the schematic diagram functions as a master drawing in the preparation of all other drawings, the lead identification will be useful later. When the inset diagram is used for lead identification, a simple outline of the bottom of the component or terminal strip is usually adequate, as illustrated in Fig. 11-25.

Alignment and Placement

The placement or orientation of the graphic symbols on the schematic has no relationship to their actual physical location or arrangement within the system. In terms of electrical function, there is no correct right side or left side of any symbol. Standard practices have been developed and should be followed to make it easier to interpret the diagram. For example, by convention the ground symbol is normally pointed down and is located near the bottom of the drawing or at least near the bottom of a functional unit. If a number of transistors are used in a circuit, they should all be arranged in a consistent manner. In addition to having similar orientations, components should be aligned horizontally and vertically. The flow paths or interconnecting lines should be as short as possible, showing the most direct routes without diagonals or curves. Figure 11-26 shows a portion of a larger circuit and maintains good horizontal and vertical alignment.

Spacing dimensions will undoubtedly vary with individual company practice. If no specific rules are listed in drafting room manuals, the following specifications can be used. Components should be at least .25 in. from corners or crossovers, as shown in Fig. 11-27 (A). There should be a .5-in. minimum between two components located on the same horizontal or vertical line [Fig. 11-27 (B)]. These are only suggested minimums, and more space may be required depending on the placement of reference designations or other notes and identifiers.

The most important characteristics of good spacing and placement are consistency and clarity. In addition to being functionally or electrically correct, a good schematic diagram should be esthetically pleasing to the eye.

The size of the drawing paper that is required is determined by the number of components that can be found in the circuit. Table 11-1 shows the size of drawing paper that should be used according to usable space when the average size of the graphic symbols is 2 in.[2] The usable space on a drafting sheet is the amount of drawing area that is left after borders, title blocks, and notes are placed on the sheet.

When space is required for revisions and parts lists, only about half of the number of symbols can be placed on each sheet size. Table 11-1 can be used only for estimating the size of paper required for a drawing. For complex circuits that include many interconnecting lines, fewer symbols may be counted in the same amount of space. When diagrams are subdivided into functional units, more space is required, so there are fewer symbols per paper size. When components appear too close together or lettering may be too crowded, a larger size of paper should be used. If reductions are standard practice

FIGURE 11–19
Layout and placement according to functional grouping. (Courtesy of Motorola, Inc., Semiconductor Products Sector)

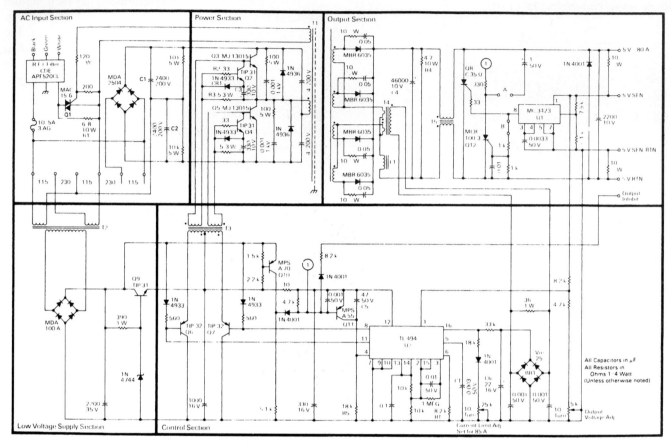

FIGURE 11–20
Adequate space around components for lettering.

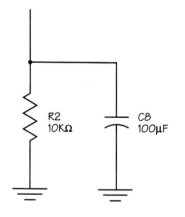

FIGURE 11–21
Common number of squares for components on an 8 × 8 grid.

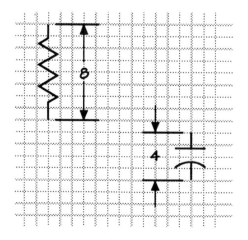

FIGURE 11–22
Spacing of at least five squares between components on
8 × 8 grid.

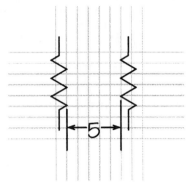

FIGURE 11–23
Diode bridge rectifier in diamond shape.

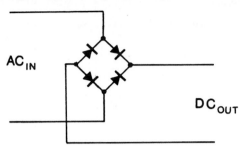

FIGURE 11–24
Inset diagram showing lug
numbers and switch orientation.

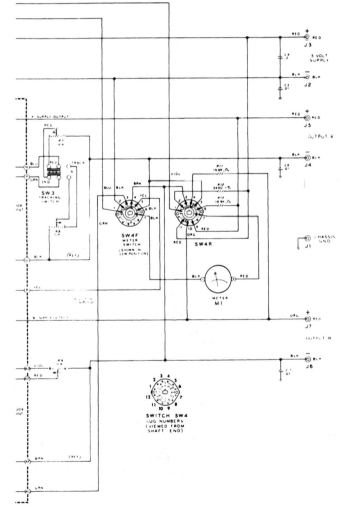

FIGURE 11–25
Inset diagram showing component bottom for lead identification.

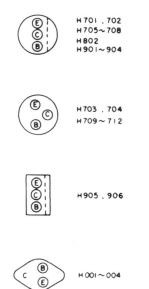

FIGURE 11–26
Sample of good horizontal and vertical alignment.
(Courtesy Joint Industrial Council)

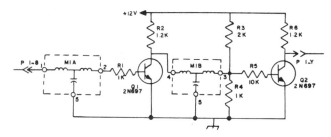

FIGURE 11–27
Space between components and corners or between components: (A) minimum .25 in. (6.35 mm); (B) minimum .50 in. (13 mm).

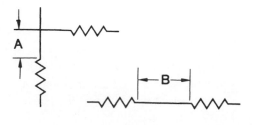

TABLE 11–1
Estimated Number of Components That Can Be Placed on Standard-Sized Papers

Standard Size, in.	Usable Area, in.²	Number of Symbols
B 11 × 17	140	60–70
C 17 × 22	300	130–150
D 22 × 34	650	300–325
E 34 × 44	1340	625–670
F 28 × 40	1008	480–500
H 28 × 50	1260	600–630
J 34 × 50	1530	725–765
K 40 × 50	1800	850–900

or company policy, then additional space allowances must be made so that the drawing will remain clear and legible when reduced.

CONSTRUCTION OF THE DRAWING

For more complex circuits, first divide the circuit into functional units. Lay out each of these units in a trial sketch to help determine the overall required space. Check the trial sketch for any alterations that can be made to eliminate crossovers or crowding. If necessary, make another trial sketch. To prepare the sketch, start with symbols that will be aligned horizontally in the finished drawing. These will normally be the active components, such as transistors. If special attention is given to the placement of the transistors in the sketch, like the placement in Fig. 11–28, additional components can be easily connected to the transistors, as shown in Fig. 11–29. Then add the interconnection lines, as shown in Fig. 11–30.

When going from the trial layout to the finished drawing, you can make slight alterations without making an additional trial sketch. Make light pencil construction lines on the paper to establish the general overall shape and placement of the finished drawing. These lines will give you a last chance to make small changes in placement or spacing.

Skills are acquired after much practice. Layout spacing, placement, and alignment, as any other skill, will become easier as you gain more experience with complex circuits.

The Finished Drawing

If necessary for photoreproduction, the finished drawing may be done in ink, with special appliqués, or with other

FIGURE 11-28
Placement of active components first.

FIGURE 11-29
Additional components in relation to transistors.

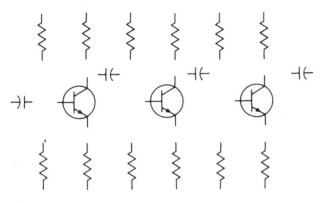

FIGURE 11-30
Connecting lines with interrupted lines identified.

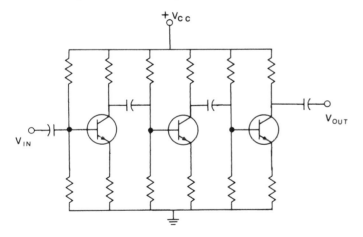

dry transfer methods. The appliqués should be placed on the drawing first. After they are applied, if the final drawing appears cluttered or crowded, these symbols can be removed with very little difficulty. After all the transfer symbols have been placed on the paper, additional symbols should be added. These symbols should be drawn with a standard template and ink. When all components have been drawn or placed on the finished drawing, the interconnection lines should be added. The

final step is to add reference designations and other notes.

You may follow these steps when preparing a schematic diagram:

1. Obtain the original sketch and check for accuracy and all data.
2. Make a sketch on grid paper with trial spacing and alignment to estimate paper size and space requirements.
3. Make any alterations or changes necessary on the trial sketch. This might include making additional trial sketches.
4. When satisfied with the circuit layout, place the finished drawing paper over a grid underlay of the same grid size as the sketch. Lightly place construction lines on the drawing using the grid spacing as the placement guide.
5. Begin the finished diagram by placing dry transfer symbols on the paper or by drawing the individual component symbols with the aid of templates.
6. Finish the schematic diagram by adding all flow lines, terminal markings, labels for interrupted lines, notes or reference designators, and other lettering.
7. Check the finished drawing for technical accuracy and clarity before submitting it as complete.

DRAWING VERIFICATION

Checking a schematic diagram is every bit as important as the initial preparation of the drawing. A print is made of the final drawing so that it may be marked on as the checking process is done. The best method of checking is a line-by-line, symbol-by-symbol check. The final drawing should be checked against the original sketch. Any questionable details must be made clear. Obviously, it is better to eliminate any misunderstandings long before the finished diagram is prepared. You can use a colored pencil or marker to indicate the portions that are being checked off, tracing over a copy of the finished drawing and the original sketch as you check it. If you find any errors, clearly mark them in red so that they may be fixed on the final drawing. In addition to checking interconnecting lines and connection points, check the following items:

1. Component symbols
2. Reference designations, including values, tolerances, and polarity markings, if needed (Do reference designations clearly identify the correct symbol?)
3. Identification of numbered terminals or other leads
4. Title blocks, including spelling and punctuation

5. All additional notes added to the drawing, including correct spelling, technical accuracy, and placement away from crowded areas

Once the checking is completed, the marked-up checking print will be returned to the drafter so that corrections can be made to the finished drawing. After corrections have been made, another checking print should be made, and the process should be repeated. When the checking process produces no errors, the finishcd drawing is initialed and dated. Then the required number of prints is made. The finished drawing will be the master for associated drawings.

ELECTRONIC SYMBOLS IN CAD

In traditional drafting methods, each **symbol,** or graphical representation of standard parts, must be drawn individually using a template or other construction methods. The use of CAD allows the drafter the opportunity to draw the symbol once and store it in memory for recall later. Once this image is created, it can be scaled, rotated, or mirrored (Fig. 11–31) and used in any circuit drawing. When attributes and properties are stored with the symbol, parts lists can be automatically generated and other calculations can be performed. The collection of stored symbols is called the *symbol library* (Fig. 11–32).

Creating and Defining Symbols

Drawing a symbol begins with the formation of the connect nodes (base points), which must be created as permanent points, or handles that are used to place and position the symbols. The symbol origin is specified and used as the reference point for future placement of the symbol. The **attributes** are text notes used for additional information about the symbol. These notes can be collected into a disk file and processed by applications (software programs) designed to automatically generate parts lists. The attributes can be a very complete description of the symbol, including component, style, designation, capacity, code, cost, and company or vendor.

Symbols are stored as an individual block or subpart of a larger graphic. New symbols can be created by nesting together individual symbols. Once a symbol is created, it may be recalled, changed, and restored. This modified symbol may be saved as a new symbol similar to an existing one or may simply replace the old symbol. If a symbol in the symbol library is changed, the drawing may be updated with the new version by simply executing a command. There is no longer any need for many tedious hours of redrawing entire schematics to update one symbol or figure. When the update command is used, only the active drawing is changed. All other drawings remain unaltered.

CAD-GENERATED DIAGRAMS

In general, the creation of an electronic diagram (Fig. 11–33) on a CAD system is very similar to preparing the diagram manually on a drafting board. The primary difference, when using a CAD system, is the capability to recall instantly all the needed symbols from the library and place them on the drawing at the desired location and in proper position. A schematic diagram provides a means of capturing, concisely and accurately, all the data required to describe a particular circuit. As such, it forms an essential basis for any electronic design project and is used throughout the design process from development, through printed circuit board layout, to inclusion in service handbooks.

Diagram symbols can be edited by using the standard editing commands of the CAD system. Symbols can be moved or reoriented if necessary. Symbol attribute text can also be added, deleted, or modified. Interconnects can be added, rerouted, or deleted as required. When two interconnecting lines cross but are not intended to be electrically connected, the old method of adding a semicircular bridge is no longer used. Instead, the dot/no dot method is preferred. An interconnect may also be broken, if desired, without affecting its electrical continuity.

Diagram Data Retrieval

CAD systems provide the capability to store and later extract information such as part number, material, vendor, and cost. These data will always accompany the part when it is used on a diagram and can easily be tallied; a report can then be generated by the system. The use of CAD and symbol libraries eliminates the need to constantly re-create the same information.

If attributes are attached to the standard symbols, data can be extracted and used to control NC/CHC equipment, including drilling machines, board profilers, automatic component-insertion machines, and automatic test equipment (ATE). CAD systems today allow for the complete design cycle with drawings, simulation, design documentation, parts listings, and engineers' reports.

FIGURE 11–31
Drawing of a circuit on a CAD
system. (Courtesy California
Computer Products, Inc.)

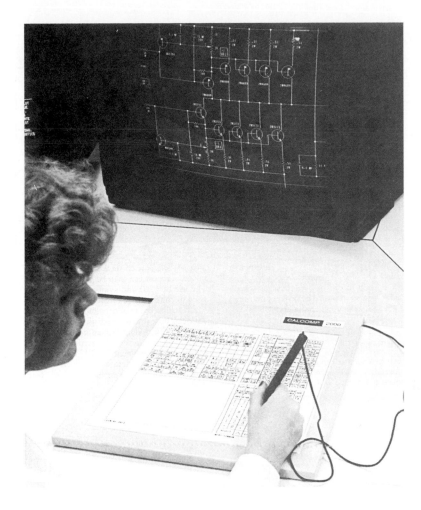

FIGURE 11–32
CAD library. (Courtesy Prime-
Computervision)

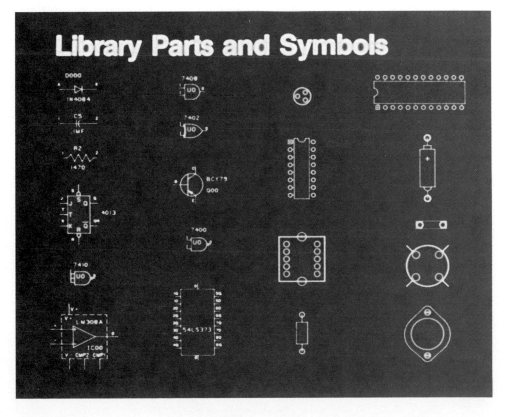

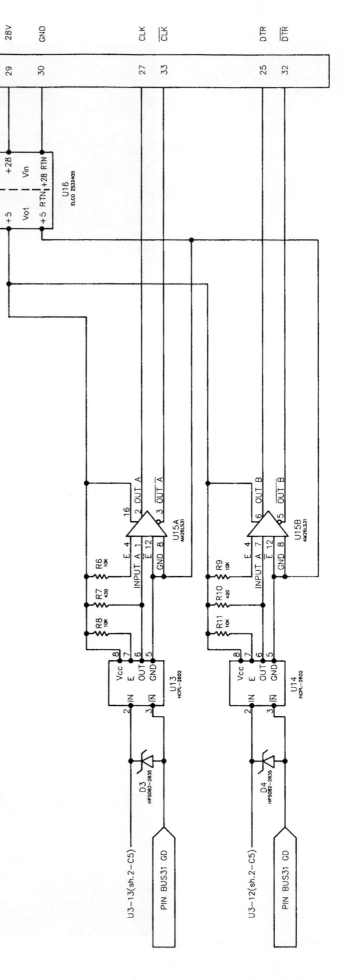

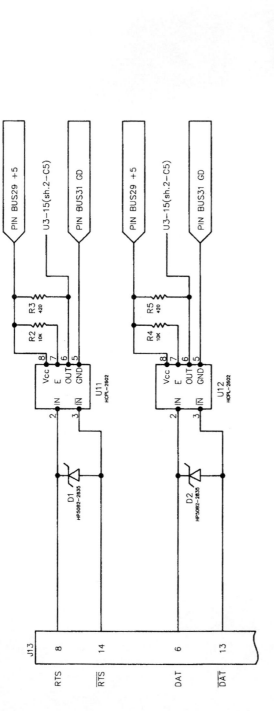

FIGURE 11-33
Electronic diagram.

REVIEW QUESTIONS

11–1 What is the primary purpose of the schematic diagram?

11–2 Discuss the primary differences between a schematic diagram, wiring diagram, and a pictorial diagram. For what would each be used?

11–3 What is a switch table and for what is it used?

11–4 Why are reference designations required on schematic diagrams?

11–5 What size drawing paper would be required for a circuit with 100 different components?

11–6 Why are inset diagrams used on the schematic?

11–7 What kind of information is usually included in the notes found on schematics?

11–8 When are diagonals used and why are they recommended in some cases and avoided in others?

11–9 When would interrupted lines be used?

11–10 Who has primary responsibility for the *correctness* of the final drawing?

11–11 Why is it important to try to have input on the left and output on the right?

11–12 When does the schematic have sections that are enclosed in boxes and what purpose do the boxes serve?

11–13 Why can multiple-layer switches be drawn in pieces in various parts of the circuit?

11–14 Why are power supplies drawn separate from the main circuit?

11–15 How many trial sketches are necessary before the final drawing can be done?

11–16 Why are trial sketches made?

11–17 Explain the dot/no dot method of showing connections and crossovers.

11–18 What is the physical relationship between the size of the drawing and the size of the actual circuit?

11–19 What items should be checked before the drawing can be considered finished?

11–20 Why are graphic symbols used on schematics instead of pictorials?

11–21 In drawing the schematic, what parts should be done first and why?

11–22 What could be considered the primary advantage of using a CAD system to generate the final schematic?

11–23 Does the use of a CAD system eliminate the need for trial sketches and other preliminary placement considerations?

PROBLEMS

When the following problems ask for a trial sketch or preliminary placement sketch, complete the exercise on grid paper. (Ten squares to the inch is recommended.) For drawings that are to be completed as the finished drawing, complete them on plain drawing vellum. (Use of a grid underlay is recommended.) Complete all graphic symbols or components using templates. Lettering should be done by hand; practice good lettering techniques, which were covered in earlier chapters. For adequate practice some exercises should be completed in ink, with Leroy lettering and Leroy symbol templates, as assigned by your instructor.

11–1 Draw a trial sketch of the circuit shown in the figure for problem 11–1.

PROBLEM 11–1
Rough sketch.

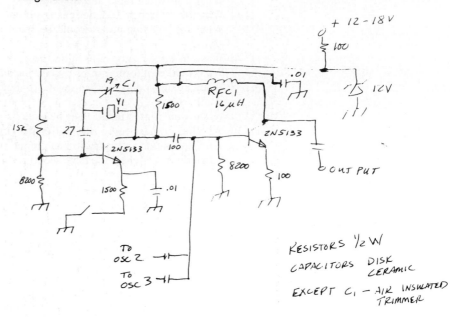

11–2 Replace the boxes shown in the figure for problem 11–2 with the following components:
1. DC battery, 9 V
2. 100-Ω resistor
3. 200-Ω resistor
4. NPN transistor
5. DC battery, 15 V

PROBLEM 11–2
Simple transistor amplifier circuit.

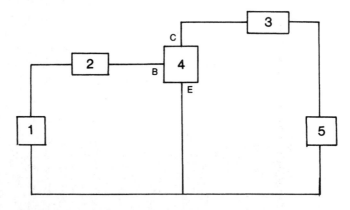

11–3 Given the original sketch in the figure for problem 11–3, draw a new trial layout.

PROBLEM 11–3
Rough sketch.

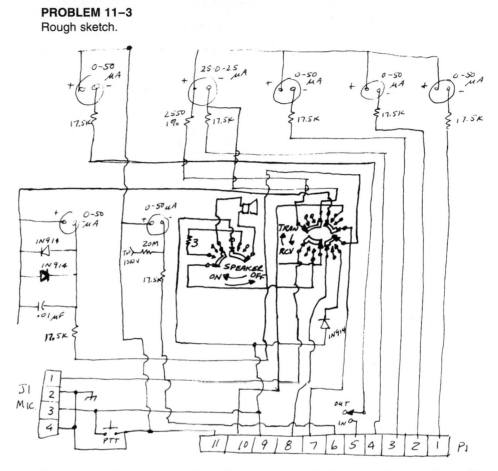

11–4 For the circuit drawn in problem 11–3, prepare a finished drawing complete with parts list.

11–5 Make a check print of the finished drawing completed in problem 11–4 and demonstrate how to perform the final check using the line-by-line, component-by-component method.

11–6 On a B size sheet, make a schematic diagram of the circuit shown in the accompanying figure.

PROBLEM 11–6
Power supply.

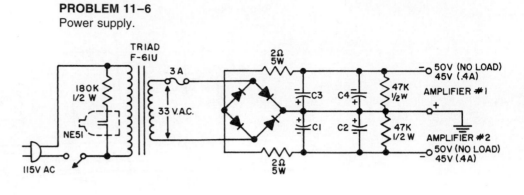

```
C1 — 1500 μf, 50 VOLTS
C2 — 1500 μf, 50 VOLTS
C3 — 1500 μf, 50 VOLTS
C4 — 1500 μf, 50 VOLTS
SILICON BRIDGE – FOUR – 1N1115
```

11–7 On a B size sheet, make a schematic diagram of the circuit shown in the accompanying figure.

PROBLEM 11–7
Discrete flip-flop.

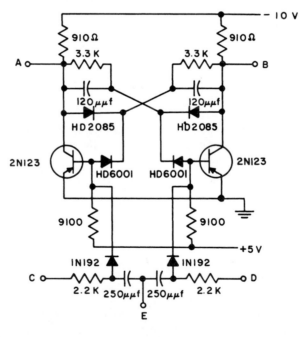

11–8 On a C size sheet, make a schematic diagram of the circuit shown in the accompanying figure.

PROBLEM 11–8
AC power input/regulated power supply output.

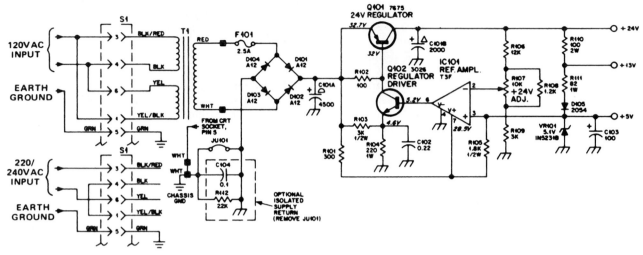

AC Power Input/Regulated Power Supply Output

11–9 On a C size sheet, make a schematic diagram of the circuit shown in the accompanying figure.

PROBLEM 11–9
Electronic circuit.

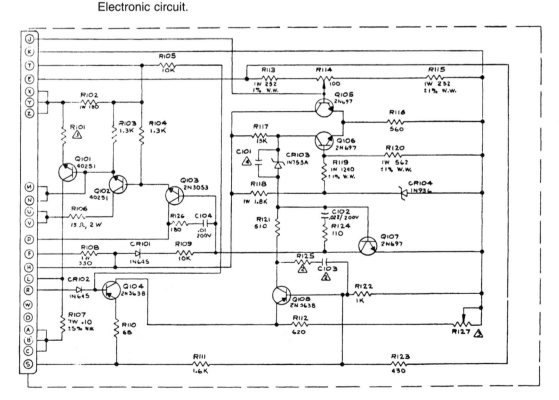

11–10 For the two-stage RC coupled amplifier shown in the figure for problem 11–10, the following components are missing:

1. 50-μF ceramic	9. 24K
2. 68K	10. 6.2K
3. 22K	11. 8.2K
4. 5.6K	12. 10-μF ceramic
5. 12K	13. 50-μF electrolytic
6. 50-μF electrolytic	14. plus 20 V
7. 10-μF ceramic	15. ground
8. 75K	

Both transistors are NPN 2N3904.

Redraw the schematic diagram, including the components listed. Place reference designations beside every device and include a note that says all resistors are tolerance of 10% and ½ W.

PROBLEM 11–10
Two-stage transistor amplifier.

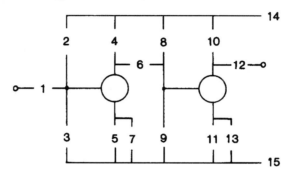

11–11 Redraw the circuit in problem 11–11 and modify it to include interrupted lines for the ground, positive voltage, and offset horizontal alignment for the transistors.

11–12 (Advanced problem) From the printed circuit board in the figure for problem 11–12, develop the schematic diagram for the power supply. Use standard graphic symbols and note all reference designations and component identifiers.

PROBLEM 11–12
Power supply printed circuit board.

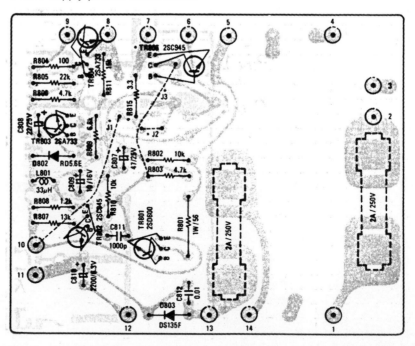

11–13 Redraw the schematic diagram shown in Fig. 11–33.

11–14 and 11–15 Create a schematic from the PCB. Lay out the drawings on an A size sheet.

PROBLEM 11–14
Printed circuit board.

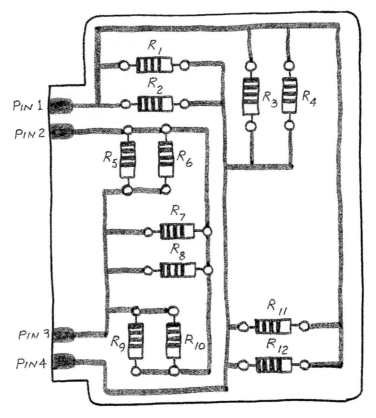

PROBLEM 11–15
Printed circuit board.

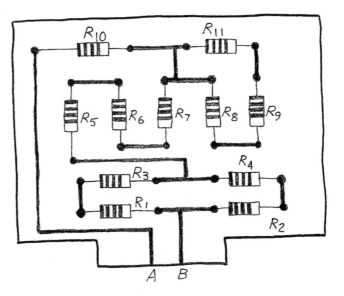

11–16 through 11–27 Using the correct symbols and designation, draw the given problems.

PROBLEM 11–16
Schematic diagram.

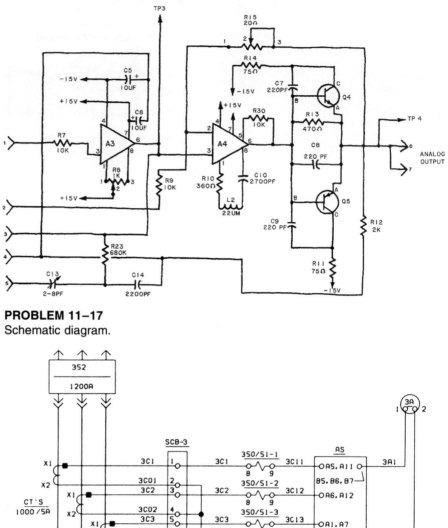

PROBLEM 11–17
Schematic diagram.

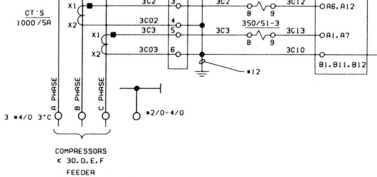

PROBLEM 11–18
Schematic diagram.

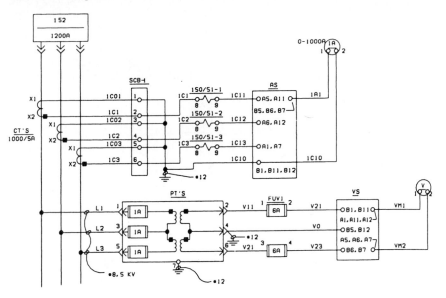

PROBLEM 11–19
Left- and right-channel separation circuits.

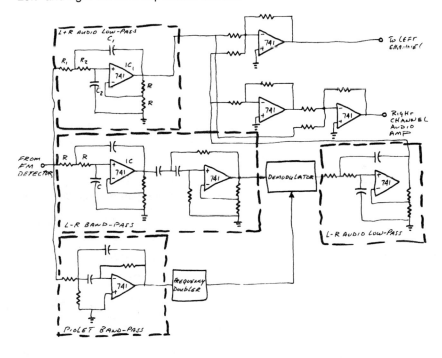

PROBLEM 11–20
Schematic diagram for a function generator.

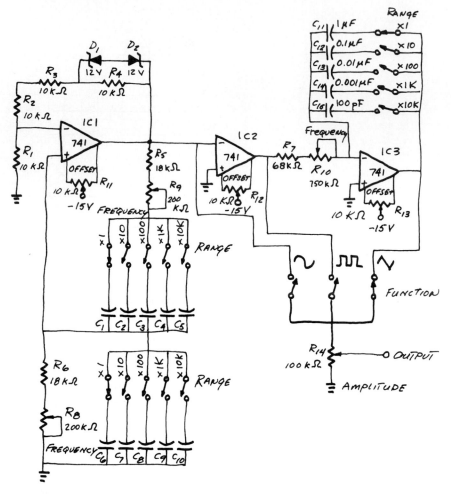

PROBLEM 11–21
Schematic diagram of a bandstop filter.

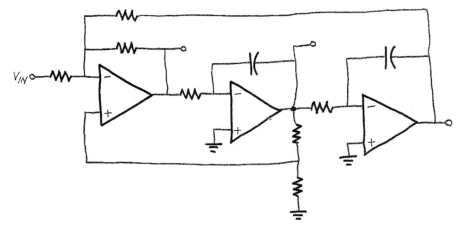

PROBLEM 11–22
Bandstop filter circuit.

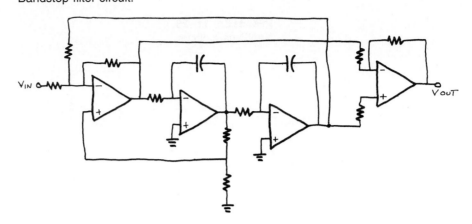

PROBLEM 11–23
Schematic diagram.

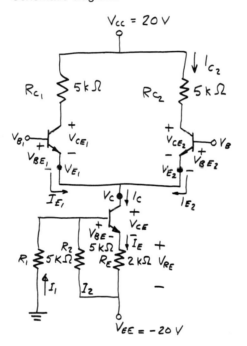

PROBLEM 11–24
Schematic diagram for a temperature-sensitive heater.

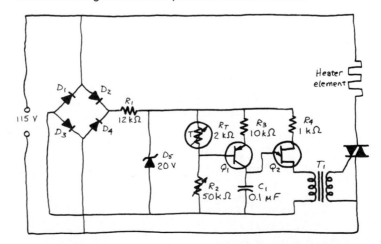

PROBLEM 11–25
NAND gate circuit.

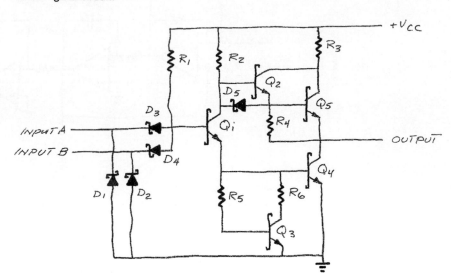

PROBLEM 11–26
Schematic diagram.

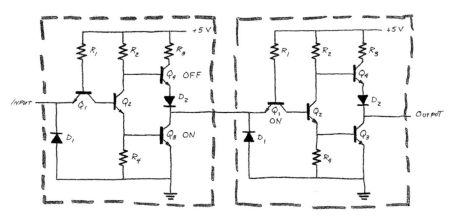

PROBLEM 11–27
ECL circuit.

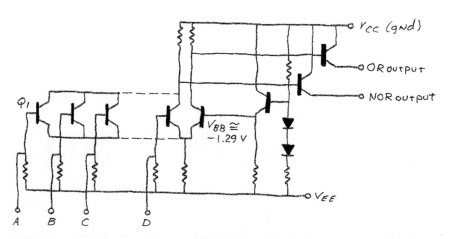

12

BLOCK DIAGRAMS

INTRODUCTION

A **block diagram** is a simplified diagram of a system, program, or process. Labeled geometric figures and interconnecting lines represent the functional relationships and flow of information between units or sections. Block diagrams are the easiest form of diagram to draw and read, which makes them ideal for conveying information to people with limited technical knowledge. Technical and sales manuals for electronic and computer equipment make frequent use of block diagrams like the one in Fig. 12–1.

Block diagrams are used in all areas of business, education, and industry for representing both technical and nontechnical information. Social, psychological, economic, historical, engineering, and technical data can easily be presented in the form of a block diagram.

The geometric shapes used on block diagrams are normally kept as simple as possible, although this form can be more artistically rendered than diagrams with strict layout, construction, and symbology specifications. Block diagrams are sometimes drawn in pictorial form. An example of a **pictorial block diagram** is provided in Fig. 12–2, a serial I/O interface to a common backplane.

The electronic and computer industries use block diagrams to convey a variety of technical information.

In electronics, block diagrams are used where it is unimportant to show specific circuit connections or the exact electronic component. Each individual block represents a stage, substage, or **unit** of the total system as an independent function. Individual blocks may represent aspects of a process, as in **flow diagrams.** Block diagrams are the first step in the design process of a system. The engineer or designer uses the block diagram to describe a particular system with all its related units. The function and purpose of a system can be more readily conveyed and understood when a block or flow diagram is used. The actual composition of each stage or unit of a system, including its connections, circuitry, and components, is not shown on the block diagram.

Blocks may represent individual **modules,** units, or removable portions of a total system made up of multiple sections. A block may represent something large, such as a complete **chassis** unit, or something small, for instance, a group of integrated circuits.

BLOCK DIAGRAMS

A block diagram—as its name implies—is a diagram composed of block shapes. Although a majority of block diagrams are composed of square and rectangular

FIGURE 12–1
Block diagrams of memory
structures. (Courtesy Intel
Corporation)

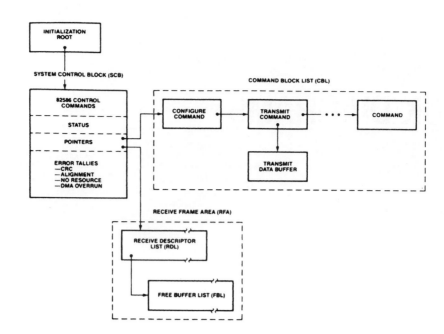

shapes, many companies use other simple geometric forms, such as ellipses, circles, and triangles. Specialized electronic symbols for antenna, ground, switch, and so on, may also be incorporated into a typical electronic block diagram. Fig. 12–3 is a typical example of a block diagram. Note that block diagrams show only the functional relationship between specific stages of a total process.

The term **functional sequence** or functional relationship is extremely important in understanding the construction and interpretation of block diagrams. The functional sequence of an operating device determines the *flow* between units of a system. The term flow signifies that something is moving from one place to another.

Therefore, a flow must start at one point and end at another; in other words, it must have an **input** and an **output.** The input for most block (and other) diagrams is placed on the left of the drawing and the output is on the right, as in a book the words are read from left to right. Whenever possible, electronic diagrams are constructed with this left-to-right sequence. Note that most *computer flow diagrams* are normally drawn with a top-to-bottom flow direction. Some electronic block diagrams use a top left-side input and a bottom right-side output, although the left-to-right sequence is preferred.

The single-sideband transmitter shown in Fig. 12–4 provides a simple example of the left-to-right flow se-

FIGURE 12–2
Pictorial block diagram of serial backplane connection. (Courtesy Intel Corporation)

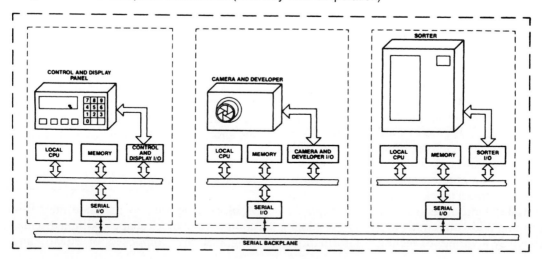

FIGURE 12–3
Block diagram.

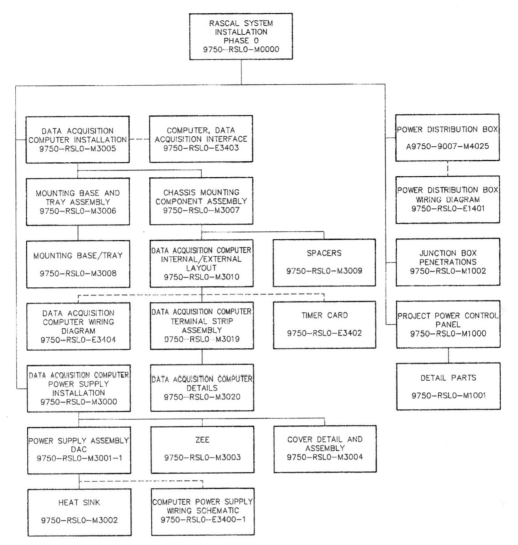

FIGURE 12–4
Single-sideband transmitter.

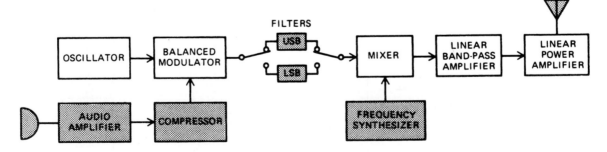

quence. The flow starts on the left with the audio amplifier and ends on the right side at the linear power amplifier. It is then coupled through the antenna system. Note that the flow from left to right is common, but some systems will *feed back,* or send a signal back through the system. This feedback may move from right to left or in a loop fashion.

Since the purpose of a block diagram is to depict the functional flow between related units of a system, each successive unit depends on the preceding unit. If one unit precedes another, it must be assumed these units are also related in time. One comes before another; therefore, the preceding unit is *before* the next unit in terms of real time. For all practical purposes the exact measurement of the time between units is zero, except when you attempt to understand the operation of a particular piece of equipment. Here, time and functional sequence are critical.

It must also be understood that if something flows from one point to another, through a sequence of steps, units, modules, or operations, that same movement precludes backflow, or the reversal of a signal (except when a system is designed to have a specific feedback capability).

Block diagrams of a system can be established on several levels. They can be used to describe the func-

tional relationship between major units, for instance, the sequence of operation for a stereo system: turntable to tuner–amplifier to speaker. A block diagram can also describe the relationship of specific subunits of just the amplifier. Component-to-component block diagrams are seldom used.

In summary, a block diagram can be more accurately defined as a series of geometric shapes depicting the functional relationship between individual units or subdivisions of that system. The functional sequence of a block diagram is based on a flow from one unit to another. This flow may be thought of as time related; one unit functions before each successive unit even if the exact time cannot be realistically measured. Block diagrams can be used to represent technical data, computer program or equipment flow processes, or electronic device or equipment operation sequence.

Flow Diagrams

Flow diagrams are used primarily for the presentation of computer program data, as shown in Fig. 12–5, or operation instructions and processes, as in Fig. 12–6. In both of these figures the flow direction is from the top down. Note the use of a variety of geometric shapes. Flowcharts such as these are laid out along a main cen-

FIGURE 12–5
Computer flow diagram.
(Reprinted by permission of
HEATH COMPANY)

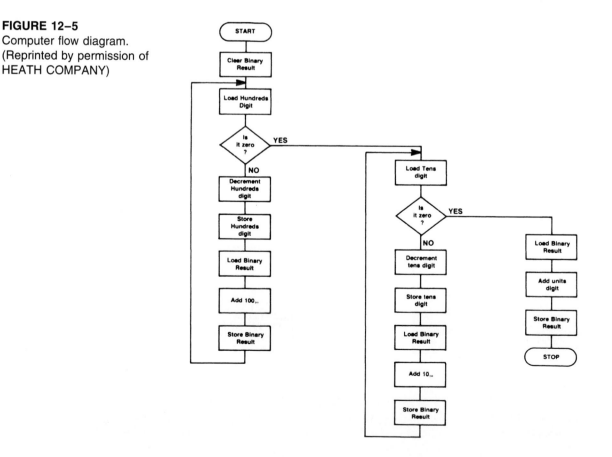

FIGURE 12–6
Flow diagram. (Courtesy International Business Machines Corporation)

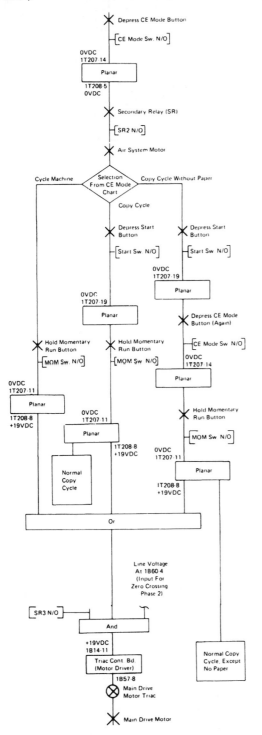

terline with the blocks evenly spaced to form a pleasing, easily interpreted diagram. More than one centerline is used for diagrams that need to depict double or multiple flow. The actual connecting **flow lines** between the geo-

metric shapes are always drawn vertically and horizontally (perpendicular). Angled and curved flow lines are not recommended.

Legend, title, and function words are lettered in the same style and are normally centered within the block (Figs. 12–5 and 12–6). These figures use both upper- and lowercase **centered lettering.** Note that, when two or more lines of lettering are required, **symmetrical lettering** should be placed around a common centerline. The size of lettering is determined by the smallest possible reduction of the drawing. Block size is determined by the space needs of the largest title, and all similarly shaped blocks should be the same size.

Computer Diagrams

Block diagrams are used throughout the computer industry to describe the organization of computer systems, as illustrated in Fig. 12–7. Each block in this **computer organization diagram** represents a series of complicated circuits and components. When preparing computer diagrams, try to maintain a uniform block size to represent similar devices or functions.

In Fig. 12–8 a diagram is used to depict a single-component, microcomputer chip. This component was formed by large-scale integration (LSI) processes and is an entire circuit in itself. A block diagram is used to represent the circuit functions on this single chip. Figure 12–9 is a photograph of an actual microcomputer chip. It includes blocks drawn around the functional areas of this chip. A functional block diagram for the chip would be very similar to Fig. 12–8.

Electronic Block Diagrams

The JIC defines electronic block diagrams like this:

> EL2.7.1 Block Diagram. The block diagram of the complete system indicates the basic purpose or function of each portion of the electronic system and its relationship to the overall system. Each block shall be identified and cross-referenced in a manner that the internal circuitry may be found readily on the elementary and related diagrams.

Preparation of a block diagram, and any other diagram, must be made to give the utmost in clarity. Good drafting practices, lettering, and linework, as discussed in Chapters 2, 3, and 4, must be applied to the drawing of all diagrams.

Electronic block diagrams originate with the design engineer in the form of a sketch. The finished block diagram is drawn by the drafter. It is the responsibility of the drafter to create a balanced, simple, and precise layout of the final drawing. The block diagram in Fig. 12–10 is an example of a well-designed and well-laid-

FIGURE 12-7
Block diagram showing
computer organization.
(Courtesy Intel Corporation)

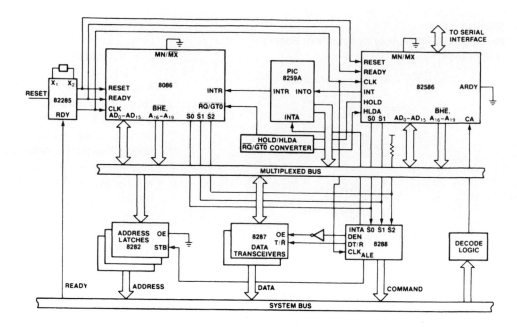

FIGURE 12-8
Block diagram of single-
component microcomputer.
(Courtesy Intel Corporation)

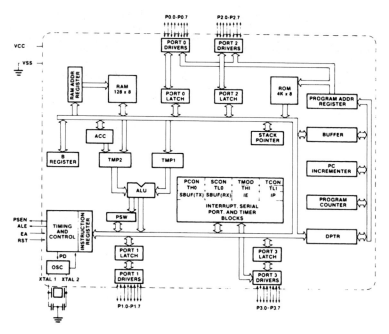

FIGURE 12–9
A microcomputer chip. (Courtesy American Microsystems, Inc.)

out block diagram of a stereo tuner–receiver. Note the use of an area corresponding to the right channel drawn with phantom lines at the bottom of the figure. Phantom lines are used to show future circuits or repetitious circuits that need not be redrawn. Missing units may also be shown blocked in with phantom lines. Note that future, missing, or repetitious units blocked in with phantom lines are still connected to the overall diagram in the proper sequence. Dashed lines are also used to show future or auxiliary units.

Block diagrams are drawn with the course of the flow circuit following the most direct path. Blocks should be aligned horizontally and vertically. The number and shape of different blocks should be limited.

After the engineer sketches the block diagram, the drafter lays out the total number of blocks, determines the proper block sizes and configuration, and provides sufficient space between blocks for flow lines. (Suggested specifications for block diagram layout are provided later in the chapter.) The drafter may need to resketch the diagram a number of times to try various layouts. Grid-lined paper or a grid-pattern underlay are excellent tools for this stage in the process. Counting the total number of flow lines and the required block series will help determine the paper size requirements for the diagram. Some trial and error is normal in this process.

Figure 12–11 shows a drafter's trial sketch. This layout sketch was drawn on grid-lined paper. Note that the original engineer's sketch would be much rougher and labels would be mostly abbreviations. The trial sketch is essential to determine the size of blocks, space needs, and lettering requirements. The drafter's finished drawing is shown in Fig. 12–12. When a block diagram is drawn with a CAD system, blocks and symbols may be entered in the drafting library and recalled for placement on the diagram. This process eliminates the need to redraw each repetitious block and symbol.

In Fig. 12–13 (A) a schematic sketch of a simple radio receiver is shown. The schematic has been broken into functional units as a block diagram sketch in Fig. 12–13 (B) and as a finished drawing in Fig. 12–13 (C).

FIGURE 12-10
Block diagram of multiplex stereo tuner-receiver. (Used with permission from Radio Shack, a division of Tandy Corporation, Fort Worth, TX 76102)

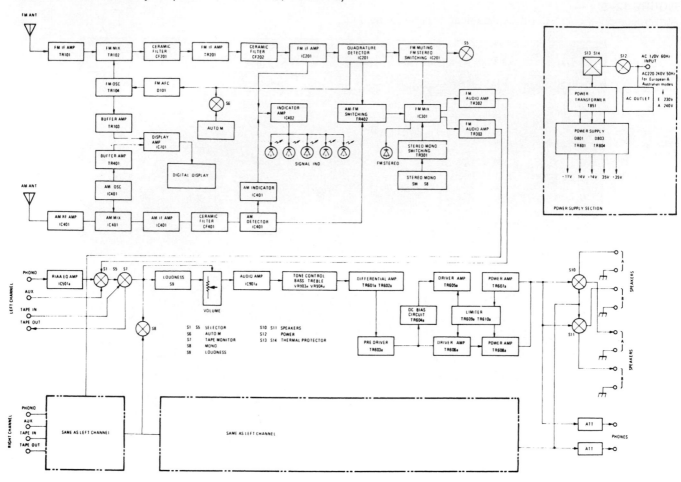

FIGURE 12-11
Trial layout sketch of the FM stereo receiver block diagram shown in Fig. 12-12.

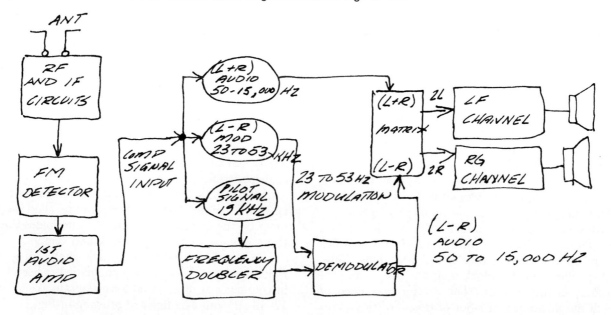

FIGURE 12–12
Completed block diagram of FM stereo receiver. (See sketch in Fig. 12–11.)

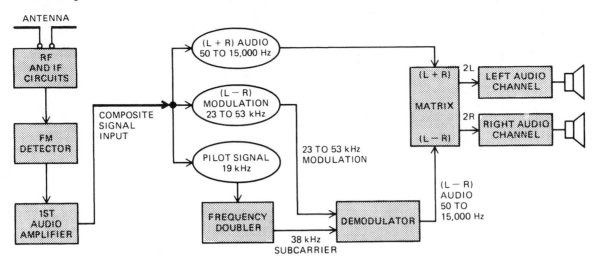

The shaded area defines the limits of each unit. In (C) the antenna is incorporated into the first block.

CONSTRUCTION OF THE BLOCK DIAGRAM

Block diagrams are made of four major parts: blocks, symbols, flow lines, and lettering. This section highlights each of these areas.

FIGURE 12–13
Simple radio receiver: (A) schematic diagram sketch; (B) block diagram sketch; (C) simplified block diagram.

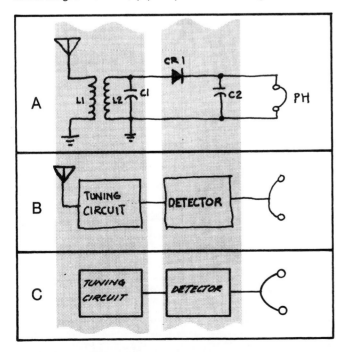

Blocks

A variety of simple geometric shapes is used for block diagrams, although rectangular forms are the most common, as shown in Fig. 12–14. Diamond, tubular, circular, elliptical, and triangular forms are used throughout the electronics and computer industry, as illustrated in Fig. 12–15. Computer flow diagrams make use of a wider range of forms than electronic diagrams, which stress square and rectangular shapes. Special electronic symbols are also used for electronic block diagrams.

Typical block proportions are shown in Fig. 12–16. Blocks are normally constructed as squares with proportions of 1 : 1 or rectangles with proportions of 2 : 1, 2.5 : 1, and 3 : 2. There is no rule governing the proportions of blocks used for diagrams. Blocks can be placed vertically or horizontally, depending on the needs of the diagram. Most blocks are drawn horizontally, as shown in Figs. 12–10 and 12–14. Fig. 12–12 has one of its blocks vertical.

The choice of size and proportion, as well as whether to use a single size of block throughout the entire diagram, depends on the length of the longest legend or title for the diagram. In some instances, abbreviations are used for legends and titles. Abbreviations are not used when the diagram is meant for nontechnical users. The full title and legend must be included in this case.

Rectangular blocks are sometimes placed with their long sides vertical to accommodate multiple inputs and outputs. The choice of block ratio may depend on the paper or underlay grid size. Grid patterns expedite the arrangement and layout of block diagrams.

In general, a block diagram should contain as few different sizes of blocks as possible. Some books suggest

FIGURE 12–14
Block diagram.

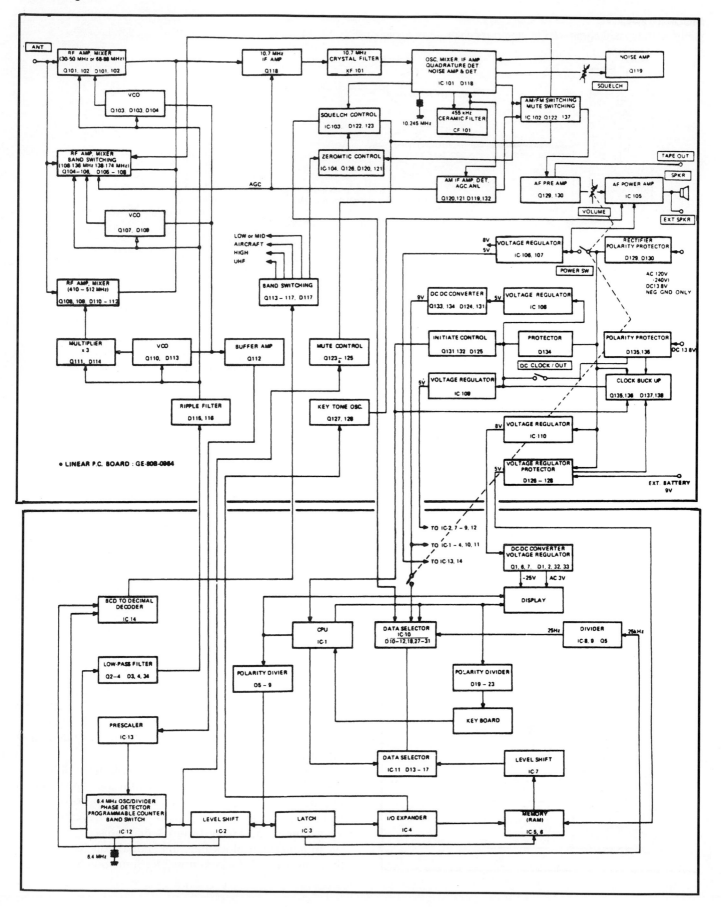

FIGURE 12–15
Common flow diagram and block diagram geometric shapes.

FIGURE 12–17
Block diagram of signal-to-noise ratio (SNR) test setup. (Courtesy TRW LSI Products)

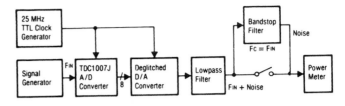

only one size per drawing; others say that two sizes is the maximum. After reviewing this chapter and the many block diagrams provided by industry, you must come to the conclusion that seldom if ever is this *rule* used in the real world. In Fig. 12–17, there are five different sizes of blocks out of a total of seven. The

complicated block diagram in Fig. 12–14 uses seven sizes of blocks, although a majority are a single size.

Blocks are drawn with a medium lineweight unless emphasis is required. Blocks drawn with dashed lines are auxiliary or future units. Symbols, blocks, and flow lines use the same alphabet of lines.

FIGURE 12–16
Common block size ratios.

Symbols

A few common electronic and electrical symbols are used on block diagrams. They are usually functional units in their own right and are normally depicted with aspects of their real-life configuration. Antennae, earphones, speakers, and switches are four such symbols. Fig. 12–18 shows eight typical **block diagram symbols.**

The tuned frequency receiver shown in Fig. 12–19 uses symbols and blocks in combination. Two grounds, one antenna, and a set of earphones are connected in sequence with the main units of the system. Note that symbols are usually of support or peripheral equipment. Terminal units, such as grounds, antennae, and speakers, are normally shown as symbols.

Flow Lines

Blocks and symbols are connected by flow lines according to the functional sequence of the system. In Fig. 12–19 the signal depicted by the flow line goes from the

FIGURE 12–18
Common symbols used on block diagrams: (A) ground; (B) single-cell battery; (C) terminal; (D) switch; (E) speaker; (F) two-conductor jack; (G) earphones; (H) antenna.

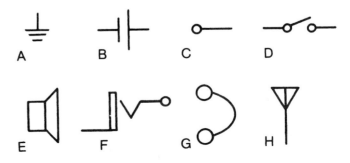

FIGURE 12-19
Tuned frequency receiver.

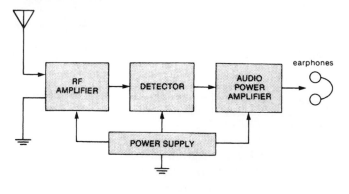

power supply into each of the three units. The standard left-to-right flow sequence connects the antenna to the RF amp to the detector to the audio power amp. Flow lines are drawn as solid, medium-weight lines (see Chapter 3). Dashed lines are used for future or auxiliary lines. Note that dashed lines also represent a mechanism connection or linkage between units. The thickness of the lines should be the same as that used for the blocks. Heavy lineweights are used for lines and blocks where emphasis is needed. Normally, the lineweight is the same for all lines on the diagram.

Diagrams should be constructed to minimize flow line **crossovers.** All flow lines are drawn with horizontal and vertical lines and make sharp 90° corners. Do not use diagonal lines or rounded corners unless this is a company standard.

Crossovers and **connections** must be clearly presented for the drawing to be properly understood, as shown in Fig. 12-20. Connecting lines are drawn with as minimum a number of changes in direction as

possible. An attempt should be made in the layout stage to limit the amount of corners and crossovers to the absolute minimum. In Fig. 12-20 (A) a crossover is shown. In (B) an acceptable method of drawing a multiple connection is shown, although (C) and (D) are the preferred method for showing this form of connection or junction.

You can designate a connection by placing a dark solid dot about three line widths in diameter at the connecting point, as shown in (F). When a line terminates at another line, it is unnecessary to use the dot, (E). When an array or bank of parallel and perpendicular lines requires connection points, the dot method is used for multiple junctions, as in (G). When a line terminates at a perpendicular line in a single connection, the dot may be eliminated, (G).

These specifications for connections and crossovers can be applied to the layout and construction of schematic, logic, wiring, and other types of electrical and electronic diagrams.

Figure 12-21 shows an example of a crossover. The dashed lines for the one block mean that the block is an auxiliary or future unit.

Spacing Flow Lines

Spacing of parallel flow lines must take into account the ultimate reduction size of the drawing. Flow lines are arranged in groupings of two and three by the functional relationship of the lines. Fig. 12-22 shows the suggested spacing for flow lines. Individual lines should be .40 in. (10 mm) apart, and .50 in. (13 mm) between groups of lines will provide sufficient reduction compatibility. The American Society of Mechanical Engineers (ASME) suggests a minimum of .06 in. (1.5 mm) spacing on the drawing's smallest reduction.

FIGURE 12-20
Crossovers and connections: (A) crossover; (B) multiple connection (dot method); (C) preferred multiple connection (no dot); (D) preferred multiple connection (dot method); (E) single connection (no dot); (F) single connection (dot method); (G) using dots for array of lines (note that single connections require no dots).

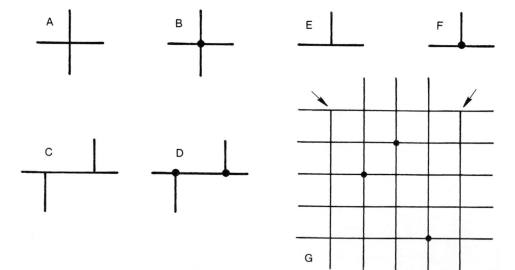

FIGURE 12–21
Crossovers and dashed lines.

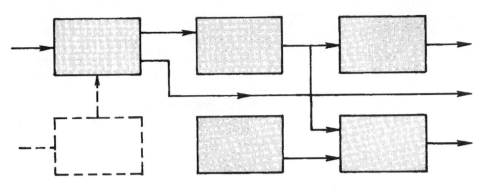

Multiple parallel lines should be spaced according to the same dimensions, as shown in Fig. 12–23. Parallel lines are to be grouped in twos and threes when four or more parallel lines are together.

The placement of flow-direction arrows on multiple parallel flow lines should be done to eliminate confusion (Fig. 12–22). In general, **flow arrows** may be positioned in a geometric or staggered pattern for more than two parallel lines. Arrows can be aligned for two lines (Fig. 12–22).

Arrowheads for lines terminating at blocks are normally placed against the left side of the block (Figs. 12–21 and 12–23). Some companies place the arrowhead halfway between two units. When a flow line is long, it should have a series of evenly placed flow arrows along its length (Fig. 12–21).

Some suggested spacing dimensions are shown in Fig. 12–24. Note that a variety of in-house standards are used by individual companies. The suggested spacing of flow lines presented here is provided as a general guideline for diagram construction. There are no standardized construction or layout specifications for the drawing of electronic diagrams. *The preceding suggested specifications for flow lines may be applied to logic, schematic, and wiring diagram layouts.*

Block Layout

Most block diagrams are laid out so that the flow direction is from the left toward the right. In Fig. 12–25, the input originates on the left, and the output is on the right. Auxiliary (AUX) units can be placed above or below the unit block to which they have input. Note that the three main units in Fig. 12–25 are drawn in a row and are evenly spaced.

Blocks are aligned vertically and horizontally for most diagrams. The following simplified specifications can be used to set up a block diagram:

☐ Vertically align blocks regardless of flow direction, as in Fig. 12–26.
☐ Align blocks in rows and columns, as in Fig. 12–27.
☐ Maintain the vertical and horizontal alignment when blocks are offset, as in Fig. 12–28.

FIGURE 12–22
Layout of multiple (parallel) lines and variations in arrowhead placement: (A) .40 in. (10 mm) between parallel lines, minimum of .06 in. (1.5 mm) after reduction; (B) .50 in. (13 mm) between groups of lines.

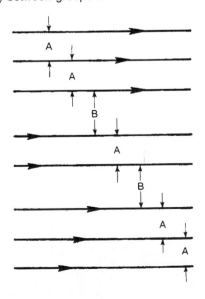

FIGURE 12–23
Line spacing in groups: (A) .40 in. (10 mm) between lines, minimum of .06 in. (1.5 mm) after reduction; (B) .50 in. (13 mm) between groups of lines.

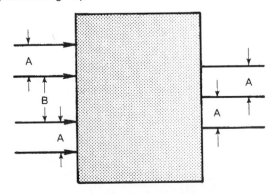

FIGURE 12–24
Line spacing; A = .40 in. (10 mm).

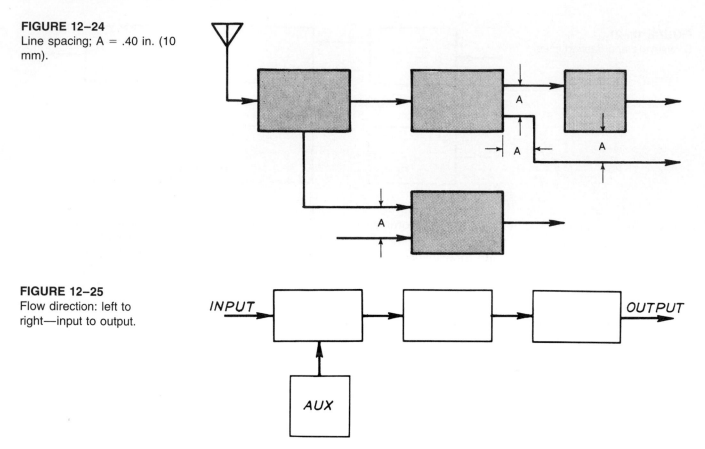

FIGURE 12–25
Flow direction: left to right—input to output.

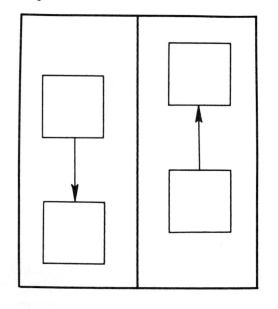

FIGURE 12–26
Vertical alignment of blocks.

FIGURE 12–27
Vertical and horizontal block alignment.

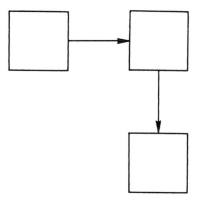

FIGURE 12–28
Offset blocks.

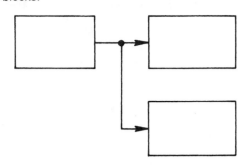

FIGURE 12–30
Block spacing: full- and half-block spaces.

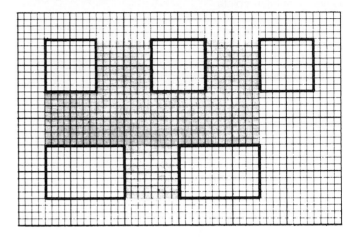

When the total length of the diagram is too long because of numerous sequential units, the scanning sequence method shown in Fig. 12–29 should be used. The last block in the first row is connected to the first block in the second row. The square blocks of Fig. 12–29 are positioned in vertical columns and horizontal rows. Obviously, it is easier to use the vertical and horizontal method if all the blocks are the same size.

The first step in laying out a typical block diagram is the positioning of blocks on grid paper. The space between the blocks should be kept as even as possible, as in Fig. 12–30. The best method is to allow one full-block space between units when the blocks are squares. Half-block spacing can be used for rectangular blocks.

Sufficient space must be provided for flow lines connecting blocks and for flow lines between rows and columns of blocks, as in Fig. 12–31. Fig. 12–32 shows a typical instrument block diagram. As can be seen, there are many shapes and sizes of blocks; not all blocks are evenly spaced or aligned according to the preceding guidelines. However, the diagram is pleasing to the eye, casily interpreted, and fairly well balanced.

Lettering

Lettering is the last step in the completion of a block diagram. Lettering within blocks and other geometric shapes should conform to specifications presented in Chapter 2, and in ANSI Y14.2M. The height of the lettering is determined by the eventual reduction size. Most lettering is symmetrical and centered, as in Figs. 12–32 and 12–33. *Left-justified lettering* is shown in Fig. 12–34, but centered lettering is very easy to execute when using a computer to generate the drawings.

Block diagrams are seldom hand lettered since they are used for printed manuals and sales literature. Most block diagrams are phototypeset. Freehand lettering is necessary for the original sketch and the layout sketch. Regardless of the type or form of lettering, it should be of a consistent size and style. The height of the lettering is normally the same throughout the drawing. When using any particular form of lettering—regular, compressed or expanded—use only one style throughout the drawing. In Fig. 12–34, two styles of lettering were mixed; this is a poor practice and should be avoided.

FIGURE 12–29
Scanning sequence used for long and complex diagrams.

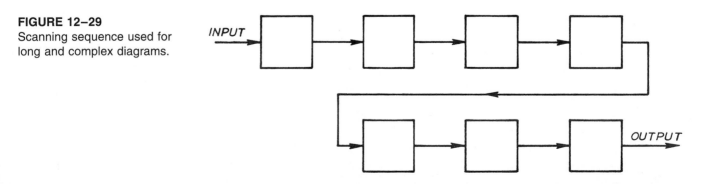

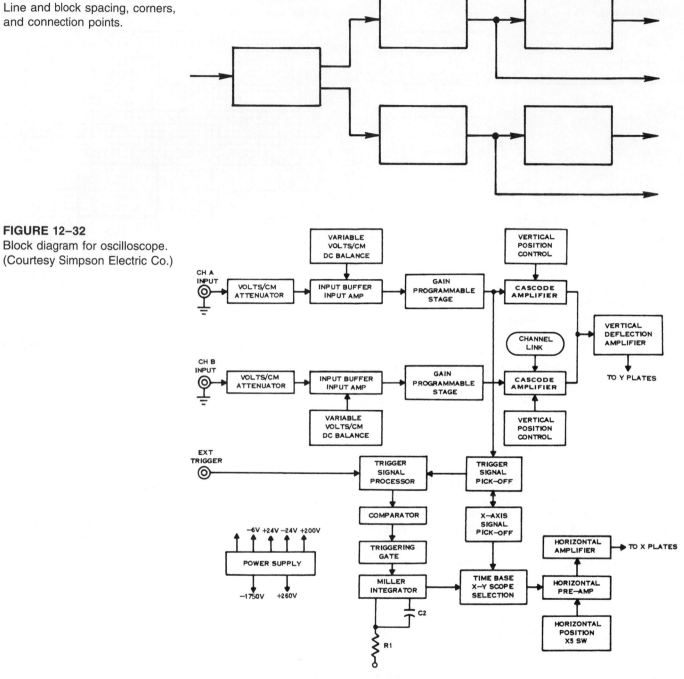

FIGURE 12–31
Line and block spacing, corners,
and connection points.

FIGURE 12–32
Block diagram for oscilloscope.
(Courtesy Simpson Electric Co.)

FIGURE 12–33
Block diagram for microcomputer support system. (Courtesy of Motorola, Inc.,
Semiconductor Products Sector)

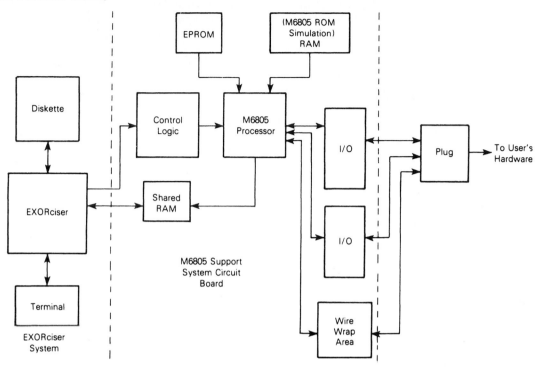

FIGURE 12–34
Block diagram of FM RF, IF, and MPX alignment connection. (Used with permission
from Radio Shack, a division of Tandy Corporation, Forth Worth, TX 76102)

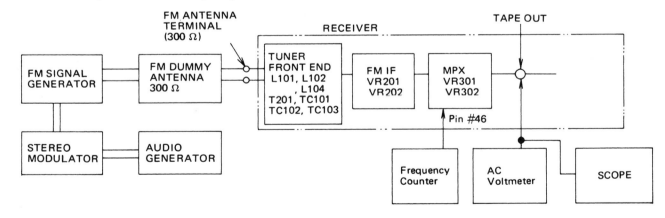

REVIEW QUESTIONS

12–1 Describe one method of determining the size of a block to be used throughout a diagram.

12–2 What are the differences between flow diagrams and block diagrams?

12–3 Name three uses for block and flow diagrams.

12–4 How should arrowheads be placed on multiple parallel lines?

12–5 Blocks should be placed in vertical (   ) and horizontal (   ).

12–6 What is a block diagram?

12–7 Name five important aspects and considerations in the layout of a block diagram.

12–8 What is the suggested spacing between parallel flow lines on a block diagram?

12–9 Describe the typical specifications for block diagram lettering. What standard governs lettering?

12–10 Why would a line or block be drawn thicker than other parts of a diagram?

12–11 What does a dashed line represent? What does a phantom line represent?

12–12 What is a flow path?

12–13 Name five typical symbols used on block diagrams.

12–14 Describe the sequence of steps used to construct a block diagram.

12–15 Name six common geometric shapes used for diagrams.

12–16 Give four typical proportions for blocks.

12–17 Name two ways of showing line junctions or connections.

12–18 What is a crossover?

12–19 How are multiple connections shown?

12–20 What is a title or legend and how is it normally placed?

PROBLEMS

12–1 Redesign the block diagram shown in Fig. 12–12. Use only block shapes and standard symbols. Align the blocks in rows and columns.

12–2 Redraw Fig. 12–17. Draw the diagram pictorially using oblique pictorial blocks.

12–3 Redraw the diagram shown in Fig. 12–3, Fig. 12–5, or 12–6.

12–4 Redraw the block diagram in Fig. 12–17 using a consistent block size and even spacing.

12–5 Redesign the diagram shown in Fig. 12–32. Use only one block size and shape.

12–6 Draw a block diagram of a simple radio: antenna, receiver, amplifier, detector, speaker.

12–7 Draw a block diagram of a stereo system: inputs from AM tuner, FM tuner, tape deck, and turntable. Each input is channeled into the preamp, which is connected to the power amp and output into a speaker and earphones.

12–8 Redraw the computer diagram in Fig. 12–7.

12–9 Draw Fig. 12–2 as a simple block diagram.

12–10 Redraw the following block diagram using one consistent block size and equal spacing. Use symmetrical lettering.

PROBLEM 12–10
Block diagram. (Courtesy TRW LSI Products)

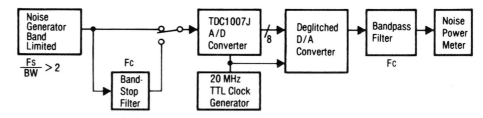

12–11 Draw from scratch a block diagram of a transmitter RF section starting from number 1 and ending with number 8. Use a consistent block size, and letter the title using vertical, centered lettering. The diagram must read from left to right and is connected in sequence: (1) oscillator, (2) buffer 1, (3) buffer 2, (4) predriver, (5) driver, (6) power amplifier, (7) antenna, and (8) power supply.

12–12 Using the specifications from problem 12–11, draw a block diagram of a superheterodyne receiver. Using the following sequence, construct the diagram: antenna, RF amp, mixer, IF amp, detector, AF amp, speaker. Place a filter below the detector. A signal goes from the detector to the filter. The filter is fed back into the IF amp, mixer, and RF amp.

12–13 Draw the interactive graphics system using the following titles for their corresponding blocks: (1) MC6809 MPU and RAM (stack and user), (2) display RAM 6K of MCM2214s, (3) MC6847 VDG, (4) MCM2716 EPROM, (5) MC1372 color TV modulator.

PROBLEM 12–13
Computer diagram.

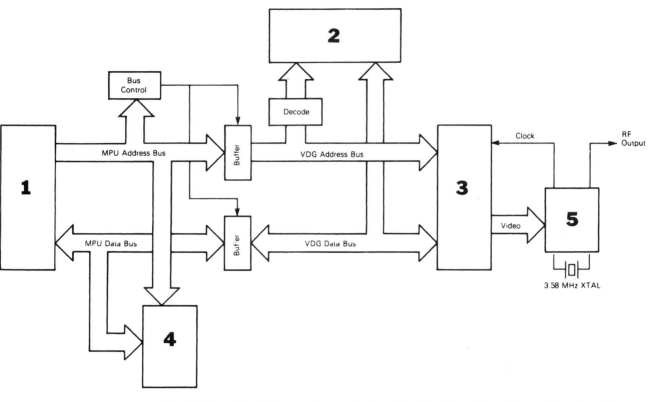

12–14 Draw the AM transmitter using the following titles: (1) oscillator, (2) buffer RF amp, (3) intermediate RF power amp, (4) final RF power amp, (5) microphone (place outside of symbol), (6) audio amp, (7) compressor amp, (8) audio power amp.

PROBLEM 12–14
Block diagram.

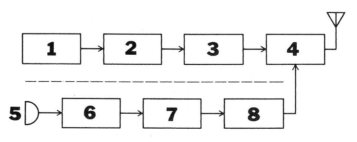

12–15 Redraw the following block diagram using uppercase, centered, and symmetrically placed lettering. Use one consistent block size with equal spacing.

PROBLEM 12–15
Block diagram. (Courtesy TRW LSI Products)

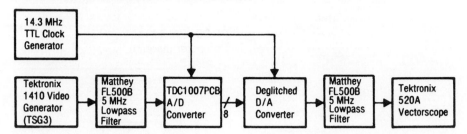

12–16 Draw the diagram of the radio-telephone telecommunication circuit. This circuit converts the voice frequency (input) into electrical impulses that modulate the output of a radio frequency oscillator (carrier) and transmits the information through the air to a receiving station. The following titles correspond to the blocks: (1) RF crystal oscillator, (2) RF buffer amp, (3) modulator, (4) RF power amp, (5) AF amp, (6) power supply. Add the speaker and earphones. Connect the power supply to all units.

PROBLEM 12–16
Block diagram.

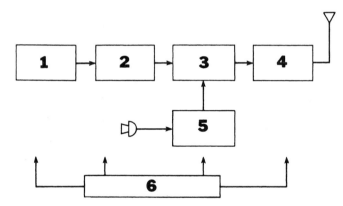

12–17 Draw the superheterodyne receiver with the following titles: (1) RF amp, (2) mixer stage, (3) local oscillator, (4) power supply, (5) IF amp, (6) detector, (7) AF amp.

PROBLEM 12–17
Block diagram.

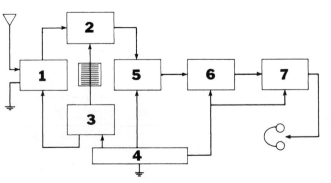

12–18 Redraw the following pictorial diagram as a simple block diagram.

PROBLEM 12–18
Pictorial diagram. (Courtesy Intel Corporation)

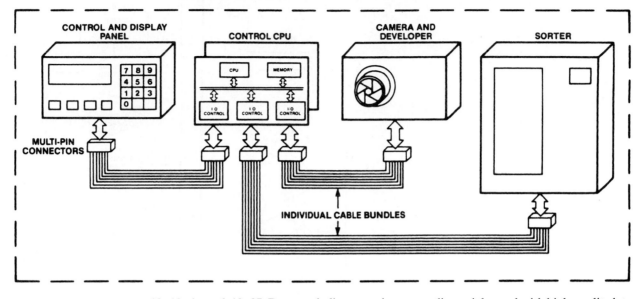

12–19 through 12–27 Draw each diagram using proper lineweights and with high-quality lettering.

PROBLEM 12–19
Flow diagram.

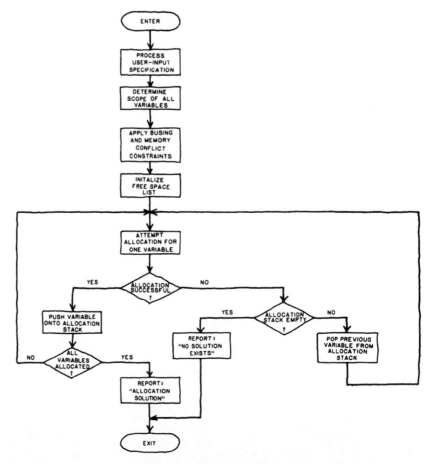

PROBLEM 12–20
Computer flow diagram.

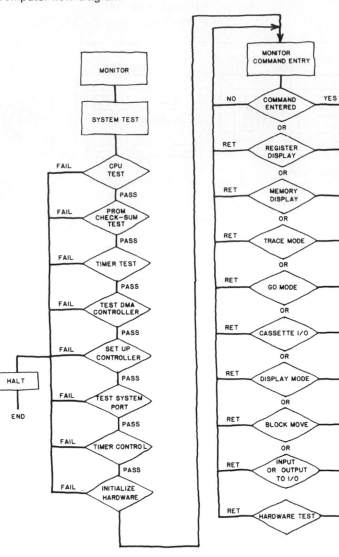

PROBLEM 12–21
Flow diagram.

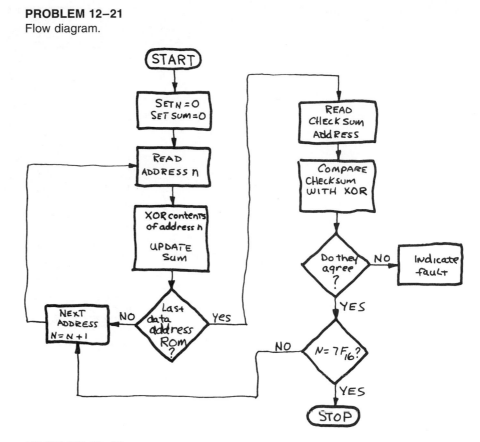

PROBLEM 12–22
Computer flow diagram.

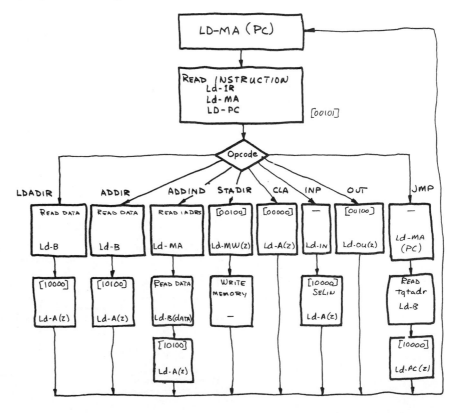

PROBLEM 12–23
Memory subsystem flow diagram.

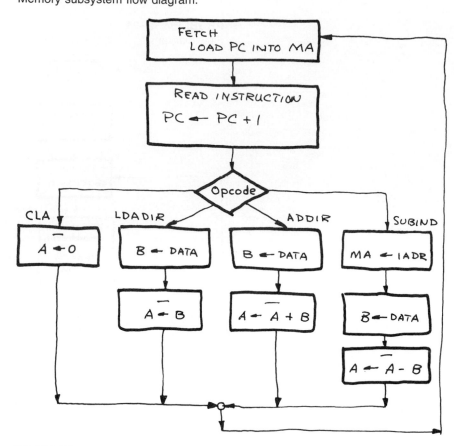

PROBLEM 12–24
Memory subsystem flow diagram.

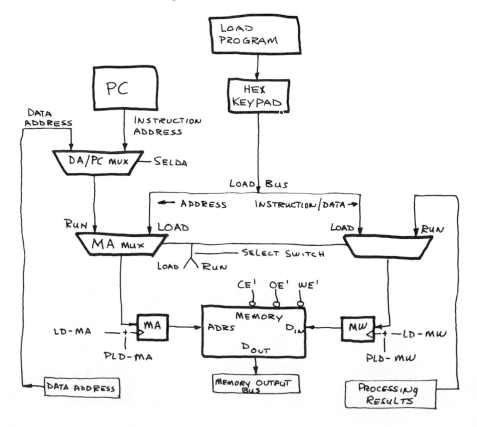

PROBLEM 12–25
Flow diagram.

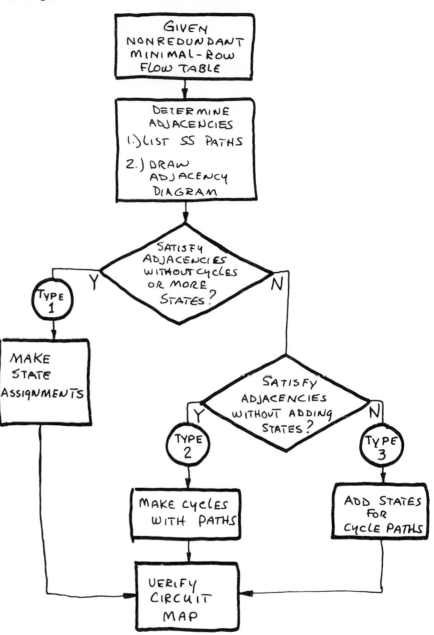

PROBLEM 12–26
Drawing tree block diagram.

PROBLEM 12–27
Power distribution box, drawing tree block diagram.

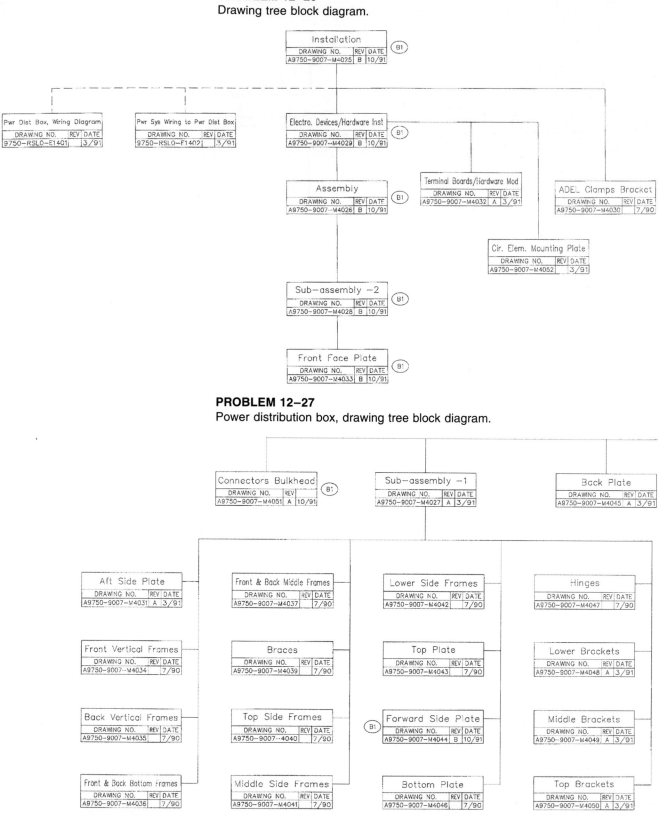

13

WIRING DIAGRAMS

INTRODUCTION

Up to this point all the drawings and diagrams discussed have primarily shown the functional or electrical relationship between parts and components. The physical relationship is also important in an electronic or electrical drawing. This type of drawing is called the **wiring diagram** and, as its name implies, is a diagram that aids in the assembly or production of the entire electronic package.

WIRE

The correct drawing or representation of wire is critical. There is not just one type of wire used for all assemblies. There are many different sizes, styles, and types of wire and many methods of wiring. Wire is made of material like copper or aluminum that will allow current to flow through it with very little resistance. Since the primary purpose of wire is to connect two points together, it should not alter the overall electrical characteristics or function of the circuit.

Connecting wire is generally either a single strand of wire or several strands fixed together as one wire. This wire can be insulated or covered with a material like nylon, enamel, polyester, rubber, or other material

that does not conduct electricity very well. The size, or cross-sectional area, of wire can vary. In addition, the size of the insulating material on the wire can vary. The insulating material can be made in many different colors to aid the technician in identifying the wire and its function.

All this information must be included on any wiring diagram, since the diagram is often used as the assembly or production drawing. This drawing may be included in a service or instruction manual, and without identification of color and type or size of wire, it would be very difficult for the technician to trace the circuit paths or repair the equipment.

Wire Gauge

The size of the wire is called the **gauge.** Wires come in standard sizes identified by whole numbers, which have been set by the American Wire Gauge **(AWG)** Standards. The smaller the number, the larger the wire is. For a complete list of available wire sizes, see Appendix A.

Wire ranges in size from about as fine as a piece of hair to as big around as a small finger, as shown in Fig. 13–1. All different sizes have specific uses, and each size has different electrical characteristics in terms of how much resistance there is in the wire or how much current can safely flow through it. Wire formed from

FIGURE 13–1
Bundle of large-size wires surrounded by coaxial cable.
(Courtesy of Western Electric Co.)

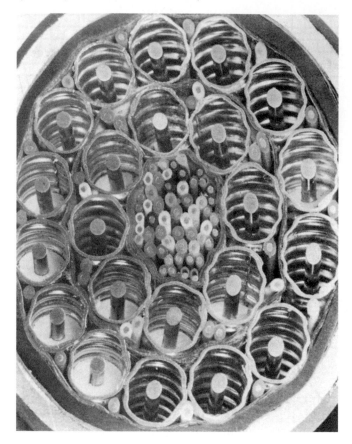

many strands of wire is not gauged in the same manner as solid wire; however, stranded wire does have similar electrical characteristics and limits. The biggest advantage of stranded wire is that it is more flexible than solid wire and should be used where the wire is likely to be moved or when vibrations can occur. The primary disadvantage is that stranded wire with the same overall cross-sectional area as solid wire can only handle about 60% of the current that solid wire can handle.

To find the equivalent solid size of stranded wire, multiply the number of strands times the cross-sectional area of each individual strand. The total cross-sectional area corresponds to an AWG equivalent size for the stranded wire on the AWG chart of solid wire. Stranded wire, however, is normally identified by the actual strand, not by the equivalent solid wire. It is identified first by the number of strands and then by the AWG number of the individual strands. The numbers are separated by a slash. For example, 8/34 means a wire made up of 8 strands of #34 wire. Since the cross-sectional area of #34 wire is 39 circular mils, this wire is equivalent to approximately AWG #25 solid wire (8 × 39 = 312 circular mils). Remember that although these strands

have approximately the same gauge or physical size as solid wire the maximum amount of current must be derated to about 60%.

Insulation

Many different types of materials are used to insulate conductors. The insulating material is necessary to provide protection to people and to make certain the wires do not make contact with other wires or the metal cases. Any time two conductors touch, an electrical connection is made. If a wire is to be used to connect two components, then it may only touch the leads of those two components. It is physically impossible to keep a long loose wire from making contact with anything else, so insulation is used to cover the length of the wire. Only the ends are exposed so that they may make contact.

Different materials have different insulating properties; therefore, the type of insulation is determined by the particular application. Insulating materials come in a number of different solid colors or in a solid color with one or two stripes of different colors, called *tracers.* The colors aid in tracing circuit connections of equipment, as well as in fabricating a system.

Method of Termination

A wiring diagram has three main labels: the size and type of wire; the size, type, and color of the insulation; and the type of connection to be made. All connecting wires will end either in bare wire that must be soldered or in terminals, clips, jacks, plugs, or other mechanical means of making connections. All this information must be included on the wiring diagram. The choice between soldering or using some other form of connection is determined by how often the connection might need to be removed, either for repair or for multiple use of the equipment. Equipment with interchangeable sections does not have soldered interconnections. A sliding contact or quick-disconnect, plug-type assembly is used instead.

PARTS OF A WIRING DIAGRAM

All wiring diagrams must include at least the components, the connection lines, and the means of connection. The components are frequently represented by a geometric shape showing physical characteristics, as in Fig. 13–2, rather than the unique symbols used for electrical characteristics. The connection lines may be represented in a number of different methods (covered in detail later in this chapter). The method of termination is designated by either a pictorial representation of the connector or simply a note identifying the type of terminals.

FIGURE 13–2
Geometric shapes showing components in wiring diagrams. (Courtesy of International Business Machines Corporation)

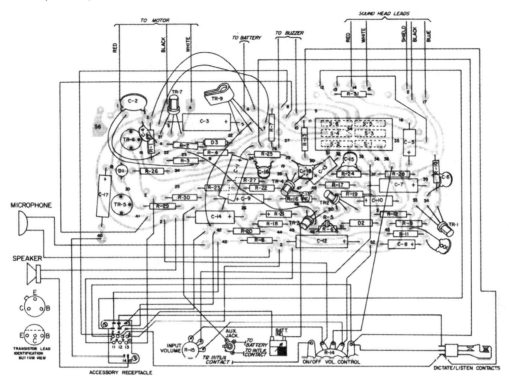

As with all other diagrams, the wiring diagram starts with a rough sketch. The circuit is divided into smaller sections based on the manner in which it will be assembled. For very simple circuits this may include just one main assembly.

The rough sketch establishes the initial space requirements for the final drawing, as shown in Fig. 13–3. All components are represented by their physical characteristics. Even though there is no proper size for these components, guidelines need to be followed. If the wiring diagram will be used in production or assembly, the picture must help locate the actual part. Components are frequently represented larger than life on the sketch. The size of each pictorial representation must show the location of parts and some proportional relationships to adjacent components.

For components with multiple leads, any identifying characteristics—such as tabs, slots, and polarity—must be shown on the drawing to ensure proper connections, as in Fig. 13–4. The pictorial representations are usually

FIGURE 13–3
Rough sketch establishing preliminary space requirements.

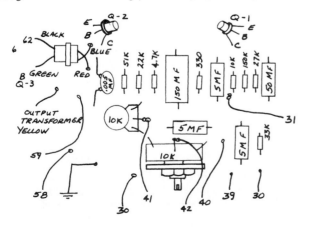

FIGURE 13–4
Component outlines, identifying all tabs, slots, and spacing.

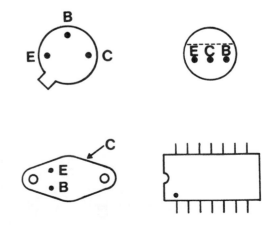

WIRING DIAGRAM

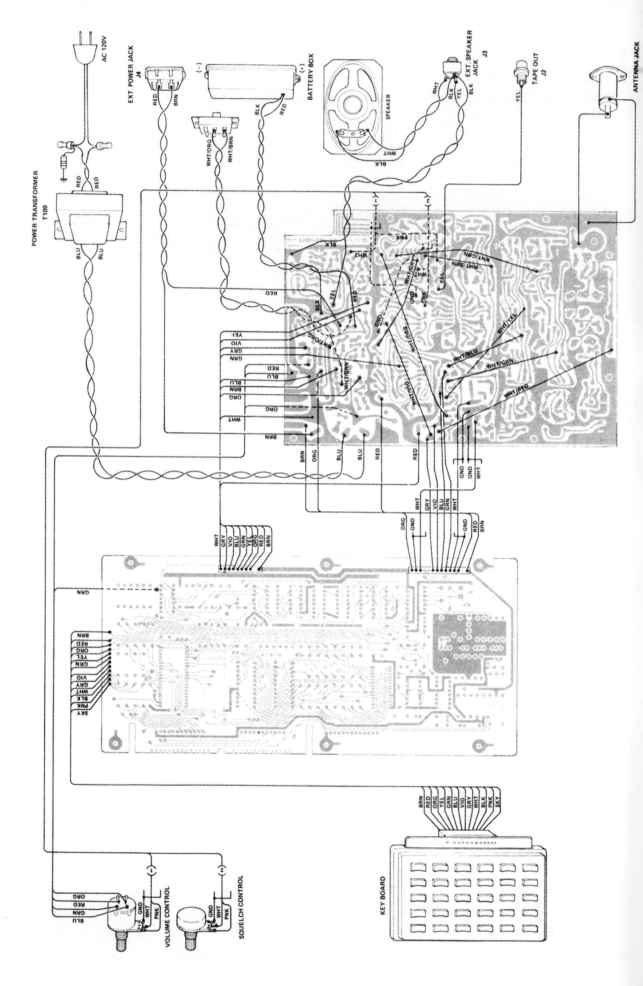

not pictorials in the sense of showing entire physical characteristics with multiviews or other pictorial methods, but are simply outlines best showing the required mounting space. If all the information cannot be shown in one view, then additional views may be required for final assembly. In many instances a single view of the bottom or actual wiring side of a chassis is sufficient to represent all components and their interconnections, as shown in Fig. 13–5.

Unfortunately, like all other electronic or electrical diagrams, there is no easy way to achieve a well-designed, well-balanced, and well-presented wiring diagram. The trial-and-error method must be used to achieve the final drawing. With practice and experience, fewer trials with fewer errors should be possible.

TYPES OF DIAGRAMS

According to ANSI there are three major categories for wiring diagrams: (1) *continuous line*, (2) *interrupted line*, and (3) *tabular* types.

Continuous Line

The **continuous line** diagram shows all connections and the location or travel of the wire from one point to another. This diagram is convenient for simple circuits not containing an excessive number of parts, where the actual assembly can be done one wire at a time. This type of drawing, illustrated in Fig. 13–6, has also been called a *point-to-point connection* diagram because it shows the drawing with the most specific details as it uses an individual line for every wire. As the circuit becomes more complex, this type of diagram can get very confusing. Showing every wire can be either a disadvantage or an advantage, depending on how many wires there are. The advantage is that it allows direct tracing of every circuit path and every component connection. The disadvantage is that too many paths are difficult to trace and may lead to mistakes or wrong connections. A trial sketch may be a good point-to-point sketch to get an idea of the overall needs of the finished drawing.

Another continuous line diagram is a modified point-to-point diagram. This diagram is frequently

FIGURE 13–6
Point-to-point wiring diagram. (Courtesy of Motorola, Inc., Semiconductor Products Sector)

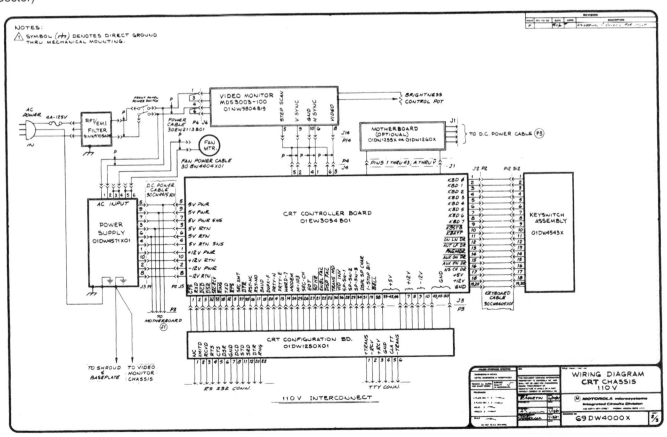

FIGURE 13–7
Highway diagram with feeder
lines joining main highway,
forming short 45° lines.
(Courtesy GTE Communication
Systems Division)

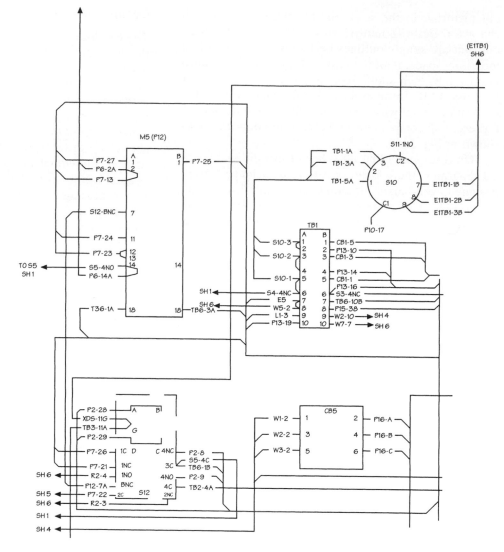

called a *highway* or *trunkline* diagram. If a group of wires flows through the circuit in the same direction, they can be represented by a single line, with each individual wire entering and leaving the main line, as shown in Fig. 13–7. As a wire enters or leaves the main trunk or highway, the connection is made by a short 45° diagonal line or a small ¼-in.-diameter circular arc, as seen in Fig. 13–8.

Each wire entering or leaving must be identified by some type of code, as shown in Fig. 13–9, or the circuit paths cannot be traced. Figure 13–10 is an example of this code, which identifies some or all of the following: wire number, size, type, and color; standard component designators; terminal numbers; harness number; origination; and destination. This code can be as simple as a number for every component in sequence, a number for all leads of the component, and a list of the destina-

tion by the component number, lead number, and color of wire.

When you draw a trial layout and many wires seem to go through the circuit in the same direction, you may decide that a highway diagram may be the most efficient method of preparing the final drawing. Keep in mind, however, that wires traveling through the assembly in approximately the same direction may not be bundled together or take exactly the same path. When a number of wires do, in fact, go through the circuit in the same direction, it may be advantageous to assemble the wires in precut lengths with the correct bends and terminations. You can then connect this group of wires as an assembly.

When a group of wires is assembled together in this manner, the assembly is called a **harness.** This is not the same as a cable. A **cable** is two or more wires or

FIGURE 13–8
Highway diagram with feeder
lines joining main highway
forming a small 1/4 in. arc.

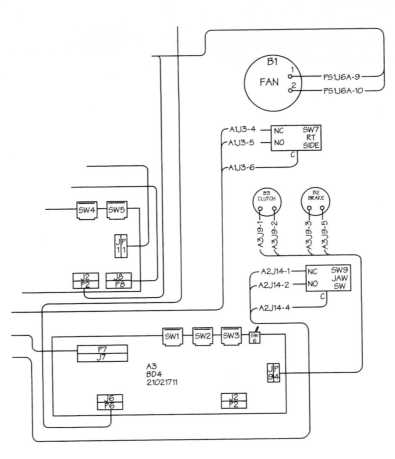

FIGURE 13–9
Wire destinations for every connection. (Courtesy California Computer Products, Inc.)

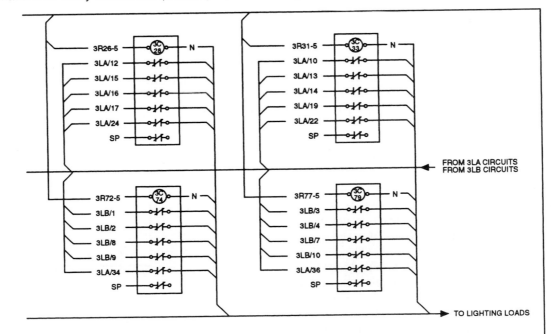

FIGURE 13-10

Number code indicating every component in sequence, all leads, and destinations. (Courtesy GTE Communication Systems Division)

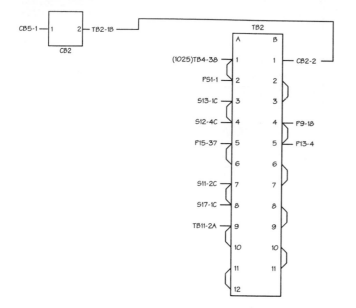

FIGURE 13-11

Typical baseline diagram.

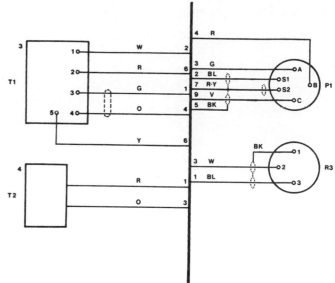

conductors insulated from each other and formed as a single unit. Cables are used when every installation is the same and when all the individual conductors originate and terminate in the same approximate location. (Harnesses and cables are covered in more detail later in this chapter.)

Interrupted Lines

Continuous line diagrams become impractical when circuits become too complex. An **interrupted line** diagram is used for complex wiring diagrams. This type of diagram is frequently called the *baseline* or *airline* method. In this method the placement of the lines may have very little to do with the actual routing of the wires. All wires go into one main line, as shown in Fig. 13-11, no matter where they terminate or what path they take to get there.

This type of diagram may not be very convenient for production or assembly, since components are not drawn in their true physical relationships. It is, however, a means of presenting interconnections of many wires in a well-organized, easy-to-follow format. This type of diagram tells very clearly the organization and termination of every wire, although it does not show the exact path in which the wire travels. Diagrams of this type are very useful for service manuals and allow large, complex circuits to be drawn in a much smaller space.

A baseline diagram consists of an imaginary main line located either vertically or horizontally. Short

feeder lines run from the baseline to every terminal. These lines are drawn in the most direct path possible and terminate perpendicular to the baseline, as shown in Fig. 13-12. This diagram should be prepared with as few changes in direction as possible. When terminal markings are located in a circle and a direct line cannot be drawn to the baseline without passing through the component, all changes in direction must be formed at 90° angles, as in Fig. 13-13.

The baseline is usually represented as a heavyweight dark line, and the short feeder lines are normally medium-weight lines. The baseline can be located through the center of the drawing with the components divided

FIGURE 13-12

Short feeder lines terminating perpendicular to the baseline.

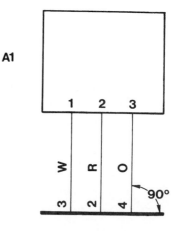

FIGURE 13–13
Changes in direction at 90° angles.

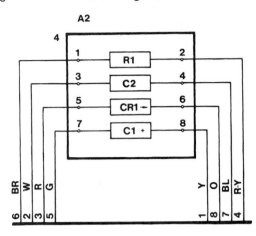

and spaced on either side, or it can be located along any side (frequently the bottom) based on the type of circuit and number of components. As shown in Fig. 13–14, the baseline extends just past all feeder lines, far enough to avoid confusing it with the highway method, which at least implies the direction of wire routing. In more complex circuits, more than one baseline can be used. This allows a feeder line to enter one baseline and leave from a different one.

Tabular Form

The third major classification of wiring diagrams is the **tabular** form. This form can be called a *lineless* diagram.

FIGURE 13–14
Baselines extending just past last feeder line.

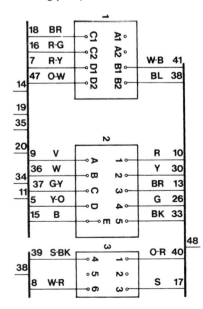

This type of presentation is the most incomplete visual representation of the interconnections, but it allows a large and complex diagram to be presented in a very limited amount of space. All connections are listed in tabular form, as shown in Fig. 13–15. This table should include some means of identifying each wire and should define the destination or endpoints. Like all other forms of identifying wires, a table can include the AWG number, color, type, and insulating material. It should definitely include the origin and destination points, as shown in Fig. 13–16.

CABLES AND HARNESSES

As stated earlier, a *cable* is a group of two or more insulated conductors confined in the same outer protection. Generally, all the conductors at either end terminate in the same approximate location. Frequently, a cable assembly terminates in a multiple connector. Typically, cable assemblies are used to connect one subassembly to another with mating connectors, as shown in Fig. 13–17. One common type of cable is the flat ribbon cable, where the conductors are side by side and the assembly looks like a flat ribbon. Although this type of cable is wide, it takes up very little routing space when used to interconnect subassemblies because it fits through a small slot between the assemblies. This can be seen in Fig. 13–18, where ribbon cables are used by Chrysler Corporation in the assembly of their travel computer.

When a group of wires is routed in the same direction, the wires can be tied together with a cord or other mechanical means, as in Fig. 13–19. This group of wires is normally formed before the wires are connected to the circuit assembly. This large group of wires is called a *harness,* or wiring harness. Unlike the cable, there does not need to be any outside protective covering, or coating over all the wires. A wiring harness can have some type of clear coating to protect it against environmental factors. In general, however, the primary difference between a cable and a harness is that the cable includes the protective outer covering, and a harness is simply a group of wires that for convenience have been bundled together. Also, harness wires can have many different lengths and can branch off at many different locations.

Both cable assemblies and harnesses need special attention and typically require individual drawings separate from the wiring or interconnection diagrams.

The preliminary step in preparing a harness diagram is to check the wiring diagram. If the wiring diagram is prepared on a highway-type drawing, the overall shape

FIGURE 13–15
Connections identified in tabular form. (Courtesy of International Business Machines Corporation)

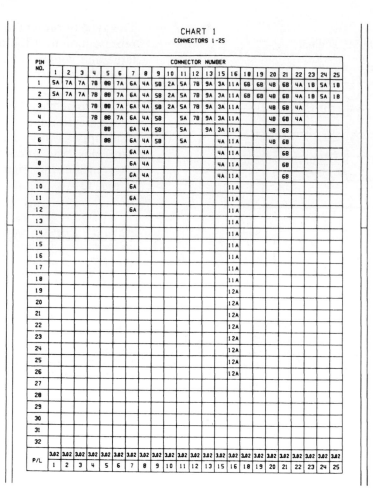

CHART 1
CONNECTORS 1-25

PIN NO.	1	2	3	4	5	6	7	8	9	10	11	12	13	15	16	18	19	20	21	22	23	24	25
1	5A	7A	7A	7B	8B	7A	6A	4A	5B	2A	5A	7B	9A	3A	11A	6B	6B	4B	6B	4A	1B	5A	1B
2	5A	7A	7A	7B	8B	7A	6A	4A	5B	2A	5A	7B	9A	3A	11A	6B	6B	4B	6B	4A	1B	5A	1B
3				7B	8B	7A	6A	4A	5B	2A	5A	7B	9A	3A	11A			4B	6B	4A			
4				7B	8B	7A	6A	4A	5B		5A	7B	9A	3A	11A			4B	6B	4A			
5					8B		6A	4A	5B		5A		9A	3A	11A			4B	6B				
6					8B		6A	4A	5B		5A			4A	11A			4B	6B				
7							6A	4A						4A	11A				6B				
8							6A	4A						4A	11A				6B				
9							6A	4A						4A	11A				6B				
10							6A								11A								
11							6A								11A								
12							6A								11A								
13															11A								
14															11A								
15															11A								
16															11A								
17															11A								
18															11A								
19															12A								
20															12A								
21															12A								
22															12A								
23															12A								
24															12A								
25															12A								
26															12A								
27																							
28																							
29																							
30																							
31																							
32																							
P/L	3.02	3.02	3.02	3.02	3.02	3.02	3.02	3.02	3.02	3.02	3.02	3.02	3.02	3.02	3.02	3.02	3.02	3.02	3.02	3.02	3.02	3.02	3.02
	1	2	3	4	5	6	7	8	9	10	11	12	13	15	16	18	19	20	21	22	23	24	25

FIGURE 13–16
Another tabular form method of showing connections.

Wire GA	COLOR	FROM	TO	LENGTH
20	ORN	SW1	J11P1	3.00
	ORN	SW2	P1	12.00
	ORN	J11P1	J10P1	2.50
	ORN	J10P1	J9P1	2.50
	WHT	T2	J11P2	4.50
	WHT	T2	J10P2	3.50
	WHT	T2	J9P2	3.00
	BLK	T1	J11P3	4.00
	BLK	T1	J10P3	4.00
	BLK	T1	J9P3	4.00
	GRN	SW3	J11P4	3.00
	GRN	J11P4	J9P4	3.50
	GRN	J9P4	J10P4	2.50
20	GRN	J10P4	P2	12.00

FIGURE 13–17
Wires connected with a mating connector. (Courtesy of Motorola, Inc., Semiconductor Products Sector)

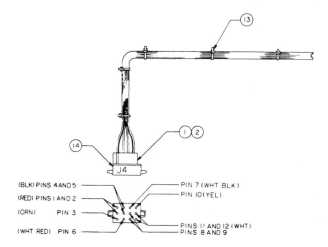

FIGURE 13–18
Ribbon cable used in Chrysler's travel computer. (Courtesy of Chrysler Corp.)

of the harness can be obtained directly from the wiring diagram. The highway diagram does show physical relationship, size, and location of all components. Usually, the harness diagram is prepared true size, whereas the wiring diagram or other electronic drawings are not. This makes sense because the harness diagram is frequently used to prepare the actual harness, and the wiring diagram is used only as a road map to trace circuit functions.

In the preparation of the actual harness, the entire assembly must be considered. Wires will not go through partitions unless holes are provided. When a harness must go through a hole, special strain relief or other protection is required. If the harness must travel around an unusually large component, the bend must be included in the preparation of the harness because the wires will not stretch to fit as the harness is being installed. The overall size of the harness must also be considered. Perhaps two or more harnesses may be simpler to install than one large one.

Once the layout of the harness is complete, the harness diagram is prepared. As shown in Fig. 13–20,

the overall outline of the harness is used, rather than a drawing of individual conductors. The conductors are individually identified where they branch out from the harness. Some identification code is necessary at that point, similar to the identifications used on other wiring diagrams. The primary use of the harness diagram is in the preparation of the harness itself; therefore, it is drawn to full size. The drawing is mounted on a board, and pins or nails are inserted into the board to form a type of track in which to lay the wires. All the precut wires are placed on this board, sometimes called a *jig,* and they are bent to the correct harness shape. Once all the wires are in place, cords or other clamps are secured around the bundle to form the harness, as illustrated in Fig. 13–21. Usually, all the individual conductors have been properly terminated before being placed on the jig because some termination methods may affect the overall length of wire.

Details for every conductor are included in a table or bill of material that includes specifications for every conductor in the harness: color, termination, origination, destination, AWG number, and type. The drawing

FIGURE 13–19

Wiring harnesses with wires bundled together and terminating in mating connectors.

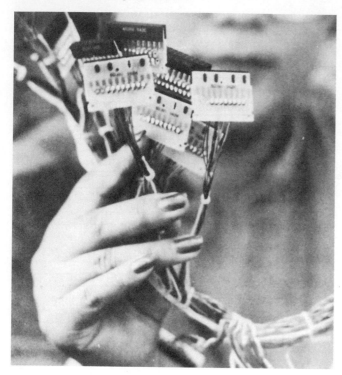

numbers of the schematic diagram, wiring diagram, and any other associated diagrams should also be included on the diagram.

DIAGRAM CONSTRUCTION

As with other diagrams, the wiring diagram must be clear, accurate, and pleasing to the eye. You can only achieve this by careful planning and using trial sketches.

FIGURE 13–20

Wire harness assembly. (Courtesy of Motorola, Inc., Semiconductor Products Sector)

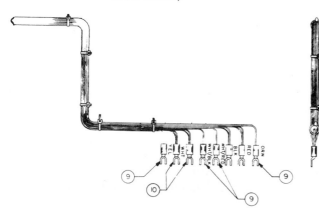

FIGURE 13–21

Wire harness assembly showing location of cable ties. (Courtesy of Motorola, Inc., Semiconductor Products Sector)

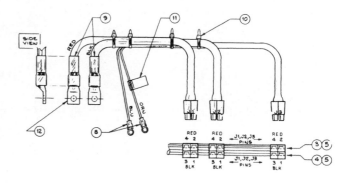

The trial sketch helps you determine the size requirements of the final drawing, the need for cable or harness assemblies, and the type of final drawing desired.

The use of grid paper or grid underlay is helpful in preparing wiring diagrams because they are generally prepared on a vertical and horizontal format. Many of the same rules must be followed that have been established with block, logic, and schematic diagrams. Crossovers and jogs should be kept to a minimum. Connection lines should be spaced in equal increments and should be no closer than $\frac{1}{16}$ in. after reductions. Unlike schematics, where connections or junctions are made at convenient locations, on wiring diagrams the connecting lines must go from one terminal to another. Junctions, joints, or splices are very seldom made in electronic assemblies.

FIGURE 13–22

Wrong identification inserted into break in line. (Courtesy of GTE Communication Systems Division)

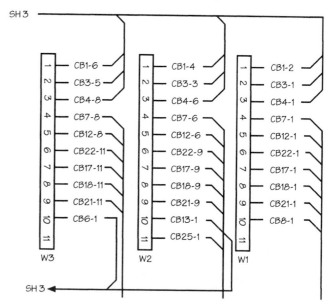

FIGURE 13–23
Typical templates used in preparation of wiring diagrams. (Courtesy of Berol RapiDesign)

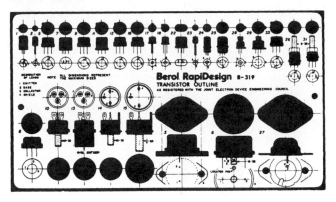

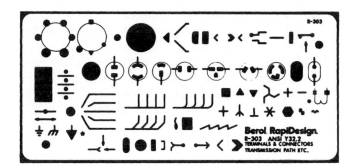

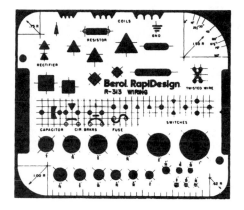

FIGURE 13–24
Data-acquisition computer wiring diagram.

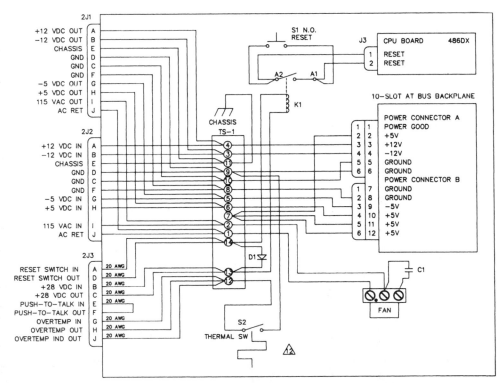

In general, all lineweights are the same on wiring diagrams and should be black, medium-weight lines. If necessary, you can use thin-weight lines for component outlines or symbols to avoid confusion and for the base-line method, where the imaginary baseline is drawn with a black heavyweight line for emphasis.

Lettering is the same for wiring diagrams as for other diagrams. The wiring diagram requires extensive lettering. In addition to including some type of identification or reference designation for every component, each wire includes identification of all specifications. The lettering can be inclined or vertical and should be aligned to be read from no more than two sides. Frequently, to avoid confusion the wire-identification lettering is done with a break in the line representing the wire. The lettering is inserted in the break, as in Fig. 13–22, so that it is in line with the wire it is describing. This eliminates any doubt as to which line the identifying code accompanies.

Templates are strongly recommended for use in the preparation of wiring diagrams. Figure 13–23 shows typical templates that are used to draw the component outlines in wiring diagrams.

FIGURE 13–25
Wiring diagram for junction box.

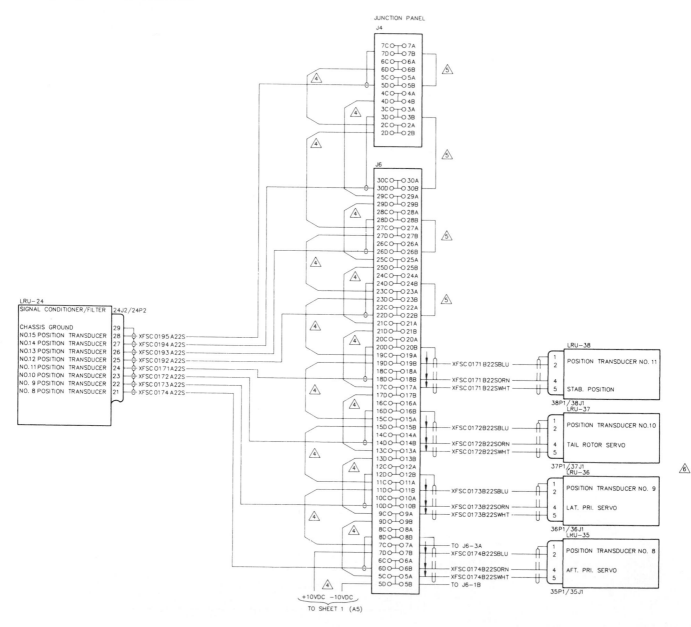

CAD WIRING DIAGRAMS

Remember that the use of CAD to prepare wiring diagrams and other drawings does not eliminate the need for a complete understanding of how these drawings are to be prepared. It is still the drafter's responsibility to ensure accuracy and completeness in all drawings. Figures 13–24 and 13–25 show examples of CAD-generated wiring diagrams.

REVIEW QUESTIONS

13–1 Why do wires come in many different sizes?

13–2 What is the primary purpose for the insulating material used on wires?

13–3 What characteristics should be considered when you choose wires?

13–4 What are AWG numbers and what are they used for?

13–5 What advantage does stranded wire have over solid wire?

13–6 Since stranded wires do not use the same type of AWG code as solid wires, how are stranded wires identified?

13–7 Why would a cable assembly be used?

13–8 When is a wiring harness most beneficial?

13–9 What are the major classifications of wiring drawings?

13–10 What kind of information is desired on wiring diagrams?

13–11 When is it most practical on a drawing to include every wire that is to be connected?

13–12 Which type of wiring diagram gives the least information of true physical relationship?

13–13 Why is it unnecessary to draw most wiring diagrams to scale?

13–14 Why should grid paper be used in the preparation of the wiring diagram?

13–15 When a harness assembly drawing is prepared, why is it often drawn full size?

PROBLEMS

13–1 Draw the geometric shapes of various component case styles shown in the following figure.

PROBLEM 13–1
Component shapes.
(Courtesy of Texas Instruments, Inc.)

DO-1

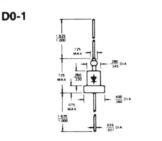

DO-5

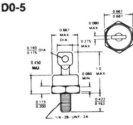

TO-41

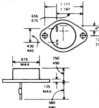

TO-98

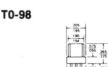

13–2 The telephone cable diagram in this figure demonstrates two types of terminations being used. Redraw this circuit on a B size sheet. Practice lettering by including the assembly instructions that are listed.

PROBLEM 13–2
Telephone cable diagram.

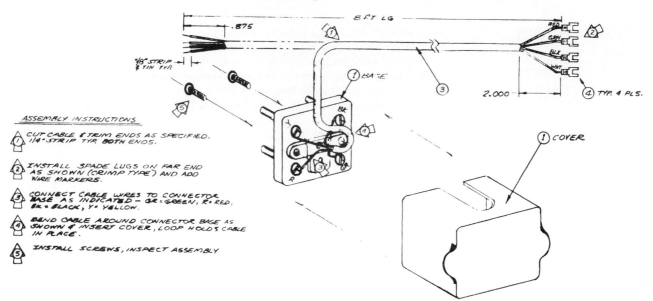

13–3 The following figure is a wiring diagram for a telephone recording attachment. Redraw this diagram on a B size sheet.

PROBLEM 13–3
Telephone recording attachment.
(Courtesy of International Business Machines Corporation)

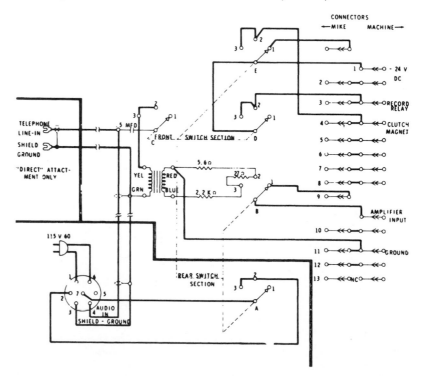

13–4 A piggyback connector is used for a foot control microphone. Draw the piggyback connector shown in this figure on a C size sheet. Complete all the lettering on this drawing with the use of some lettering tool such as a Leroy letter set.

PROBLEM 13–4
Piggyback connector.
(Courtesy of International Business Machines Corporation)

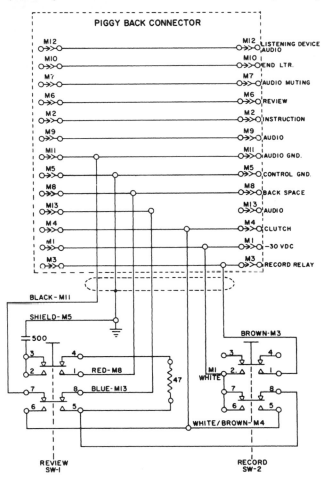

13–5 Redraw Fig. 13–10 illustrating the use of numbering codes.
13–6 Figure 13–14 demonstrates how the baseline extends just past the short feeder lines. Redraw this diagram on an A size sheet. Use ink and a lettering template.
13–7 Produce a table similar to the one in Fig. 13–16 for the baseline diagram shown in Fig. 13–14.
13–8 Redraw the wire harness assembly shown in Fig. 13–21. Include the additional views to fully describe the terminals.

13–9 (Advanced problem) A wiring diagram for a direct connection to a computer is shown in the following figure. Draw this on a C size sheet or larger.

PROBLEM 13–9
Chassis wiring diagram.
(Courtesy of Motorola, Inc., Semiconductor Products Sector)

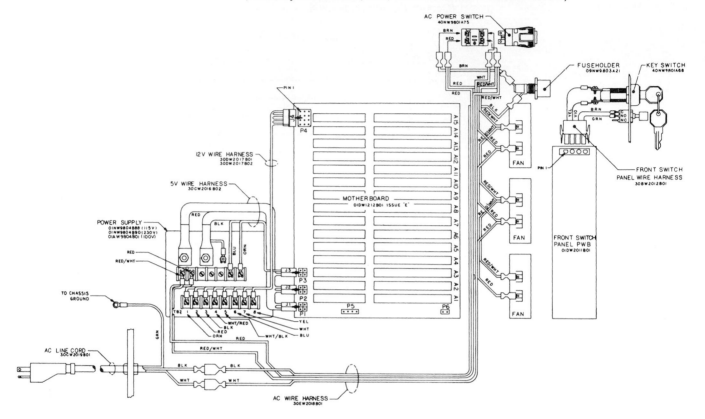

13–10 Draw the wiring diagram on a D size sheet.

PROBLEM 13–10
Wiring diagram.

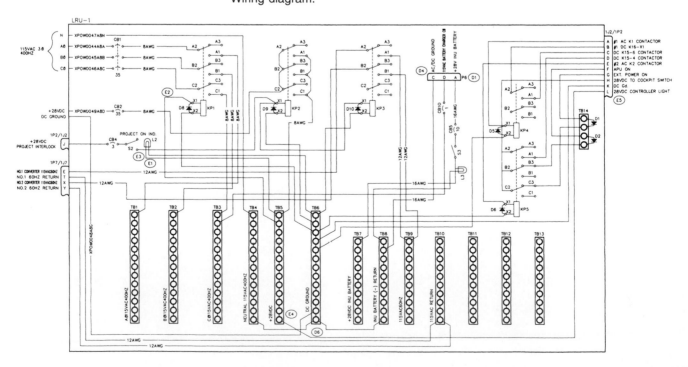

13–11 Using a C size sheet, draw the power system wiring to the power distribution box diagram.

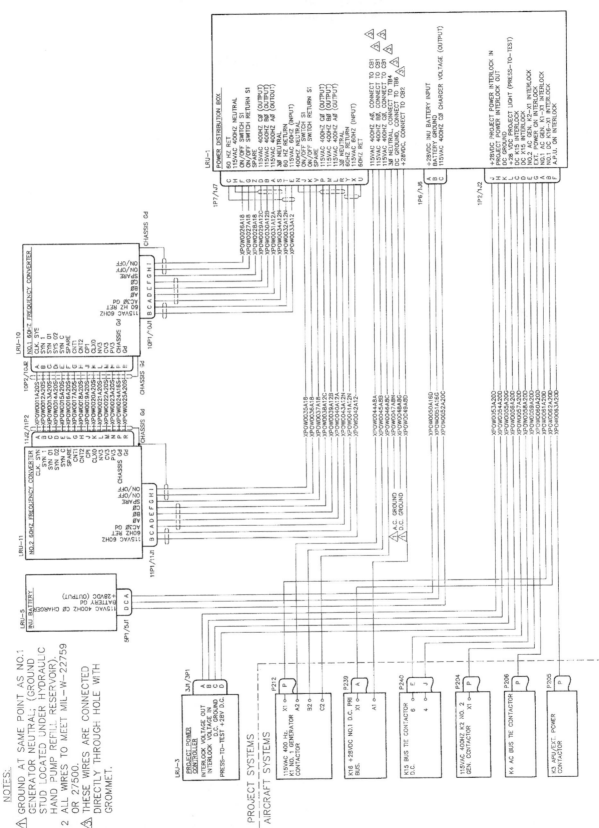

PROBLEM 13–11
Power distribution box wiring diagram.

NOTES:

⚠ GROUND AT SAME POINT AS NO.1 GENERATOR NEUTRAL; (GROUND STUD LOCATED UNDER HYDRAULIC HAND PUMP REFILL RESERVOIR).

2 ALL WIRES TO MEET MIL–W–22759 OR 27500.

⚠ THESE WIRES ARE CONNECTED DIRECTLY THROUGH HOLE WITH GROMMET.

13–12 Draw the air data system on a C size sheet.

PROBLEM 13–12
Air data system wiring diagram.

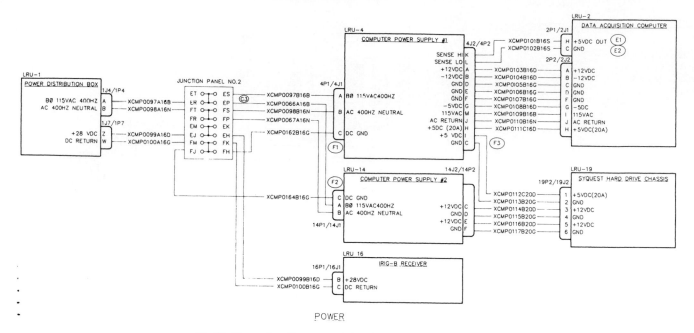

POWER

13–13 Draw the diagram of the data-acquisition computer shown in Fig. 13–24.
13–14 Draw the wiring diagram for the junction box shown in Fig. 13–25.

13-15 Draw the wiring diagram shown. Use a D size sheet. The upper-right portion of the diagram marked "Servo Control Unit" is a harness wiring diagram.

PROBLEM 13-15
Power and control wiring diagram.

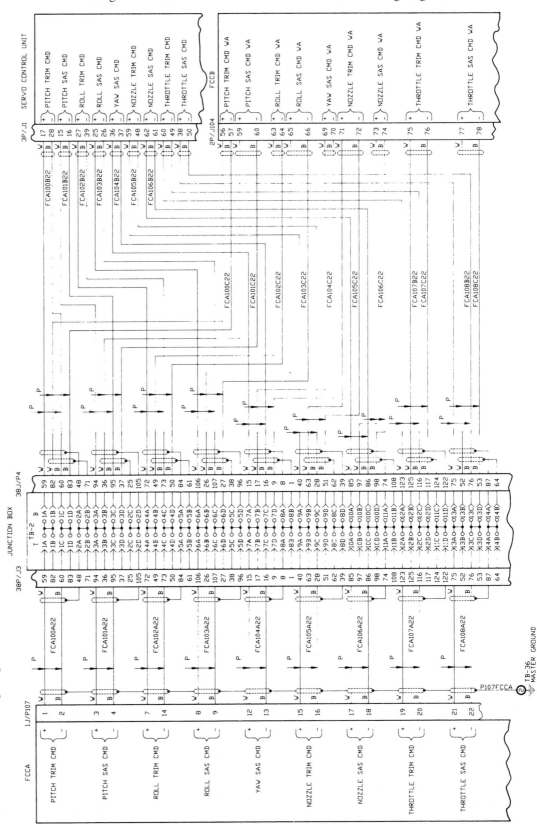

14

MOTORS
AND CONTROL CIRCUITS

INTRODUCTION

A basic understanding of motors is essential to understanding motor control and why it is necessary. Generating electrical energy and using this energy is basic to our industrial structure. Motors or electrical machinery can be found everywhere in applications often overlooked: automobiles, turntables, electric pencil sharpeners, elevators, and computer printers. Each of these applications has a different control requirement.

MOTORS

There are two major categories of motors, *AC* and *DC*, but there are many ways to construct and connect the internal windings of these motors. Generators operate on the same principle and are often included in the same discussions. **DC motors** and **DC generators** are practically impossible to tell apart by appearance. Both have the same parts—*armature core, armature windings, commutator, brushes,* and *field poles* (see Fig. 14–1). There are three basic types of DC motors, determined by how the field windings are connected: **series, shunt,** and **compound.** The selection of the type of motor to use is determined primarily by the requirements of the

load. In a series motor, the field circuit is in series with the armature, whereas the shunt motor is connected in parallel. A compound motor has two sets of field windings, one connected in series and one connected in parallel. These different types of connections are shown in Fig. 14–2. DC motors are rated by voltage, current, speed, and horsepower.

The operation of all motors is essentially the same and is based on the principles of magnetics and its interaction with conductors. Currents are caused by conductors cutting magnetic fields set up in the windings of the motor. The field poles produce this magnetic field, and the armature windings are the conductors. The field poles are either permanent magnets or electromagnets. To use the electrical energy produced by the spinning armature, a method must be developed to connect to the wires in these windings. Obviously, it would not take too many spins for the wires to become hopelessly tangled if a connection were made directly to the wire. Instead, a method must be used that makes contact but still allows the armature to turn freely. As a result, sliding contacts, or *slip rings,* are connected to the ends of the conductors. These slip rings, in turn, make contact with carbon blocks called *brushes,* which are connected to the commutator to transfer the electrical energy to the output

FIGURE 14–1
Cutaway of DC motor.
(Courtesy General Electric
Company)

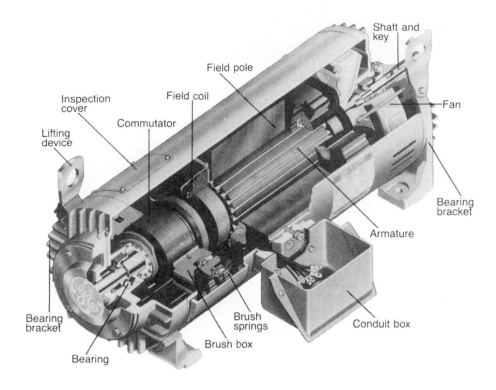

FIGURE 14–2
Connecting the field windings in series, parallel, or a
combination of the two.

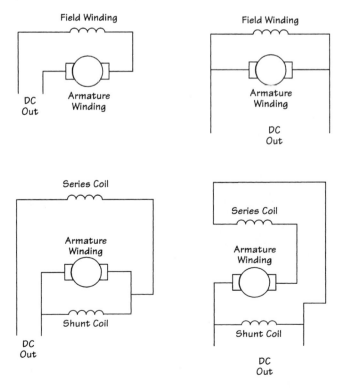

terminals. Figure 14–3 demonstrates how slip rings
are connected.

DC Motors

DC generators and motors can be grouped into many
different categories based on different characteristics of
the motors: type of winding, size, type of magnet, and
so on. The simplest type of motor uses a permanent
magnet and is sometimes referred to as a **PM generator**
or PM motor. Such a motor has the advantage of being
small and light. It is also the most efficient, since no
power is required to generate the magnetic field. The
primary applications of these motors and generators,
however, must be low-current situations. Too large a
current through the armature windings will produce a
magnetic field that could be large enough to demagne-
tize the field magnets. Electromagnetic fields can either
be externally excited or self-excited.

The way the field coil is connected will produce
different output results. For example, a series-connected
coil will produce higher-output voltage for higher-output currents. Shunt-wound motors or those with the
field coil connected in parallel produce almost constant
output voltage, regardless of the output current. This
situation is easy to understand when you realize that in
a series-connected motor the field coil is in series with
the output. If the output current increases, the field
current also must increase, producing a stronger magnetic field and creating a higher induced voltage. When

FIGURE 14–3
(A) Wound-rotor induction motor showing rheostat connections (Courtesy Electric Machinery Operations of Dresser–Rand Company); (B) rotor showing location of slip rings. (Courtesy General Electric Company)

the coil is connected in parallel, it is unaffected by output current.

Each type of motor has its own set of advantages and disadvantages. The application will determine the type of motor that needs to be used. Shunt-wound motors provide good speed regulation, easy reversibility, and ease of braking. The main disadvantage is that if the field current is interrupted while the motor is turning, but not under load, the motor increases in speed until it reaches a *runaway* state, where it can literally fly apart. Series-wound motors are probably the most common. The primary advantage is that they are small and light. Because the field coil has high current, it requires fewer turns of wire, resulting in less expense and lighter construction. Series-wound motors provide the most horsepower for their size.

A series-wound motor can also be used with either AC or DC because the polarity of the field windings and the armature windings both reverse, and the magnetic force remains in the same direction. For this reason, series-wound motors are often called *universal* motors. Series motors do not adapt easily to dynamic braking, however. Starting torque is very high; as a result, starting current is also very high. Speed regulation is poor under varying loads. Runaway can occur under no-load conditions. Series motors provide excellent results in high-speed operations, but these operations can cause considerably shorter life spans and higher maintenance requirements. For this reason, series-wound motors are usually not designed for continuous duty.

Compound-wound motors can be used to provide a combination of the characteristics of series- or shunt-wound motors. Depending on how the fields are connected, different operating characteristics are produced. Compound motors are more expensive but are desirable for high-torque, constant-speed applications. Table 14–1 gives a summary of the types of DC motors and some of their operating characteristics.

AC Motors

The majority of motors used today are AC. Many of these are the universal type that can be used for either AC or DC applications. Another type of motor that has the same advantage of low cost and is even more reliable is the **induction motor** (see Fig. 14–4). Motors constructed using induction eliminate the need for brushes and thus lower the maintenance cost. The two principles used in induction motors are rotating magnetic fields and induced voltages. Even though AC induction mo-

TABLE 14–1
Summary of DC Motors and Their Characteristics

	TYPE OF MOTOR					
Characteristic	PM	Shunt	Series	differential	Compound cumulative	low inertia
Starting current	High	Low	Very high	Low	Moderately high	High
Starting torque	High	Low	High	Low	High	Low
Speed	Varying	Constant	High and varying	Very constant	Fairly constant	Varying
Speed control	Simple	Simple	Simple	Simple	Simple	Simple
Reversibility	Easy	Easy	Easy	Easy	Easy	Easy
Dynamic braking	Yes	Yes	No	Yes	Yes	Yes
Size/weight	Smallest	Normal	Small	Large	Large	Small
Cost (relative)	Low	Moderate	Low	High	High	High
Horsepower range	<1	Any	<2	Any	Any	Very low
Uses	Portable equipment	General	Portable equipment	General	General	Instruments

FIGURE 14–4
Cutaway of a wound-rotor induction motor. (Courtesy Magnetek Louis Allis Company)

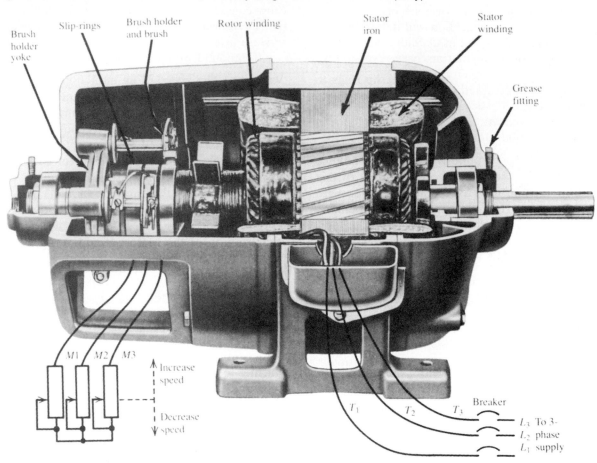

tors have parts similar to those of DC motors, the construction is slightly different. An AC motor has a stator and a rotor, rather than a field and armature. These produce very similar results, but may have different construction. The rotor is made of very thick rods, rather than the thinner windings found in other motors. These rods are connected in such a way as to resemble an exercise wheel for a small animal. Because of this, they became known as *squirrel-cage* motors (see Fig. 14–5). The stator windings are grouped much differently than the armature windings. They are divided into three equal groups 120° apart, and each of these groups is connected to a separate phase of the three-phase power. When you look at this type of motor, however, all the wires look as if they are uniformly distributed. The rotor is located in the middle of the magnetic fields. As the fields "move" past the conductors in the rotor, a current flow and an induced voltage are created. The magnetic action between the rotor and the rotating field causes a starting torque and makes the rotor rotate without

any electrical connection. This results in a very dependable and easy-to-maintain motor. If the rotor is constructed so that it is almost parallel to the rotation axis, it will provide much more uniform torque and reduce noise during operation.

The **wound-rotor induction motor** is significantly different in construction from the squirrel-cage induction motor. Each rotor conductor is insulated and lies in slots very similar to the DC armature windings. Also, the grouping of the rotor windings is very definite and can be identified by inspection. They are usually wound for three phases with a Y connection and must have the same number of poles as the stator. Wound-rotor induction motors also have slip rings to allow banks of resistance to be connected easily to each phase of the rotor. If the resistance in this bank of resistors is set to zero, the motor will operate just the same as a squirrel-cage induction motor. Adding resistance to the rotor of the wound-rotor induction motor is a method of speed control. The advantages of this type of motor include

FIGURE 14–5
Squirrel-cage motor. (Courtesy Siemens Energy and Automation)

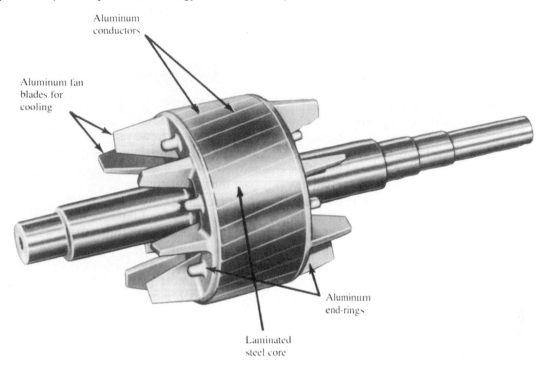

increased starting torque with less starting current. However, the cost of producing this type of motor increases significantly with the addition of brushes, insulation of the rotor windings, slip rings, and the resistance bank. These motors are less efficient with a greater power loss in the rotor and have poorer speed regulation. The speed control available through the resistance bank is not very good at low-load conditions. However, wound-rotor induction motors are very useful when high starting torque and long starting periods are required. In today's applications, the DC motor is found more and more often. The use of solid-state controls enables the use of DC motors to provide a wide range of speed control, good speed regulation, and higher starting and running torque with much lower starting currents.

Braking of induction motors can be accomplished very simply by applying DC to the field windings. This causes a stationary magnetic field, rather than a rotating one. When the rotor attempts to follow this magnetic field, the result is very quick braking. One big disadvantage of induction motors, however, is speed variation. The speed of an induction motor depends directly on the physical design and the frequency of the AC power used for the field windings. It is practically impossible to alter this frequency, since it uses the standard generated frequency for all power consumption. So, the primary method for varying speed is to make changes to the physical design of the motor itself. Increasing the number of coils per pole decreases the rotation speed of the rotor. There are trade-offs with this type of modification to the actual construction of the motor. If additional windings are added, the motor becomes larger, more expensive, and less efficient.

For applications in industry where three-phase power is available, the **polyphase induction motor** is widely used. The polyphase induction motor is typically cheaper and more efficient than a single-phase induction motor. It is also able to produce a higher starting torque with a smaller starting current. Whenever possible, a polyphase motor is preferred; however, when three-phase power is not available, a single-phase motor must be used.

To gain some of the advantages of a polyphase motor using single-phase power, a **split-phase motor** can be used. Two sets of windings are used in this type of motor, one for starting and one for running. The start winding is smaller-gauge wire, which provides a higher resistance. The run windings have more inductance, and this inductance causes a phase shift between the magnetic fields to increase the starting torque. They are designed so the starting windings will drop out after the motor has reached 70% of its operating speed. If they did not, these windings would overheat from the continuous high currents through the starting windings. One big disadvantage in the use of this type of motor is a lower starting torque than with a true polyphase motor. The low torque is due to only a 60° phase difference between the start and run coils. A larger phase difference is found in the polyphase motor.

If a capacitor is added in series with the start winding, a greater phase difference can be obtained, resulting in a higher starting torque. Such motors are called **capacitor-start motors.** Like the start winding, the capacitor is used only during the actual starting. Both the split-phase and the capacitor-start motor (see Fig. 14–6) need a method for dropping the start windings or the starting capacitor once the 70% operating speed has been reached. To disconnect the power from these, two methods are generally used: a centrifugal switch or a current relay. Unfortunately, these devices cause maintenance to increase, since they tend to fail with time and use. If a larger wire is used for the start winding, the start winding can continue to be in the circuit all the time, which eliminates the need for the switch or relay. Such a motor is called a **permanent-split-capacitor motor.** The windings are referred to as capacitor windings and main windings instead of start and run. This type of motor has the advantage of increased reliability, since the starting switch or current relay is eliminated and has smoother, but somewhat lower, starting torque. To increase the starting torque, a second capacitor can be added, which adds the expense of a second capacitor, as well as for the switch or current relay. Each type of motor has its advantages and disadvantages, and the type chosen will be the one that is optimal for the individual application. A summary of AC motors and their characteristics is given in Table 14–2.

FIGURE 14–6
Capacitor-start motor. (Reprinted with permission from James T. Humphries, *Motors and Controls.* Columbus, OH: Merrill Publishing Co., 1988.)

As you can see, the selection of a motor depends greatly on the type of application and the various different operating characteristics provided by the motor itself. The factors for choosing motors are not really within the scope of this book. A general knowledge of motors is helpful in producing drawings and understanding the documentation that is required to accompany those drawings. Motors, like many other parts encountered in the drawing experience, have ratings and standards. In this case they are provided by the National

TABLE 14–2
Summary of AC Motors and Their Characteristics

Characteristic	TYPE OF MOTOR				
	Multiphase	**Split-Phase**	**Capacitor-Start**	**PSC**	**Two-Capacitor**
Starting current	Medium	High	Medium	Medium low	Medium
Starting torque	High	Moderate	High	Moderately high	High
Reversibility	Easy, at rest or in motion	Easy, at rest	Easy, at rest	Easy, at rest or in motion	Easy, at rest
Cost (relative)	Low to normal	Normal	High normal	High normal	High normal
Horsepower range	Any	Up to 2	Up to 3	Up to 5	Up to 5
Uses	General industrial	Fans, washers	Power tools, compressors	Tape recorders	Power tools, compressors

Electrical Manufacturing Association (NEMA), and most manufacturers use this code when identifying their motors and control components.

CONTROL CIRCUITS

When motors are used in circuits, there must be a method of controlling the motors. Control can be as simple as a method for starting and stopping—a switch that provides line voltage or removes it. Typically, however, a little more than that is required.

Control circuit voltages are often lower than the voltages used for the motor. There are two primary reasons for using a lower voltage: safety and the need to standardize on values for control devices. In most control circuits, single-phase 120 V is used. To get from the line voltage to a usable control voltage, a step-down transformer is used. The symbol for a transformer is shown in Fig. 14–7. The coil connected to the line voltage is called the **primary** and the coil connected to the control circuit is called the **secondary.** Both the primary and the secondary can have a single coil or more, depending on the ratio between the two. Because the transformer is not capable of generating power, the power on the primary side must be equal to the power on the secondary side. Remember that power is the voltage times the current. So, if the voltage is being stepped down, the current will be stepped up by the same ratio.

Protective Devices

In addition to the power for the motor and the control circuit, protective devices are required. Fuses, disconnect switches, and circuit breakers are used to disconnect the energy from the circuit. The disconnect switch is often a three-pole switch ganged together so that all three lines can be disconnected at the same time with a single handle. These devices are shown in Fig. 14–8.

Protection for circuits can be provided by fuses, circuit breakers, or thermal overloads. **Fuses** and **circuit breakers** perform similar functions. They are both used to break the circuit and stop current flow if an unsafe condition occurs. Another device used to open an unsafe

FIGURE 14–7
Symbol used for a transformer.

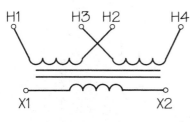

FIGURE 14–8
Disconnect switches ganged together so that they all operate at one time.

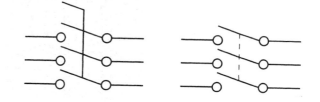

circuit is called simply an **overload device.** This device is sometimes called a **thermal overload** because the contacts will open due to high temperatures. It is often found in motors and motor starters. Fuses can be standard, one-time fuses that must be replaced after they trip or they can have a renewable fuse link. Time-delay fuses are used when heavy overloads may occur for a short period of time, such as high currents during starting. In this case, the circuit must remain closed, but if for some reason the high current remained for a sustained period, this fuse would trip and the circuit would open.

In some applications a short circuit even for a fraction of a second could cause dangerously high currents and result in damaged equipment. In these cases, the current-limiting fuse is used. The symbols used most often to represent fuses, circuit breakers, and overloads are shown in Fig. 14–9.

Switches

Circuit breakers are used for load switching as well as protection. There are a number of different kinds of circuit breakers, both nonautomatic and automatic. The nonautomatic kind is used as a disconnect or a circuit interrupter for load switching and isolation. These cir-

FIGURE 14–9
Symbols for thermal overloads, circuit breakers, and fuses.

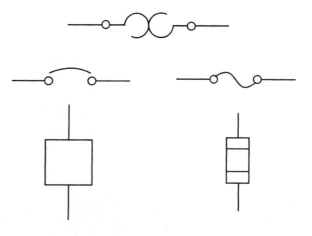

cuit breakers can have a thermal strip added for automatic tripping. A **thermal trip unit** has a time lag or delay, since it depends on temperature rise. A momentary overload would not trip this circuit breaker. If there is a short-circuit situation, a much faster means must be used to disconnect power. A **magnetic trip unit** is used to trip instantaneously on a short circuit or even a momentary overload. A combination of the two units is often desirable, since a magnetic trip works best with short-circuit high currents and a thermal trip works best with sustained overloads. Figure 14–10 shows the symbols for circuit breakers with thermal overload and magnetic trip units included. For three-phase applications, these circuit breakers can be ganged together and disconnected manually with a single handle.

Disconnect switches are used primarily to connect and disconnect the primary power lines. They are not used to start and stop motors. Start and stop switches come in a large variety of sizes and shapes, with many different types of operators or actuators. They can be manual or automatic, momentary contact, normally open, or normally closed. In some applications the color of the operator is critical. Certain colors have become associated with certain functions. For example, red is generally used for stop switches and green is used for start switches. Although the symbols are not necessarily different, the switch symbols are used more extensively in control circuits. More types of switches are used. Each different type of switch requires a slightly different symbol. All switches indicate that some contact is made and also indicate a direction of movement, or *throw,* for the switch. Switches are identified by the number of *poles,* or contacts, that they can make and the throw, or number of directions in which they can be moved. For example, a typical switch used to control the over-

FIGURE 14–11
Two-position, single-pole, single-throw switch.

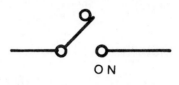

head lights is called a single-pole, single-throw switch. It is represented by the switch shown in Fig. 14–11. The switch can be opened or closed, but makes contact with the circuit in only one direction; therefore, it is single throw. This same switch symbol could be used to represent a single-pole, double-throw switch if the switch were used as a selector switch and if the off contact were connected to another line. Since most overhead lights need to have only one wire opened to stop current flow, the single-pole designation means it is connected to only one wire. Opening and closing the switch makes or breaks contact for that wire; hence, it is single pole. Switches are used to make and break contact manually, as in the light switch, automatically, as in liquid level, pressure, or flow, or by many other means. A switch that is used to indicate liquid levels, or a float switch, is not the same as a switch used to indicate pressure or temperature. Symbols for different types of switches are shown in Fig. 14–12. For a more complete list of switches available, consult Appendix F.

Automatic switches are used for controlling many different kinds of circuits. A level switch may start a pump when a fluid level is too high or too low. A limit switch may stop a conveyor belt when an object passes a certain point. A proximity switch can stop machinery when a person or object moves within a danger zone.

FIGURE 14–10
Symbols representing a circuit breaker with thermal overload and magnetic disconnect for both single-line and three-phase application.

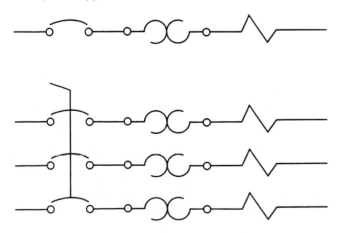

FIGURE 14–12
Methods to represent different types of switches graphically: (A) momentary contact, normally closed; (B) momentary contact mushroom head, normally open; (C) limit switch, normally closed; (D) time delay, normally open, timed closing when energized.

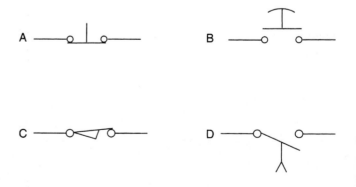

FIGURE 14–13
Mechanical connection between two contacts of a limit switch, represented by dashed line, indicates that when one is open the other must be closed.

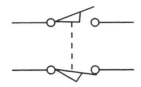

Switches are identified as having normally open and normally closed contacts. Although these are not electromechanical devices, they are mechanical devices that can be wired as normally open or normally closed; both sets of contacts can even be used. In the situation when both sets of contacts are used, the mechanical relationship or connection between the two is often indicated with a dashed line, as seen in Fig. 14–13; this line means that when one switch is open the other is closed, and vice versa.

Another important component in some control circuits is the **indicator light.** These lights are used in many applications either to indicate that power is applied to a circuit or to warn of a dangerous or alarm situation. As with the push button, different colors for different lights are generally accepted to mean certain things. Red is almost always used for dangerous or alarm situations and green or clear is used for normal operating conditions. Indicator lights are drawn as shown in Fig. 14–14. The letter in the circle represents the color of the lens, which can be glass or plastic. Sometimes the slash marks are not included, and only the letter indicating the color is used. In this case, it is important to include a comment on the diagram to identify it as a lamp and perhaps note what a lighted condition indicates.

Control Devices

Control relays, time delay relays, and *contactors* are all used in control circuits. The **relay** is an electromechanical device, a combination of electrical and mechanical functions. An electrical current passing through the re-

lay causes a corresponding mechanical action. As the name implies, the relay device consists of a coil of wire formed from a number of turns of wire wrapped around a magnetic core. When a current is caused to flow through the coil, a magnetic field is established and the coil becomes an electromagnet. The mechanical function of the relay coil occurs when this magnetic field is established. A set of *contacts* is located close to the magnetic coil, and the magnetic action causes these contacts to be either pulled together or pulled apart. These contacts are used to make and maintain a complete circuit or to break a circuit. Many relays have multiple sets of contacts. These contacts will be either normally open (**NO**) or normally closed (**NC**). When the relay is at rest or is de-energized, these conditions exist. Contacts are always drawn in their *de-energized* state so that when you are analyzing a circuit you will know that a normally open contact will close when the relay is energized. The symbol for contacts is two parallel lines, as shown in Fig. 14–15, and the symbol for the coil portion of the relay is a circle with a letter to help identify it. Many different letters are used to indicate the coil. CR is often used to identify a control relay, but it is just as common to use other letters, such as an M for motor starter, TR for timed relay, or simply F and R for forward or reverse direction of motor movement. The control relay should not be confused with a motor starter, which is discussed a little later in this chapter. Also, the control relay should not be confused with the motor itself. There is little chance for that when the entire circuit is shown. Company policy will often dictate the lettering identification needed for a set of control relays, but, if not, the two basic things to remember are to keep the identification simple and to be consistent in the method you use.

Figure 14–16 shows a simple control circuit with two relays, three normally open contacts, and an indicator light. Notice that the contacts are labeled showing to which relay they are connected; also, they are numbered sequentially. No contacts are shown in this small portion of a circuit for the second control relay. 1CR-1 is used to "hold" or "seal" the circuit closed after 1CR is energized. This is referred to as an *interlock contact* and maintains the circuit after the start button has been released. Contact 1CR-2 is used to complete the circuit to the indicator light, and contact 1CR-3 completes the

FIGURE 14–14
Symbol for an indicator light and its use in a typical circuit.

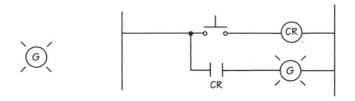

FIGURE 14–15
Symbol for a control relay and two sets of contacts, one normally open and one normally closed.

FIGURE 14–16
Simple control circuit with two relays, three sets of contacts, two switches, and an indicator light.

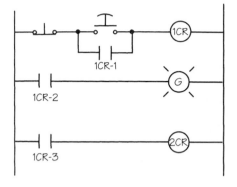

circuit so that coil 2CR is energized. It is obvious that this is only a portion of a larger control circuit, since it would be pointless to include a control relay that was not used to control anything.

In some applications a small delay may be required between the time a relay is energized and the contacts are activated. The delay could come in four different configurations. For example, it could be a delay after energizing or a delay after de-energizing. The configurations are as follows:

1. Normally open, time delay on closing
2. Normally closed, time delay on opening
3. Normally open, time delay on opening
4. Normally closed, time delay on closing

The first two represent a time delay when energized, since normally open contacts close when energized and normally closed contacts open when energized. This same principle is applied to the second two. The symbols for these configurations are shown in Fig. 14–17.

FIGURE 14–17
Symbols for various time delay configurations: (A) normally open, time delay on closing; (B) normally closed, time delay on opening; (C) normally open, time delay on opening; (D) normally closed, time delay on closing.

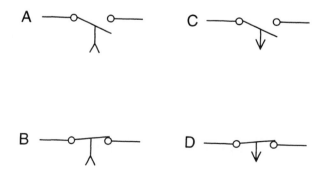

Note that the symbols used to represent the time delay contacts look like switches. Also, when discussing switches, contacts are often mentioned. The interchanging of reference to switches and contacts is understandable when you remember that each of these devices performs the simple task of opening and closing, or completing, a circuit.

The *contactor* is also an electromechanical device; it works in the same way as the control relay. A contactor also has sets of normally open and normally closed contacts. The primary difference between a contactor and a control relay is the size and type of load they are able to control. Contactors are used for much heavier currents and are capable of handling up to 2250 A. They can be found in high-resistance or heavy industrial solenoid applications. The symbols used for contactors are the same as those used for control relays. If separate overload circuit protection is available, a contactor can be used to start and stop motors.

Most often, however, an actual *motor starter* is used. A motor starter is very similar to the control relay and the contactor. It is an electromechanical device with contacts that operate when a coil is energized. The primary difference between them is that motor starters generally have built-in overload protection. These motor starters come in a range of sizes (see Fig. 14–18) and are used in single-phase, two- and three-phase, and also three-phase, four-wire applications. There are manual motor starters that provide overload protection but very limited control. Most often, motors require some remote operation and added control safety and convenience. In such cases a magnetic motor starter is used. Overload relays are added to the motor starter and are generally thermal elements. The biggest need for protection for motors is for sustained periods of heavy load. Thermal overload protection allows high currents to pass for a period

FIGURE 14–18
Full-voltage starters. (Courtesy Eaton Corp., Cutler–Hammer Products)

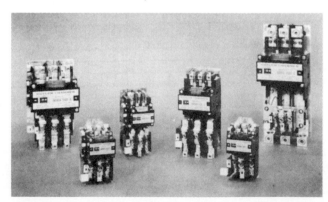

FIGURE 14–19
Simple motor and control circuits representing full-voltage, across-the-line starting.

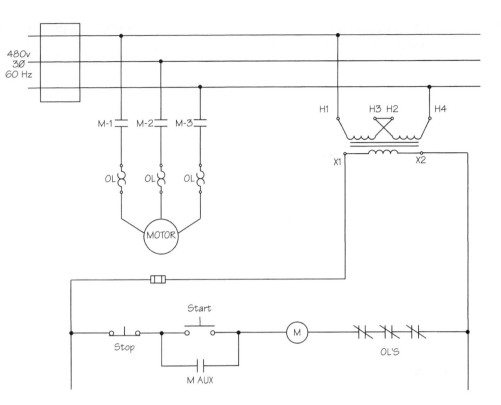

of time and then trips open and removes power from the starter coils. When the coil de-energizes, its contacts open and disconnect the motor from the live power. Remember that overload relays do not protect motors or control circuits from an instantaneous short circuit. This type of protection requires the use of circuit breakers or fuses. Most motor starters have three sets of load contacts for use with three-phase motors and an additional high-current auxiliary contact. This auxiliary contact is often used to provide the interlock contact around the start switch. The contact is often called the *seal-in* contact and, when energized, is said to seal the start circuit.

Starters

There are many different types of starters in the following categories: full voltage, reduced voltage, reversing, and multispeed.

Full-voltage or across-the-line starters are the simplest form of control and the simplest to draw. A complete circuit showing the control and motor portions is given in Fig. 14–19. Note the step-down transformer used to go from 480 to 120 V for the control circuit.

Reversing motor starters need an additional motor starter with another set of contacts to reverse the connections on two phases of the motor. Note the sets of normally closed contacts shown in Fig. 14–20

FIGURE 14–20
Simple motor control circuit representing both forward and reverse control with mechanical interlock between the two coils.

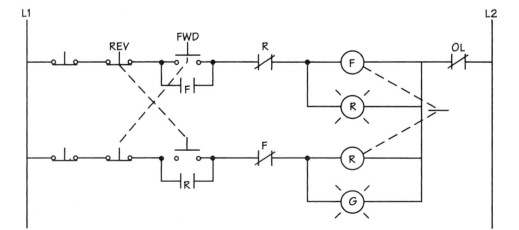

next to each motor starter. This is an electrical interlock to prevent both motor starters from being energized simultaneously. In addition, a mechanical interlock is often used. The start switch in each motor-starter branch is mechanically linked so that when one closes the other one opens. This mechanical interlock is demonstrated by the dashed lines shown in the illustration.

The terms *two-wire control* and *three-wire control* are often used to describe control circuits. Unfortunately, it is not as simple as two wires versus three wires. The concept is, however, not too difficult to understand. Figure 14–21 shows both a two-wire and three-wire control circuit. Another way to describe the two different circuits is no-voltage or low-voltage release and no-voltage or low-voltage protection. In a two-wire system, or low-voltage *release,* if for some reason the line voltage drops out the motor will stop. If the contacts on the control device, such as a pressure switch, limit switch, and so on, are still closed when the line voltage returns, the motor will restart. For three-wire control circuits, or low-voltage *protection,* when the line voltage is removed, the coil de-energizes and the interlock contact drops out. In this circuit the motor will not start again until the start button is pushed. A three-wire circuit can provide additional safeguards to keep the motor from restarting after a power loss or unsafe condition. However, the disadvantage of this type of control circuit is that some physical intervention is required to restart the circuit.

FIGURE 14–21
Two-wire control circuit versus three-wire control circuit.

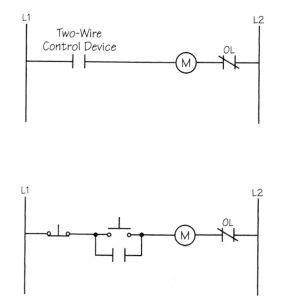

TYPES OF DIAGRAMS

Two basic types of diagrams are encountered frequently in control circuit drawings: (1) the *one-line diagram* and (2) the *ladder diagram.* The one-line diagram includes everything in a single line, which is read from the top to the bottom. The one-line diagram is most often used in power distribution drawings. It represents one phase of a three-phase system (where each phase is identical), which if all three were drawn would represent the entire distribution system. The reference to one line means that an entire function can be described in one line from the power source to ground. For motor-control circuits the ladder diagram is most often used. The diagram itself looks somewhat like a ladder, with two long vertical lines representing the primary power source and all the components connected horizontally (like *rungs*) between the power lines.

The primary difference then is the application of distribution or control functions. As can be expected, one-line diagrams can be found most often in utility or power companies to represent their distribution systems across the country. These are discussed more fully in Chapter 17. Ladder diagrams are used in industries that need some control functions performed by switches, relay coils, and motors.

Ladder Diagrams

Ladder diagrams consist primarily of motor-control circuits and include all the circuitry necessary to energize or de-energize these motors. The ladder diagram, like other circuit diagrams, is prepared so that it can be read from the upper left-hand corner, across to the right, and down the page. The diagram is constructed so that the main power lines are drawn as vertical lines down either side of the diagram. All circuit functions are drawn on horizontal lines connected between the two main lines. Frequently, control functions are produced sequentially, and the horizontal lines indicate the sequence.

The control circuit consists of two basic parts: the actual motor circuit, or *power* portion, and the *control* portion. The use of the term *ladder diagram* stems from the layout of the control portion of the circuit. Figure 14–22 illustrates that an easy visual distinction could be made between the two portions of the control circuit if the motor portion were prepared with a dark, medium to heavyweight line and the control portion were prepared with a light, thin-weight line.

Spacing requirements for the layout of ladder diagrams are determined by the amount of lettering that must be included between the horizontal lines. There should be sufficient room to ensure that the lettering

FIGURE 14–22
Control circuit showing heavy, dark lines for the motor portion and light, thin lines for the control portion.

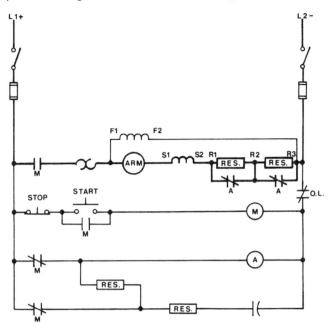

FIGURE 14–23
A single stop button to de-energize all coils at the same time.

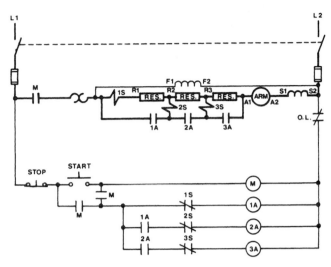

will not run into the linework even after reductions. Keep in mind that when a drawing is reduced the overall size is reduced, but the line width and the space between the lines are also reduced. A drawing that may seem comfortable to read on a large (D size) print may become impossible to read after it is reduced. Most lettering on diagrams should be from ⅛ to ¼ in. throughout. Thus, the space between the horizontal rows on a ladder diagram should be no smaller than ½ to ¾ in., which will ensure that the lettering can be included without confusion.

The length of the vertical main power lines is determined by the number of horizontal lines needed for the control portion of the circuit. The width between the vertical power lines is determined by the horizontal line that contains the largest number of devices. Once that line has been located, it is used to help lay out the rest of the circuit. It is not uncommon for the control portion of the circuit to be a different width than the motor portion. One of the most obvious reasons is shown in Fig. 14–23, where the stop button needs to de-energize all the coils at the same time. The only way to be certain that this will be done is to make all the control lines connect to the side of the stop button that will place the button between the coils and the main power line.

As for all other circuits, it is highly desirable to present a well-balanced and carefully laid out diagram. When placing the devices and components on each hori-

zontal line, be sure that they maintain some form of alignment. For example, many times a lengthy control circuit contains several rows of sequential circuit functions with a pair of contacts and a coil. All the pairs of contacts should be aligned vertically, and the centerlines for the circles used to represent coils should also be aligned vertically, as shown in Fig. 14–24.

These coils and pairs of contacts may not line up with the main contactor and pairs of contacts in the first line of the control circuit. Generally, this line contains more devices than the other lines because it usually includes the start and stop switches and overload devices that other coils do not need. When this is the case, the first line should be well balanced and nicely spaced. Equal distance between each device gives a well-balanced drawing. This may be a little difficult to achieve without some practice. The width of each device is dif-

FIGURE 14–24
Vertical alignment for coils and contacts.

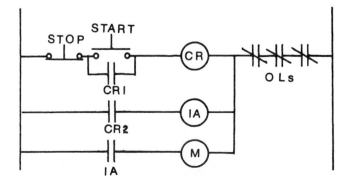

FIGURE 14–25
Different devices maintaining consistent size relative to each other: circles, ½ to ⅝ in.; contacts, ¼-in. lines, ³⁄₁₆ in. apart; overall switch length, ½ in.; and resistor rectangle, ³⁄₁₆ × ⅝ in.

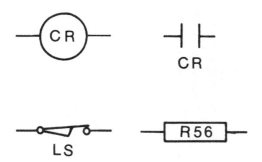

FIGURE 14–26
Very simple motor circuit with manual ON/OFF control and single operating speed.

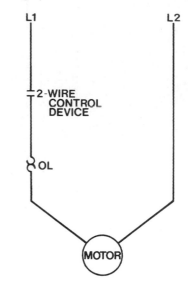

ferent and has to be taken into account when the overall spacing is determined.

As shown in Fig. 14–25, a good size for the circles is ½ to ⅝ in. diameter, which leaves enough room to add the letter or reference inside the circle. The pairs of contacts should be drawn with ¼-in. parallel lines located approximately ³⁄₁₆ in. apart. All the other devices should be drawn relative to these sizes. For example, the switches should be about ½ in. overall, with the rectangles for the resistors drawn about ³⁄₁₆ by ⅝ in., which allows for ⅛-in. lettering to be included within the rectangle.

DC CONTROL CIRCUITS

Control circuits must provide a number of control functions pertaining to the operation of the motor. The type of function is determined by the actual use of the motor. For example, the control functions of a motor used in an elevator are much different from the control functions of a motor used in a heating system. In both instances, however, the primary functions of starting, stopping, running, speed control, and protection all have to be considered. The simplest circuit may include only a two-position switch for the starting and stopping and a single speed and direction for the running portion, as shown in Fig. 14–26. There should be some type of circuit protection, such as overloads, on even this simple a circuit. The other extreme can include time delays, levels of speed control, forward and reverse functions, and automatic starting, as shown in Fig. 14–27.

The main power lines for the DC control circuit are shown as positive and negative. As a general rule, the vertical line on the left is used as the positive, which allows conventional current to flow from left to right across the diagram. If any switches, circuit breakers, or

fuses are needed for the main power lines, they should be drawn very near the top of the diagram. In preparing the location of all the circuit devices to be placed on the horizontal lines between the main power lines, keep in mind that switches, contacts, resistors, or other similar devices and components are placed on the line so that they are encountered first. The motor starters or relay coils are located toward the right or at the end of that line toward the negative power line, as shown in Fig. 14–28. When resistors are used in the control circuits, the reference designation and value of the resistor are normally lettered within the rectangle used to represent the resistor. Designations for the relay coils and other contactors are also lettered within the symbols used to represent these devices.

The connecting lines used in control circuits must follow the conventions for crossovers and junctions. The method of using a dot for a connection and no dot for a crossover, as shown in Fig. 14–29, is almost always used. The dot is used to indicate that an electrical connection exists between the two wires that are crossed. When a crossover is drawn and no dot appears, it simply means that the lines pass by each other and do not make any electrical connections. As on all other diagrams, crossovers and changes of direction should be kept to an absolute minimum. So that it is easier to follow circuit paths and to document the control circuit, the horizontal lines are given a number. All horizontal lines should be numbered even if they do not extend across the entire diagram from one main power line to the other, as in Fig. 14–30.

FIGURE 14–27
Portion of a larger, more complex control system.

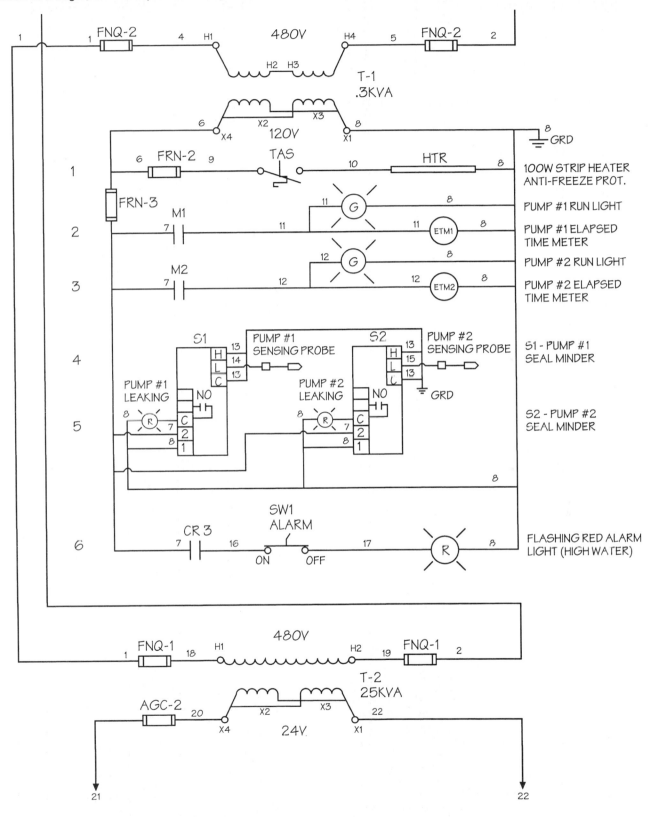

FIGURE 14–28
Contact located toward the left side of the control line and coils situated closer to the right side.

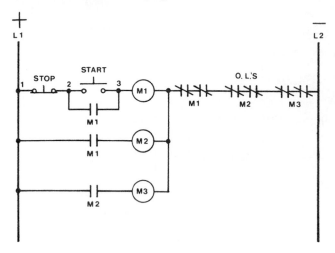

DC Motor Circuits

For the DC control circuit diagrams, the motor portion is included as additional horizontal lines connected between the two primary or main power lines. The line includes two sets of main contacts, with one on each end of the line, the armature of the motor, any starting resistors or other means of direct motor control, and all the necessary overloads. The motor circuit portion normally contains two lines, an armature, and a field line that includes the field coil, a variable resistor, and a field loss coil, which keeps the motor from running extremely fast if the field becomes open.

In Fig. 14–31, the vertical lines represent 250 V DC. Both the motor circuit and the control functions are shown on the diagram. Notice that only the lines of the control portion of this circuit are numbered. The first device is a main disconnect switch to remove all power from the circuit. Both lines, positive and negative, contain the disconnect switches and a fuse to provide overload protection. The first horizontal line includes the armature portion of the motor. In this circuit, speed control is provided and is represented by the rectangle for the resistor and contacts connected at intermediate points. Notice that this line includes the contacts for the motor starter (M). There are two sets—one on each end of the line. The second line also contains a portion of the motor circuit, frequently referred to as the *field circuit* because it represents the field portion of the motor. This line includes the field windings, the field rheostat, and the field loss relay for protection. Lines 1 through 5 contain the motor control circuit. The control

FIGURE 14–29
Connections indicated with a dot. (When wires simply cross over, there is no dot, indicating no connection.)

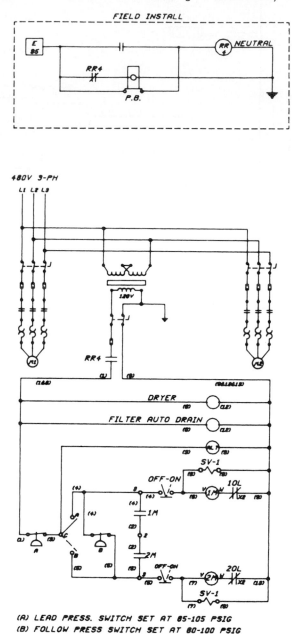

(A) LEAD PRESS. SWITCH SET AT 85-105 PSIG
(B) FOLLOW PRESS SWITCH SET AT 80-100 PSIG

AIR COMPRESSOR

portion of the circuit includes all necessary devices to provide control functions, such as starting and stopping and, in this case, speed control of the motor. Speed control is provided by the resistor in the armature circuit.

FIGURE 14–30
Control lines numbered on the left hand side for ease in tracing circuit functions.

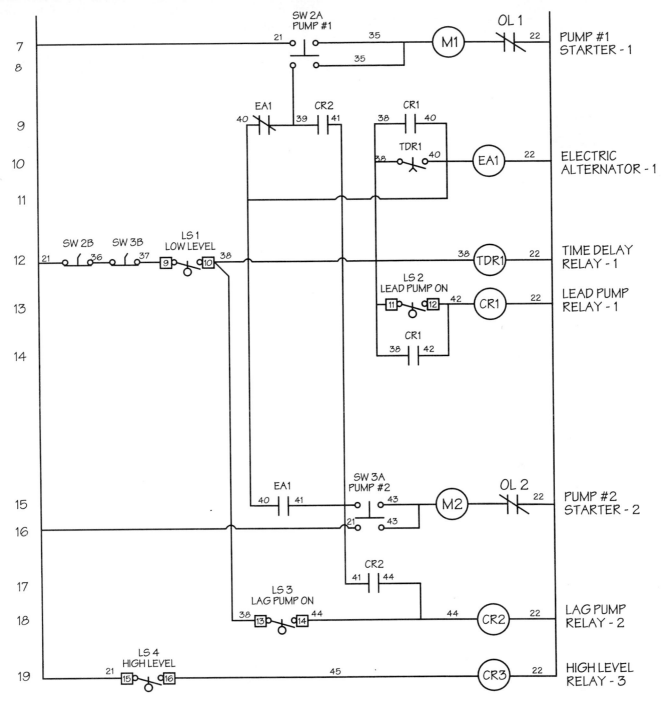

AC Motors and Control Circuits

The primary difference between AC and DC motor control circuits is in the motor itself. Either motor can be controlled by a DC control system. The difference is that the AC motor itself needs to be connected to an AC power supply, either single phase or multiple phase. Typically, an AC motor operates on a three-phase system, which is fairly easy to produce.

For this type of diagram it is not uncommon for the motor circuit to be arranged horizontally and the control circuit to be arranged vertically. Figure 14–32 shows a

FIGURE 14–31
DC control circuit and motor circuit.

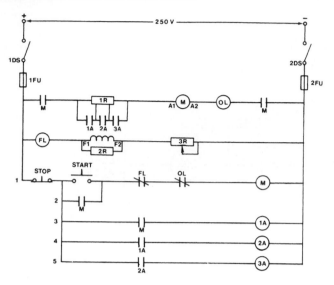

FIGURE 14–32
AC motor using AC control with push-button speed selection.

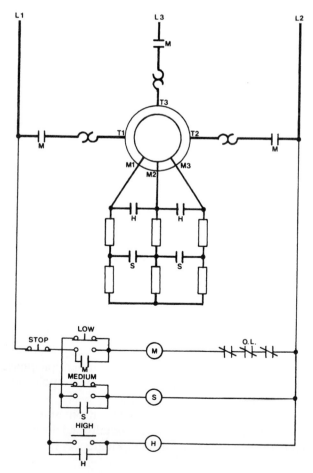

typical motor control circuit, which includes both an AC control circuit and AC motor circuit. Notice the dark, heavyweight lines used to identify the motor portion of the circuit. The three lines are identified as L1, L2, and L3. They indicate the main power supply, which should be three-phase power. Shown connected directly to the rotor windings are the resistors and contacts used in the speed control of the motor circuit. The control circuit itself is connected between L1 and L2. Sometimes the main power supply is too large to be safe for the control operations or personnel, and a transformer is connected between the main power lines and the control portion of the circuit, as shown in Fig. 14–33.

The motor portion of the circuit is not always represented with the horizontal orientation. It is just as commonly presented with the power lines as vertical lines and the control portion drawn to one side or the other, as in Fig. 14–34.

STANDARDS

The use of widely accepted standards has been suggested frequently in this text. The Joint Industrial Council (JIC) publishes standards for control circuits and, as mentioned in earlier chapters, the use of these standards by industry is strictly voluntary. Most companies, however, find it convenient to use these widely accepted standards in the preparation of all *in-house* material, as well as for materials produced for publication.

The elementary or ladder diagram shown in Fig. 14–35 is used to demonstrate drawings conforming to JIC standards. The motor circuit is included at the top of the page and is oriented horizontally with a three-phase main power supply. The control circuit portion is connected between two of these lines and is oriented vertically. Notice that this circuit includes two motors, not just one. These two motors are connected to be started with the same set of main contacts. This particular circuit contains many different types of control functions. Although this circuit is meant only to demonstrate good layout techniques, it is important to note that a more complex control circuit may be very similar to this.

The control portion of the circuit is connected by way of a transformer to step the voltage down to suitable levels. Each line shows excellent spacing and vertical alignment. All the control relays are in line vertically and are located nearest the right-hand power line. Each line in the control circuit is assigned a *line number*, which is used simply for reference and identification. Line numbers are found on the left-hand side of the drawing. On the right side, for more complex circuits a short description of circuit function is frequently included. In addition to the actual circuit and the notes

FIGURE 14–33
Transformer used to step down voltage for control portion.

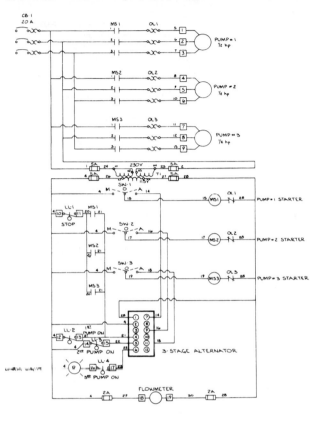

FIGURE 14–34
Vertical orientation for main power lines to motor with control portion located off to one side.

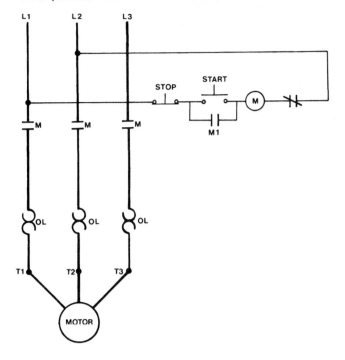

on the circuit, it is necessary to include documentation that describes the sequence of operation. This sequence should be a complete description of how the circuit operates. It is often included on the same drawing as the ladder diagram.

In addition to the ladder diagram that describes the circuit function or operation, a wiring diagram is prepared that actually demonstrates how the control circuit should be wired. Wiring diagrams were covered extensively in Chapter 13. One type of drawing discussed was the point-to-point wiring diagram. Most wiring diagrams for control circuits using electromechanical control are the point-to-point type of diagrams.

The primary difference between the control circuit diagram and the wiring diagram is in the layout. The control circuit is a functional representation and is laid out for the most convenient visual recognition of the circuit functions. The wiring diagram is most concerned with a quick visual recognition of the actual wiring lay-

out. As illustrated in Fig. 14–36, this diagram is frequently prepared to show the physical relationship between devices and connecting wires. This wiring diagram could be the physical representation for the circuit shown in Fig. 14–34.

USE OF CAD

As in all other types of drawings, it is possible to use a CAD system to aid in the preparation of control circuit diagrams. Figure 14–37 shows the typical electrical symbols as they were generated by a CAD system. These symbols can be in the library of parts and can be placed anywhere on the circuit diagram with the use of a light pen or digitizer tablet. It is the ultimate responsibility of the drafter to maintain company standards in all drawings or diagrams that are prepared.

FIGURE 14-35

Ungrounded control circuit. (Courtesy of Joint Industrial Council)

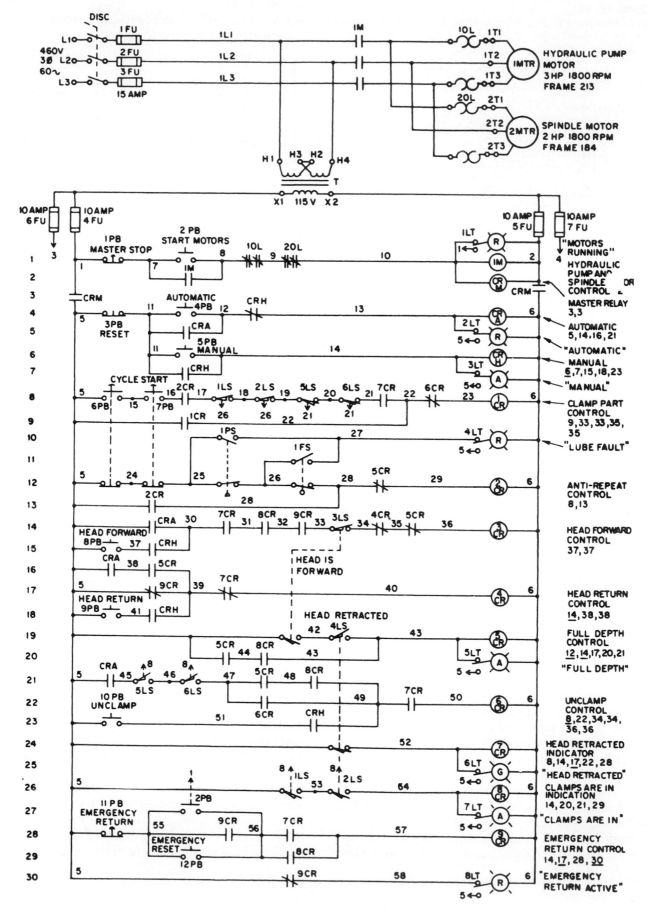

FIGURE 14–36
Typical wiring diagram showing physical characteristics and placement.

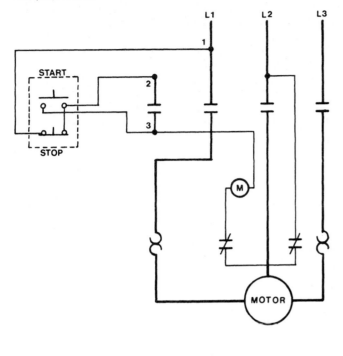

FIGURE 14–37
Electrical symbols generated by a CAD system. (Courtesy of Interactive Computer Systems, Inc.)

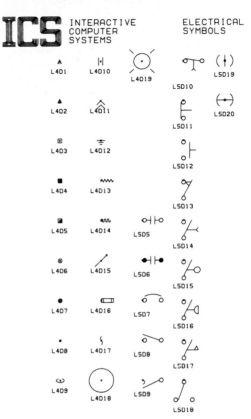

REVIEW QUESTIONS

14–1 List the basic types of drawings found in control circuits.

14–2 Why does a one-line diagram or single-line diagram usually have more than one line?

14–3 Do control circuit symbols conform to the same standards as symbols used in schematics diagrams?

14–4 How are the motor portions and control portions represented differently on drawings?

14–5 Using Appendix E, determine the meaning of the following reference designations:

    a. CB   e. R
    b. F   f. TB
    c. LS   g. T/C
    d. Q

14–6 What two basic circuits are needed to draw a motor control circuit?

14–7 How are contacts drawn on the circuit, in their energized or de-energized states?

14–8 Why are horizontal lines in the control circuit numbered?

14–9 What must be done if the control circuit needs less voltage than the motor circuit?

14–10 Why are CAD systems being used so often in industry today?

PROBLEMS

14–1 Practice drawing different types of switches. If a template is not available, use the JIC symbols located in Appendix G and construct them using straightedges and circle templates.

14–2 Draw circuit protection devices (fuses, thermal overloads, and circuit breakers) both with and without the use of a template.

14–3 Redraw the circuit presented in Fig. 14–22. Be sure to use heavy dark lines for the motor portion and lighter thin lines for the control portion. Use ink if available.

14-4 Draw a sketch of Fig. 14–23. Be sure that the motor portion is done with lines that are darker than the control lines.

14-5 Practice drawing the devices in Fig. 14–25 using the size guidelines as indicated. Use both pencil and ink and make as many as necessary to produce acceptable drawings.

14-6 Practice lettering by hand and with the use of a lettering template by redrawing Fig. 14–27. This drawing should be prepared on at least a C size sheet. Make certain that the lettering does not become crowded or confusing.

14-7 Redraw Fig. 14–34 using a horizontal orientation for the motor portion of the circuit and locate the control portion in a more typical configuration horizontally across the bottom.

14-8 Draw the wiring diagram showing physical location in Fig. 14–36. Prepare this on a B size sheet. If a template is not available, construct each symbol using the guidelines presented in this chapter.

14-9 through 14–13 Draw each of the control circuit diagrams on a C size sheet.

PROBLEM 14–9
Control circuit.

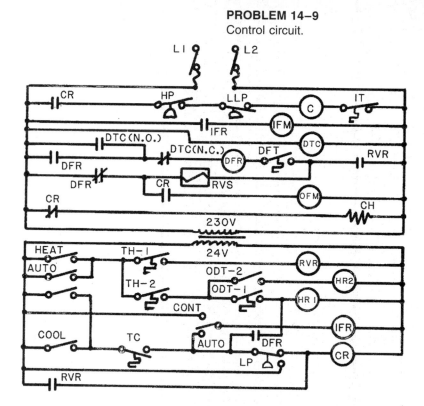

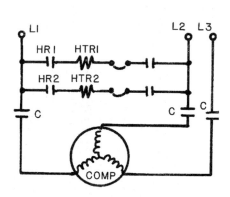

PROBLEM 14–10
Thermostat control circuit.

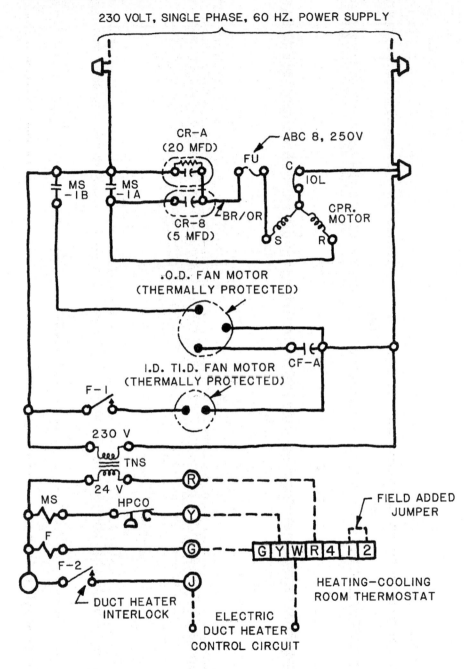

PROBLEM 14–11
Control circuit.

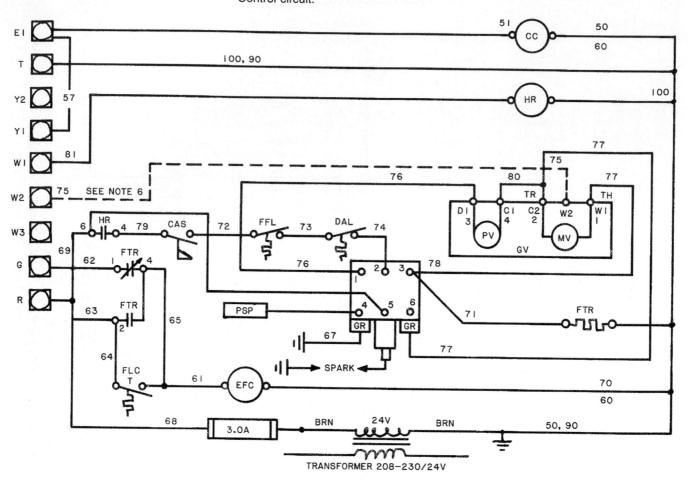

PROBLEM 14–12
Control wiring diagram.

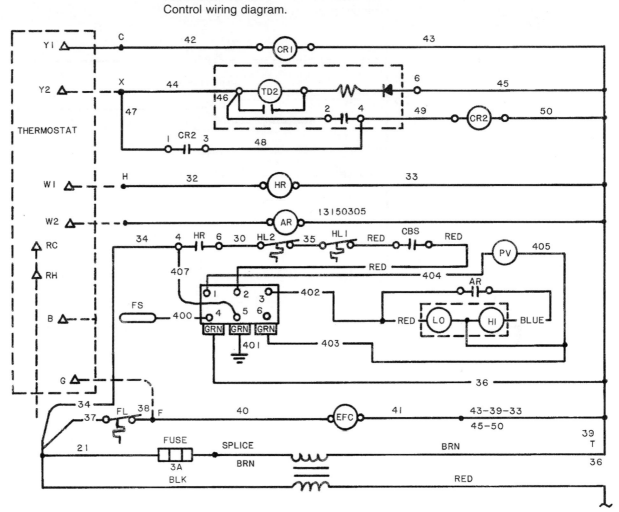

PROBLEM 14–13
Heat pump control circuit.

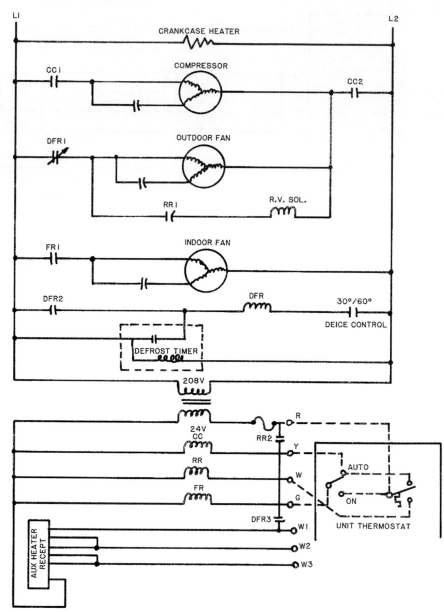

15

LOGIC DIAGRAMS

INTRODUCTION

A **logic diagram** is a simplified diagram of a digital system, program, or process in which special shapes or geometric figures and interconnecting lines are used to represent the relationship of parts. The use of special shapes simplifies the diagram because the actual shape represents the *logic function* that is being illustrated.

Unfortunately, not all logic functions can be represented by graphically unique shapes. Frequently, block diagrams must be used with all the circuit functions described within the block or surrounding it. However, even when a circuit function can be represented by a unique or single shape, the electrical circuit itself can be rather complex.

The actual internal circuitry has become significantly unimportant. Each function discussed in this chapter can be contained in a single integrated circuit or IC. This package may contain a few components or as many as thousands, and it will still be represented by the basic logic symbol or block and the logic function. For this reason, digital logic circuits or drawings are more like block diagrams than schematic diagrams, which show individual components. Sometimes, however, logic diagrams contain wiring information and other interconnection information concerning both

logic and nonlogic functions. For this reason it is not uncommon to find diagrams labeled *Logic Schematic Diagram,* as in Fig. 15–1.

The logic functions generally have one or more inputs and one or more outputs. The primary purpose of each type of logic circuit is to cause an output that is either *on* or *off*. This output is frequently referred to as *open* or *closed;* hence these types of logic functional circuits are called **logic gates.**

LOGIC DIAGRAMS

Different types of logic diagrams are used to portray logic circuits and functions. The most elementary is the *basic logic diagram,* which shows the functional logic relationship using simple logic symbols. This type of diagram makes no reference to physical relationships. Remember that the logic circuits themselves may be contained in ICs, but the basic logic diagram simply uses symbols representing the logic functions.

A *detailed logic diagram* includes the symbols not only for logic functions but also for nonlogic functions, connection points, pin numbers of individual circuit parts, and any other information necessary to describe the circuits, both functionally and physically.

FIGURE 15–1
Logic schematic diagram.
(Courtesy of Motorola, Inc.,
Semiconductor Products Sector)

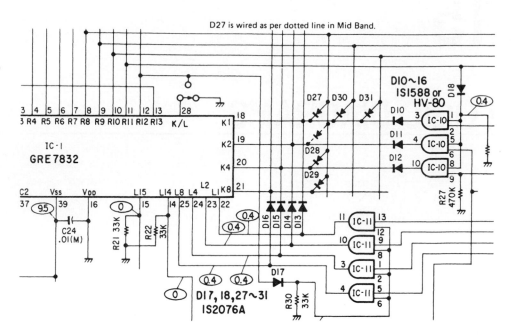

Both these types of diagrams can be drawn with the unique shapes of common logic gates. Less common functions can be represented by rectangles that are labeled describing the function. An earlier method of portraying both logic and nonlogic functions was done with rectangles for all functions. It was necessary then to describe each block in the circuit diagram. The advantage of using unique symbols is evident in that it cuts down on lettering time and allows quick identification of circuit functions.

Logic Rules

Logic means valid and efficient decision making. If we apply the definition of logic to the study of digital electronics, it also means that the decision is displayed as the output of the functional circuit. The logic circuit has one or more inputs and is generally limited to one output. The function of the circuit, then, is to consider the inputs, make a valid and efficient decision, and provide an output.

Basic **logic rules** apply to gates. These gates are thought of as switches, and like switches have only two states that are to be considered: *ON* or *OFF*. The same type of logic can be applied to a ceiling light in your home that can be controlled from either end of the room. To activate the light (output), either switch can be *on* or *off* (inputs). If a switch is *on,* then the output results in the lamp being on.

Logic gates also have two states, *ON* and *OFF.* These states are not produced by switches opening or closing, but by the application or removal of a voltage

to one or more of the inputs. To keep all supporting data as simple as possible, the binary, or two-digit, number system has been applied to digital logic circuits. To describe the voltage level of the input, a 1 or a 0 is used. If the *ON* state is a +Vdc, then it is represented by the binary 1. If the *OFF* state is 0 Vdc, then it is represented by the binary 0. This same system is used to represent either an ON or OFF output state. These states are also frequently referred to as **high** and **low.**

Truth Tables and Logic Symbols

One way the logic function has been described is with a table showing what the possible combinations of inputs are and what outputs result from the different input combinations. This type of table is called a **truth table.**

A two-input gate can have four unique combinations of high and low input states. A truth table shows all the possible input combinations and the output produced by the various input combinations. The output is a function of the type of logic circuit being described. A discussion of the most basic logic gates should include the following: **AND, OR, NAND, NOR,** and **EXCLUSIVE-OR.** Each of these gates has two or more inputs and only one output. The output is a direct result of the circuit function and the combination of inputs.

The *AND* gate has an output of 1 *if and only if all* inputs are 1. If two inputs are considered, then the truth table shown in Fig. 15–2 describes this gate. The inputs are labeled A and B and the output is X. Also shown in Fig. 15–2 is the unique symbol used to indicate an AND gate. To analyze the information supplied in this

FIGURE 15-2
Symbol and truth table for the AND gate.

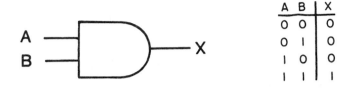

A	B	X
0	0	0
0	1	0
1	0	0
1	1	1

figure, it should be apparent that when a 1 (+Vdc) is applied to both A AND B inputs, then a 1 (+Vdc) appears at X output. For any other combination of inputs, the output at X is 0 (0 Vdc). Although often included when AND or any other gate is drawn, the input and output lines are not really a part of the unique symbols.

The *OR* gate has a 1 output *if either or both* of the inputs are 1. This truth table and symbol are shown in Fig. 15-3.

Sometimes it is necessary simply to change the state of a signal. Since this is not making a valid decision, this functional circuit is frequently not called a gate. However, more often than not engineers refer to the logic circuit that simply inverts a signal as a **NOT GATE.** Technically, it should be called an *inverter,* since the name given generally reflects the logic function being performed.

The truth table and symbol for the inverter are shown in Fig. 15-4. This gate is very easy to understand. If the input is 1, the output is 0; likewise, if the input is 0, then the output is 1 (or the input is inverted). In this case, *inverted* means changed to another state. Because there are only two states, there are only two possible combinations. The symbol contains the triangle, which simply implies direction, and the circle, which indicates inversion. This circle is always used to represent the function *inversion.*

The primary reason to be concerned with the NOT gate or the inverter becomes more apparent as we proceed to the explanation of the *NAND* and *NOR* gates. Functionally, these gates can be thought of as AND gates and OR gates with their outputs inverted. This

FIGURE 15-3
Symbol and truth table for the OR gate.

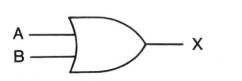

A	B	X
0	0	0
0	1	1
1	0	1
1	1	1

FIGURE 15-4
Symbol and truth table for the inverter or NOT gate.

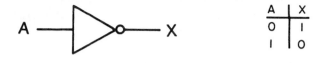

A	X
0	1
1	0

combination of gates then is described as ***NOT an AND*** or ***NOT an OR,*** which are shortened to *NAND* and *NOR.* Because these functions are required frequently, gates have been manufactured to perform these functions, thus alleviating the need to actually wire the inverter and the AND gate to make a NAND.

The truth tables and symbols for these gates are shown in Fig. 15-5. As you can see, the primary difference between the AND gate symbol and the NAND gate symbol is the small circle that is added to the output side. Similarly, the only difference between the OR gate symbol and the NOR gate symbol is the small circle added at this output. By comparing the truth table of an AND and NAND, you can see that the different combinations of inputs produce exactly opposite outputs for these gates.

The last gate frequently referred to as one of the basic gates is an *EXCLUSIVE OR (EX-OR).* This gate has a 1 output *if one or the other* inputs are 1 *but not both.* The symbol and the truth table are demonstrated in Fig. 15-6.

To aid you in preparing logic diagrams, a number of templates is available that includes the distinctively shaped symbols, as well as various rectangles and squares used to represent other logic and nonlogic functions. One such template is shown in Fig. 15-7. (Other examples are found in Chapter 1.) In addition to various

FIGURE 15-5
Unique symbols and truth tables for NAND and NOR gates.

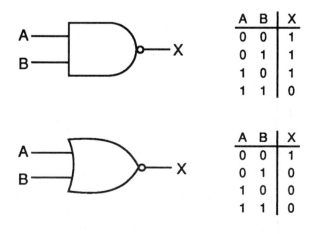

A	B	X
0	0	1
0	1	1
1	0	1
1	1	0

A	B	X
0	0	1
0	1	0
1	0	0
1	1	0

FIGURE 15–6
Symbol and truth table for the EX-OR gate.

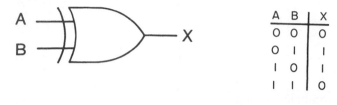

A	B	X
0	0	0
0	1	1
1	0	1
1	1	0

FIGURE 15–8
Logic diagram stick-on symbols. (Provided courtesy of Bishop Graphics, Inc.)

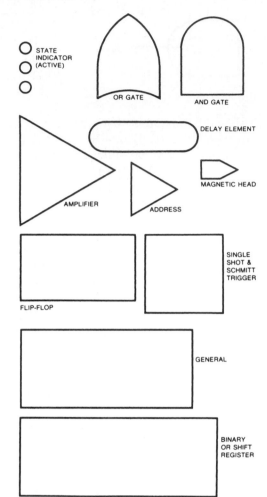

templates, dry transfer symbols of various logic symbols or circuit functions are also available, as shown in Fig. 15–8. These symbols generally are stick-on or rub-on kinds and produce a clean, crisp drawing when reduced.

In addition to the two-input gates just shown, many of these basic gates can have more than 2 inputs. It is not uncommon to have 3, 4, or even as many as 8 or 16 inputs for any of these gates. Remember that additional inputs represent additional possible combinations of inputs, but the basic definition remains the same.

For an AND gate the output is 1 *if and only if all* inputs are 1. Therefore, there is still only one possible combination that would result in a 1 output for the AND gate. Similarly, the OR gate produces a 1 output *if any* of the inputs are 1 *or if all* of the inputs are 1. The only case when the output will not be 1 is if all inputs are 0. This is the same no matter how many inputs there are. An example of a four-input AND gate is shown in Fig. 15–9.

Constructing Logic Symbols

Sometimes it is necessary to draw the logic symbols without the aid of either a template or dry transfer

FIGURE 15–7
Logic template. (Courtesy of Berol RapiDesign)

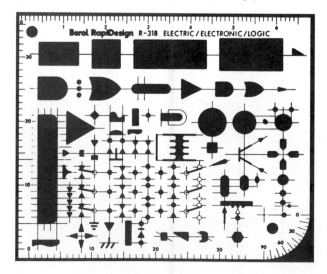

shapes. Like the circuit components shown in Chapter 8, logic symbols have no required size. Guidelines can be found in MIL-STD-806B for recommended sizes. If a template is used for drawing these symbols, there is no need to worry about a *correct* size because the size is determined by the template itself. Any drafter who regularly draws logic circuits should make certain to have these templates.

FIGURE 15–9
Four-input AND gate.

There are, however, certain requirements for the proportions used in forming arcs and lines in logic symbols. This means that there is a required ratio or relationship between the height and width and the radii of the arcs. If you need to construct these logic symbols, keep in mind the following steps. For an AND gate using an overall length of 1 in., the overall height should be .8 in. The values given in Fig. 15–10 are proportions. For example, if the overall length needs to be 2 in., the overall height will be 1.6 in. to maintain the ratio of .8/1.0 between the height and width. Centerlines need to be located and an arc drawn to finish the AND gate. It should be apparent that constructing each gate individually is a time-consuming task, so use a template for this task. Review the use of templates for inking or pencil in Chapter 1. When it becomes necessary to actually construct the logic symbols, try to keep the ratios or proportions demonstrated in Fig. 15–10.

Sometimes the number of inputs indicated for a gate requires more room than is convenient for the desired size of gate. To make certain that enough room is left for identification of the inputs, various methods are used. The least desirable—though frequently seen—method includes making bends in the input connecting leads. The method approved by ANSI is to include extensions on the edge of the gate. These different methods are shown in Fig. 15–11. The eight-input AND gate and OR gate each have extensions added to the input side of the gate symbol. A four-input AND gate is also shown with the input leads bent to allow more room for lettering; however, this method is not recommended.

FIGURE 15–10
Typical proportions to be used in the construction of logic gates.

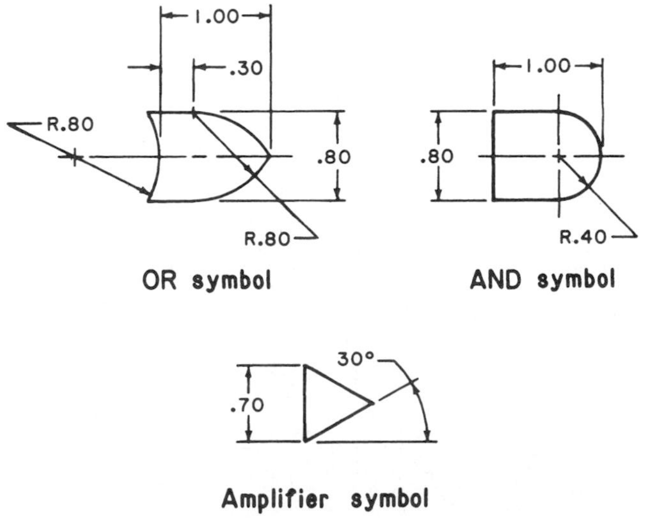

OR symbol

AND symbol

Amplifier symbol

FIGURE 15–11
Additional inputs to extend the symbol for adequate space for lettering.

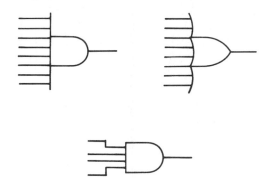

Combinational Logic

Up to this point only individual gates and their logic functions have been discussed. It is very seldom that any one of these gates is used alone to form a complete circuit. The use of the term circuit implies a complete logic system, where individual gates perform functions within the system. When many different gates are connected together to form a system, the result is called **combinational logic.** When you prepare diagrams of combinational logic, follow the same layout rules as for block diagrams. The flow of the circuit from input to output should be from left to right. Because the logic functions are performed sequentially, their physical position should indicate the required sequence in the left-to-right flow.

All the logic gates discussed so far are *asynchronous* logic gates. This means that as soon as a change appears on the input a logical decision is made by the gate and, if necessary, a change occurs at the output. This happens almost instantaneously, on the order of nanoseconds. If this output acts as the input to the next group of logic gates or next stage, then that input is acted on, any changes in the output occur, and a change at the input flows through the circuit. Since each stage depends on the changes occurring in the prior stage, it is important to physically locate these gates to demonstrate this sequential action or flow. Since it is typical to read from left to right the sequential flow is from left to right. Figure 15–12 shows an example of sequential logic.

Combinational logic must be thought of as a series of logic gates representing the individual units or parts of a larger functional system. Each individual gate has a unique function. If these functions are combined, a decision can be made based on more conditions or more combinations of inputs than is allowed by a single or individual gate with multiple inputs. For example, if a dangerous situation arises in a boiler when several

FIGURE 15–12
Two-stage sequential logic.

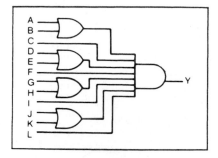

different combinations of conditions exist, it will be necessary for an alarm to ring if any one of the combinations is present. If either of the following conditions occurs, a dangerous situation is thought to exist and an alarm should ring: the temperature is too high and the water is too low or the temperature is too high and the pressure is too high. The engineer could look at the possible dangerous conditions and determine an appropriate logic circuit that would produce the correct outcome, which in this case is an alarm ringing. Assign labels or letter codes for all the variables:

$$T = \text{high temperature}$$
$$W = \text{no water}$$
$$P = \text{high pressure}$$

To determine the combinational logic, it is frequently necessary only to read the initial statements.

If high temperature **AND** no water

OR

If high temperature **AND** high pressure

The statements tell which logic gates are needed. The two AND statements make the first stage. They can stand alone as individual AND gates, but then two alarms are needed, as shown in Fig. 15–13. If it is more

desirable to have only one alarm, then there is a second level represented by the connecting OR statement. This says if either of the conditions exists then the alarm should sound. Since this is a decision that is made based on the outcome of two other decisions, then this must be a second-level, or **second-stage,** gate. In this particular example it is more important to know that any dangerous situation has occurred so that the boiler may be turned off. If it is necessary to know exactly which dangerous situation has occurred, then individual alarms or lights are required. Figure 15–14 illustrates the basic logic diagram representing this logic circuit. This particular illustration is fairly simple to lay out. More complex circuits would take a little more patience and perhaps a number of trial layouts. It is ultimately the responsibility of the drafter to produce a balanced, precise, and easy-to-read final layout and drawing from an engineer's sketch.

Figure 15–15 is an example of a well-done layout of a logic diagram with a number of levels or stages. Remembering a few simple guidelines should make the drafter's job a little easier. Logic diagrams are prepared with a general left-to-right and top-to-bottom flow from input to output. Symbols or individual logic gates should be aligned both horizontally and vertically. Gates that are in the same level are normally aligned on the same vertical line.

After the engineer produces a rough sketch of the desired logic diagram, the drafter must lay it out, determine the proper shapes and sizes for the gates, and provide sufficient space between gates for flow lines or other supporting data. (Specific suggestions are made later in the chapter for specifications in the final drawing.)

Grid-lined paper or a grid-pattern underlay makes an excellent tool for the preparation of all electronic drawings. So many drawings or diagrams need to be aligned both vertically and horizontally that a grid pattern is an excellent guide and a wise investment. Flow lines or input lines, although not representing true length or physical dimension, must be spaced properly for a more appealing look, for ease in circuit tracing,

FIGURE 15–13
Individual logic circuit used for each unsafe condition.

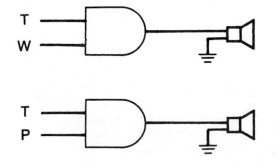

FIGURE 15–14
Sequential logic and a single alarm to indicate unsafe conditions.

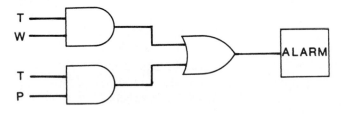

FIGURE 15–15
Well-designed logic circuit with multiple stages.

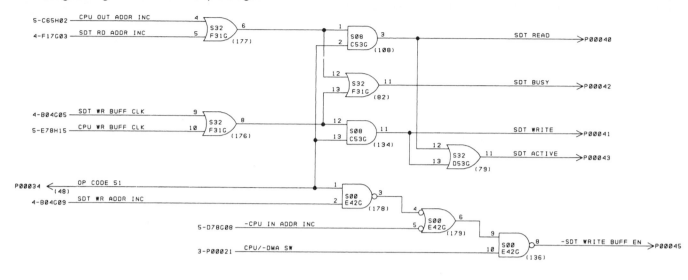

and for notes or other required lettering. Always assume that a final layout may be achieved only after some trial and error. In general, the more complex a circuit with more gates and levels, the more difficult and time consuming a final layout will be. All the guidelines for preparing block diagrams are valid in the layout of logic diagrams. Refer to the related information in Chapter 12.

LAYOUT

Since the unique shape used in logic diagrams represents a logical function, it is extremely important to accurately depict the desired shape. In some cases, the complete circuit may include additional discrete or individual components like the transistors or resistors shown in Fig. 15–16. Many times these are **logic schematic diagrams.** In general, they are found in the input or output areas of the diagram. Most illustrations in this chapter are limited to drawings that contain either unique shapes only or a combination of unique shapes and rectangles.

Keeping in mind that some logical decisions must occur sequentially will help in preparing the drawing and locating these gates. If the output of one gate becomes the input of another gate, it will be necessary to locate the gates showing that sequence. Many logic circuits, then, are time related or contain sequential operations. The diagram must be able to quickly demonstrate that sequence. Since it is normal to read from left to right the sequence can usually be followed from left to right, although some diagrams read both left to right and right to left, as seen in Fig. 15–17. Logic diagrams

should be aligned both vertically and horizontally. The vertical alignment is frequently determined by the timing or sequence of events. Figure 15–17 shows an example of sequential logic with both vertical and horizontal alignment.

Flow Lines

The logic symbols have interconnecting lines called *flow lines.* Flow lines for logic diagrams are drawn as solid, medium-weight lines, just as they are for block diagrams. Auxiliary or optional functions are represented by dashed lines. The thickness for the flow lines should be the same as the thickness used in drawing the symbol or rectangle in the diagram. Sometimes, when special emphasis is needed, a portion of the diagram is done in heavier and darker lines. Unless this special emphasis is needed, all lines on the diagram should be of the same line weight.

Too many crossovers on a diagram may confuse the reader. It is good practice to choose a layout scheme that allows for an absolute minimum number of crossovers. Flow lines are drawn as vertical or horizontal lines with sharp 90° corners, as shown in Fig. 15–18. Diagonal lines or rounded corners are not generally used except where not using them would cause more complicated or confused drawings. When crossovers and connections are required, be consistent and follow established standard guidelines. (Detailed methods of handling crossovers and connections were discussed in Chapter 12.) This method, accepted by ANSI, includes the dot/no-dot method of marking a connection (dot) and crossover (no dot). Fig. 15–19 shows an example of handling connections and crossovers.

FIGURE 15–16
Complete logic circuit with transistors, resistors, and other components in input and
output. (Courtesy Motorola, Inc., Semiconductor Products Sector)

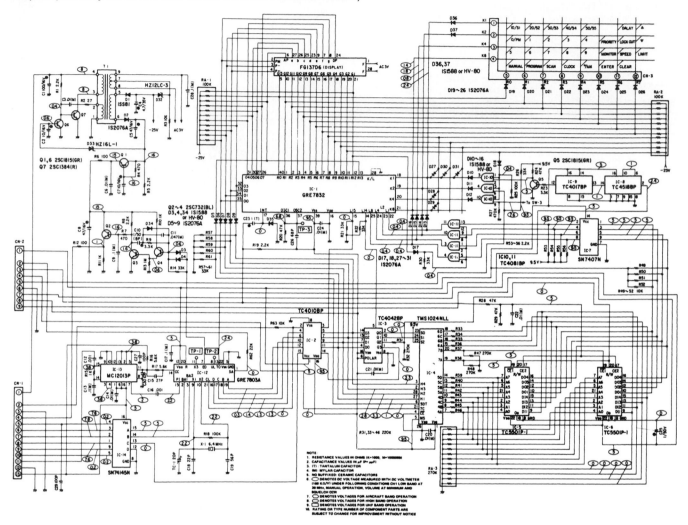

FIGURE 15–17
Logic circuit with good horizontal and vertical alignment.

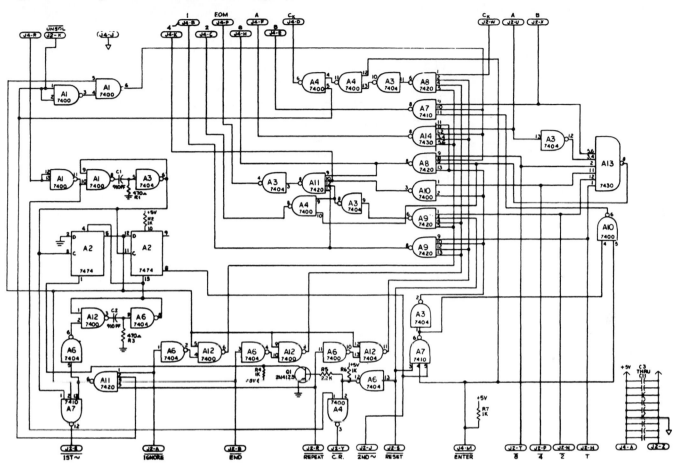

FIGURE 15–18
Changes in direction of flow lines in 90° angles. (Courtesy of Interactive Computer Systems, Inc.)

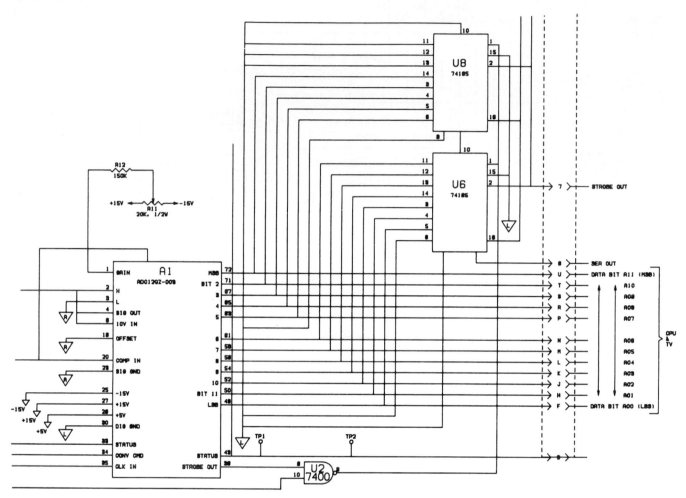

FIGURE 15–19
Representing connections and crossovers.

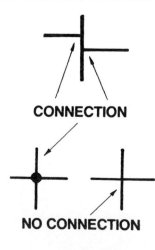

During the trial-and-error stage of layout, try to limit the changes in direction. Changes in direction require more space and tend to distract from the main flow of the circuit. With more complex circuits it is necessary to include some changes in direction, and this must be accomplished with sharp corners and a minimum number of crossovers.

Spacing Flow Lines

Spacing of the interconnecting lines must take into account any reductions that must be done. Fig. 15–20 shows the suggested spacing between individual lines. When reduced, a drawing should have no less than .06 in. between two lines.

In logic circuits the use of arrowheads placed on flow lines—although not restricted—is not generally

FIGURE 15–20
Line spacing: (A) .40 in. (10 mm), not less than .06 in. after reduction; (B) .50 in. (13 mm).

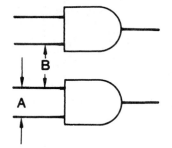

found. The use of the unique shapes gives the direction of input and output. However, if arrowheads are used, they should be placed centrally on the flow lines, not touching any of the logic symbols.

Some additional spacing dimensions are shown in Fig. 15–21. These should be regarded as suggested spacing and are good guidelines. If company practice requires different standards, it is the drafter's responsibility to maintain company policy.

Lettering

Lettering is the last step in the completion of most drawings. Most of the lettering on logic diagrams is used to identify inputs, outputs, or other control connections, as illustrated in Fig. 15–22. Some lettering includes part numbers or actual pin numbers, as shown in Fig. 15–23. When a rectangle is used, the logic function to be performed is lettered preferably inside the block or box, as shown in Fig. 15–24. All lettering, whether freehand or with lettering aids, must be consistent in height, line density, and style.

Reduction consideration cannot be stressed enough. All lettering must be readable at the greatest reduction. Also, space between components and lines must be maintained as the drawings are reduced.

FIGURE 15–21
Line spacing; A = .40 in. (10 mm).

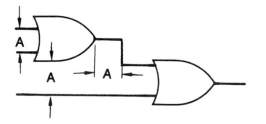

FIGURE 15–22
Control connections with labels. (Courtesy of TRW LSI Products)

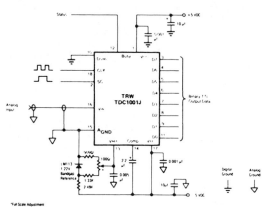

FIGURE 15–23
Lettering indicating part numbers and pin numbers on ICs. (Courtesy of Texas Instruments, Inc.)

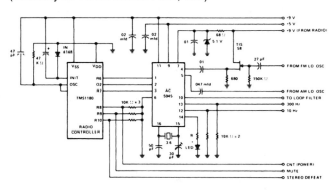

FIGURE 15–24
Lettering indicating circuit functions. (Courtesy of Texas Instruments, Inc.)

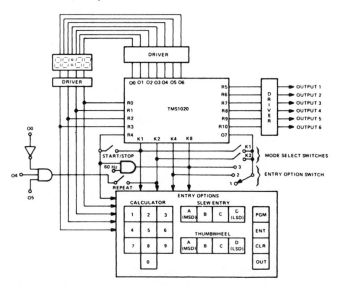

FIGURE 15–25
Arrowheads indicating flow direction when it is not standard left to right. (Courtesy of Motorola, Inc., Semiconductor Products Sector)

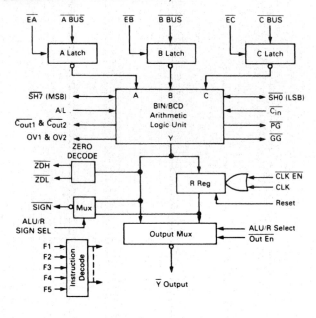

FIGURE 15–26
CAD-generated logic diagram. (Courtesy of T & W Systems)

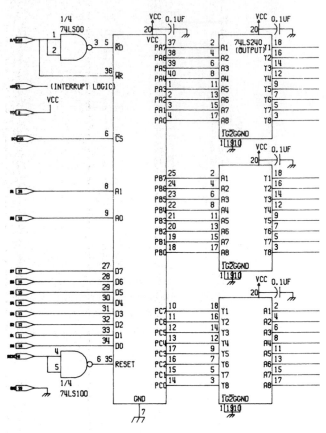

Logic and Computers

With the rapid growth in integrated circuits, more and more logic circuits are being developed as logic or microprocessor modules. These diagrams are almost always prepared with rectangles because the module contains a number of larger circuit functions that have no unique shapes. Rectangles are used and identified as counters, memory, input/output (I/O) unit, or central processing unit (CPU). They are interconnected by many flow lines that can travel in either direction. Attempts are made to keep the flow from input to output from the left to the right. However, when this is not possible, arrowheads are used to indicate signal flow, as shown in Fig. 15–25.

USE OF CAD

As previously stated, the use of CAD systems in the preparation of any drawings or diagrams can save time. It is extremely important for you to be aware of all rules identifying layout, spacing, or flow, because placement is still determined by the drafter. The system can eliminate some of the repetition of the drawing, but the knowledge of good drafting practice is essential.

Figure 15–26 is an example of a CAD-generated drawing of a logic circuit. The overall appearance is relatively pleasing, and good vertical and horizontal alignment has been followed. One of the primary advantages of CAD systems is that shapes can be repeated. Once you have entered a shape into the library of parts, you can call it up and place it anywhere on the drawing simply by using the cursor or light pen and a digitizer tablet, as shown in Fig. 15–27. Keep in mind that it is essential to understand good drafting practice because the final drawing—whether generated by machine or by hand—still reflects your knowledge and ability as a drafter.

FIGURE 15–27
Logic design on P-CAD.
(Courtesy P-CAD)

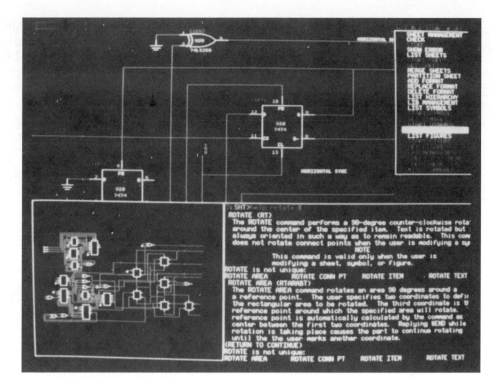

REVIEW QUESTIONS

15–1 Why are unique shapes used rather than rectangles for logic functions?

15–2 What is the difference between a basic logic diagram and a detailed logic diagram?

15–3 Why is the binary number system so useful in describing logic functions?

15–4 What is a truth table?

15–5 Define the logic function performed by an AND gate.

15–6 What is the primary difference between an EX-OR and a regular OR gate?

15–7 What is the small circle in a logic diagram used for and what does it mean?

15–8 What does the term asynchronous mean when applied to a logic gate?

15–9 What is combinational logic?

15–10 Why are rectangles used in logic circuit diagrams?

PROBLEMS

15–1 Practice drawing the unique symbols for the AND gate. Draw at least two for each gate with combinations of two, three, four, and eight inputs. These should be done with the use of a template and should be completed in pencil or in ink as assigned by your instructor.

15–2 Construct the symbol for an AND gate using a compass and straightedge following the guidelines specified in the chapter. Refer to Fig. 15–10 for typical dimension ratios. Complete as many as necessary to produce an acceptable symbol.

15–3 Construct the symbol for a NAND gate using a compass and straightedge following the guidelines specified in the chapter. Refer to Fig. 15–10 for typical dimension ratios. Complete as many as necessary to produce an acceptable symbol.

15–4 With the use of a template, construct two, three, four, and eight input OR gates. Remember to draw extension lines on the input side of the symbols for added space of flow lines.

15–5 To represent the following set of conditions, draw a circuit using the appropriate logic gates. The fasten seat belt warning should buzz if the driver OR the passenger belt is not connected AND the key is turned on.

15–6 Redraw the logic symbols illustrated in Fig. 15–8 without the use of a template. Taking measurements from the figure, produce these at 2× size.

15–7 This figure illustrates the integrated circuit to produce the decoding function to change BCD to decimal. On the right side of the figure is a representation of the internal logic portion of this IC. Redraw this circuit using all the suggested guidelines for flow line spacing. Do not attempt to find a different layout. Produce this as a B size drawing with the use of templates in pencil or ink as assigned by your instructor.

PROBLEM 15–7
Logic diagram.
(Courtesy of Motorola, Inc., Semiconductor Products Sector)

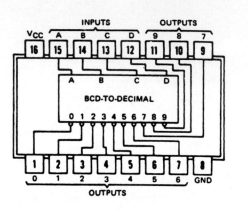

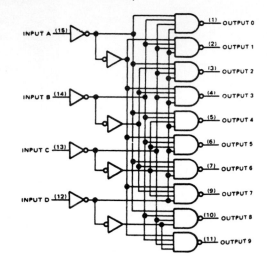

15–8 Redraw the logic circuit of problem 15–7, changing all the NAND gates to AND gates so that an inverted output can be obtained.

15–9 A universal timer logic diagram is shown in this figure. Redraw this circuit in ink on C size or larger paper. This problem will provide practice for lettering, linework, and a few component symbols.

PROBLEM 15–9
Universal timer diagram.
(Courtesy of Texas Instruments, Inc.)

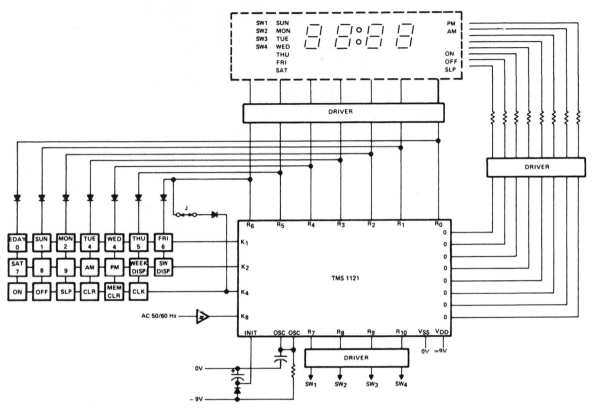

15–10 A traffic light controller is shown in this figure. Redraw this circuit in ink on C size or larger paper. This problem will provide practice for lettering, linework, and a few component symbols.

PROBLEM 15–10
Traffic light controller.
(Courtesy of Texas Instruments, Inc.)

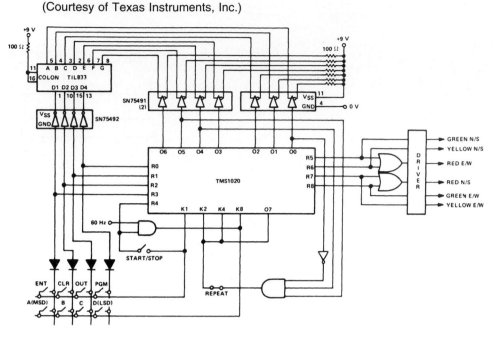

15–11 through 15–27 Redraw each assigned project on the appropriate sized paper.

PROBLEM 15–11
Logic diagram using JK flip-flops.

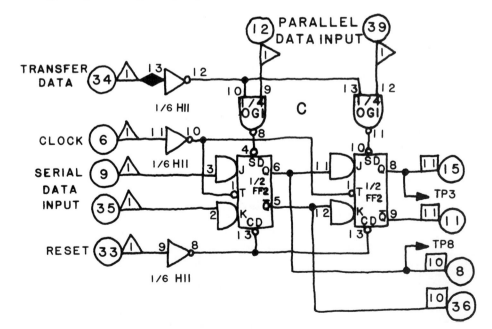

PROBLEM 15–12
Partial logic diagram.

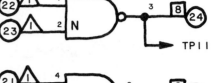

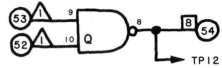

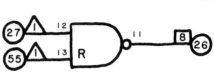

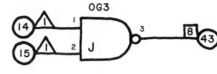

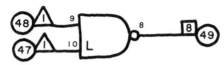

PROBLEM 15–13
Logic diagram.

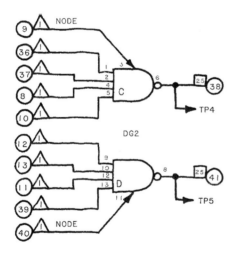

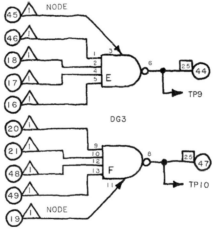

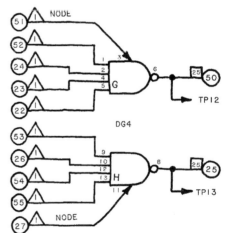

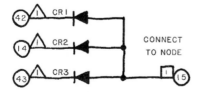

PROBLEM 15–14
Logic diagram.

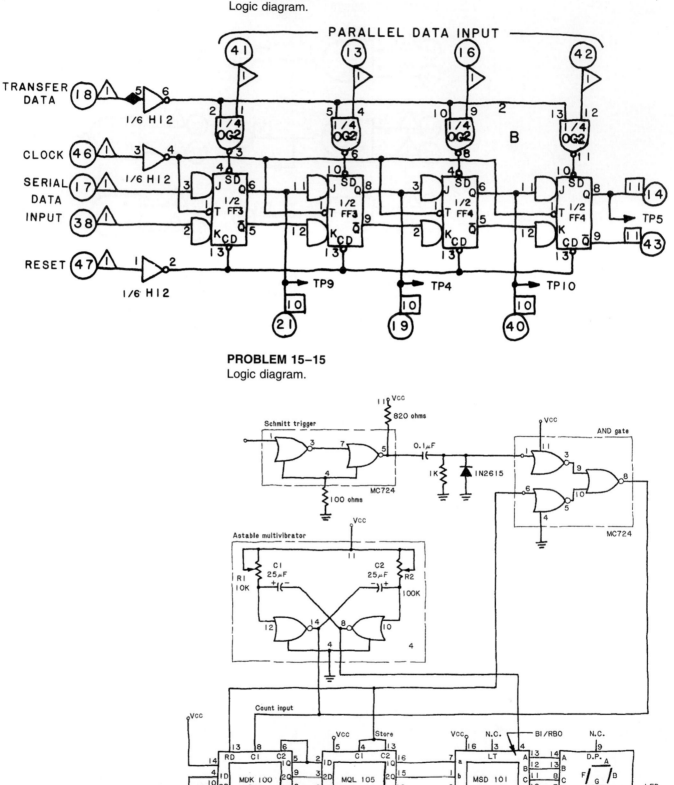

PROBLEM 15–15
Logic diagram.

PROBLEM 15–16
Setup for switching regulator logic diagram.

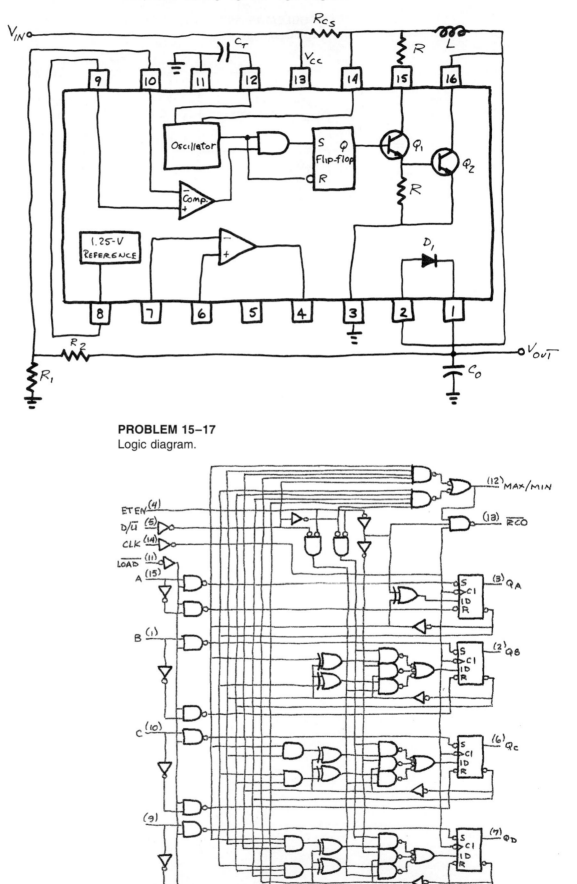

PROBLEM 15–17
Logic diagram.

PROBLEM 15–18
Logic diagram.

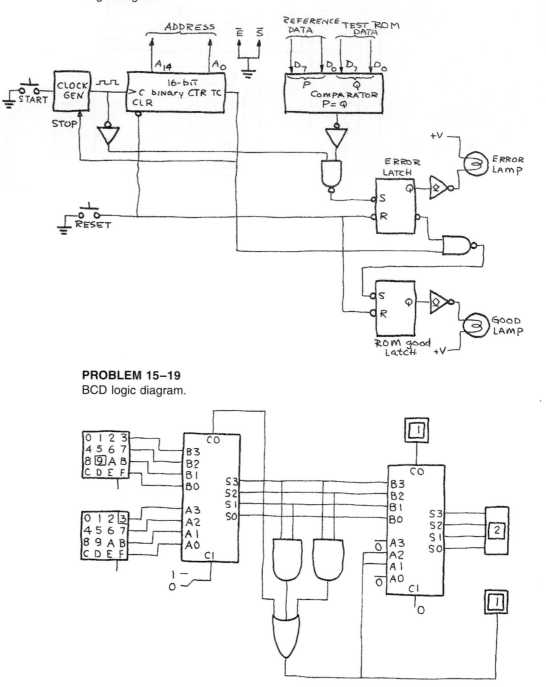

PROBLEM 15–19
BCD logic diagram.

PROBLEM 15–20
Logic diagram.

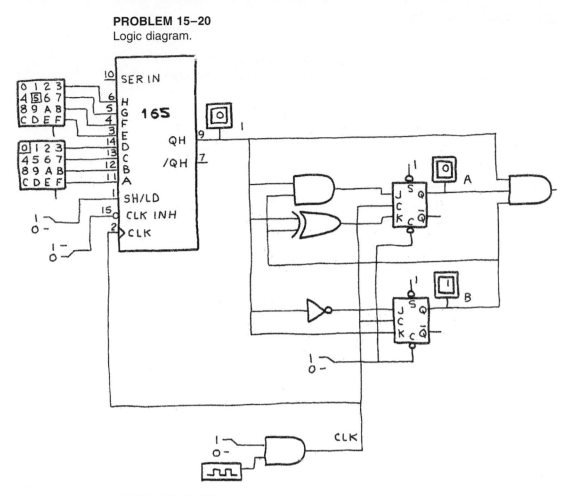

PROBLEM 15–21
Logic diagram for an even-parity check machine.

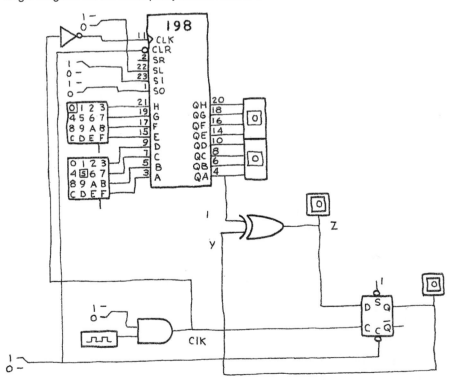

PROBLEM 15–22
Controller output decoder logic diagram.

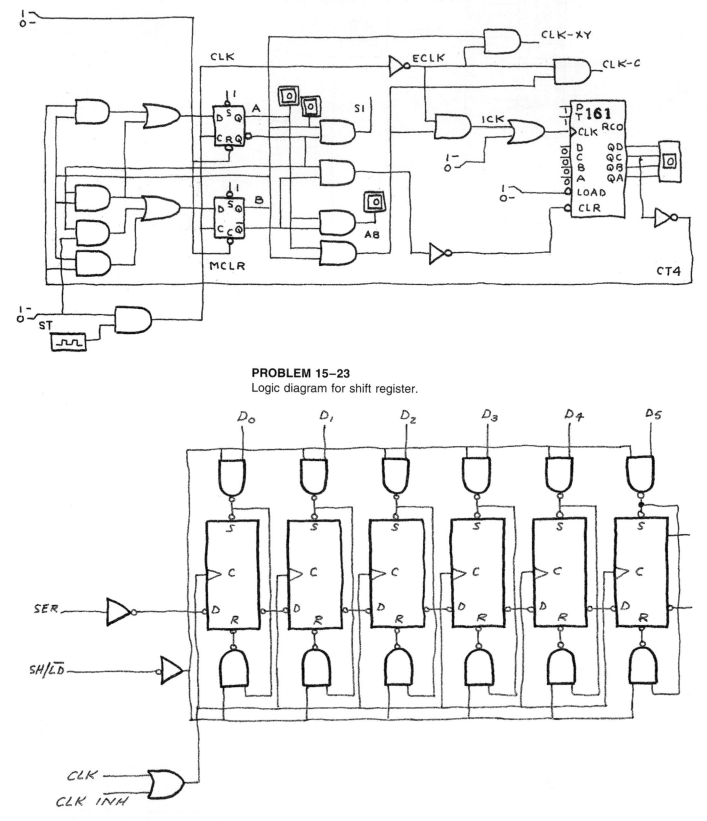

PROBLEM 15–23
Logic diagram for shift register.

PROBLEM 15–24
Sequential logic circuit.

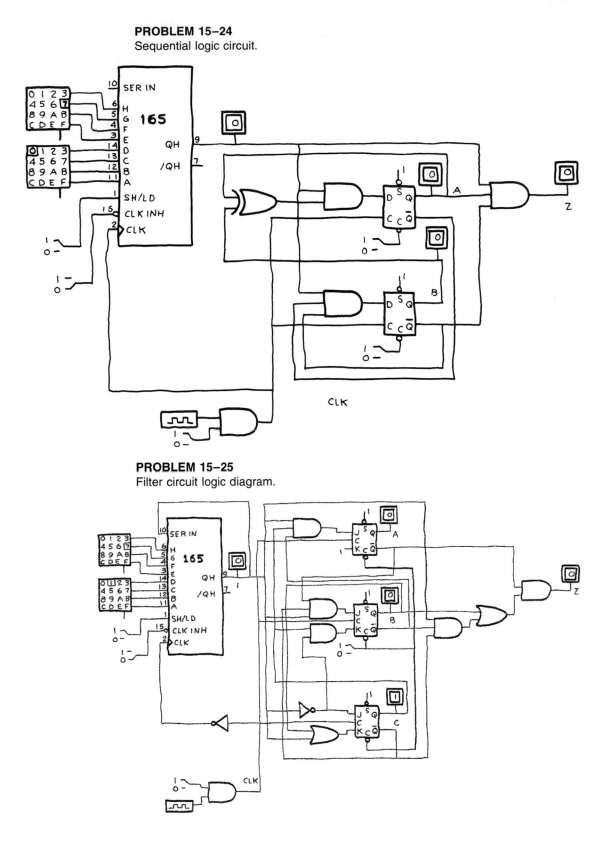

PROBLEM 15–25
Filter circuit logic diagram.

PROBLEM 15–26
Logic diagram of a converter.

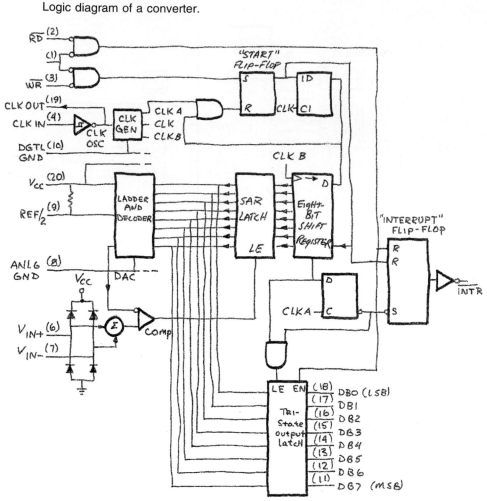

PROBLEM 15–27
Logic diagram.

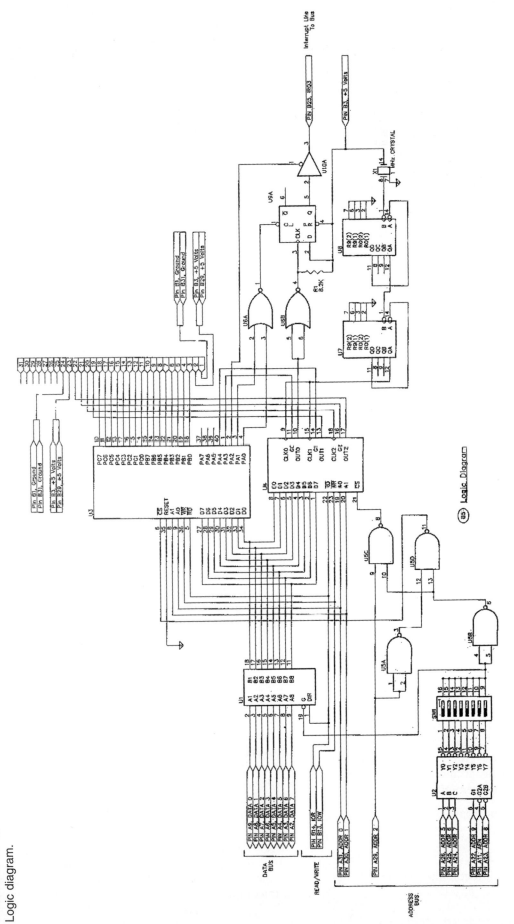

16

PROGRAMMABLE CONTROLLERS AND ROBOTICS

INTRODUCTION

Industry's desire to increase productivity has been the main catalyst for increased automation. This automation is used to control not only individual machines but an entire process. A **control system** must consider input information, output, and the processing itself.

As early as the late 1960s, manufacturers tried alternatives for control circuits. Minicomputers were used, but they were not designed for industrial life, they had no input or output capabilities, and they generally had to be programmed in assembly or machine language.

The evolution to programmable controllers from standard or conventional minicomputers occurred through the late 1960s and early 1970s. Some credit General Motors with initially conceiving the programmable controller in 1968. Engineers at Cincinnati Milacron also worked on the development of a programmable controller. The early programmable controllers had to meet the following requirements:

1. Easily programmed and reprogrammed
2. Easily maintained
3. More reliable
4. Physically small
5. Capable of handling data communication
6. Cost competitive

7. 115 V AC input/output
8. Expandable in size and memory

Although the shutdown times were reduced because reprogramming was less time consuming than with traditional hard-wired methods, there were still some major problems with these early programmable controllers. The programming language was still very complex and required training in programming to understand. The first units were very large and costly and did not save floor space.

During the development phase of computer-controlled systems, the only desire was to have a method of replacing magnetic relays. The introduction of the microprocessor in 1978 led to smaller, faster, and less expensive computers. These new designs brought improvements to the programmable controller and made it what it is today. In order for programmable controllers to be used by persons knowledgeable in ladder diagrams and relay logic, several manufacturers use high-level programming languages based on more conversational style. These languages include PROTEUS from Westinghouse, BICEPS from Honeywell, and PCL from Hitachi. Unfortunately, some of these high-level languages are still implemented only on mainframes and minicomputers.

PROGRAMMABLE CONTROLLER

The programmable controller is defined by the National Electrical Manufacturers Association as a digitally operating electrical apparatus that uses a programmable memory for the internal storage of instructions for implementing specific functions such as logic, sequencing, timing, counting, and arithmetic to control machines or processes through digital or analog input or output modules. A digital computer used to perform the functions of a programmable (logic) controller is considered to be within this scope. Excluded are drum and similar mechanical-type sequencing controllers. The definition of a programmable controller as a replacement for solid-state logic controls or electromechanical relay controls must also indicate that it can be programmed or reprogrammed and maintained by persons unskilled in the use of computers. A person who is already knowledgeable in relay logic systems should be able to master most programmable controller functions in a few hours.

A programmable controller system controlling a simple process may be housed in one or more units. For a more complex process, the entire system may consist of three to five or more units connected together. Connecting these units together is relatively simple and is usually done with single cables. The connections for the input and output modules present a still more complex task. The overall programmable controller system always consists of the following: the CPU (central processing unit), the programmer/monitor, and the input/output modules; it may include peripherals such as a printer and a program recorder/player. Figure 16–1 shows a block diagram representation of a programmable controller.

A CPU is a computer centered around a single microprocessor chip. Its operation is similar to a desktop or personal computer, although it is generally limited to industrial control applications, whereas a personal computer would be used for all kinds of personal productivity. The CPU includes the microprocessor, fixed memory, alterable memory, and usually the power supply. A typical programmable controller is shown in Fig. 16–2. Although the input and output modules are included, the keyboard and the display unit are not shown. They are not required for the operation of a programmable controller, but are necessary only for the actual programming steps.

The input modules receive either binary or analog signals provided by transducers. In early programmable controllers, only digital or binary inputs were used. This provided only ON/OFF control. However, today it is not uncommon to use programmable controllers that provide proportional, integral, and derivative (or PID) control. Fig. 16–3 shows standard input modules.

Transducers

Various instruments are used to measure variations in temperature, pressure, current, or other variables. These variable measurements can be changed into electrical impulses. Another variable that is often measured is position. The measurement of position can be either absolute or incremental. **Absolute transducers** measure position with respect to a fixed or datum point. An **incremental transducer** measures distance moved from the last measured point. This would result in an inaccurate measurement after a power failure.

Proximity detectors may be defined as solid-state limit switches. Figure 16–4 shows such a device. Many

FIGURE 16–1
Block diagram of a programmable controller.

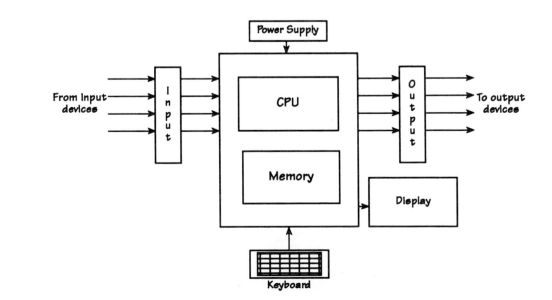

FIGURE 16–2
Typical programmable controller with CPU, memory, power supply, I/O slots, and communication processor. (Courtesy of Modicon, Inc.)

of these use optical detection for measurement. Measurement of weights is very important for economic and often legal applications. The simplest method for weight measurement is a balance arm with a known weight used to balance an unknown weight. A strain weigher requires a secondary transducer, which is used to measure the amount of deflection caused when a weight is measured by exertion of a force on a spring.

Devices are available to measure almost any variable. To make these measurements in control circuits requires the use of transducers. In the strictest sense, a transducer is a device that converts one physical quantity into another. The term transducer is often used interchangeably with the sensor itself, rather than to refer to the complete process.

DESIGN USING PROGRAMMABLE CONTROLLERS

Keep in mind that ON/OFF control, although simpler than PID, should not be thought of as an inferior form of control. Many processes will require only ON/OFF control systems. The output from the output module might be used to actuate pumps, motors, pistons, relays, or other devices.

When programmable controllers are used in control circuits, the following design steps should be followed:

1. Define the process.
2. Sketch the operation using block diagrams or flowcharts.
3. Give a written description of each step.
4. Show where sensors are required.
5. Show where manual controls, such as switches, are required.
6. Add master switches for stopping and starting the operation.
7. Consider fail-safe operation to protect personnel.
8. Draw a ladder diagram representing all steps of the operation.
9. Create a parts list of sensors, relays, and the like.
10. Create a wiring diagram showing the interconnection of various modules.

The final step should include a series of "what could go wrong" questions and what steps may be necessary to correct faults.

FIGURE 16–3
Programmable controller with three standard input/output units. (Courtesy of Modicon, Inc.)

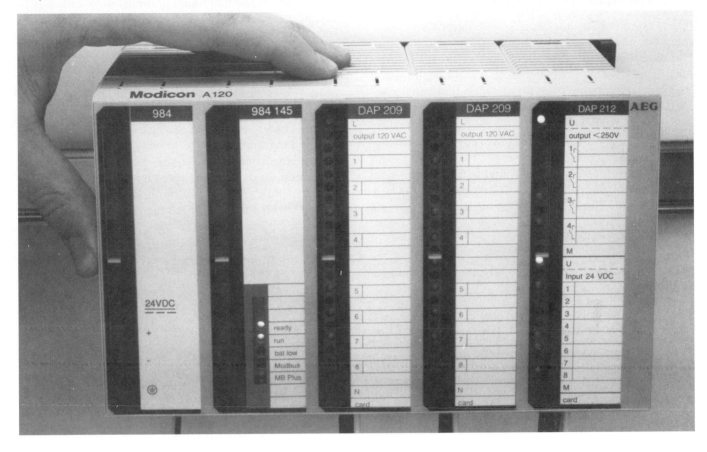

FIGURE 16–4
(A) Simple limit switch and (B) proximity detector. (Courtesy Eaton Corporation, Cutler–Hammer Products)

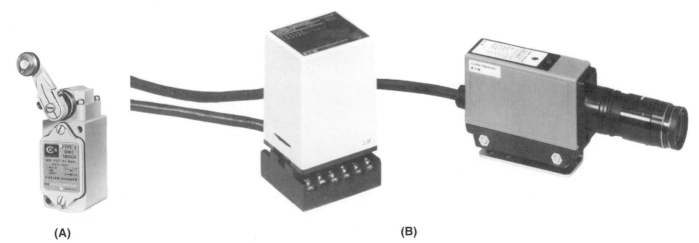

(A)   (B)

Programming the Programmable Controller

Once an operation has been fully described and documented, the programmable controller can be programmed. Each manufacturer may use a slightly different format for its programmable controller; however, the principles are the same for all types. Individual manuals will include the required screen format, numbering schemes, limitations, and other important information for each controller. A representative page from a manufacturer's manual is shown in Fig. 16–5.

Major differences can be found between hand-held units and large monitor units. The large-monitor-type unit generally displays a number of ladder rungs at a time, whereas the hand-held unit may show only a portion of one or more rungs. Two types of controllers are shown in Fig. 16–6. There is a trade-off for the large-monitor type in cost factor and, although the view of the complete circuit is helpful during programming, it is of no use during operation. Some programmable controllers, with an appropriate interface, allow a standard microcomputer to be used as the keyboard and display unit for programming. Using a microcomputer only during the programming stage provides the advantage of a full-screen display without the cost associated with a large monitor dedicated only as an output display. The microcomputer can be used in other applications when not needed for programming the controller. Each pro-

FIGURE 16–5

Page from programmable controller manual demonstrating programming steps and screen format. (Courtesy of Eaton Corporation, Cutler–Hammer Products)

4.6 Three-Wire Example

Let us program the controller with a basic control arrangement:

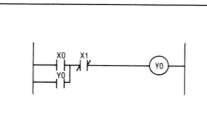

Three wire Circuit

This is a simple holding circuit which demonstrates how output Y0 can be turned ON and OFF through the operation of the two inputs X0 and X1. X0 and X1 are the two NO pushbuttons connected to the controller input. The following figure shows the required keystrokes and the resultant screen display.

Program First Screen — begin with a Normally Open X000 contact:

[F1] [X] [0] [Enter]

Add Normally Closed X001 contact:

[F3] [X] [1] [Enter]

Add Coil Y000:

[F5] [Y] [0] [Enter]

Add the Y000 holding contact:

[F1] [F2] [Y] [0] [Enter]

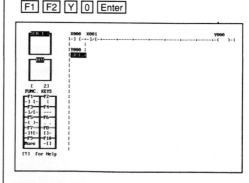

Write screen to computer memory:

[F9] [F1]

FIGURE 16–6
Hand-held and small graphic programmers. (Courtesy Eaton Corporation,
Cutler–Hammer Products)

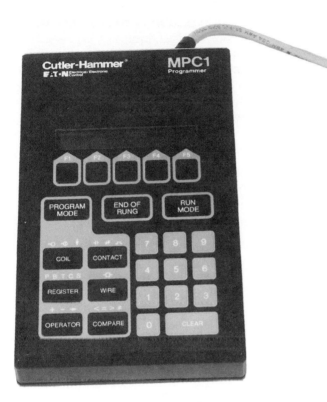

grammable controller provides codes for incorrect programming or problems during operation. On a large screen, the error message may include the error code as well as some explanatory phrase. On a smaller unit, only the code may appear, requiring the operator to check the reference manual for a description of the error. For example, 24 might appear on the display of a small system; this number refers to a memory overflow, so the words *memory overflow* would probably appear on the screen of a large-monitor system.

Each programmable controller has format limitations that must be observed when programming ladder diagrams. If the format is not correct, the CPU will not accept the program into memory. Figure 16–7 shows a page of error messages and what these messages might represent for a typical programmable controller.

The following limitations are typical for most programmable controllers:

1. A contact must always be inserted in the first slot of the first rung.
2. A coil or output must be inserted first.
3. A limited number of outputs and contacts can appear on the screen at one time.
4. Only one output can be connected to one contact.
5. Flow must always go from left to right.
6. All contacts must run vertically.
7. A coil or contact must be inserted last.

Some common relay symbols, including normally open contacts, normally closed contacts, and coils or outputs, are used no matter what type of programmer is used. Also, there is a limit to the number of parallel

FIGURE 16–7
Page from installation manual of Cutler–Hammer D500 PCL indicating errors. (Courtesy Eaton Corporation, Cutler–Hammer Products)

6.0 Troubleshooting

The prime diagnostic device is the LED's on the CPU and on each module. Regular observance of these LED's will alert the user to potential or actual trouble. Other diagnostic information is available through peripheral devices like the programmer.

6.1 Table of Graphic Programmer ALARM Messages

Alarms are displayed in the message area of the graphic programmer display panel. Following are the alarms which may appear.

BATT	—	Battery condition or low voltage alarm.
TL	—	EasyNet error.
LINK	—	Abnormal occurrence in Computer Link.
DIAG	—	An error has occurred during diagnosis.
DOWN	—	The processor is in the ERROR DOWN state, an Error Reset command is required to HALT the system.

6.2 Table of Graphic Programmer ERROR Messages

Error Location	Item	Error message	Cause	LED indicator Corresponding LED	LED indicator Status Change	Remedy
Software	Programming or operation	! NO END ERROR ! MC/JC ERROR ! OPERAND ERROR	No END instruction No pair instructions (MCS/MCR or JCS/JCR) Operand error	RUN, PRG	Turns off	Insert missing instructions and try again.
		! ILLEGAL INST ! W/D TIMER ERROR ! SCAN TIME OVER	Illegal instruction Watchdog timer (300 ms) error First scan time over 200 ms	RUN, CPU RUN, CPU RUN	Turns off Turns off Turns off	Cycle power OFF-ON after programming correctly. Cycle power OFF-ON after programming correctly. Change program or select higher capacity model.
Hardware	CPU	! CPU ERROR ! E-POWER FAIL	System error Power supply voltage low dc voltage low	CPU RUN, I/O RUN, CPU, POWER	Blinks Turns off Turns off	Replace CPU unit. Check power supply to I/O unit, replace I/O unit. Cycle power OFF-ON. Check CPU unit.
	I/O	! I/O UNMATCH ! I/O NO SYNCHRO ! I/O BUS ERROR	Mismatch of setting/registration with mounted I/O unit No answer from I/O unit I/O bus error	RUN, I/O RUN, I/O RUN, I/O	Turns off Turns off Turns off	Change type of I/O unit or I/O setting. Check expansion unit for proper mounting. Check I/O units for proper mounting.
	PROM	! ROM ERROR	ROM error	RUN, PRG	Turns off	Replace PROM Module.
	Programmer	! GP COMM ERR ! PC COMM ERR ! COMM TIMEOUT	Data error to programmer Data error to PC No data to programmer	COMM	Remains off	Check transmission line between controller and programmer, try again.
	Data transmission	TL LINK	EasyNet Talker/Listener error Computer link error	LED indicator unit		See
	Battery	! BATTERY FAIL	Low battery voltage	BATT	Flickers or unlit	Replace battery.

6.3 Interpreting LED Displays

All eight LED indicators on the basic unit are lit or blink during normal operation. Should one or more LEDs not light, a malfunction is indicated.

LED indicator	Meaning
POWER	Lit when power supply voltage is normal.
BATT	Lit when battery voltage is normal.
COMM	Blinks when communicating with programmer.
HOLD	Lit when stopped.
RUN	Lit when running.
CPU	Lit when CPU is normal.
I/O	Lit for normal I/O.
PROG	Lit for normal program.

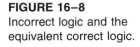

FIGURE 16–8
Incorrect logic and the
equivalent correct logic.

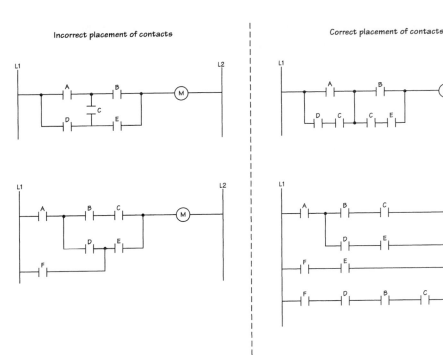

branches or lines and the number of contacts or other logic symbols that can be included on any line. Some programmable controllers limit the number of nested branches allowed. Figure 16–8 illustrates some correct and incorrect ladder diagrams used in programmable controllers. During operation, a programmable controller must scan every rung from top to bottom in order. In a relay logic system, any event occurring anywhere results in an immediate output, but in a programmable controller, output will not occur until that rung is scanned. This must be considered when a ladder diagram is drawn to ensure that the right sequence of events occurs.

Methods of Control

Open-loop control is the simplest form of control used. It uses no feedback, as in a closed-loop system. The feedback system can provide many kinds of control action and correction. Proportional control is the simplest

form, but it is rarely sufficient by itself. In addition, derivative and integral calculations are also used. This type of feedback and control is based on the rate of change of the error (the derivative) plus the integral of the sum of the errors, as well as the size (proportional) of the error. This combination of error correction used in control systems, called PID, is one of the most widely used forms today. Figure 16–9 shows a general flow diagram using feedback.

Relationship of Digital Logic to Relay Logic

Most larger programmable controllers do not require digital gate logic principles. They are programmed by keying in lines, connections, contacts, coils, timers, or functions. However, many smaller programmable controllers have keyboards with LED displays and digital logic symbols on the keypad, such as AND, OR, and NOT, and a dot, +, −, 0, or =. These are Boolean algebra operators used with logic terms.

FIGURE 16–9
Block diagram of general flow
with feedback.

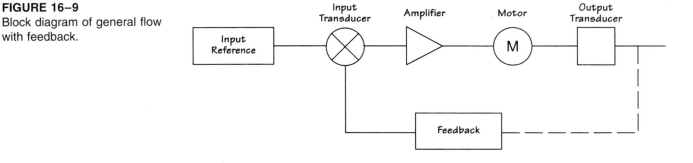

The voltages used to operate programmable controllers are in the 5- to 15-V DC range, but the process signals are generally much greater. The input/output modules provide the interface between the controller and the actuators and transducers. For NEMA size 4 or larger motor starters, a standard control relay is required to connect the output module. When used this way, the relay is called an *interposing relay*.

Converting from Relay or Logic Diagrams

The following examples demonstrate how gate logic can be translated to relay logic and then implemented by a programmable controller.

A conveyor system is to be controlled by several inputs. It is not important at this time how those inputs are detected, because we are interested only in the logic and control portion of this circuit at present. The inputs in this example are simple momentary-contact push buttons. The conveyor will run if any one of three switches is activated, and it will stop if any one of three other switches is activated. Figure 16–10 demonstrates the gate logic to implement this problem. This same circuit can be implemented by the relay logic shown in Fig. 16–11. The relay ladder diagram would be converted to a programmable controller, as shown in Fig. 16–12. This circuit requires any one of the start buttons to continue to be depressed for the conveyor to continue running. As soon as the start button is released, the conveyor will stop. The use of momentary-contact push buttons for this application is not too realistic for this type of circuit. To allow the conveyor to continue to run after any of the three start switches is pressed, a sealing contact from the motor can be added to the circuit. This contact is shown in Fig. 16–13.

The actual wiring for relay logic circuits may become rather complex, although the programmable controller wiring remains fairly simple. For example, the

FIGURE 16–10
Gate logic.

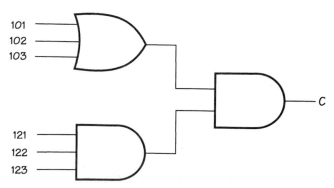

FIGURE 16–11
Relay logic.

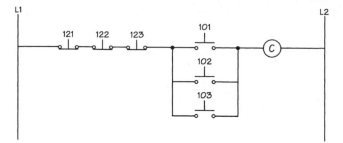

reversing motor control circuit shown in Fig. 16–14 is fairly standard. The dotted lines connecting the REV and FWD push buttons indicate that these are interlocked; when one is closed, the other must be open. For the programmable controller, all the interlock and sealing contacts are a part of the program portion, and the only wiring required includes the start and stop buttons for forward and reverse, the motor connections for forward and reverse, and the red and green indicator lights. This wiring diagram is shown in Fig. 16–15. The programmable controller logic would be similar to that shown in Fig. 16–16.

Sequential Logic Control

When various devices such as pumps, motors, or timers need to be activated in a particular order and one step must be completed before the next, the system is referred to as a **sequential system**.

In the beginning, these systems were designed mostly by intuition, experience, and trial and error. Such a design was very impractical for large and complex systems, so a more systematic method of design became necessary. Flow tables, merged flow tables, and Karnaugh maps are used to design what is referred to as an **asynchronous system**. However, this system can become very impractical for even medium-sized circuits. Also,

FIGURE 16–12
Programmable controller logic.

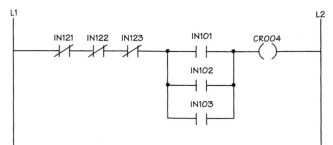

FIGURE 16–13
Addition of a sealing contact.

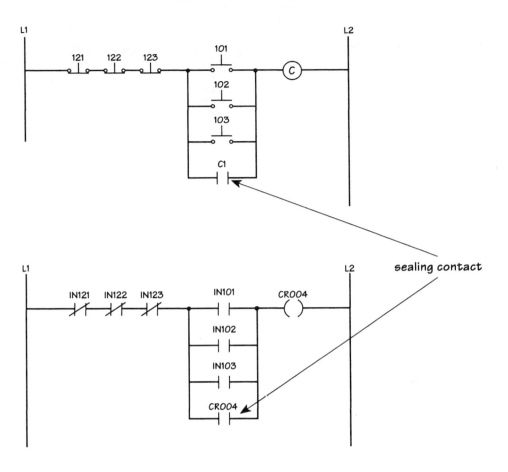

sealing contact

if only one small change is made, a whole new solution must be derived.

One solution to this problem is to use commercially available sequential-system programmers. Such programmers involve either punched-card or punched-tape readers. These systems are extremely flexible but rather expensive. Programmable sequence controllers can provide the greatest flexibility and convenience. Also called logic counter programmers, they can be assembled from standard, commercially available logic elements that are

FIGURE 16–14
Reversing motor relay logic control.

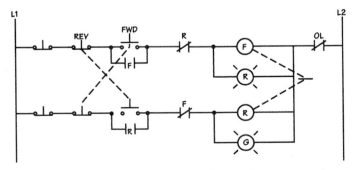

FIGURE 16–15
Wiring of a programmable controller.

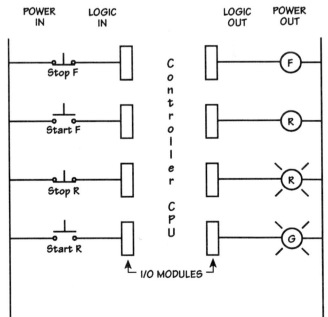

FIGURE 16–16
Programmable controller logic.

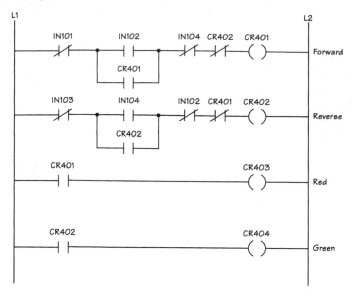

completely compatible with existing logic elements. Logic counter programmers keep track of program steps by counting them.

The use of programmable logic controllers allows the individual required program steps to be stored in memory, rather than using individual switching elements for the steps. The programming steps for sequential logic are very similar to those of combinational logic

from earlier discussions:

1. Describe the process.
2. Create a flowchart or function diagram.
3. Identify conditions and convert to Boolean equations.
4. Translate the Boolean equations into ladder logic.

Relay ladder diagrams or logic circuits are not normally used to represent sequential logic circuits, since the diagrammatic layout would become far too complex and difficult to follow. Flowcharts can also become unwieldy. Consequently, function charts or diagrams tend to be used most often. A function diagram is shown in Fig. 16–17. Note that the function diagram is laid out much like a flowchart.

Limitations of Ladder Programming

Ladder programs are ideal for combinational and simple sequential tasks. Full documentation is required for small programmers, since they have no means of adding comments to the circuit. Ladder diagrams are much too cumbersome to be used with most complex sequential tasks. They are difficult to design as well as troubleshoot.

APPLICATIONS OF PROGRAMMABLE CONTROLLERS

Programmable controllers are often used in robotics as either the controllers for individual robots or as overall

FIGURE 16–17
Function diagram.

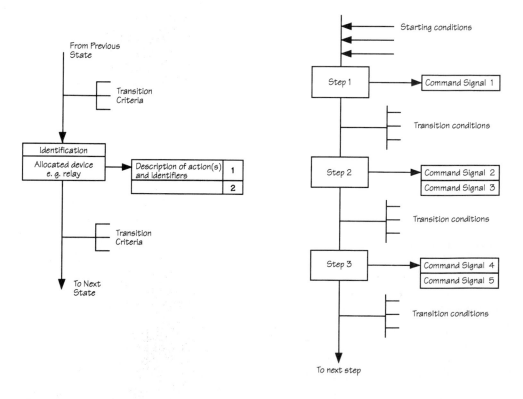

system controllers for a number of robots. More often, a microcomputer is used to control a robot to provide the necessary and adequate processing speed. Programmable controllers are used to process large numbers of input and output signals when only simple logic functions are required on the inputs. Programmable controllers are a very reasonable and cost-effective choice for point-to-point and limited-sequence robots.

The next step is to link and group programmable machines into flexible manufacturing cells, which might include robots, CNC machines, a conveyor system, and a programmable controller.

Robots

The definition of a robot as adopted by the Robotic Industries Association is as follows:

> A robot is a reprogrammable multifunctional manipulator designed to move material, parts, tools, and specialized devices through variable programmed motions for the performance of a variety of tasks.

Industrial robots can be categorized by a number of different features. For example, they can be servo and nonservo controlled or they can have different degrees of freedom. A **degree of freedom** is an axis of motion for the robot. Robots with six or more degrees of freedom are classified as medium or high technology. As more degrees of freedom are added, complexity and cost can quickly become prohibitive in many applications. Most industrial tasks require from three to six degrees of freedom, or axes of motion. Keep in mind that limited axes of motion do not mean inaccurate or limited usefulness. In robots, more is not necessarily better. A robot should be selected to perform only the needed tasks; extra capabilities can simply result in additional and unnecessary costs.

A two-axis robot can have both linear and rotary movement. Basic robot elements include some sort of mechanical arm and either computer control for the most flexibility or a fixed-sequence controller. The three basic components of a robot include the **controller,** the **manipulator,** and the **end effector,** or gripper. Figure 16–18 identifies the six possible axes of rotation in a complex robot. The controller (Fig. 16–19) includes both the hardware and software needed to control the movements of the robot. The manipulator is the base and the arm itself. It is usually internally powered. Each individual joint in a robot arm can have up to three axes of movement, and each axis has its own actuator. These actuators are either mechanical, electrical, or pneumatic. Each of these presents its own advantages and disadvantages. The gripper, or the robot hand, is used for grasping, moving, spray painting, welding, and

FIGURE 16–18
Robot with six axes of movement. (Courtesy Staübli Unimation, Inc.)

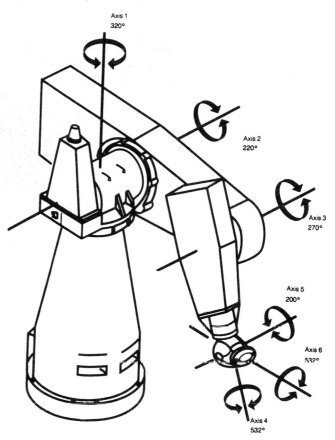

other similar tasks. The hand can have fingers that are mechanical, magnetic, or vacuum devices or can just be some type of tool.

The simplest of robots are typically two-position (sometimes called bang-bang) robots, the positions being up and down or in and out; these robots perform very limited tasks. Programmable controllers can often be used to control these low-technology robots. The programs for these robots usually contain only time delays or sequential events. Very sophisticated robots can perform many tasks and have very flexible movements. A robot such as Cincinnati Milacron's T3® is one of these remarkable units. It allows for six degrees of freedom and is used in many industrial applications.

There are four primary classifications of the functions performed by robots:

1. Pick and place
2. Point to point
3. Continuous path
4. Assembly (which combines continuous path with machine tools such as drills or lathes)

FIGURE 16–19
Controller for PUMA 761/762 robot. (Courtesy Staübli Unimation, Inc.)

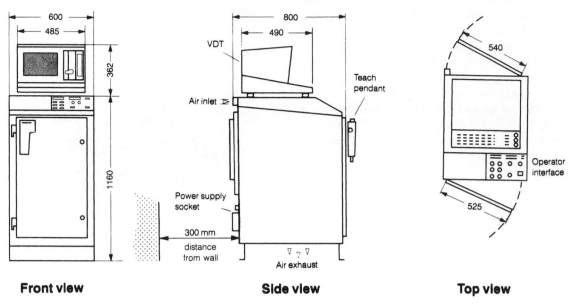

Front view   **Side view**   **Top view**

The more flexible the robot, the more complex the control system has to be. A control system is used to cause a process to occur and may be able to adjust for disturbances that will affect the final product of this process. A block diagram of a general control system with feedback is shown in Fig. 16–9. The feedback may be a signal from a meter reading, a limit switch, a pressure gage, or any other device used to measure a disturbance.

Each different robot has a different operating envelope or space needed for clearance of the arm movement based on the dimensions of the arm, gripper, base, and so on, as well as the directions of movement. Figure 16–20 shows various types of operating areas or envelopes required. A robot with a simple linear in-and-out arm movement does not require any side-to-side clearance or up-and-down clearance.

The programming of a robot includes designing the work pattern that must be performed so that the robot can operate without human intervention. When a person programs a robot to perform a task, that person must generally be skilled in performing that task. There are several methods of programming a robot, and the method used has a good deal to do with the complexity of the robot as well as the task to be performed. Many robots have an attached pendant or training box that allows the operator to move the arm through a series of steps and points. This situation requires the operator to move each axis for the robot until it is in exactly the right place. The movements are not done in real time, so the operator should make certain that the series of points or movements, when connected, do not require

the arm to travel through the base of the robot or to make too large a movement at once. If the arm needs to move over a large distance, it is better to create a series of small moves so that the final movement is not too rapid or too large an individual swing. Most robots have built-in self-preservation. If you do attempt to move the arm from one point to another in a straight line that would cause the arm to crash into the base, the robot will not perform that task. Some robots are "trained" by actually grasping handles and moving the robot arm through the task to be performed. This is sometimes called *teach by showing* or *walkthrough.* It can also be called *continuous path,* since you are moving the robot on a continuous path rather than from individual point to individual point. This method requires a great deal of memory and can be a problem when the manipulator arm is quite heavy. As the arm is moved through a sequence of steps, a sample of the arm's position is made and entered into the control program. Offline programming can also be used to control the robot's movement. This is sometimes referred to as the *controlled-path* method, and it takes advantage of the computational ability of the computer interface. One way of producing the controlled path is for the operator simply to move the arm with the pendant; the computer can then generate the path to produce that movement.

Software for controlling robots is as diverse as popular programming languages for general-purpose computers. Some of the more popular languages available are VAL (Westinghouse), AML (IBM robotics), RAIL (Automatix), and T3 (Cincinnati Milacron). Also, BA-

FIGURE 16-20
Working range, or operating envelopes, of various robots.
(Courtesy ABB Robotics, Inc.)

IRB 2000 ROBOT

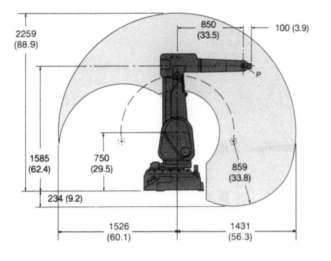

IRB L6E ROBOT

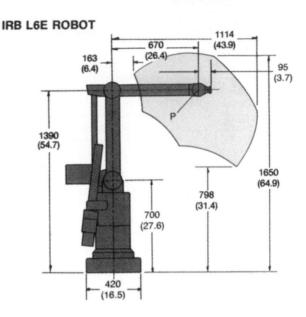

IRB 1000 ROBOT

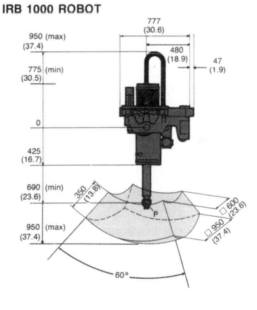

SIC and PASCAL can be used for real-time robot control. The basic function of each of these languages is to move the robot arm through a series of tasks. Robots can be found performing in the following work environments:

1. Spray painting
2. Parts transfer and assembly
3. Electronic and PCB assembly
4. Plastic molding
5. Machining, welding, and finishing
6. Loading and unloading machines or machine tools

INSTRUMENTATION

Very few control systems exist without instrumentation used as either instrumentation-monitoring or instru-

TABLE 16–1

Identification Letters

	FIRST LETTER		SUCCEEDING LETTERS		
	Measured or Initiating Variable	**Modifier**	**Readout or Passive Function**	**Output Function**	**Modifier**
A	Analysis (5, 19)		Alarm		
B	Burner, combustion		User's choice (1)	User's choice (1)	User's choice (1)
C	User's choice (1)			Control (13)	
D	User's choice (1)	Differential (4)			
E	Voltage		Sensor (primary element)		
F	Flow rate	Ratio (fraction) (4)			
G	User's choice (1)		Glass, viewing device (9)		
H	Hand				High (7, 15, 16)
I	Current (electrical)		Indicate (10)		
J	Power	Scan (7)			
K	Time, time schedule	Time rate of change (4, 21)		Control station (22)	
L	Level		Light (11)		Low (7, 15, 16)
M	User's choice (1)	Momentary (4)			Middle, intermediate (7, 15)
N	User's choice (1)		User's choice (1)	User's choice (1)	User's choice (1)
O	User's choice (1)		Orifice, restriction		
P	Pressure, vacuum		Point (test) connection		
Q	Quantity	Integrate, totalize (4)			
R	Radiation		Record (17)		
S	Speed, frequency	Safety (8)		Switch (13)	
T	Temperature			Transmit (18)	
U	Multivariable (6)		Multifunction (12)	Multifunction (12)	Multifunction (12)
V	Vibration, mechanical analysis (19)			Valve, damper, louver (13)	
W	Weight, force		Well		
X	Unclassified (2)	X axis	Unclassified (2)	Unclassified (2)	Unclassified (2)
Y	Event, state or presence (20)	Y axis		Relay, compute, convert (13, 14, 18)	
Z	Position, dimension	Z axis		Driver, actuator, unclassified final control element	

mentation-actuated controls. The link between instrumentation and control circuits becomes more obvious when one considers using automatic control in response to measurements. For example, a process that needs water cooling could actuate water flow after a temperature indicator signals a high temperature. Instruments are required to measure many different variables. From an electrical point of view, there are only two broad classes of instruments: analog and digital. Analog instruments generally require a power supply and often require some wiring to obtain the necessary signals for operation. Often, digital instruments used to measure flow, pressure, or temperature are included in discussions of instrumentation, because they perform the task of measurement. But they should be included in discussions of control circuits, because one of their primary functions is that of control. The need for instrumentation introduces a unique set of drawings for industrial control. These drawings are called **balloon drawings,** and they are found only in instrumentation diagrams. There have been two sets of standards for symbols used in the area of instrumentation. The most widely used in electrical areas is ANSI/ISA-S5.1-1984, *Instrumentation Symbols and Identification,* from the Instrumentation Society of America, which was approved on November 5, 1986. The other set is from Scientific Apparatus Makers Association (SAMA). These standards use a method of functional diagramming to describe functions with blocks and designators. In addition to these standards, the key elements of ISA-S5.3, *Graphics Symbols for Distributed Control/Shared Display Instrumentation, Logic, and Computer Systems,* were incorporated into ISA-S5.1 with the intent of withdrawing that document after the approval of the revision of ISA-S5.1. The standards cover instrumentation used in many areas, such as chemical, petroleum, power generation, air conditioning, and metal refining processes, to name a few. The emphasis here is on power generation.

Looking up the definition of balloon in the standard gives the following:

Balloon Synonym for *bubble.*
Bubble The circular symbol used to denote and identify the purpose of an *instrument* or *function.* It may contain a tag number. Synonym for *balloon.*

The tag numbers or other identification include information such as plant area designation or the function of the instrument itself. Each set of tag numbers and identification can be translated using designations from the standard, as shown in Table 16.1. The numbers in this table refer to notes in the standard. The standard should be consulted when preparing instrumentation diagrams. The location of a particular letter in the tag string is important. The first letter is the identifier, and all other letters are for additional readouts or functions. For example, the use of the letter V as the measured or initiating variable means vibration or mechanical analysis. But using V as a succeeding letter indicates a valve, damper, or louver. The first letter, then, is determined by the measured value or a functional identification. The succeeding letters indicate additional readouts, recordings, or output functions of the instrument. In addition, special modifier letters may be used to modify the initial variable or output variables. For example, the use of T indicates temperature, but the use of TD indicates temperature *differential,* or the difference in temperature. These are two entirely different variables. Figure 16–21 shows two samples of the use of this identification method. Figure 16–21 (A) pictures an indicating voltmeter connected to a turbine–generator pair. The line cutting across the circle shows that the primary location of this instrument is normally accessible to the operator. The tag in Fig. 16–21 (B) includes the use of a modifier. This instrument measures the temperature and engages a switch when that temperature is low. This action causes one output to split, sounding a low-temperature alarm and tripping the circuit.

FIGURE 16–21
Examples of primary element bubbles.

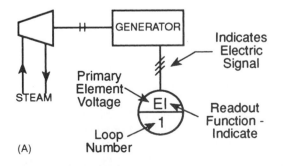

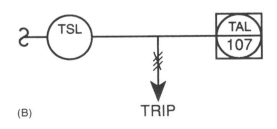

FIGURE 16–22
Instrumentation diagram. (Courtesy Brown and Caldwell Consultants)

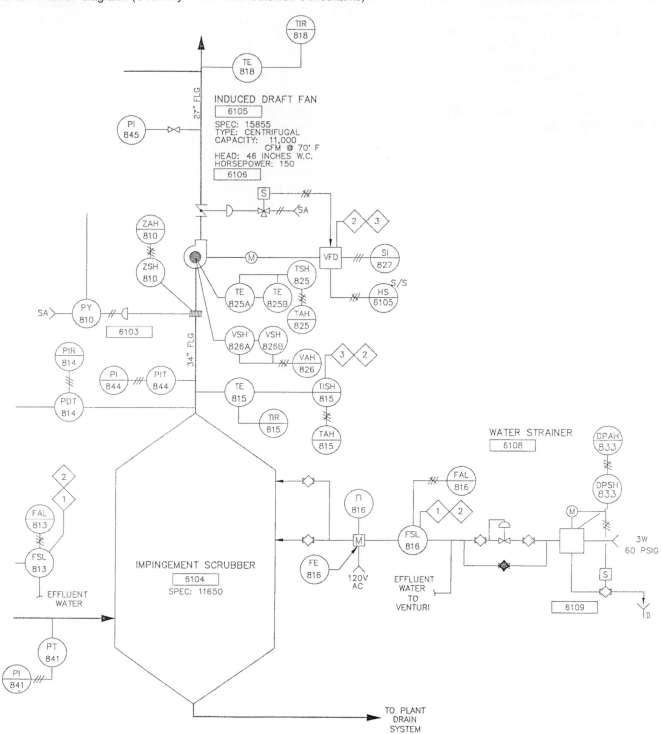

The numbering of the loop that contains the instruments may be done with either a parallel numbering system or a serial one. The use of a parallel numbering system requires the same starting numerical sequence for each different function or first letter, such as FT-100, TE-100, and TIC-100. For serial numbering, a single numerical sequence is used regardless of the function or first letter, such as, FT-100, TE-101, and TIC-102. The most important thing to remember is that the numbering system used should be as simple as possible.

Actual drafting practices for the use of tagging bubbles are covered in the standard. In general, good drafting practices should be followed for neatness and legibility. Adequate space must be left between symbols for notes and other information. Symbols may be drawn with any orientation, but the tag numbers should always be horizontal. Electrical, pneumatic, and power supplies are not shown unless they are required for a complete understanding of the operation. The arrangement of the diagram will reflect the functional flow and not necessarily indicate correct interconnections. These interconnec-

tions are shown on another diagram. The sizes of the tagging bubbles will vary, depending on final reductions of the finished drawing and sizes of other symbols used to represent equipment on the diagram. A suggested starting size for circles and squares to be used on large diagrams is $7/16$ in. The most important consideration is consistency throughout the entire drawing. The degree of detail used in the final drawings will vary for each instance. A simple functional representation indicating what instruments are used or what variables are measured is shown in Fig. 16–22. This diagram indicates the kind of hardware and types of signals that will be measured.

REVIEW QUESTIONS

16–1 What three things must be considered for a complete control system?

16–2 What type of functions are performed by programmable logic controllers?

16–3 *True or false:* To program or maintain a programmable logic controller, one must be a computer expert.

16–4 Both the programmable logic controller and the personal microcomputer contain a microprocessor. What makes them different?

16–5 What is a transducer, and what is its primary purpose?

16–6 List the primary parts of a programmable logic controller, and discuss the function of each part.

16–7 Discuss the limitations found in most programmable logic controllers.

16–8 Why is the sequence of operations a more important consideration in programmable logic controllers than for hard-wired relay logic?

16–9 What is the difference between an open-loop and a closed-loop control system?

16–10 Why is wiring for an actual relay logic circuit so much more complex than for programmable logic controllers?

16–11 List some common applications of programmable logic controllers.

16–12 What are the three basic components of a robot?

16–13 List two methods of classifying industrial robots.

16–14 What is a degree of freedom?

16–15 What is an operating envelope?

16–16 List and describe different methods of programming, or training, robot movements.

16–17 Why is instrumentation an important part of a control system?

PROBLEMS

16–1 A motor is to be started and stopped from any of three locations. Each location has a start and stop button. Draw a logic circuit representing this. Convert the circuit drawing to a ladder diagram using relay logic.

16–2 A grinding machine (G) must have lubricating oil (O) whenever it is running. Both the grinder and the oil reservoir use three-wire control, and O must be running before G is started. Also, if O stops, G must stop immediately. Draw the logic circuit representing this and convert it to a ladder diagram using relay logic.

16–3 Convert the logic figure shown to both relay logic and programmable controller logic.

PROBLEM 16–3
Gate logic circuit.

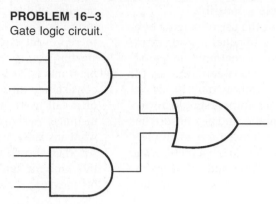

16–4 Convert the logic figure shown to both relay logic and programmable controller logic.

PROBLEM 16–4
Gate logic circuit.

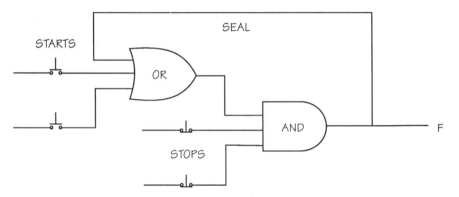

16–5 Convert the relay logic circuit shown to a gate logic circuit.

PROBLEM 16–5
Relay logic circuit.

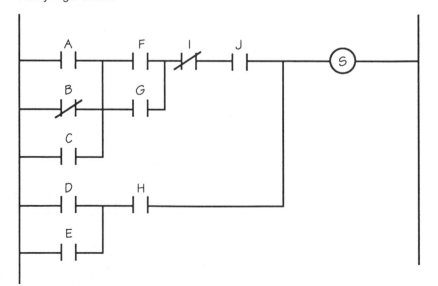

16–6 On a size B sheet, draw the programmable controller circuit shown here.

PROBLEM 16–6
Programmable controller circuit.
(Courtesy of Eaton Corporation, Cutler–Hammer Products)

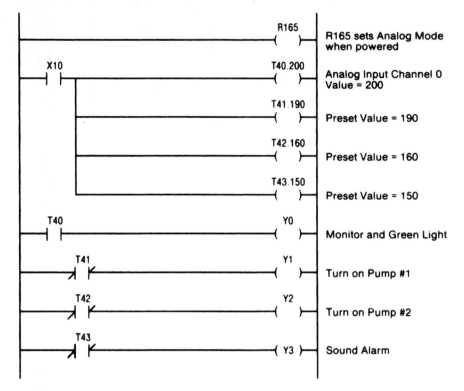

16–7 Shown is the chassis for the D100 programmable logic controller. On a size B sheet, draw these views at one-half scale, showing all dimensions.

PROBLEM 16–7
D100 programmable controller chassis.
(Courtesy of Eaton Corporation, Cutler–Hammer Products)

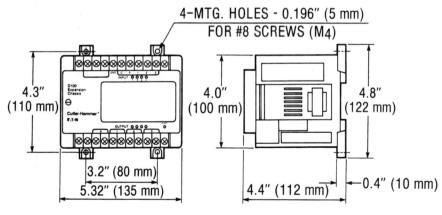

16–8 Shown is a chassis for a programmable controller. On a size B sheet, draw these views at one-half scale, showing all dimensions.

PROBLEM 16–8
Programmable controller chassis.
(Courtesy of Eaton Corporation, Cutler–Hammer Products)

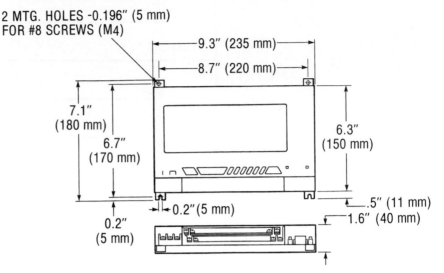

2 MTG. HOLES -0.196″ (5 mm)
FOR #8 SCREWS (M4)

9.3″ (235 mm)
8.7″ (220 mm)
7.1″ (180 mm)
6.7″ (170 mm)
6.3″ (150 mm)
0.2″ (5 mm)
.5″ (11 mm)
1.6″ (40 mm)
0.2″ (5 mm)

16–9 On a size C sheet, draw the PLC enclosure shown.

PROBLEM 16–9
PLC enclosure.
(Courtesy of Brown and Caldwell Consultants)

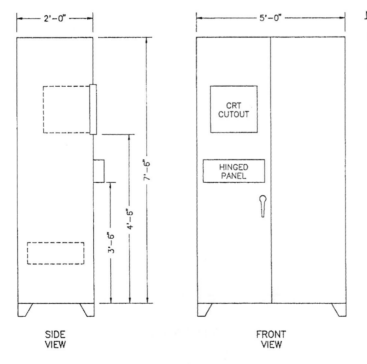

NOTES

1. CRT SHALL BE MOUNTED AND SUPPORTED INSIDE THE PANEL, FLUSHED WITH DOOR PANEL.

2. KEYBOARD SHALL BE MOUNTED AND FASTENED ON A HOLD DOWN TRAY ATTACHED TO THE PLC CABINET BY HINGED PANEL. TRAY SHALL HAVE PROVISION FOR PAD LOCKING TO THE CABINET.

2′-0″
5′-0″
7′-6″
4′-6″
3′-6″

CRT CUTOUT

HINGED PANEL

SIDE VIEW

FRONT VIEW

PLC ENCLOSURE

16–10 On a size B sheet, draw the operating envelopes shown for the IRB 3200 robot. Use metric dimensions.

PROBLEM 16–10
Robot operating envelope.
(Courtesy of ABB Robotics, Inc.)

IRB 3200 ROBOT

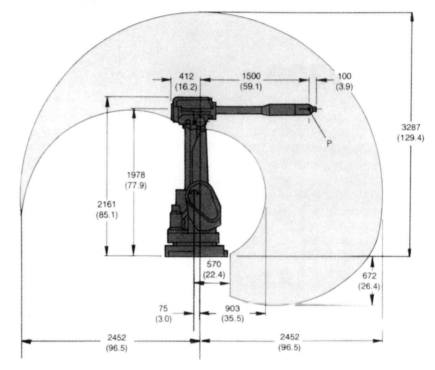

16–11 On a size B sheet, draw the operating envelopes shown for the IRB 3000 robot. Use metric dimensions.

PROBLEM 16–11
Robot operating envelope.
(Courtesy of ABB Robotics, Inc.)

IRB 3000 ROBOT

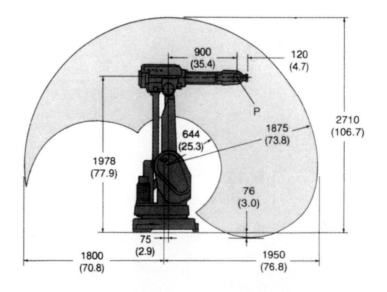

16–12 On a size C sheet, draw the wiring diagram for the programmable controller shown.

PROBLEM 16–12

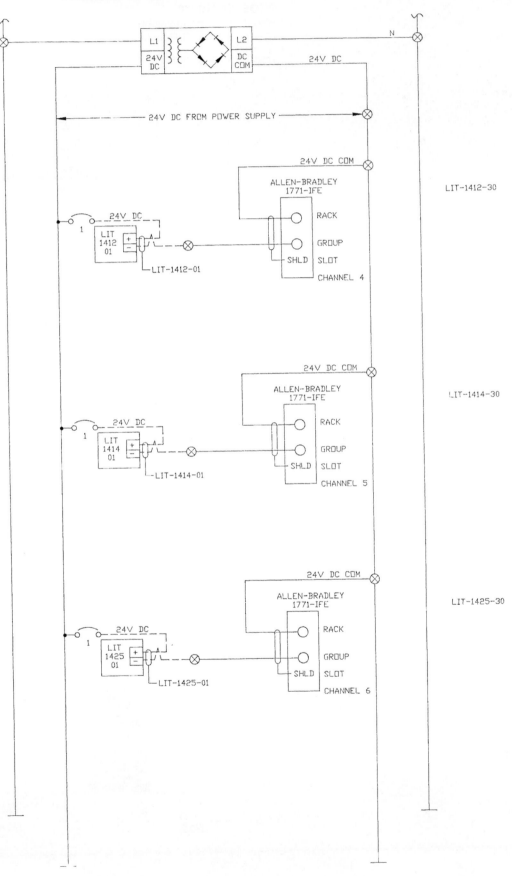

16–13 On a size C, sheet draw the Packer I/O detail shown.

PROBLEM 16–13

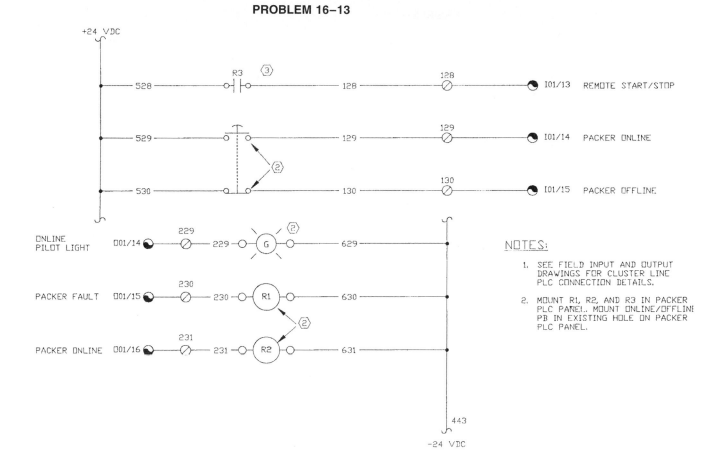

NOTES:

1. SEE FIELD INPUT AND OUTPUT DRAWINGS FOR CLUSTER LINE PLC CONNECTION DETAILS.

2. MOUNT R1, R2, AND R3 IN PACKER PLC PANEL. MOUNT ONLINE/OFFLINE PB IN EXISTING HOLE ON PACKER PLC PANEL.

PACKER I/O DETAIL

16–14 On a size B sheet, draw the instrumentation circuit shown in Fig. 17–14.

16–15 On a size C sheet, draw the ladder diagram shown.

PROBLEM 16–15

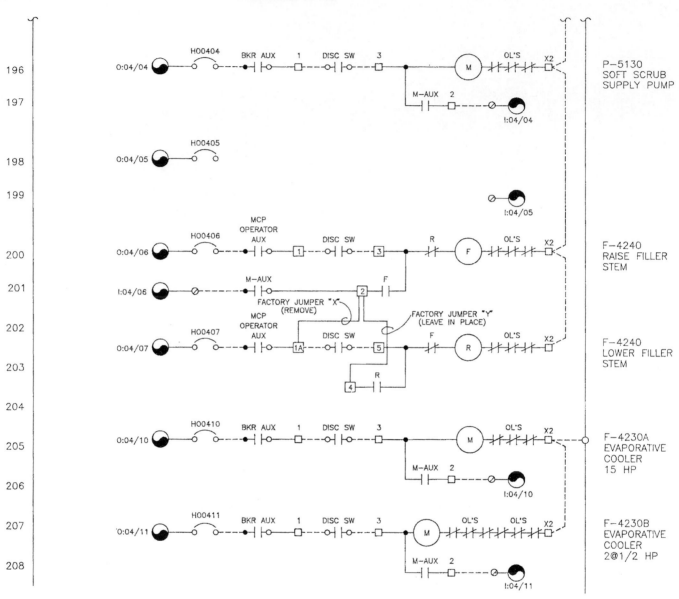

16–16 (Advanced Problem) On a size C sheet, draw the schematic diagram shown.

PROBLEM 16–16

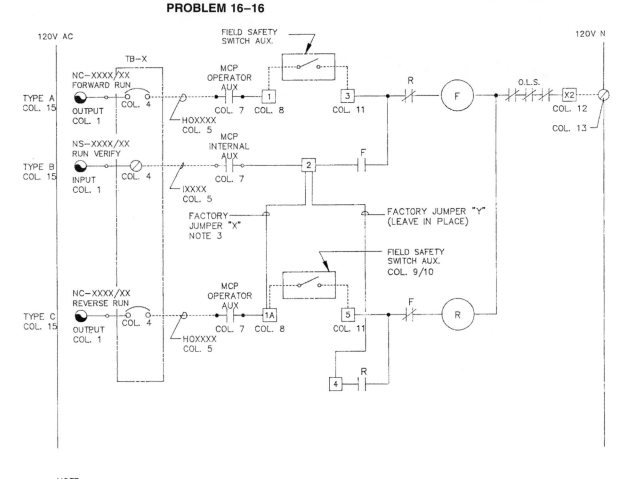

NOTE:
ALL "COL. X" CALLOUTS PRETAIN TO WIRING POINT–TO–POINT LIST.

FULL VOLTAGE REVERSING COMBINATION MOTOR STARTER
TYPICAL I/O SCHEMATIC

16–17 (Advanced Problem) On a size D sheet, draw the instrumentation diagram shown in Fig. 16–22.

17

POWER DISTRIBUTION

INTRODUCTION

There are many definitions of electrical power and distribution systems, but, in general, electricity is a form of energy, and a distribution system includes all machines, equipment, and cabling used in generation, transmission, and distribution of this energy, which must be available at any location when it is needed. Three things are required: production, transmission, and utilization. Many methods are used in producing electricity. In the beginning the biggest source was water power. Water is stored at a high elevation and released. The energy that results in this water movement is capable of doing work. Work in this case is the moving of turbines, which in turn rotate generators, thereby generating electricity.

To be useful to the consumer, this electricity must be transformed to a usable level and delivered to the home or business. The transmission and distribution of electricity is generally accomplished two ways: overhead or underground.

PRODUCTION OF ELECTRICITY

In addition to falling water, steam and nuclear power are sources of energy useful in generating electricity.

Although wind and solar power have been used to generate electricity, they still account for a very small percentage of all electricity generated today. A block diagram of electrical generation is shown in Fig. 17–1. Steam can be produced in a boiler using coal, oil, or natural gas. Any source of energy that can be used to produce rotational movement can be used to turn a generator. A typical generator is shown in Fig. 17–2.

A generator produces electricity at a low voltage, which must be raised by the use of a transformer so that it can be transmitted economically over long distances. The distribution of electrical power can be described as a *tree structure*. The roots are the generating station. The trunk is the main transmission line. The large branches from the main substation are sub-transmission lines. Finally, the smaller branches are the primary distribution lines, or feeders, with the smallest branches being the secondary distribution lines going to the leaves, or individual consumers.

This transmission and distribution system requires the use of transformers to step the voltages up and down, meters and other instruments to measure the quantity, and regulating devices, protective relays, and circuit breakers to control and ensure the stability of this electrical power. Figure 17–3 is a flow diagram representing a typical electric system.

FIGURE 17–1
Block diagram of electrical generation.

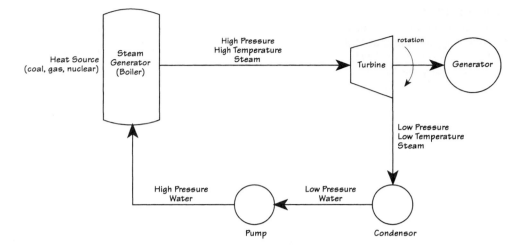

Drawings made for the electrical power field can be made either for the generator plant or for an individual commercial consumer. These drawings can be the most complex, because they include all types of drawings: schematics, mechanical, control, and logic, as well as some drawings that are unique to the power industry, such as the single-line drawing.

JOB PREPARATION

The basic system, even though it is generally a three-phase system, is often represented by a single-line drawing, which shows the major equipment, interconnection, ratings, loads, and possibly instrumentation. Standard scale for electrical drawings is generally either ⅛ in. =

FIGURE 17–2
Cutaway of a typical generator. (Reprinted with the permission of Merrill, an imprint of Macmillan Publishing Company, from *Motors and Controls* by James T. Humphries. Copyright © 1988 by Merrill Publishing Company.)

1.0 ft or ¼ in. = 1.0 ft. The number of drawings required is determined by how large a system is represented and how large the area to be covered is. If a project is very large, it requires a number of drawings to cover individual details. A numbering scheme is used to tie these drawings together, and related diagrams are listed on each drawing, as shown in Figs. 17–4 and 17–5. The detail drawings required for a power substation include the following:

1. Plot plan
2. Single line
3. Substation

FIGURE 17–3
Flow diagram of electric system.

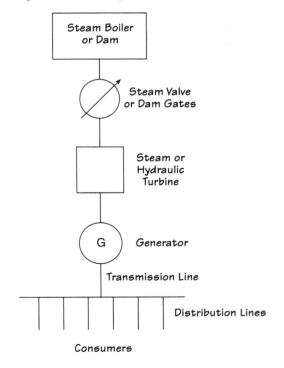

FIGURE 17–4
Numbering system for cross referencing multiple drawings related to the same project. (Courtesy Georgia Power Company)

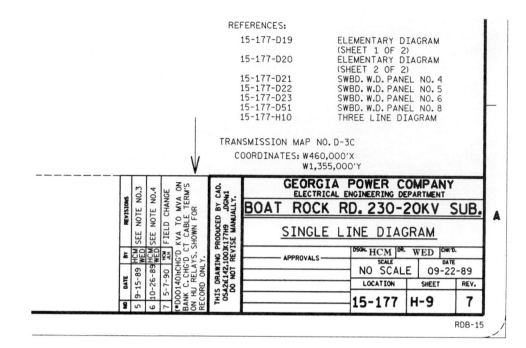

REFERENCES:
15-177-D19   ELEMENTARY DIAGRAM (SHEET 1 OF 2)
15-177-D20   ELEMENTARY DIAGRAM (SHEET 2 OF 2)
15-177-D21   SWBD. W.D. PANEL NO. 4
15-177-D22   SWBD. W.D. PANEL NO. 5
15-177-D23   SWBD. W.D. PANEL NO. 6
15-177-D51   SWBD. W.D. PANEL NO. 8
15-177-H10   THREE LINE DIAGRAM

TRANSMISSION MAP NO. D-3C
COORDINATES: W460,000'X
W1,355,000'Y

GEORGIA POWER COMPANY
ELECTRICAL ENGINEERING DEPARTMENT
BOAT ROCK RD. 230-20KV SUB.
SINGLE LINE DIAGRAM
APPROVALS — DSGN. HCM DR. WED CHK'D.
SCALE NO SCALE DATE 09-22-89
LOCATION 15-177 SHEET H-9 REV. 7
RDB-15

4. Underground conduits
5. Aboveground conduits
6. Equipment layout
7. Control and instrumentation
8. Area lighting

The list of drawings may not be limited to these items, and not all of these items are required for every project.

The first distribution system was direct current and low voltage and was installed underground. This system was pioneered by Thomas Edison. Widespread expansion of electrical systems did not occur until the adoption of alternating current and the application of transformers. Also, overhead construction and transmission can be a much more economical method of distributing elec-

FIGURE 17–5
Numbering system for cross referencing multiple drawings related to the same project. (Courtesy Georgia Power Company)

REFERENCES:
15-177-D21   SWBD. W.D. PANEL NO. 4
15-177-D23   SWBD. W.D. PANEL NO. 6
15-177-D51   SWBD. W.D. PANEL NO. 8
15-177-H9   SINGLE LINE DIAGRAM

TRANSMISSION MAP NO. D-3C
COORDINATES: W460,000'X
W1,355,000'Y

GEORGIA POWER COMPANY
ELECTRICAL ENGINEERING DEPARTMENT
BOAT ROCK RD. 230-20KV SUB.
THREE LINE DIAGRAM
APPROVALS — DSGN. HCM DR. WED CHK'D.
SCALE NO SCALE DATE 9-15-89
LOCATION 15-177 SHEET H-10 REV. 4
RDB-15

TABLE 17–1
Typical Voltages for Power Distribution

Main Transmission	Subtransmission	Primary Distribution	Distribution Secondary
69,000	13,800	2,400	120
138,000	23,000	4,160	240
220,000	34,500	13,800	240
330,000	69,000		480

trical power. The first electric power station, Pearl Street Electric Station in New York City, went into operation in 1882, and the electric utility industry was born. Planning for power distribution must start at the customer level, considering demand, type, load factor, and other load characteristics. All these will determine the type of distribution system necessary.

Types of Power Lines

There are two major categories of power lines: transmission and distribution. Transmission lines are the high-voltage (115 to 800 kV) lines that connect the main substations to the generating plants. Distribution lines include medium-voltage (2.4 to 69 kV) lines used primarily to tie load centers to the main substations and low-voltage (< 600 V) lines leading directly to the consumers. Although these categories indicate a range of voltages, there are voltages preferred for most uses and recommended by ANSI or some of the other standards boards. Standardization also helps to reduce the cost of the distribution equipment. Industries do not have to produce equipment capable of supplying or operating at different voltage and frequency levels. Some typical voltages are shown in Table 17–1. Throughout the process of transmission and distribution, the voltage must be raised or lowered to meet the particular requirements. A transformer is shown in Fig. 17–6. This transformer will step up or step down the voltages from one level to another, which is one of the main purposes of the substation. Substations also contain circuit breakers, fuses, and lightning arrestors. These protective devices keep high levels of current from leaving the substation and endangering the consumer, but they also serve as protection for the equipment itself. Typical circuit breakers and lightning arrestors are shown in Figs. 17–7 and 17–8.

Conductors need support to get from one place to another. These supports may be towers, poles, or other structures. Usually, steel towers are used for transmission lines, and wood and concrete poles are used for distribution circuits. In general, steel towers are used where exceptional strength and reliability are required. Given proper care, a steel tower lasts indefinitely.

Two factors are considered in choosing poles: length and strength required. The length depends on the required ground clearance and the number of crossarms or other equipment that will be attached to the poles. Poles are generally between 25 and 90 ft high. Required pole strength is determined by the weight of the crossarms, insulators, wires, transformers, and other equipment, as well as by ice and wind loading.

Line conductors are insulated from each other, as well as from the pole, by insulators. The most practical of these are made from glass or porcelain. A pin insulator is mounted on a pin, which is installed on the crossarm. These insulators can weigh anywhere from $\frac{1}{2}$ to 90 lb. The most commonly used insulators are the pin, or post, type and the suspension, or hanging, type. A third type is a strain insulator, which is a variation on the suspension type and is designed to sustain extraordinary pulls. These insulators are shown in Fig. 17–9. The main advantage of the pin type is that it is cheaper. The higher the voltages are, the more insulation that is needed. In such a case, the pin type becomes so large it is impractical, and suspension insulators are used.

Line conductors may vary in size according to the rated voltage. Copper, aluminum, and steel are the most commonly used conductors. Of these three, copper is the best conductor. Aluminum is used because of its weight, and steel is used because of its strength. ACSR (aluminum conductor, steel reinforced) is used for long transmission spans. Copperweld or alumoweld, a clad-steel combination, is used for rural distribution and for guy wires.

One-Line Diagram

Used most often in the power field, the **one-line diagram** represents all the major components of an electrical plant or system in one line. Generally, the system is a three-phase system, but only one line (or phase) is necessary to describe the system fully. As for all other drawings or sets of drawings, certain rules must be followed.

1. Drawings typically are vertical.
2. Highest voltages should be placed at the top of the drawing.

FIGURE 17–6
Typical transformer. (Courtesy ABB Power T&D Company, Inc.)

3. Adequate distance between components must be maintained for reference designations and other notes.
4. Standard recognized reference designations and abbreviations must be used.

Depending on the use of the diagram, additional information listing actual part numbers and descriptions may be included on the diagram, rather than on a separate bill of materials.

A checklist for the drafter in the power field should include the following:

1. Ratings of all devices, cables, and wires
2. Ratios of instrument transformers and phasor diagrams

3. Fuse and circuit-breaker ratings
4. Title of drawing and correct names of substations, buses, generators, and the like.
5. Neatness of drawing, spelling, and proper abbreviations
6. Minimum line crossings and space to avoid crowding
7. Lettering uniformity
8. Notes that are conclusive and readily understood

The use of a template is highly recommended for drawing these symbols. However, if a template is not available, each symbol can be completed using triangles, straightedge, and compass. Figure 17–10 shows some typical construction characteristics that must be followed.

FIGURE 17–7
Typical circuit breaker.
(Courtesy Dennis Cherry/
Columbus Southern Power
Company)

FIGURE 17–8
Typical lightning arrestor. (Courtesy Dennis Cherry/
Columbus Southern Power Company)

Preparing Single-Line Drawings

The power distribution field presents a whole new vocabulary, including **switchgear, substation, voltage regulator, pothead, air break switch, circuit breaker,** and **lightning arrestor.** All these devices are represented graphically in circuit diagrams. In addition to the single-line drawing, many other diagrams are found in the power distribution industry: three-line diagrams; logic diagrams; general-arrangement diagrams, showing physical arrangement of equipment; connection diagrams, showing physical connection of controls, meters, and so on; power distribution plans, showing the actual distribution and sometimes routing for service routings inside plants; and detail drawings.

Often standard detail drawings are produced and bound in sets of standards for an individual power company (Fig. 17–11 and Fig. 17–12). Once these detail drawings are made, they can be included by simply adding a copy of the standard to a project, rather than redrawing it each time it is needed.

Control circuits are also found in the electrical distribution field. Generating stations may have control circuits for water pumps or other applications. Since these were covered in detail in Chapter 14, they are not included in this chapter.

In a large company, a drafter may get the information or a single-line drawing in the form of a rough sketch from a development or project engineer. The finished drawing is produced by taking all the information from that sketch and laying it out following company policies. It is important for the drafter to check

(A)

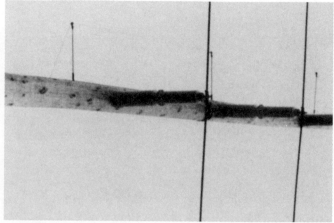

(B)

(C)

FIGURE 17-9
Insulators: (A) pin-type; (B)
stand-off; (C) strain. (Courtesy
Dennis Cherry/Columbus
Southern Power Company)

FIGURE 17-10
Typical construction
characteristics.

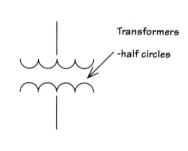

Transformers

-half circles

Disconnect Device

-both triangles are equal length
-clear space between the points
-legs are 30° from vertical

30°

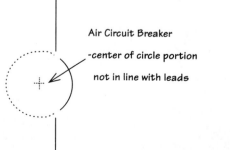

Air Circuit Breaker

-center of circle portion

not in line with leads

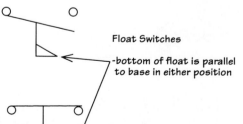

Float Switches

-bottom of float is parallel
to base in either position

FIGURE 17-11
A typical standard from a bound set. (Courtesy Georgia Power Company)

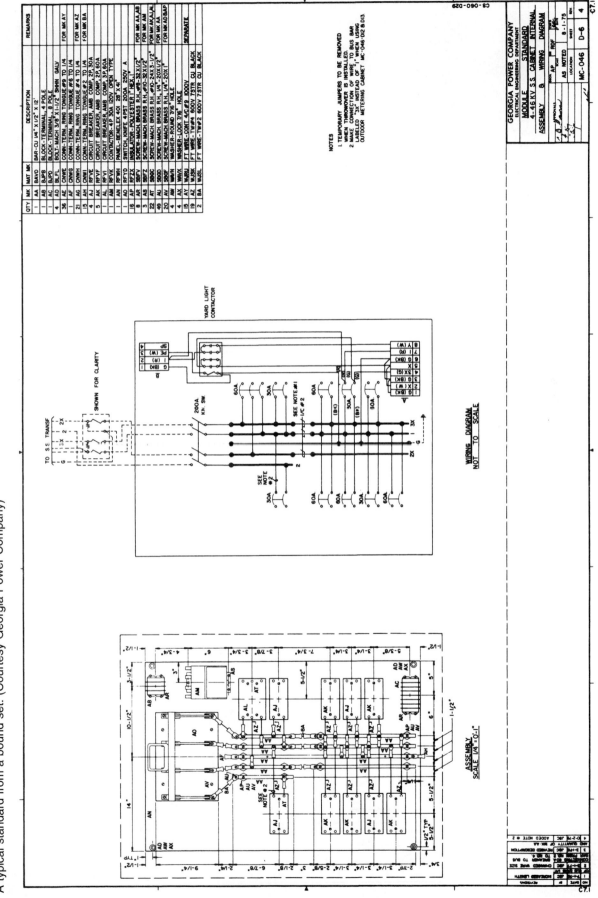

FIGURE 17-12

A typical standard from a bound set. (Courtesy Georgia Power Company)

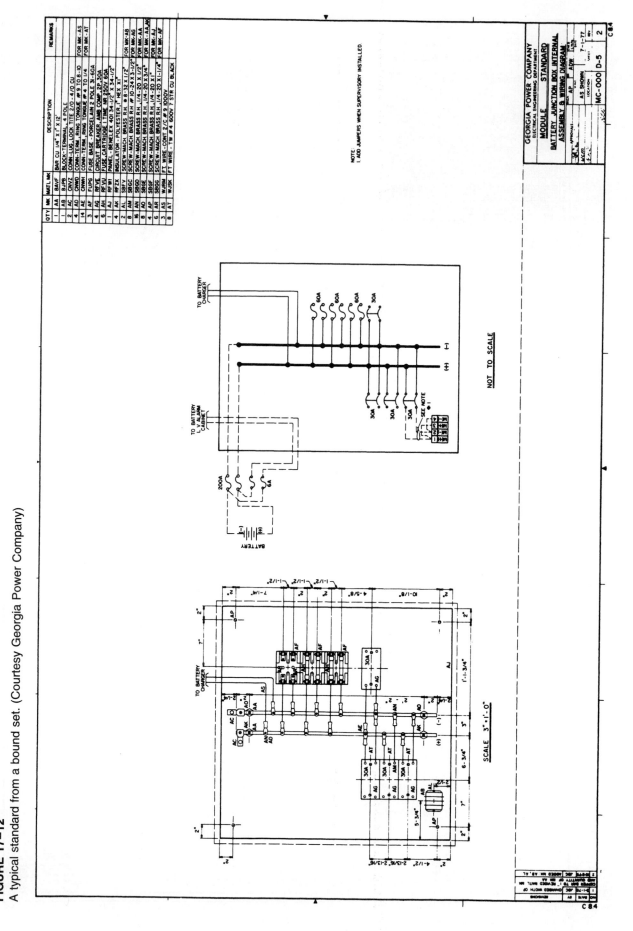

the specifications of all equipment and cabling: cable characteristics and ratings of power transformers, circuit breakers, fuses, connected loads, meters, and other miscellaneous devices such as lightning arrestors or other protective devices. This information must be included on the final drawing if the drawing is to represent the system and fully describe all its components.

Reference designations for the power industry are published by standards boards, as are the symbols used in previous chapters. These are shown in Appendix F. Unless the company dictates otherwise, it is always good and acceptable practice to use these approved symbols and designators. A typical one-line diagram is shown in Fig. 17–13.

As with all other diagrams, a primary consideration is that the drawing be concise, simple, and easy to understand. To achieve these goals, the drawing should not be cluttered, and there should be adequate room between all components for notes and reference designations. Crossovers should be kept to a minimum. This can sometimes be done by simply relocating an item on the drawing, as physical relationships are not critical and are demonstrated on other drawings.

For the many drawings used to represent power distribution areas, it is important that the data presented not be duplicated and used in more than one place. When future revisions are necessary, it would be difficult for the person doing the revision to track down all references to these data and ensure that they are all changed. Even if the same person is doing the revisions, it may be impossible to remember that the same information is found on multiple drawings. Often notes can be used to identify devices standard to the industry, stating that all devices used are the same size, rating, and so on, unless otherwise noted. It is important to include all known data as well as projected future elements. Sometimes future items can have a bearing on the size of cables or the current rating of devices used in the initial installations. These future items should be drawn with dotted lines and identified as planned and used in calculations but as a future item. See Fig. 17–14.

For large customers of a utility company, high voltages may be provided directly to the installation. These customers then have their own switchgear, which functions in exactly the same way as a utility company substation. Remember, the primary function of a substation is to take the feeder lines and step down and distribute power at lower voltage levels. A *pothead* may be used to terminate the cable and allow it to be connected to the switchgear. A pothead is used to connect an underground cable and an overhead wire. It can be a single or multiple conductor. Figure 17–15 shows a cross section of a three-conductor pothead. In this pothead, three separate cables come in and are wrapped into one

cable. The three conductors pass individual porcelain-encased terminals. These terminals look very similar to insulators and perform the same function. The cable is attached to the pothead by a wiped joint or a clamping device. Often the pothead is filled with a liquid insulating compound. This must be cooled before the pothead can be connected to the overhead wires.

Relays and meters are required for monitoring purposes in the switching equipment. Often these are mounted directly in the doors of the switchgear itself, as shown in Fig. 17–16. The meters provide billing information and may be typical kilowatt-hour meters. In addition to use, peak demand may be monitored in order to determine if additional billing is necessary.

Cables running from the main or service entrance switchgear of the customer installation distribute the power to various load zones or unit substations. Often load zones are fed by two feeder lines. A typical customer distribution is shown in Fig. 17–17. This distribution provides maximum service in case of power failures. Unit substations are custom made for every installation. Although this may seem cost prohibitive, in reality, individual elements are standard. They include fuse sections, transformers, breakers, relays, meters, and switches. Installations for all customer-site unit substations are covered in the National Electrical Code. Unit substations may have very limited metering, used only to monitor the operation of the unit. Feeders from this unit provide power for the customer site. Feeder size is determined by present load, predicted future load, cost, length of feeder, and short-circuit requirements. The feeders generally go to the panelboard, which is used to actually divide the feeder into individual circuits, such as lighting. Although **riser diagrams** are used most often in plumbing or heating and air conditioning, they may be included in the electrical area. Figure 17–18 shows electrical service in a multistory building. Riser diagrams typically show the equipment located in the building and how it is connected. They present a picture of how this equipment is wired and are often thought of as interconnection diagrams.

One-line diagrams are not restricted to power distribution. They can be used any time a simplified representation of a larger or more complex system is required. Typically, much of the detailed information is left out, and the circuit function is represented, rather than the individual components. The diagram is completed using thick dark lines for the main connection lines and medium lines to connect to instruments or potential and current sources. Only major equipment is typically represented. Ratings of major components are included only if they are essential to the understanding of the operation. Detailed ratings are included in a connection diagram or a schematic. The rating of a main generator

FIGURE 17-13

A typical one-line drawing. (Courtesy Georgia Power Company)

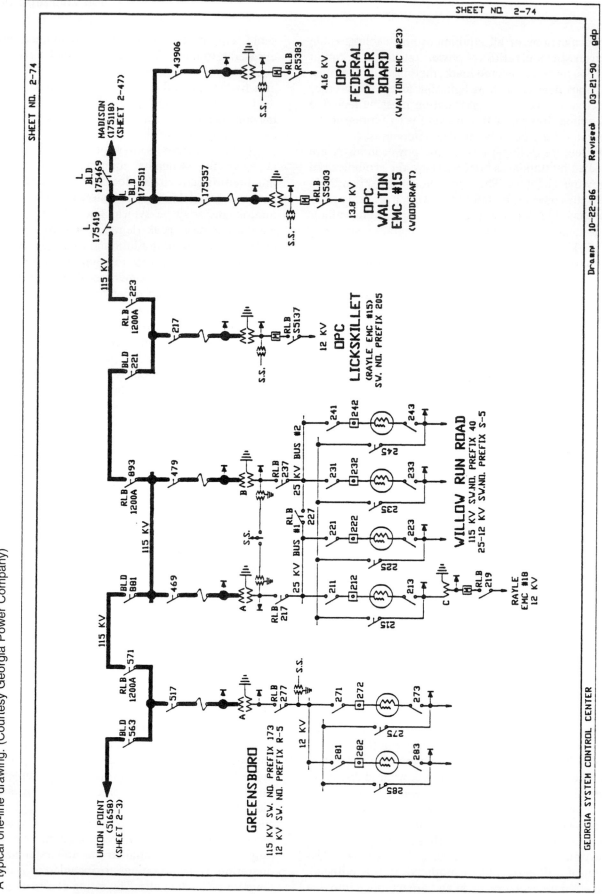

FIGURE 17–14
Future items are indicated with dotted lines.

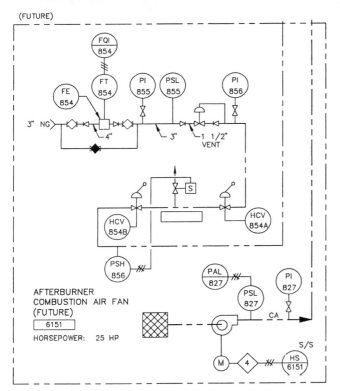

may be included. It is much more complicated to understand simple circuit functions using the complete, or three-line, drawing.

Typical transformer connections can be made either in a delta or a wye connection. Typically, the high side of the transformer is wired using wye connections, but the low-voltage side is delta connected. These winding connections should be indicated on the drawing, as shown in Fig. 17–19. The transformer was introduced in Chapter 14.

The wye connection often includes a neutral or ground wire connected to the center point. The physical representation is shown in Fig. 17–20. Also shown are the neutral, or ground, connections for the transformer. Instruments and meters are also included in the drawing. These are generally represented by a small circle and a device function designation.

The primary guide in this drawing, as in all others, is clarity. The diagram must clearly represent the circuit operation and functions. The symbols should conform first to company standards; if there are none, the guidelines in ANSI or other national standards sources should be followed.

All lines should be either horizontal or vertical, with as few crossovers as possible. There should be adequate spacing between any pairs or sets of wires. Also, line spacing must be adequate even after the maximum reductions have been made to the drawing.

The one-line diagram in Fig. 17–21 is a good example of circuit functions contained in this type of diagram.

FIGURE 17–15
Cross section of a three-conductor pothead.

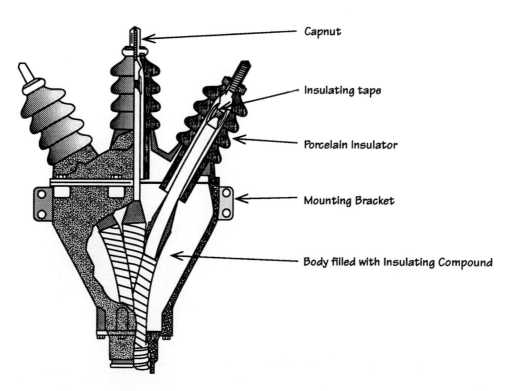

FIGURE 17-16

Meter mounted in door of switchgear. (Courtesy of Georgia Power Company)

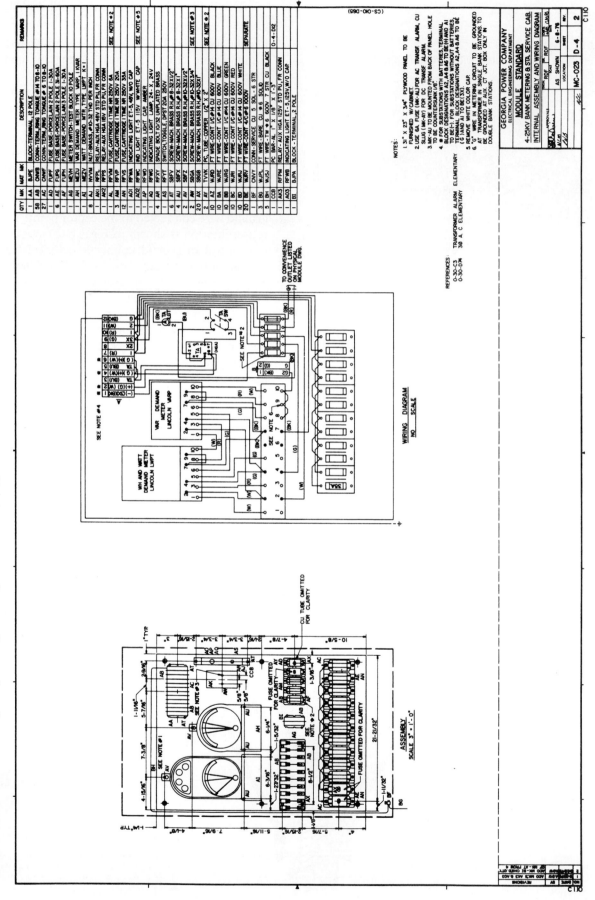

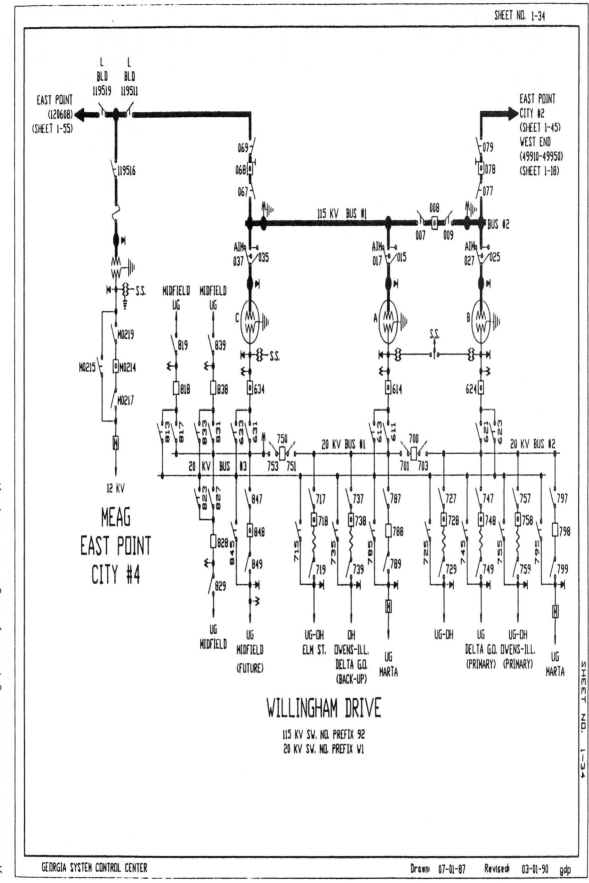

FIGURE 17-17
Typical customer distribution drawing. (Courtesy Georgia Power Company)

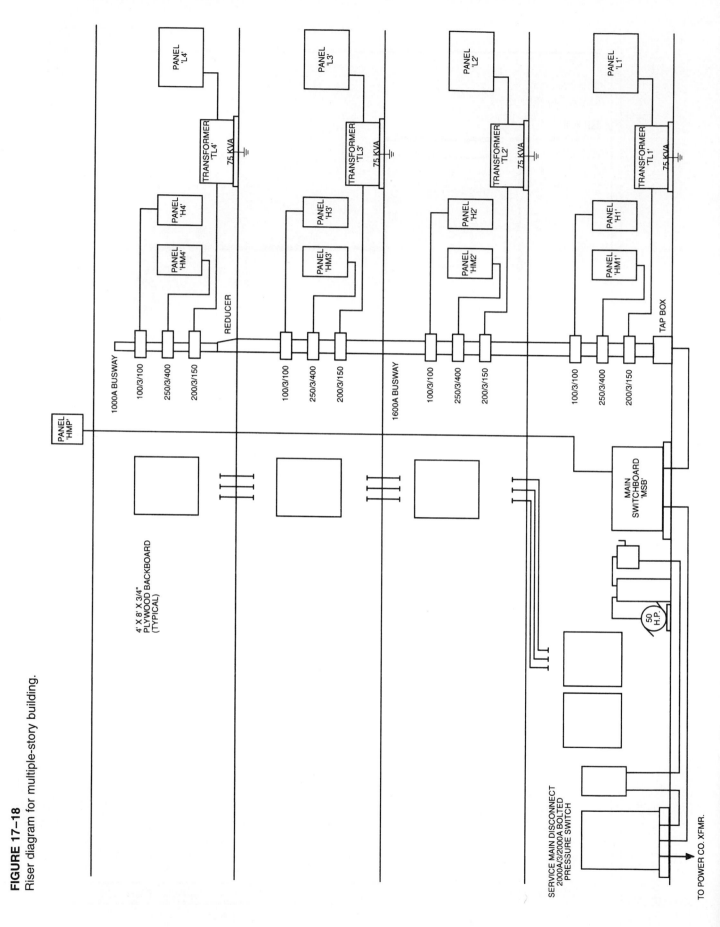

FIGURE 17–18
Riser diagram for multiple-story building.

394

FIGURE 17-19
Winding connections indicated
on drawings. (Courtesy Georgia
Power Company)

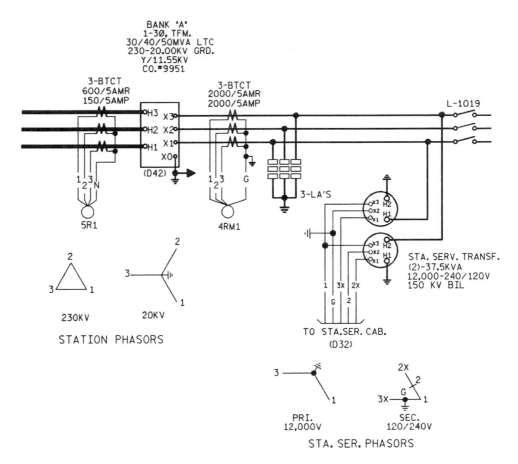

The short horizontal lines are connected to the main lines by way of current transformers and are used to provide power for such additional functions as protection and metering. Starting at the top of the diagram, the main power is represented as 5 kV, 1200 A, three-phase, 60 Hz. A disconnect device is the first device that should be encountered. This circuit does not include one, but a high-voltage disconnect should be located between the high-voltage supply and the main transformer. This is represented by either a square with main circuit breaker noted or the unique symbol shown in

FIGURE 17-20
Physical connection shows actual windings.

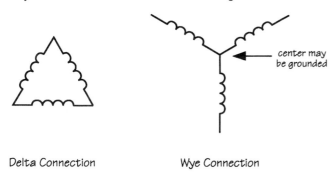

Delta Connection   Wye Connection

Fig. 17-22. Typically, one-line diagrams include a transformer to bring the high voltage down from the main power source to a standard distribution voltage. Current transformers are connected at this point to provide power for the metering circuits. Meters connected to this current transformer should be protected with a fuse, and the meter itself must include an indication of the type of meter and the range. In this example, the meter is an ammeter, so a circle is used to indicate a meter, and an A is included in the circle to indicate an ammeter. The horizontal line from the current transformers provides an auxiliary function to the main line.

Three-Line Drawings

Although one might think of the **three-line drawing** as an expansion of a single-line drawing, in reality it is a supplemental drawing. The single-line drawing is used to represent the function and operation of the entire circuit; the three-line drawing, on the other hand, places most of its emphasis on metering and relaying circuits. The three-line drawing has also been called a *schematic diagram* for the power plant. In addition, it usually includes information about connections and interconnections between devices.

FIGURE 17–21
One-line diagram showing circuit functions. (Courtesy of Interactive Computer Systems, Inc.)

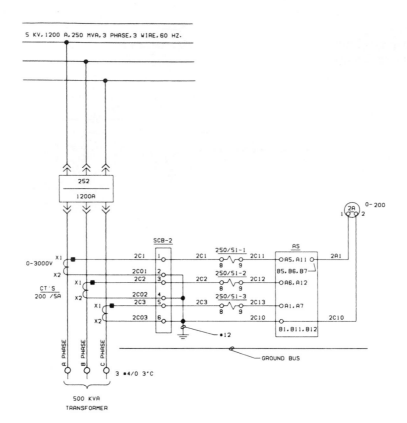

Three-line diagrams are presented in the same manner as one-line diagrams. The primary or power conductors are typically found across the top or along the left side of the drawing, with the primary or source voltage being in the upper left-hand corner. The "flow" of electricity follows the normal reading flow, because this is intuitively easier to grasp.

The ratings for devices are shown following prescribed drafting methods. Placement of the ratings, device designations and other notes should be consistent throughout the drawing. As in earlier chapters, these designations should not be split; if possible, they should be located to the right or below the component. The primary, or power conductor, lines should be thick dark lines, whereas all those used for meters or control circuits should be medium-weight, thinner lines.

Device terminal numbers and wire numbers are shown on this diagram for use in demonstrating interconnections and the physical relationship between the devices. It is important to be very consistent in the numbering scheme used to represent the wires and the terminals so that they are very clear and concise. For example, Fig. 17–23 shows a current transformer. The small numbers 1 to 8, which are shown close to the rectangle used to represent the secondary side of a trans-

FIGURE 17–22
Use of rectangle or unique symbol for circuit breaker.

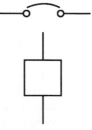

FIGURE 17–23
Current transformer represented by a rectangle showing all terminal connections and wiring numbers.

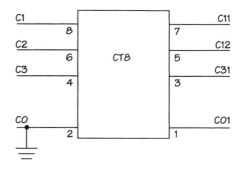

former, are, in fact, the terminals of this transformer. These numbers are located very close to the rectangle and below the wire. The numbers located on top of the lines and away from the rectangle are the wiring numbers.

Three-line drawings may not contain all the detail and device ratings found on a single-line drawing, but there will be meters and other devices that cannot be represented on any other drawings. It is necessary to follow good practices in spacing, layout, and accuracy. When drawings are prepared as D or E size drawings, you must keep in mind that the drawings may need to be reduced. In the reduction process, not only are the devices reduced, but so are lineweights, text size, and even the space between lines and devices. Therefore, you must consider the maximum reduction when laying out such a drawing.

Logic Diagrams

The generation of electrical power requires the use of many control circuits. These circuits can be analyzed and prepared in exactly the same way as those shown in Chapter 15 using contacts and relays. Logic symbols can also be used to represent these circuit functions. The logic symbols shown in Chapter 15 are often reserved for use in digital logic circuits. Although they do represent the operational function, their use is often associated with low-voltage DC logic applications. A different set of symbols representing these same functions can be found in the electric power industry. The primary requirement in the use of any symbol is clarity in representation and consistency. Figure 17–24 shows typical symbols used to represent logic functions.

Keep in mind that *symbols* represent the function of a circuit. They are not actual devices. They represent contacts and relays that occur in the actual physical application. By using a combination of a logic diagram to quickly and simply represent the functions and a connection or interconnection diagram to demonstrate the actual devices used, a complete picture can be presented. When representing control circuits, remember that all devices are shown in their de-energized or deactivated stages. If a float switch is used to activate a sump pump, it is represented in the at-rest state. It could be a normally open or a normally closed switch, depending on the design of the circuit, so it is impossible to simply say that all switches should be shown as open.

In addition to the logic and the interconnection diagrams, a schematic representation of the circuit may be used. By actually representing the contacts, switches, and relays used, ratings and specifications for these devices may be added to the diagram. Since the logic diagram is only a representation of circuit function and not circuit components, it is impossible to note device ratings and specifications on the logic diagram.

General Arrangement

The general arrangement of all equipment in a power-generating plant or a utility substation is an essential part of a complete set of drawings. Such a drawing is similar to an interconnection diagram in that it demonstrates the physical relationship between different pieces of equipment. This drawing—or, in some cases, set of drawings—identifies the location of switchgear, transformers, meters, disconnect switches, lightning arrestors, and so on. However, this diagram does not show the actual wiring. Often the general-arrangement dia-

FIGURE 17–24
NEMA symbols used to represent logic functions in power applications.

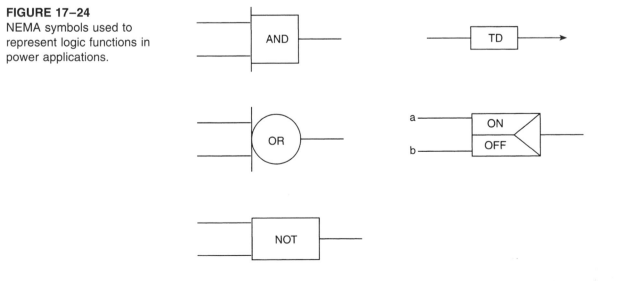

gram appears to be a mechanical drawing representing "packaging," since the enclosures for many of the devices are represented. This drawing demonstrates the total size, shape, and space required. So a set of drawings may include an elevation diagram, details, sections, or a number of other drawings. Figure 17–25 shows a general arrangement drawing.

Wiring Diagrams

Connection and **interconnection diagrams** are a critical part of installation diagrams. The connection diagram is used to represent the connections within a piece of equipment and is often used to show wiring for meters or auxiliary devices (Figs. 17–26 and 17–27). It may or may not show the main or power circuits but most often does not. The interconnection diagram, on the other hand, shows wiring external to an individual piece of equipment. This drawing is used to demonstrate the wiring required between major units, subassemblies, equipment, and the like. The types of diagrams used are the diagrammatic and the tabular types. **Diagrammatic diagrams** include continuous-line (point-to-point or highway) and interrupted-line types. These diagrams are covered in detail in Chapter 13. Refer to that chapter as you complete the exercises. In the preparation of these diagrams, the following practices should be followed.

The use of squares, rectangles, or circles is permissible to represent the individual items. The view should be completed as though all devices and connections are in one plane. The arrangement of the items must provide utmost clarity and interconnection information. Remember, this diagram is a representation of the physical

FIGURE 17–25
Typical section drawing showing general arrangement of equipment.

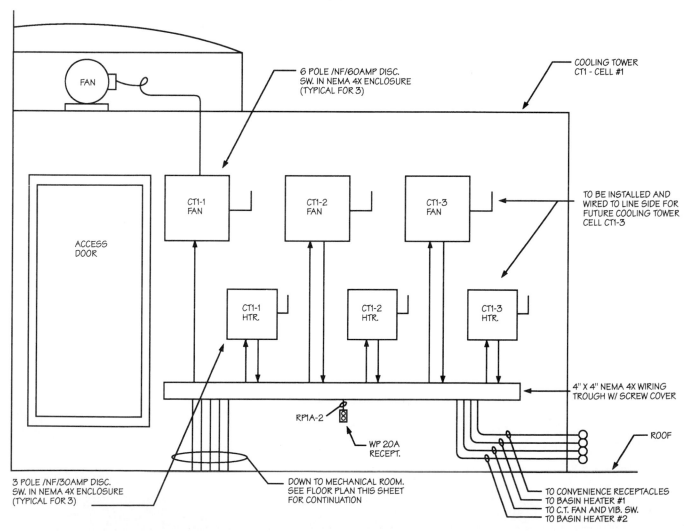

FIGURE 17-26

Detail point-to-point connection diagram. (Courtesy Georgia Power Company)

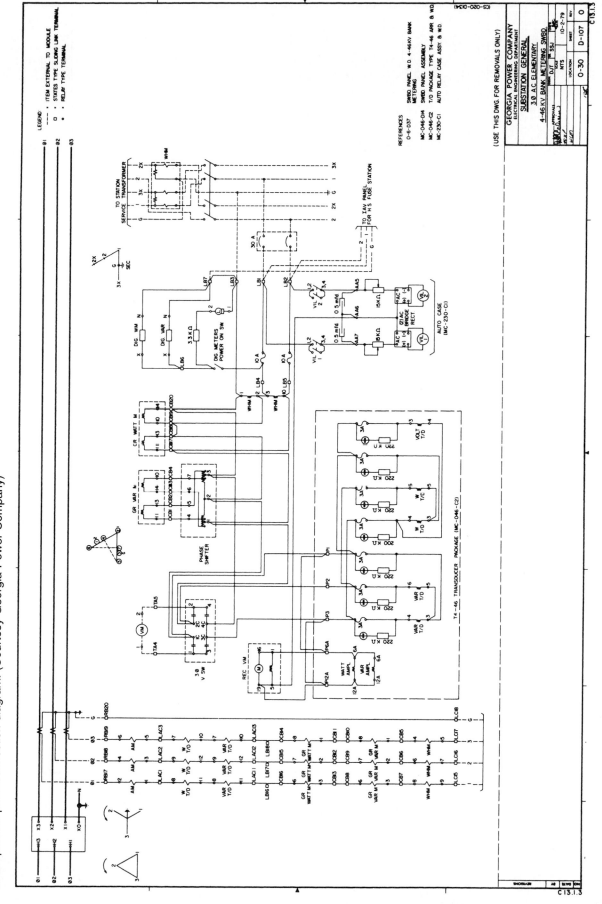
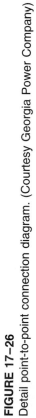

FIGURE 17-27

Detail highway connection diagram. (Courtesy Georgia Power Company)

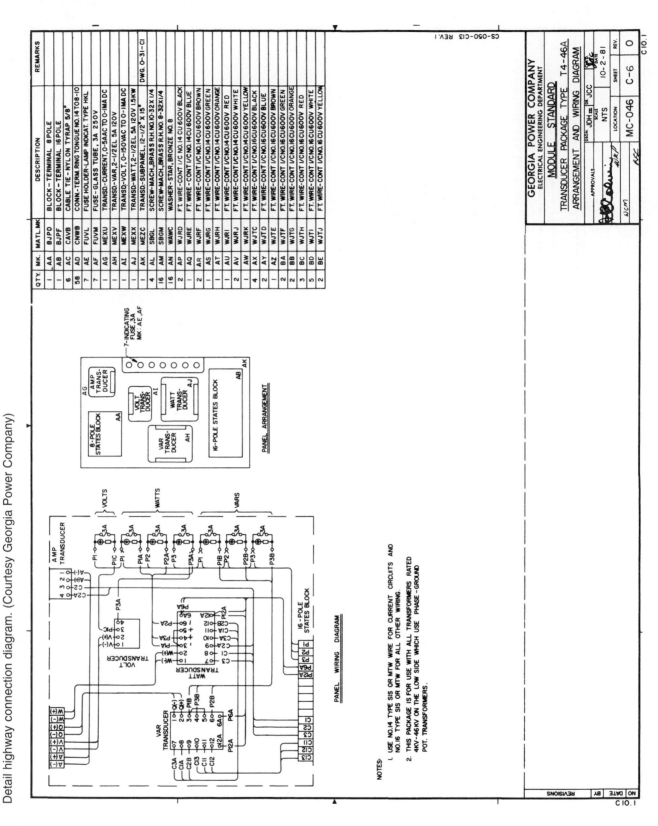

relationship between items and not necessarily the physical location. Relative location of all terminal boards, connectors, and so on, must be clear. Often a diagram showing the physical location is included in a set of drawings. Connecting lines must be spaced so that at the maximum reduction the space will be no less than .06 in. Longer parallel lines should be grouped, and routing should follow the most direct and logical path.

A code may be used in the identification of each wire run. This code may be anything from a simple number to any combination of letters and numbers. Figure 17–28 shows how numbers may be used in an interconnection diagram from the motor-control center to various devices.

It is very important in a connection diagram representing the physical location of cabling and wiring in an individual piece of equipment to actually demonstrate wiring routing.

Power Distribution Plans

Power distribution plans are required to show the actual power distribution in a building or manufacturing plant. A distribution plan includes all electrical service from the incoming line to the building, to the distribution center, and through all the service panels in the structure. The first consideration, then, is the incoming line. If this line is underground service, it requires only a notation showing the service as it enters the distribution center. If these lines are overhead, then there must be some detail showing the poles and the connections to the utility company service from the customer's point of view. These service feeds should never be routed over buildings, as they would interfere with building maintenance, security, and emergency functions.

As the distribution plans are developed, certain items need to be included: special load requirements

FIGURE 17–28
Wire numbering codes in interconnection diagram for switchgear. (Courtesy of Brown and Caldwell Consultants)

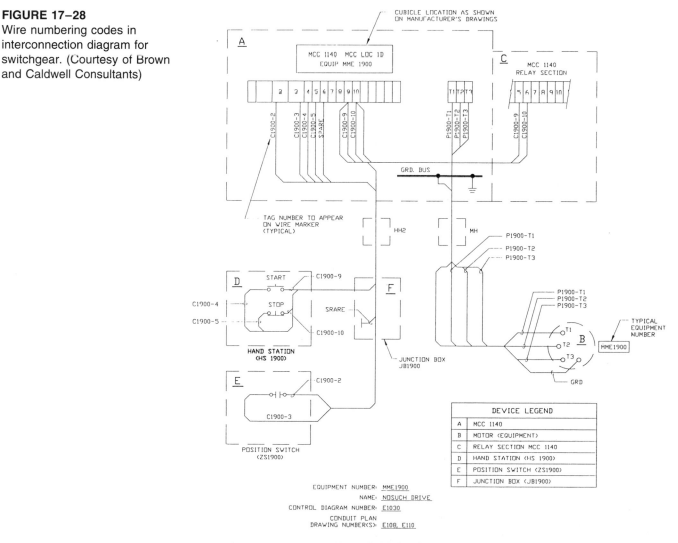

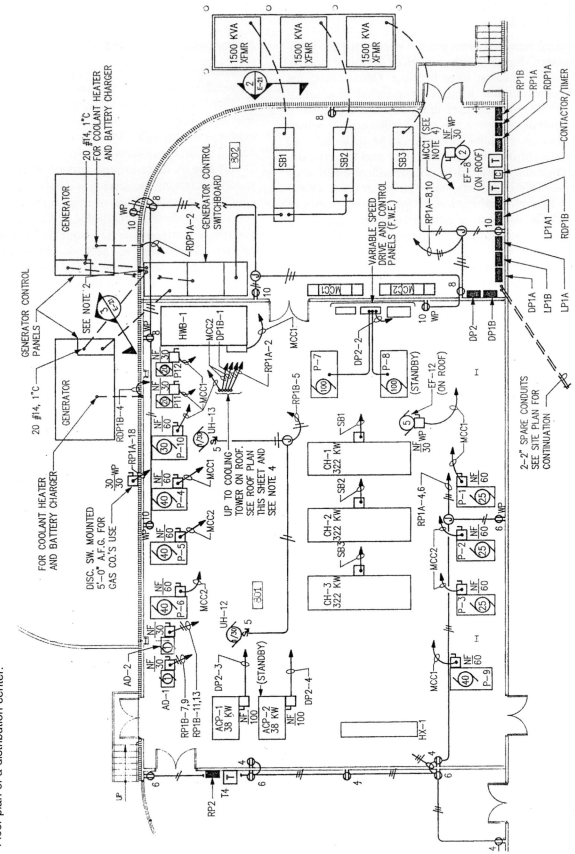

FIGURE 17–29
Floor plan of a distribution center.

402

such as machines and computers, size of motors, various voltage-level requirements, peak demand, and any other special requirements. The other area to consider is where the service will be located, what the space requirements are for meters, and what future changes are anticipated in either the service or the load requirements. Typically, in the past most office buildings were supplied with the well-known 208/120-V system. Today, however, it is more common to find a 480/277-V, three-phase, four-wire grounded wye being used to supply manufacturing customers as well as office parks. The load requirements will determine the service required. In the preparation of these details it is critical to note which areas of the drawing will be completed by the service utility provider.

Load Centers

The **load** or distribution **center** of a building can be located either inside or outside the building. The determining factor is the voltage level present. For example, if the primary side of the incoming service is 15 kV or less, the center could be enclosed in a structure. This distribution center is often called a substation because of the function of the distribution center. The voltage is stepped down and the power is distributed to individual points within the building. The floor plan of the distribution center (Fig. 17–29) should be drawn to proper scale. This is one way to ensure that all equipment will fit in the desired structure with the appropriate clearances.

A schematic diagram should be completed for the distribution plan. This will demonstrate the flow from the transformer through the building. Often a backup system is included as part of the original design. Using two transformers, the incoming service is split. If there is a loss of power in either cable, the system is designed so that one transformer and feed will supply the total load requirement. Grounding is essential in this substation, and all the switchgear, enclosures, and gates are grounded. Specification data for the various components are often listed on separate specifications pages, as data inclusion on the schematic would cause clutter and confusion.

Another drawing used in power distribution is the *riser diagram.* This diagram, as seen in Fig. 17–18, represents the electrical service lines throughout the building. Conceptually, it is similar to an interconnection diagram showing physical relationship, path of wires, and location of lighting panels. The riser diagram is generally limited to representing the wiring from the supply power to a panel box and does not include all the wiring for lighting, outlets, and the like. This information is found in architectural drawings. The emphasis for riser diagrams is on the distribution path. It has been suggested that the diagram is called a riser diagram because it shows the electric service rising through a building.

Detail Drawings

Certain installations, connections, or raceways are not covered in any other drawing. These detail drawings are added to complete the set of drawings for power distribution. The drawings may be part of a set of standards. For example, all raceways or wire channels are the same. This detail is drawn once and included in the set of standards to be used when necessary. A special drawing is required for a one-of-a-kind installation or connection. All utility companies have multiple volumes of standards. Often they are reduced and compiled in a loose-leaf binder. This allows easy updating should any standards be revised. Figure 17–30 shows a typical "standard" detail included in a set of standards.

SUMMARY

Each utility company substation or large commercial installation requires the preparation of a great number of drawings. Specifications for these drawings must also be done to cover installations, rating, devices, loads, and so on. Some of the specifications may include requests for bids on contracted work, installations, equipment and materials, legal requirements, and standards to be maintained. Drawings for the power industry require all portions of electrical drafting covered in previous chapters: symbols, reference designations, control circuits, schematics, logic diagrams, and wiring diagrams, as well as mechanical drawings for layout plans or enclosures.

Many changes will be seen in drawings for the power industry. More logic diagrams representing circuit functions will be used rather than single-line or schematic diagrams for control circuits. Operational representations will become more common and individual details will be covered in small detail drawings. Complete wire routing for equipment is being replaced by less cluttered terminal and wire numbers. Very little work is done manually in the power field. This is certainly advantageous for revisions and neatness. The lineweights are controlled and the notes and references are all lettered consistently, clearly, and neatly. The use of computer-generated drawings produces good-quality drawings. However, unless good drafting practices are known and used, the computer can also be used to generate very poorly laid out and cluttered drawings that are printed in a very high quality fashion. The computer is not a

FIGURE 17–30
Detail drawing from company standards. (Courtesy Georgia Power Company)

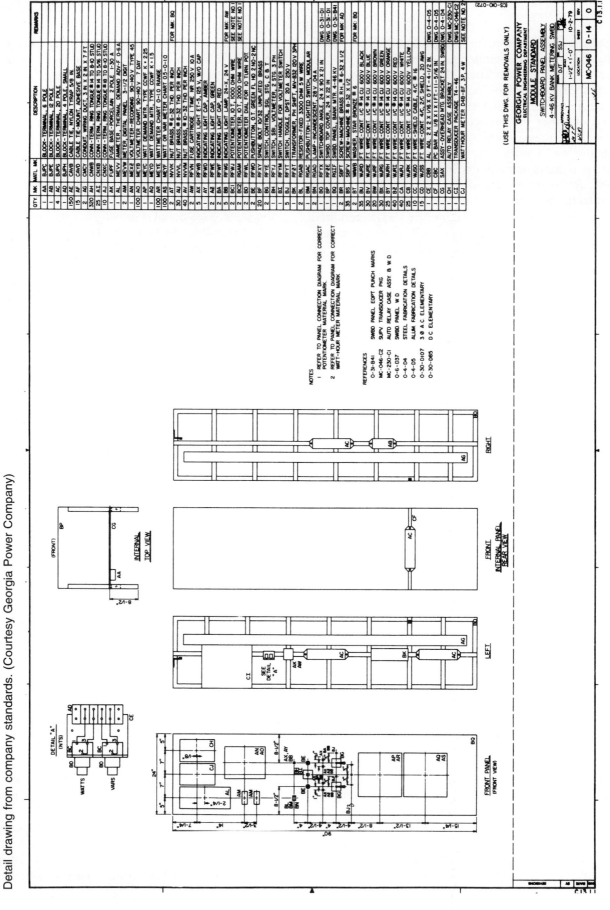

replacement for knowledge of good drafting practices; it is simply a replacement for the use of a straightedge, angle, and drafting pen.

One trend that may affect the need for a great number of drawings, however, is the use of computers to control the power-generating facilities. As a facility becomes more computerized, the number of complicated drawings required could be reduced.

REVIEW QUESTIONS

17–1 What three things are required for a power distribution system?

17–2 What is the primary source of energy used in the production of electricity?

17–3 What is the main function of a transformer?

17–4 Why are the drawings for the electrical power field considered to be complex?

17–5 What are the two major categories of power lines?

17–6 What is the main purpose of a substation?

17–7 What two factors must be considered when choosing support poles?

17–8 What is the main advantage of a pin-type insulator?

17–9 Define the following terms: switchgear, pothead, circuit breaker, and lightning arrestor.

17–10 What is the advantage of bound sets of standards in a company?

17–11 What is the primary function of a substation?

17–12 What is a riser diagram?

17–13 What is the primary emphasis of a three-line drawing?

17–14 Why is a different set of logic symbols often used in power applications?

PROBLEMS

17–1 On a size A sheet, make an assembly drawing of the panelboard and transformer shown.

PROBLEM 17–1
Panelboard and transformer surface mounting assembly. (Courtesy of Brown and Caldwell Consultants)

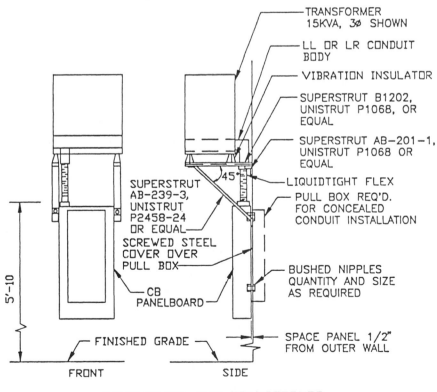

PANELBOARD AND TRANSFORMER

SURFACE MOUNTING ASSEMBLY

17–2 On a size B sheet, redraw the elevation and plan for the floodlight detail shown.

PROBLEM 17–2
Elevation and plan of floodlight.
(Courtesy of Brown and Caldwell Consultants)

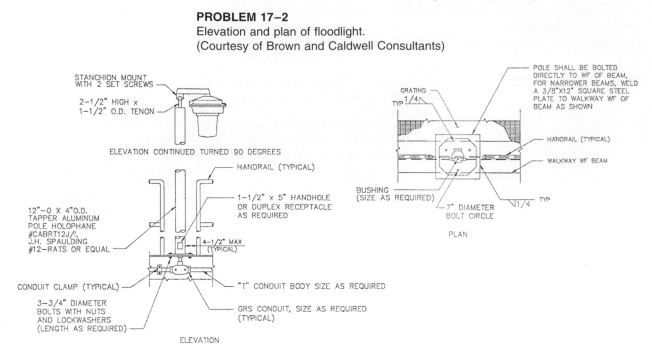

17–3 An elevation for a motor control center is shown. Draw this elevation on a size C sheet.

PROBLEM 17–3
Elevation of motor control center.
(Courtesy of Brown and Caldwell Consultants)

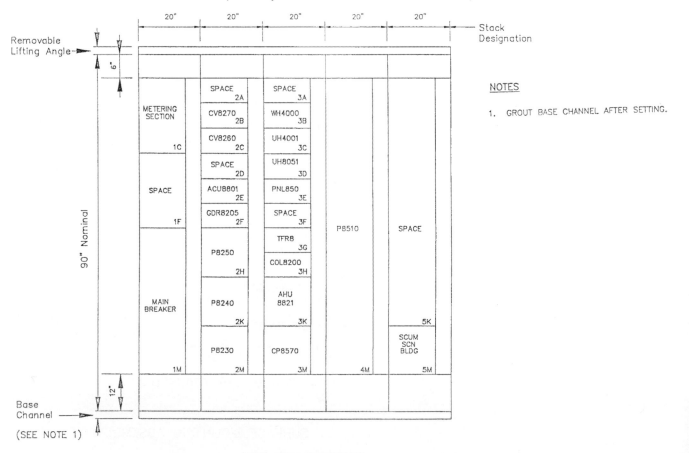

MCC 801 ELEVATION

17–4 Practice drawing the symbols for transformers, circuit breakers, fuses, disconnects, and other symbols encountered in this chapter.

17–5 On a size B sheet, draw the internal assembly and wiring diagram shown in Fig. 17–11.

17–6 On a size B sheet, draw the single-line distribution drawing shown in Fig. 17–13.

17–7 On a size C sheet, draw the partial circuit shown in Fig. 17–19 representing the transformers and wiring connections.

17–8 Practice drawing the logic symbols used in power applications, as shown in Fig. 17–24.

17–9 On a size C sheet, draw the section drawing found in Fig. 17–25.

17–10 On a size B sheet, draw the wiring diagram shown in Fig. 17–26.

17–11 On a size B sheet, draw the wiring diagram and notes shown in Fig. 17–27.

17–12 On a size B sheet, draw the detail drawing shown, which illustrates the location of a motor feed line.

PROBLEM 17–12

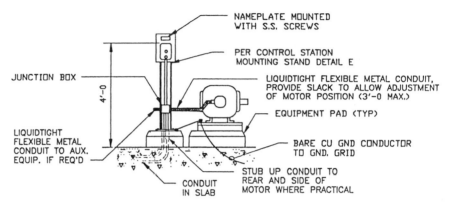

MOTOR FEED FROM BELOW

17–13 (Advanced Problem) On a size C sheet, draw the interconnection diagram shown in Fig. 17–28.

17–14 (Advanced Problem) On a size C sheet, draw the distribution floor plan shown in Fig. 17–29.

17–15 (Advanced Problem) On a size E sheet, draw the single-line diagram, the MCC elevation, and the MCC schedule shown.

PROBLEM 17–15

Single-line diagram, motor-control center elevation, and motor-control center schedule.

(Courtesy of Brown and Caldwell Consultants)

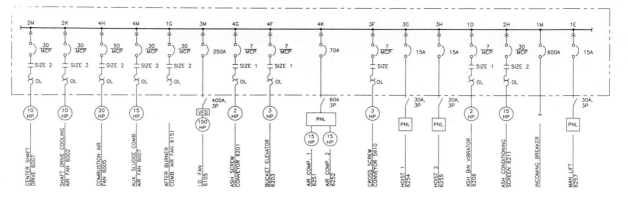

MCC "M" SINGLE LINE DIAGRAM (PARTIAL)

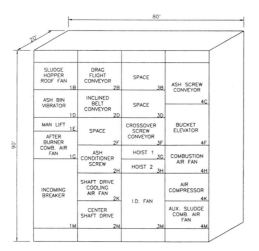

MCC "M" ELEVATION

NOTES

1. THE CONTRACTOR SHALL PROVIDE AND INSTALL A NEW MCC "M" IN PLACE OF EXISTING MCC. THE CONTRACTOR SHALL REUSE THE EXISTING MOTOR STARTER AT LOCATION 1B FOR SLUDGE HOPPER ROOF FAN. INSTALL AND REWIRE THE MOTOR STARTER AT LOCATION 1B IN THE NEW MCC "M". THE EXISTING MCC "M" SHALL BE RETURNED TO THE OWNER.

MOTOR CONTROL CENTER M: 480 VOLT, 3PH 60HZ
HORIZONTAL BUS: 600 AMPS; VERTICAL BUS: 300 AMPS
MCC FAULT CAPACITY: 25000 AMPS (SYM)

LOC.	EQUIP. NO.	EQUIPMENT DESCRIPTION	HP OR KVA	INTERRUPTER TYPE	P	AMPS	STARTER TYPE	SIZE
1B	–	SLUDGE HOPPER ROOF FAN	–	–	–	–	–	–
1D	6206	ASH BIN VIBRATOR	2	MCP	3	7	FVNR	1
1E	6257	MAN LIFT	2	CB	3	15	–	
1G	6151	AFTER BURNER COMB. AIR FAN	25	MCP	3	50	FVNR	2
1M		INCOMING BREAKER	–	CB	3	600		
2B	5910	DRAG FILGHT CONVEYOR	15	CB	3	30	–	
2D	5903	INCLINED BELT CONVEYOR	10	MCP	3	30	FVNR	2
2F		SPACE						
2H	6211	ASH CONDITIONING SCREW	15	MCP	3	30	FVNR	2
2K	6002	SHAFT DRIVE COOLING AIR FAN	10	MCP	3	30	FVNR	2
2M	6007	CENTER SHAFT DRIVE	10	MCP	3	30	FVNR	2
3B		SPACE						
3D		SPACE						

FOR CONTINUATION SEE NEXT COLUMN

LOC.	EQUIP. NO.	EQUIPMENT DESCRIPTION	HP OR KVA	INTERRUPTER TYPE	P	AMPS	STARTER TYPE	SIZE
3F	5910	CROSSOVER SCREW CONVEYOR	3	MCP	3	7	FVNR	1
3G	6254	HOIST 1	2	CB	3	15	–	
3H	6255	HOIST 2	2	CB	3	15	–	
3M	6105	I.D. FAN	150	CB	3	250	–	
4C	6201	ASH SCREW CONVEYOR	2	MCP	3	7	FVNR	1
4F	6203	BUCKET ELEVATOR	3	MCP	3	7	FVNR	1
4H	6005	COMB. AIR FAN	20	MCP	3	50	FVNR	2
4K	6251 6252	AIR COMP. 1 & 2	30	CB	3	70	–	
4M	6007	AUX SLUDGE COMB. AIR FAN	15	MCP	3	30	FVNR	2

MCC "M" SCHEDULE

17–16 (Advanced Problem) On a size E sheet, draw the three-phase AC elementary diagram shown.

PROBLEM 17–16

Three-phase AC elementary diagram.
(Courtesy of Georgia Power Co.)

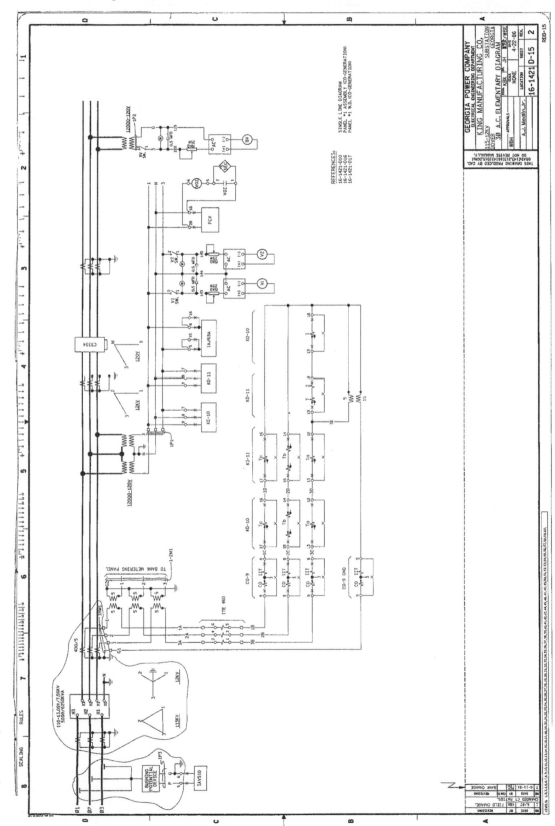

17–17 (Advanced Problem) On a size E sheet, draw the single-line diagram shown.

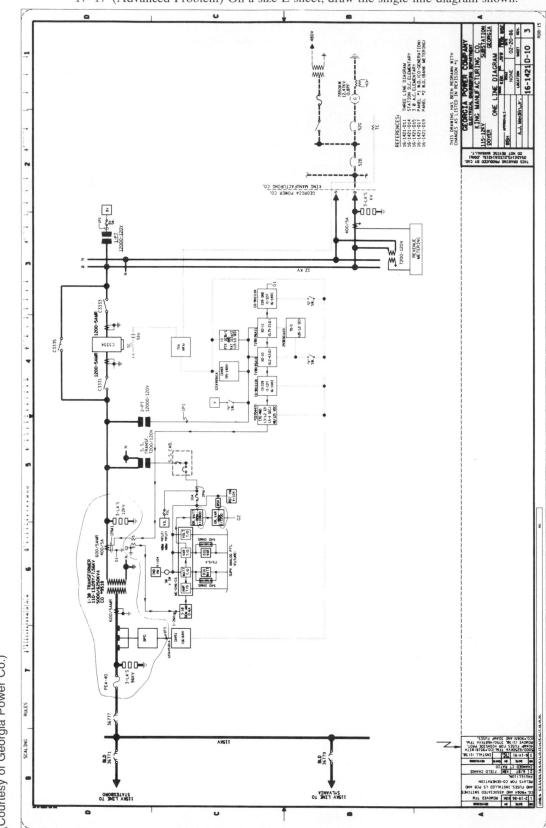

PROBLEM 17–17
Single-line diagram.
(Courtesy of Georgia Power Co.)

17–18 (Advanced Problem) On a size E sheet, draw the three-line diagram shown.

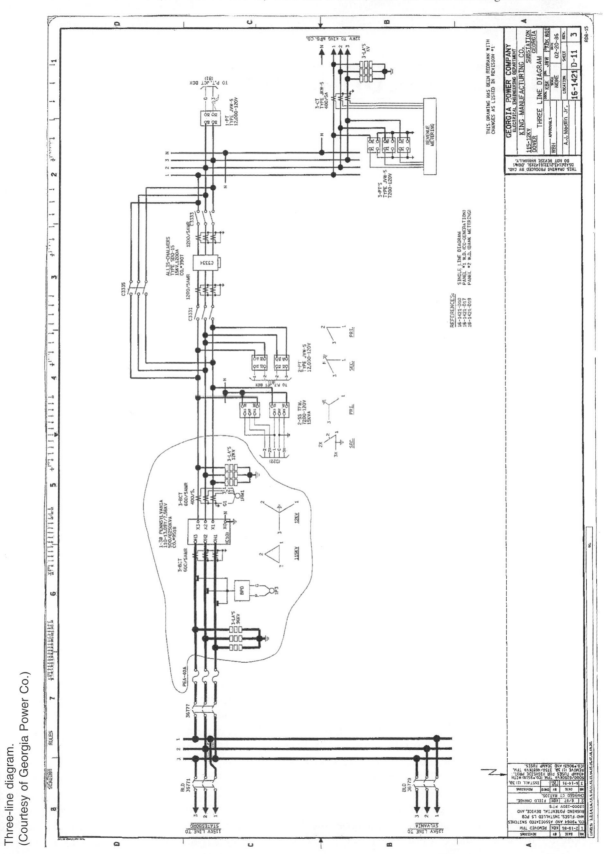

PROBLEM 17–18
Three-line diagram.
(Courtesy of Georgia Power Co.)

18

PRINTED CIRCUIT BOARDS

INTRODUCTION

A printed circuit (PC) board is a baseboard made of a laminated insulating material that contains ICs and other components, along with the connections required to implement one or more electronic functions. The connections of a PC board are a thin layer of conductor material. The conductor pattern is established on the board by one of two main processes. The additive process involves depositing the conductor material on the base in its required conductor pattern. The subtractive process uses an etching solution to eat away the nonconductor path areas on the board after the conductor path pattern is coated or printed with an etch-resisting paint or ink. The etching method is the most common process used to make PC conductor patterns.

Printed circuit boards provide a rigid and modularized circuit package, as shown in Fig. 18–1. Each board or module is connected to the system with terminal connections, cables, or other simplified connectors that allow the circuit board to be plugged into the assembly. Note that where the board does not require frequent removal it is rigidly mounted and connected electrically with cables or other forms of wiring.

Printed circuit boards get their name from the fact that the circuitry is connected not by wires but by cop-per-foil lines, paths, or traces that are actually *photo-printed* or etched onto the board.

The *printed* conductor path pattern eliminates the need for wires to interconnect the components. The use of printed or etched conductor paths has led to extremely accurate miniature circuity in the electronics industry.

Printed circuit boards are also referred to as printed wiring boards. There is some confusion as to the use of these two terms. Some manufacturers or users define a printed wiring board as a board whose conductor pattern provides connections for the components mounted on the board. A printed circuit board is said to provide the same function as well as a circuitry function such as capacitance. Others suggest that *printed wiring board* is an obsolete term used to define boards used to mount components and their accompanying wiring before the introduction of *printed wiring* (printed or etched conductor circuits).

Be aware that both *printed wiring* and *printed circuits* are terms used throughout industry interchangeably. Both terms refer to a laminated board where components are mounted and where conductor patterns are fixed for connections between components. Note that the conductor patterns replace movable wires; however, some wires may be required where jumpers are needed.

FIGURE 18–1
Printed circuit board layout and design. (Courtesy Gerber Scientific Instrument Co.)

FIGURE 18–2
Circuit boards in various shapes and sizes.

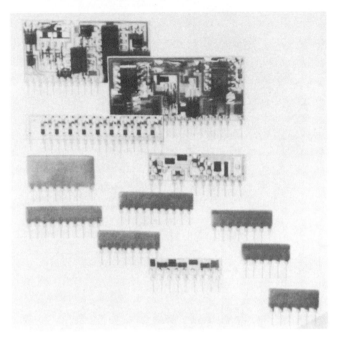

(This will be explained later in the chapter.) Printed circuit board, or simply **PCB**, is used in this text since it appears to be the most accepted term in industry.

A PCB may have all the components mounted on one side and the circuitry (conductor pattern) on the other (single-sided board), components on one side with circuitry on both sides (double-sided boards), or circuitry layered on a number of bonded boards (multilayer boards). Components are sometimes mounted on both sides of the board, although this practice is discouraged.

A PCB can be designed with manual or automated processes. Figure 18–1 showed three boards in front of the CAD terminal on which they were designed. Manual methods are frequently used for prototype boards or where the boards' tolerance and accuracy specifications are less stringent. Single-sided and double-sided boards are more apt to be manually prepared than multilayered boards with their high density and critical alignment of layers.

The board or laminated material for a printed circuit is typically square or rectangular, as shown in Fig. 18–2.

The size of the board ranges from 1×1 in. to 15×20 in., with the most common about 5×7 in. or 10×12 in. Even though the most common shape is rectangular (Fig. 18–1), other shapes are used, such as the circular board shown in Fig. 18–3. Current trends are toward larger boards and greater circuit density (more ICs or other components in smaller areas). While single-sided or two-sided boards with components mounted on only one side seem to be the most common, multilayered boards are being used more frequently as circuits become more complex in computers or communication systems. The circuit board shown in Fig. 18–4 is used in the computerized control portion of the robotics system shown in Fig. 18–5.

PROCESSES AND MATERIALS

Two basic processes are used in manufacturing printed circuit boards: the subtractive (etched) method and the additive method. At the present time the additive process is the most common. A printed circuit board is shown in Fig. 18–6.

Subtractive Process

In the subtractive process, a laminated board is clad with a conductive foil, normally copper. The material of the board itself depends on its use and specification.

FIGURE 18–3
Alternative shapes for PC boards. (Courtesy Western Electric Co.)

FIGURE 18–4
Circuit board plus components for robot computer numerical control. (Courtesy Cincinnati Milacron, Inc.)

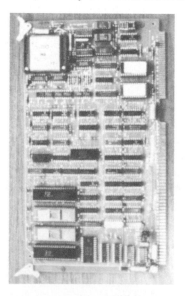

FIGURE 18–5
Hydraulic robot and computer. (Courtesy Cincinnati Milacron, Inc.)

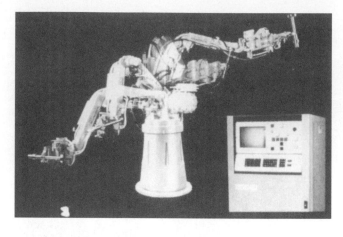

Paper and epoxy or glass and epoxy are common materials for PC boards. Before the conductive pattern is etched on the board, tooling holes are drilled to position the artwork layout. The holes are used for correct alignment or *registration* throughout the production sequence.

The artwork pattern is positioned on the board in alignment with the tooling holes. The artwork pattern is the full-scale drawing of the conductor pattern. The artwork is prepared at an enlarged scale, normally 2 : 1 or 4 : 1, and then is photoreduced. When a CAD system is used to generate artwork, the drawing is prepared full scale (Fig. 18–7). No enlargement or reduction is necessary (Fig. 18–8).

The conductor patterns (traces) are applied to the clad board with a silkscreen or appliqué process that deposits an acid-resisting coating (called a *resist*) to the board. A photographic emulsion process can be used instead. The light-sensitive emulsion is applied to the board. The pattern develops on the board after it is exposed to an intense light source with the artwork negative in contact with the board.

Regardless of how the pattern is applied to the board, the etching process is the same. An acid etching solution like ferric chloride is applied to the board. The etching process removes (eats away) all copper conductor areas that have not been coated with the resist. Only the conductor pattern is left. After the etching is complete, the resist is removed from the copper conductor pattern with a solution bath.

There are many variations in the production of PC boards. Methods are refined, simplified, automated, and invented daily. The state of the art is constantly changing and adapting to new needs and specifications. The etch-

FIGURE 18–6
Printed circuit board. (Courtesy Prime-Computervision)

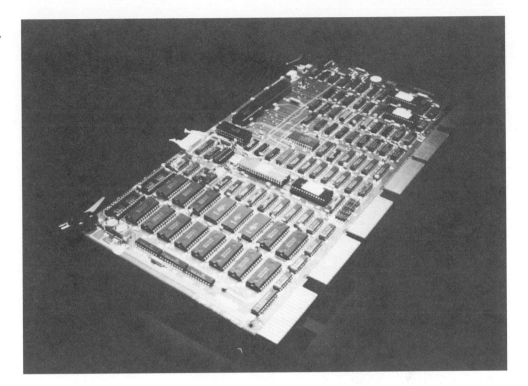

FIGURE 18–7
PCB art generated with CAD system.
(Courtesy Prime-Computervision)

FIGURE 18-8
PCB design on CAD terminal. (Courtesy Hewlett-Packard Co.)

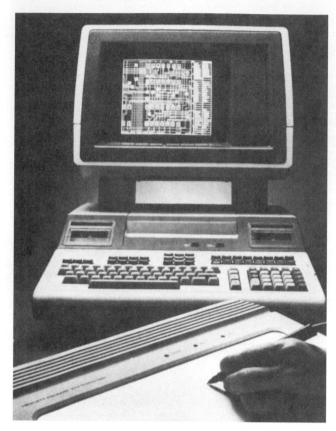

FIGURE 18-9
PCB production.

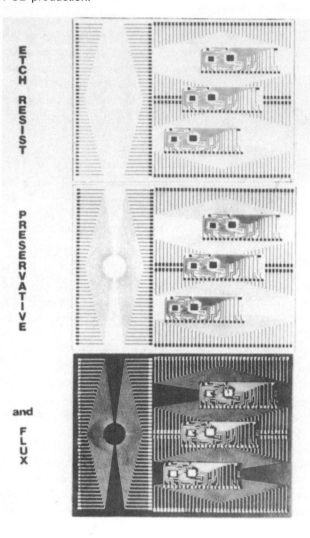

ing process shown in Fig. 18-9 is just one of the many variations found in industry.

After the board is cleaned, plating the conductive pattern with a metal like silver or gold may be required. Better contact can be made with gold as the conductor. However, it would not be very economical to use a gold clad board and etch away all but the desired conductor traces and patterns. As a result the traces are produced with copper, and then a layer of gold is deposited on the traces through a plating process. This gives the advantage of gold conductors and still remains fairly economical.

Holes required for conductor leads are drilled manually or with the aid of automated NC equipment.

The last step in the PC process involves assembly of the board components, including soldering the component leads to the conductive pattern. Leads can be hand or wave soldered (explained more fully later in this chapter), depending on the level of automation.

The process just described was for a single-sided PC board. Single-sided boards have the conductive pattern on one side, the **solder side,** and the components mounted on the opposite side, the **component side,** as shown in Fig. 18-10. Double-sided boards have a conductive pattern on both sides of the board.

Multilayer boards consist of individual sandwiched circuit boards. A multilayer board may have 3 to 20 or more layers. Each layer has a conductive copper-foil clad to its related laminated board. Each layer is etched separately. The individual layers are bonded by heat, pressure, and glass epoxy resins called *prepreg*. A multilayer board allows for more components to be mounted in smaller spaces and a much greater wiring conductor density without making conductor paths unreasonably narrow. *Plated-through holes* are used for interconnections between layers.

Additive Process

The additive process starts with an adhesive-coated laminated board (without conductive plating). Tooling

FIGURE 18–10
Component side and solder side of a PCB.

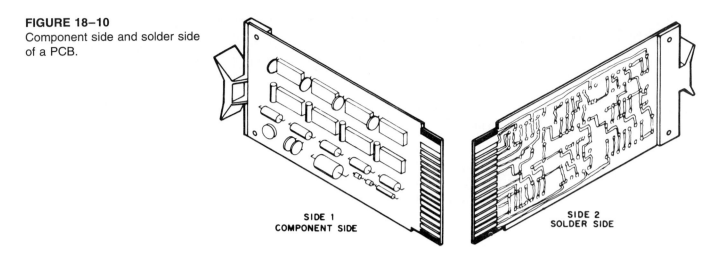

SIDE 1
COMPONENT SIDE

SIDE 2
SOLDER SIDE

holes are drilled or punched first. A reverse-pattern mask is applied to the board and the conductive pattern deposited through electrolysis. The reverse pattern mask is stripped from the board. This, however, is an optional process since the mask can remain on the board and not interfere with the conductor pattern. A solder mask may be applied to the pattern and a protective coating added as the last step.

The additive process is used for production of PCBs. This process is also used on very large scale integrated circuits. The additive process is necessary for complex miniature circuitry, where the subtractive process is inadequate. (See Chapter 10 for a more detailed explanation of the additive process as it is used in the development of microcircuits.)

Materials

A variety of base and conductor materials is used in the manufacturing of printed circuit boards. In addition, plating, finishing, coating, and other materials are added to the board both during production and after final assembly. The final board is inspected for dimensional correctness, proper conductor areas, plating, and so on. In Fig. 18–11, the gold plating on contact points on a printed circuit board is inspected.

Two standard laminate grades are used as the base material for a printed circuit board: FR-4 and G-10. FR-4 is a glass epoxy material that is chemically resistant and flame retardant and that has low moisture absorbtion. G-10 is a glass epoxy similar to FR-4, except that it is somewhat less flame retardant.

Regardless of the board laminate material, with its copper conductor layer added, the etching process is usually the same. The choice of materials is the responsibility of the design engineer.

Table 18–1 lists the primary base materials used for PC boards.

PRINTED CIRCUIT DOCUMENTATION

Whether a PC board is to be a one-of-a-kind prototype or a high-volume production article, it will have some informative graphic documentation describing how to produce the complete PC board. A printed circuit drawing package may include the drawings shown in Figs. 18–12, 18–13, and 18–14. Exactly how much documentation is needed and how it should be prepared varies with company practice, application, and method of generating the documentation (manual or automated).

FIGURE 18–11
The gold plating on contact points on printed circuit boards, inspected under 3X magnification. (Courtesy Western Electric Co.)

TABLE 18–1
Materials for PC Boards

NEMA Grade Properties	Material	Advantages
PC	Paper base/phenolic resin	Moisture resistant
P	Paper/phenolic	Good mechanical characteristics
FR-2	Paper/phenolic	Flame retardant
FR-3	Paper base/epoxy resin	Flame retardant, good mechanical/ electrical characteristics
FR-4	Glass fabric/epoxy resin	Chemical resistant, flame retardant
FR-5	Glass/epoxy	Flame and temperature resistant
G-3	Glass/phenolic	Good dimensional and strength characteristics
G-5	Glass/melamine	Good impact strength
G-9	Glass/melamine	Similar to G-5
G-10	Glass/epoxy	Similar to FR-4
G-11	Glass/epoxy	Similar to G-10
G-30	Glass/polymide	Flame retardant
GPO-1	Glass/polyester	General purpose
GPO-2	Glass/polyester	Similar to GPO-1

Adequate documentation conveys to the user the basic electromechanical design concept, the type and quantity of parts and materials required, special manufacturing instructions, and up-to-date revisions.

Too little documentation results in misinterpretation, information gaps, and loss of uniform configuration. Manufacturing becomes dependent on individuals rather than on documentation, often resulting in expensive rework and lost time.

Too much documentation can result in increased drafting costs and decreased manufacturing productivity because of time-consuming interpretation of overly complicated and confusing drawings.

Printed circuit documentation may be divided into three classifications:

1. *Minimum:* Use for prototype and small-quantity runs.
2. *Formal:* Use for a standard product line and/or production/quantities (similar to Category E, Form 2, per MIL-D-1000 without source or specification support documentation).
3. *Military:* Complies with government contracts specifying procurement drawings for the manufacture of identical items by other than the original manufacturer (Category E, Form 1, per MIL-D-1000).

Regardless of the number of design steps, method of design, or type of documentation, the end result is the assembled functional PC board, like the module in Fig. 18–15.

The schematic drawing (Fig. 18–12) is the first step in this process. After the schematic is drawn, it is used as the basis for all subsequent steps in the design process. Using the schematic, the designer begins to *lay out* the PC board. The **layout drawing** is used in arranging the components and their interconnecting patterns for the physical configuration of the board.

The layout drawing is used to create the PC board *artwork drawing,* or drawings if the board is to have more than one layer. The *marking artwork* is then created as a scaled drawing of all marking configurations to be printed on the board. A silkscreen or stencil process is used for the marking artwork to show the component shapes, reference designations, and any other information needed to assemble the board.

The PC board **master drawing** (Fig. 18–13) is normally the next drawing to be completed. The PCB master details the board outline (geometry), the connector pattern, and tooling holes, along with dimensions for the entire configuration. Notes on the master drawing define the board material, plating instructions, finishing requirements, and any other information needed to produce the board.

The **assembly drawing** (Fig. 18–14) shows the board with all components mounted. Assembly instructions and dimensions are shown on this document. A complete parts list can also be shown on this drawing.

Other drawings may be required for the fabrication of the board, such as *drilling drawings* for machining the board. Drilling drawings can be generated from the previous documents. Figure 18–16 shows a PCB without components mounted. A complete parts list can be either a separate document or part of the assembly drawing.

The drawing sequence and type of documentation may differ among companies. The PC board projects and problems in this chapter will provide you with experience in modern printed circuit board drafting and de-

FIGURE 18–12
PCB design and fabrication. (Provided courtesy of Bishop Graphics, Inc.)

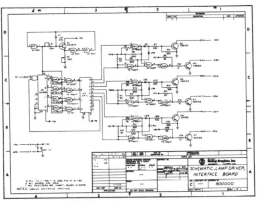

SCHEMATIC DIAGRAM: Consists of graphic symbols indicating the interconnections and functions of an electronic circuit. It is the basis for a printed circuit design. It is also used to test, evaluate and troubleshoot the completed circuit board.

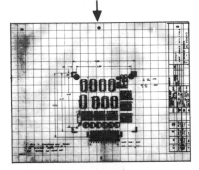

LAYOUT: The conceptual intermediate link between the schematic diagram and the master pattern.

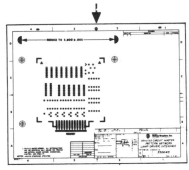

ARTWORK: Accurately scaled configuration of the printed circuit pattern from which the master pattern is photographically produced.

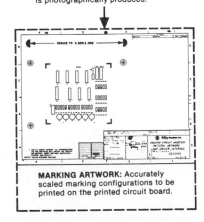

MARKING ARTWORK: Accurately scaled marking configurations to be printed on the printed circuit board.

sign procedures. Each of the steps and drawings is explained in detail in this chapter.

PRINTED CIRCUIT BOARD DESIGN

This section is a compilation of the basic data you need to prepare a printed circuit board design. Because the state of the art is constantly producing more sophisticated components, tooling, materials, and processes, we recommend that you consult the latest specifications and standards available from the standards organizations.

Printed Circuit Board Specifications

The following list introduces general specifications for PC board design and layout.

1. Keep board design, component layout, and conductor traces as simple as possible. Avoid odd-shaped boards, multiaxis component arrangements, and complicated conductor paths.
2. Attempt to restrict maximum board area to 50 in.^{2}, and keep the board outline dimensions as nearly rectangular as possible.
3. Lay out the board on a toleranced grid medium.
4. Draw artwork at an enlarged scale of at least 2:1; use 4:1 for greater tolerancing. CAD-generated art can be 1:1.
5. Provide a minimum of two (preferably three) registration marks for artwork layering.
6. Lay out the board as viewed from the component side.
7. Limit components to one side of the board.
8. Locate components for easy assembly and servicing. All components should be removable without disturbing other aspects of the board.
9. Keep the number of layers to a minimum.
10. Keep the board outline (geometry) as simple as possible.
11. Attempt to limit the number of axes for component placement to one or two. Keep components on an X–Y axis whenever possible. Avoid multiaxis arrangements, especially for boards that are to be assembled mechanically.
12. Place IC DIP packages on only one axis.
13. Keep conductor traces as short as possible.
14. Keep conductor traces on grid axes whenever possible. Use 45° for angles.
15. Run long conductor traces along the same axis that the board will travel during wave soldering, usually parallel to the long side of the board.
16. Keep conductors from the board edge; use .025 in. (.635 mm) as a general limitation.
17. Break up ground plane areas (when the ground plane is on the solder side of the board) with grid

FIGURE 18–13
Printed circuit board assembly master drawing.

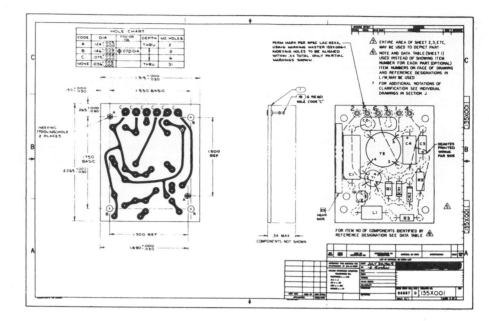

FIGURE 18–14
PCB assembly drawing. (Provided courtesy of Bishop Graphics, Inc.)

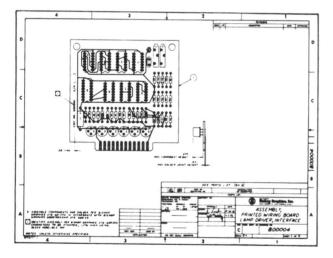

FIGURE 18–16
PCB without components mounted. (Courtesy of Motorola, Inc., Semiconductor Products Sector)

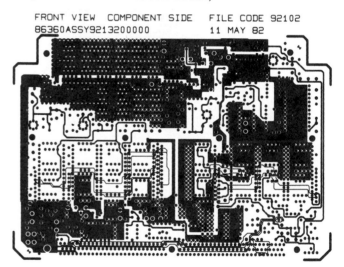

FIGURE 18–15
PCB module. (Courtesy Texas Instruments, Inc.)

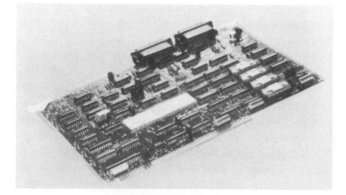

patterns to prevent board warp or blistering during wave soldering.

18. Lay out the board so that the major axis on the solder side is parallel to the major component-lead clinch direction and perpendicular on the component side. This will minimize bridging if the board is to be wave soldered.

19. Locate terminal pads, lands, and areas at grid intersections.

20. Keep terminal pads a minimum of .05 in. (1.27 mm) from the board edge.

21. Where component lead spacing does not conform to grid spacing, place at least one lead on a grid

intersection or place the pattern center on a grid intersection.

22. Keep the total number of terminal area diameters, hole sizes, and pad configurations to a minimum.
23. Keep the number of different conductor trace widths to a minimum.
24. If possible, conductor corners or angles should not be smaller than 90°.
25. Identify all layers of a multilayer board.
26. Indicate polarity and component orientation on the layout.
27. Lay out adjustable components and test points where they are easily accessible.
28. Place heat sink components dissipating more than 2 W directly to the chassis.
29. Provide heat dissipators where necessary.
30. Place heavy components around the board edge for extra support.
31. Position components so that they do not interfere with module packaging and mounting requirements.
32. Keep components away from the board edge according to design requirements.
33. Mount components so that they do not form moisture traps.

Grid System

The use of a grid system is essential in laying out and preparing the master pattern artwork. It facilitates the placement of components, terminal areas, and conductor traces on the PC layout, as shown in Fig. 18–17. A grid system is essential if automated circuit board artwork preparation, manufacturing, or assembly techniques are being considered, because the machines that perform these functions are coordinated to basic grid location systems. An increasing number of multiple lead components, such as dual-in-line packages, conform to standard grid increments.

A grid is a two-dimensional rectangular network consisting of a set of equidistant parallel lines superimposed on another set of equidistant parallel lines, with one set of lines perpendicular to the other. The line intersections provide the basis for an incremental location system. The standard increments are .100 in. (2.54 mm), .050 in. (1.27 mm), and .025 in. (.635 mm) in order of preference. If more finite increments are needed, MIL-STD-275D specifies any multiple of .005 in. (0.127 mm), while the International Electrotechnical Commission (IEC) specifies 0.5 mm (.0197 in.) and 0.1 mm (.0039 in.).

Grid layout sheets are available on many different types of materials. Their accuracy falls into three main classifications: *precision, semiprecision,* and *nonprecision.*

Precision **grid patterns** are used for manually printed circuit layout and artwork preparation to ensure maximum accuracy on the final circuit board. Precision grids are usually photographically prepared on dimensionally stable materials, such as glass or polyester film with accuracy tolerances of ±.002 in. (.051 mm) over 36 in. (91.4 cm) at specific environmental conditions. Grids produced on glass are the most stable and, therefore, the most preferable. However, they are more expensive and fragile to handle than film.

Semiprecision grids may be used during layout of circuit boards when the master pattern is to be generated by computer-controlled equipment, which can compensate for minor inaccuracies in the grid pattern. They can also be used for layout and artwork preparation of single-sided circuit boards where close tolerances are unnecessary. They are generally manufactured by printing or photo techniques on polyester or acetate film with accuracy tolerances of ±.010 in. (0.254 mm) to ±.020 in. (.508 mm) over 36 in. (91.4 cm).

Nonprecision grids are not recommended for printed circuit master artwork layouts. However, nonprecision grids may be used economically for sketches or drawings with less critical applications. They are usually in roll form. Their lack of precision may cause a severe dimensional disparity between the X and Y axis coordinates. Tolerances are usually not stated, but variations of ±.062 in. (1.57 mm) over 36 in. (91.4 cm) are typical with nonprecision grids.

Scale

All printed circuit board artwork and layout should be prepared at an enlarged scale. Drafting errors, such as terminal pad misalignment with grid centers, conductor spacing variances, and imperfections in drafting aids, are reduced proportionately with the reduction of the artwork to actual finished board size.

The scales most universally used are evenly multiplied factors of 10×, 4×, 2×, and 1×.

FIGURE 18–17
Grid pattern for preparing and laying out master pattern artwork. (Provided courtesy of Bishop Graphics, Inc.)

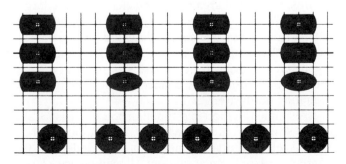

Selection of scale should be based on the relationship of artwork drafting error to finished board tolerance requirements within practical limitations. Scale can be estimated with the following formula:

$$\text{Artwork scale} = \frac{DE}{FL}$$

where DE = drafting error, usually .015 in. (.381 mm) diametric true position for manually prepared artwork and .0025 in. (.064 mm) diametric true position for artwork prepared by automated techniques

FL = tolerance required for location of circuitry with respect to true position on the finished board

For example, .015 in. (.38 mm)/.005 in. (.127 mm) = 3× scale.

The reason for using standard scales (10×, 4×, 2×, 1×) is the ready availability of artwork and layout drafting aids and templates in these scales. Other limiting factors in scale selection are ease of artwork handling, availability of standard film sizes, and capacity of camera copy boards [usually 30 × 40 in. (76.2 × 101.6 cm) or 30 × 44 in. (76.2 × 111.8 cm)].

Avoid using 1× scale for manually laid out boards whenever possible since there can be no reduction in drafting error. 1×-scale artwork should be used only in the following instances:

□ When there is absolutely no reduction facility available
□ When there are no limiting tolerances on the circuit board (i.e., large conductor spacings and oversized pads)
□ When speed and cost are important for prototype work
□ When the master pattern is prepared with automated phototechniques, ensuring a minimal drafting error

A 4× scale greatly reduces drafting errors and should be used only when extremely accurate dimensions are required on the printed circuit board. However, due to size limitations of the light table, the artwork itself, camera copyboards, and material costs, 4× scale often is not the practical choice.

With the advent of precision prespaced artwork symbols, a 2× scale can produce the exacting dimensional requirements, yet maintain lower artwork material costs and ease of handling. A 2× scale is commonly preferred for all-around use.

Board Size and Number of Layers

Determining the size and number of layers of a circuit board is an essential preliminary to the actual artwork layout.

Printed circuit boards can be designed and manufactured in three basic configurations: single layer, multilayer, and multilayer sandwich. Single-layer boards contain all printed wiring on one side, with the components on the opposite side. The plastic laminate serves as an insulator. Multilayer boards have printed wiring on both sides, with the bulk of the circuitry on one side and the components on the other. Multilayer sandwich boards are actually many thin boards laminated together, with the components on one, or sometimes both, of the external layers. There may be as many as 20 conductive layers in a multilayer sandwich board.

Cost is a basic consideration in determining board size and number of layers. For single-board applications, small boards are generally less expensive than larger boards. Single-layer circuit boards cost less than multilayer boards, which are less costly than multilayer sandwich boards. However, a few large multilayer or multilayer sandwich boards can be less expensive than many single-layer PC boards in a large system. For instance, testing is usually less expensive on a few large boards than on many small ones. The larger boards require less material and either reduce or totally eliminate costs associated with interconnecting many small boards.

Mechanical and design limitations also require consideration. Large boards have a greater tendency toward warping and bowing. Single-layer and multilayer sandwich boards often require more design time than multilayer boards. The maximum capacity of design equipment should be considered for the use of very large boards. The mechanical design may cause size restrictions by requiring use of a certain standard connector or package enclosure.

Once the basic design limitations have been identified, the specific board size can be established. Ideally, the circuit board should be just large enough to contain its components and interconnections and still be economically produced.

One method of estimating finished board size is to compute the square area required to contain each type of component, including lead terminations, and to multiply by the number of like components. For integrated circuits, additional area must be included for the larger amount of circuitry required. This can usually be estimated from other previously designed circuit boards. Total all the component areas, and divide by the usable layout area on a proposed board size. Do not include the area to be utilized by connectors, board edges, and mounting hardware.

A simpler method in estimating board size is to take actual scaled cutouts or symbols, like Bishop's PUP-PETS®, of all the components to be included on the board and place them at random, allowing space around them for interconnections. Be sure to include extra area for integrated circuit connections.

Conductor Width and Spacing

Conductors can be divided into **signal traces, ground traces,** and **voltage traces.** Signal traces are all conductor traces on a board other than ground or voltage traces. Signal traces generally have low voltage and current and therefore are thinner and more closely spaced than the heavier ground and voltage traces. All traces on a board are laid out with **slit tape.** Since manually designed boards are drawn at an enlarged scale, 2:1, 4:1, and so on, the tape width used on the layout is scaled accordingly. One of the more common signal trace widths at 2:1 scale uses a .040-in. (1.016-mm) tape width. This gives a finished board trace width of .020 in. (.508 mm). The air gap between adjacent signal traces will be the same width as the signal trace.

Ground or *voltage traces,* as well as air gaps and spaces, are frequently .100 in. (2.54 mm) wide. These traces have a greater voltage and current requirement and are therefore wider than signal traces.

Careful consideration must be given to conductor or *trace* widths and spacing when a circuit board is designed. If a conductor width is too small, discontinuity of circuitry or heat problems on a finished board may result. Narrow spacings often cause short circuits (arcing). Widths and spacing that are too large can waste valuable space and increase cost. Conductor width and thickness should be determined on the basis of the required current-carrying capacity. (See Appendix A, "Conductor Thickness and Width.") The maximum size within the available area, consistent with minimum spacing requirements, should be maintained for ease of manufacture and durability in usage. Avoid using conductors larger than .500 in. (12.7 mm). If larger conductive areas are desired—for example, ground planes—relieved areas should be incorporated to prevent blistering and warping during soldering. (See the "Ground Planes" section later in this chapter.) A nominal conductor width of .050 in. (1.27 mm) or .062 in. (1.57 mm) is recommended for low-voltage applications where space permits. (See Appendix A, "Conductor Spacing Minimums," for a summary of recommended minimums.) Figure 18–18 shows good conductor spacing, which should be determined from peak voltage between conductors, altitudes at which the circuit board will be in use, and conformal coatings to be applied to the board. A nominal spacing between conductors of .031 in. (.79 mm) or .050 in. (1.27 mm) is recommended for low-voltage applications when space permits.

Selection of conductor widths and spacings should always allow for manufacturing process variations, such as

FIGURE 18–18
Conductor spacing. Arrows indicate air gaps.

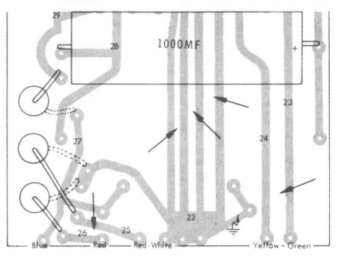

☐ Exposure of the master pattern
☐ Type of resist
☐ Variations in etching or plating

Conductor width adjustments that allow for processing variations are detailed in Table 18–2.

Final conductor width dimensions after etching depend on the following.

☐ Photographic opacity of type
☐ Tape width tolerance
☐ Over- or underdevelopment in photoreduction
☐ Copper laminate thickness
☐ Manufacturing process (photo direct or silkscreened, and intensity of etch)

Component Terminal Holes

A printed circuit board should have a separate mounting hole for each component terminal and for each end of a jumper or interconnection wire, unless the leads or wires are to be soldered to standoff type terminals.

Two basic types of **terminal holes** used on printed circuit boards are unplated holes and plated-through holes, as shown in Fig. 18–19. Unsupported holes contain no conductive material, plating, solder, or any type of reinforcement. Figure 18–20 shows etched clearance holes in a logic card.

Holes may either be drilled or punched into a circuit board, or they may be etched as in commercially available cards (Fig. 18–20).

Plated-through holes begin as unsupported holes. Conductive material is then electrically deposited, or *plated,* on the inside walls, forming an electrically conductive connection between layers of the circuit board. The plating usually consists of tin lead solder over electrodeposited copper (Fig. 18–19).

TABLE 18–2
Conductor Width Processing Tolerances[a]

Plating	Conductive Material	Conductor Width in.	mm
Unplated	2 oz copper	±.005	±.127
	1 oz copper	±.003	±.076
	½ oz copper	±.001	±.025
Panel plated copper .001 in. (.025 mm) minimum	2 oz copper	+.005 −.008	+.127 −.203
	1 oz copper	+.003 −.005	+.076 −.127
	½ oz copper	+.001 −.003	+.025 −0.76
Pattern plated copper .001 in. (.025 mm) minimum	2 oz copper	±.005	±.127
	1 oz copper	±.003	±.076
	½ oz copper	±.0015	±.038

[a] Dimensions are based on reduced 1:1 artwork.

Plated-through holes provide a reliable layer-to-layer interface connection and should be used on all production quantity, multilayer sandwich boards. Whenever requirements dictate that some holes be plated-through, all holes should be specified as plated-through to reduce drilling costs. Boards with both types of holes require two drilling operations. (Plated-through holes are drilled before the etching process.) The second drilling operation (after etching and plating) is much slower because the solder plate binds the drills.

To determine the minimum diameter of an unsupported hole, use the following formula:

(Minimum hole diameter) = (maximum lead diameter) + (minimum drill tolerance)
(Adjust to next larger standard drill size)

FIGURE 18–19
Cross section of plated-through hole. (Provided courtesy of Bishop Graphics, Inc.)

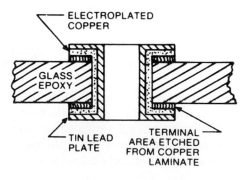

FIGURE 18–20
Back side of universal logic card 4112-4, with overall ground plane and inset showing etched clearance holes in plane. (Courtesy Vector Electronic Co.)

Suggested hole sizes and tolerances are listed in Appendix A, "Recommended Terminal Areas for Unplated Holes" and "Recommended Terminal Areas for Plated-through Holes." For plated-through holes, add the plating thickness plus the maximum plating tolerance. Usually, plating thickness is specified as a minimum, with an accepted tolerance of minus 0, plus 100%.

An unsupported hole should be no larger than .020 in. (.51 mm) greater than the minimum lead to be inserted. For plated-through holes, this may be increased to .028 in. (.71 mm).

The number of different-sized holes on a circuit board should be kept to a minimum. As the number of different hole sizes increases, so does the cost and difficulty of manufacturing the board. Usually, three to five different sizes can accommodate most components. See Appendix A for some suggested standard hole sizes.

There should be enough space between holes so that the terminal areas surrounding the holes meet minimum conductor spacing requirements. The space between holes should be equal to or greater than the board thickness or hole diameter, whichever is smaller.

Terminal Area

A **terminal area (pad)** is a portion of a printed circuit used for making electrical connections between a component or wire and part of the conductive circuit pattern. Three types of terminal areas are shown in Fig. 18–21. These are also called *pads* or *lands*.

Pad size is based on the hole diameter, and the hole diameter is determined from the lead diameter. Calculate the hole size first and then the pad size. The hole size can be found under the maximum lead diameter in the manufacturer's catalog. For radial lead components, the manufacturer's catalog also lists the lead spacing, which is then used as the pad spacing dimension.

There should be a separate terminal pad for each component lead or wire attachment. Generally, the pads should completely surround the mounting holes. Exceptions are when *flat pack* components with flat *ribbon* leads are to be mounted on the board surface or when an offset terminal is used adjacent to the mounting hole in conjunction with a clinched lead. When offset terminal areas are used, as in Fig. 18–22, they should be placed far enough from the mounting holes for the component leads to be clipped off before they are soldered. Offset terminals should be used on the circuit side of the board when very close lead spacing patterns make use of a regular terminal area difficult or impossible or when stress relief is desired for the component lead at termination. A hole is drilled at the indicated circle (Fig. 18–22). The component lead is inserted into the hole through the board. The lead is bent over, clinched to the terminal area (pad), and then soldered.

Terminal area shapes vary with designer preference, as shown in Fig. 18–23. Certain shapes, however, have specific design advantages and disadvantages. Square or rectangular terminal areas provide maximum adhesion of the copper pad to the circuit board. They are useful when a large component hole is required where there is a minimum of usable terminal area space. On the other hand, terminal areas with straight sides, like squares, rectangles, and ovals, when placed close to traces or other pads can contribute to solder bridging problems during **wave soldering.** Round and elliptical

FIGURE 18–21
Terminal areas. (Provided courtesy of Bishop Graphics, Inc.)

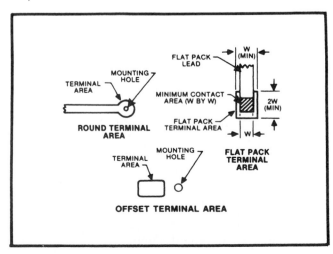

FIGURE 18–22
Offset terminals. (Provided courtesy of Bishop Graphics, Inc.)

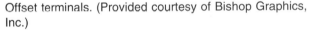

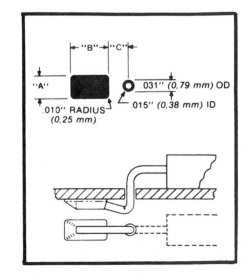

FIGURE 18–23
Terminal area *pad* or *land* shapes.

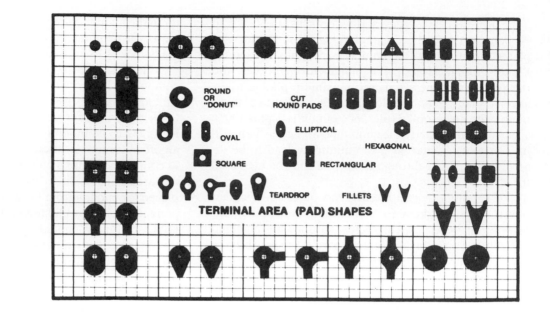

pads reduce solder bridging under the same circumstances.

Some designers prefer the use of teardrop-shaped terminal pads. The fillets where the trace meets the terminal pad provide a stronger mechanical connection, while providing a base for clinched leads. Teardrop drafting aid pads are difficult to align and are not usable with **pad master** artwork techniques unless separate fillet drafting aids are used. With the improved quality and increased density requirements of circuit boards, most designers find the use of teardrop-shaped terminal pads to be unnecessary.

Terminal area size should be as large as practical while maintaining design density consistent with minimum spacing requirements. The minimum terminal area should be based on the following criteria:

☐ Maximum hole size
☐ Hole location tolerance
☐ Terminal area location tolerance
☐ Conductor processing tolerance
☐ Minimum required annular ring

There should be a concerted effort to establish a common pad size that can be used throughout the board or for as many terminals as possible. Try to keep the number of different pad sizes to a minimum. You can establish a common pad size by listing maximum lead sizes for all components. When the list is complete, consult the table for hole and pad sizes to see if a particular pad size can be used for groups of holes that fall within an acceptable size range. The minimum required annular ring is the smallest part of the circular strip of conductive material surrounding a mounting hole that will be sufficient for design requirements. MIL-STD-

275D specifies .015 in. (.38 mm) for the minimum annular ring surrounding an unsupported hole. For plated-through holes, a minimum of .005 in. (.13 mm) on external layers and .002 in. (.05 mm) on internal layers of multilayer sandwich boards are specified.

The formula for calculating the minimum terminal area is

$$TA = H + L \text{ tol} + CP \text{ tol} + 2AR$$

where TA = minimum terminal area
   H = maximum hole diameter (see "Component Terminal Holes" section)
   L tol = locational tolerances (double-sided and multilayer: 2× hole location tolerance or 2× feature location tolerance, whichever is larger; single-sided hole location tolerance plus feature location tolerance)
   CP tol = conductor processing tolerance (the *minus* conductor width tolerance expressed as a *plus* amount)
   AR = minimum annular ring requirement

"Recommended Terminal Areas for Plated-through and Unplated Holes" in Appendix A lists suggested terminal areas along with corresponding doughnut pads in 1×, 2×, and 4× scales for standard unplated and plated-through hole sizes based on the criteria explained above. Also see Appendix A for "Printed Circuit Design Criteria: Dimension and Tolerance Considerations."

Component Mounting

Component mounting is one of the most important steps in the design of a PC board. Component type, mounting style, and orientation on the board are all factors in the placement and layout of the total board. Standards and specifications concerning component spacing and mounting instructions are available to the designer from manufacturers of components and through standards organizations. Spacing can also be determined by specific formulas that have been developed. Another factor in the design and layout of component mounting is how the components will be inserted, manually as in Fig. 18–24 or with automated insertion equipment.

In general, determining how to mount components involves the following steps:

1. Total number of board-mounted components required as shown on the schematic diagram
2. Total amount of space needed for the components
3. Pad spacing for each component, from manufacturers' dimensions for radial lead components and from a pad spacing formula for axial lead components
4. Spacing requirements between components
5. Hole sizes for the components
6. Component pad size
7. Heat sink or the heat dissipator requirements of the board and individual components, or both

FIGURE 18–24
PC board assembly and component wiring. (Courtesy Western Electric Co.)

8. Method of component insertion and board assembly
9. Method of soldering the board, hand or wave
10. Method of applying conformal coating, if needed

Note that steps 5 and 6 have already been discussed in "Component Terminal Holes" and "Terminal Area" sections.

Component Mounting Rules. Components should always be mounted on the side of the printed circuit board with the least amount of printed circuitry. For single-sided boards, components are mounted on the noncircuit side. Whenever possible, components should be placed so that their major axis is parallel to a board edge, as shown in Fig. 18–25, and to the flow of cooling air if applicable. In addition, an effort should be made to place components parallel (preferable) or perpendicular to each other, as in Fig. 18–26, to provide an orderly appearance, but not at the expense of good functional design. Components should also be located so that their identification or value codes can be read from the same direction, preferably from top to bottom and left to right, with polarity markings visible if possible.

No part of a component should project over the board edge unless required by its function. Typically, a minimum clearance of .062 in. (1.57 mm) is maintained between a component and a board edge, card guide, or other mounting hardware. All components should be mounted so that they do not restrict the removal or insertion of any other component or mounting hardware, as shown in Fig. 18–27.

Components weighing ¼ oz (7.1 g) or more per lead should always be mounted with clamps or other means of support so that the soldered joints are not relied on for mechanical support.

All parts dissipating 1 W or more should be mounted so that the body of the part does not come

FIGURE 18–25
Component mounting. (Provided courtesy of Bishop Graphics, Inc.)

FIGURE 18–26
Vector plugboard accommodates 14-, 16-, 24-, and 40-pin DIPs for use in a variety of circuit applications. (Courtesy Vector Electronic Co.)

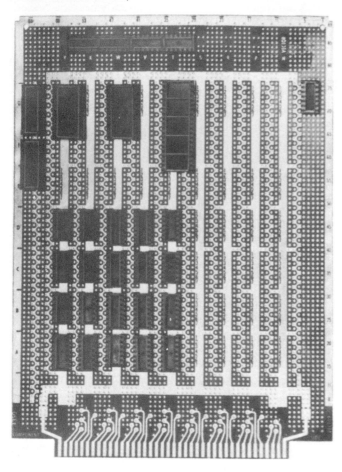

into direct contact with the circuit board unless heat-dissipation devices are used, such as heat sinks or thermal ground planes. Heat dissipators that are mounted on the component body, as in Figs. 18–28 and 18–29, are available for a variety of component types. In Fig. 18–28 the heat dissipator is designed to come in contact with both the top and bottom of the package. This type of heat dissipator requires slightly more board space than when just the component package is mounted. The heat dissipators in Fig. 18–29 are attached to the top of the component and may not require extra board space, although their extra height must be taken into account when the board housing is designed since more clearance may be required.

Component leads or parts with conductive cases should be mounted a minimum of .062 in. (1.57 mm) from the conductive pattern (if consistent with spacing, voltage, altitude, and coating data; see Appendix A). If adherence to minimum spacing requirements is not practical, insulation can be used between conductors and part leads or cases.

Horizontally mounted axial lead components, shown in Fig. 18–30, should be attached so that the body of the part is in contact with the circuit board, Fig. 18–30 (2). However, components should not be placed in contact with more than one conductor unless the board surface is suitably protected from moisture traps.

When axial lead components are mounted vertically [Fig. 18–30 (1)], they should be spaced a minimum of .015 in. (.38 mm) above the board surface to allow for good solder joints and adequate cleaning [Fig. 18–30 (1), dimension D]. The highest point of the top lead should not extend more than .550 in. (13.97 mm) above the board surface, dimension C. The bottom lead should extend straight into the board, whereas the top lead should be bent 180° around the component body and down into the board. The top lead should be insulated to prevent contact with other conductive elements [Fig. 18–30 (1) E]. The straight portion of the lead should be a minimum of .0156 in. (.4 mm) before the radius bend is formed. The minimum bend should be equal to the lead diameter.

Radial lead components should be mounted within 15° of being perpendicular unless a large case size makes this impractical. In this instance, they should be mounted with a side surface in contact with the board and the leads bent down at a 90° angle. If the vertically mounted components have coating extending down the leads from the body, as in Fig. 18–31, they should be mounted so that the coating is a minimum of .060 in. (1.52 mm) above the board surface to prevent interference with the solder joint.

Components in transistor packages may be mounted vertically, as in Figs. 18–32 and 18–33 (A through D), or horizontally. When mounted horizontally, the cases should be secured to the board with clips or other mechanical fasteners, Fig. 18–33 (E). For vertical mounting (the most common practice), the package should be spaced .015 in. (.38 mm) to .125 in. (3.18 mm) above the board to permit flux removal. The base of the component should be parallel to the board surface within .050 in. (1.27 mm). Transistor components can be mounted straight through with or without a spacer (A and D), offset (B and C), horizontally with a clip (E), or inverted (F). Flat packs, Fig. 18–33 (G), can be surface mounted or inserted. Dual-in-line (DIP) packages, Fig. 18–33 (H), can be mounted in the standard *through-the-board* manner, or they can be plugged into a receptacle.

Pad Spacing. Pad spacing or lead spacing is the distance between the centers of one lead to the center of the other lead, as shown in Fig. 18–34 (C). Pad spacing must be calculated for axial lead components. Pad spacings for radial components, transistors, and dual-in-line packages are taken from manufacturers' specification

FIGURE 18–27
Component mounting instructions. (Reprinted by permission of HEATH COMPANY)

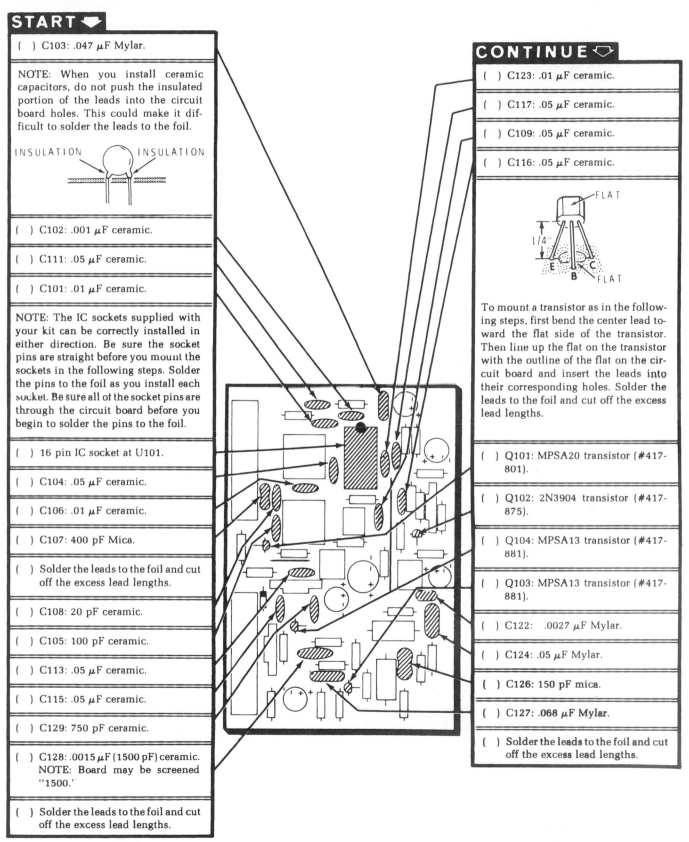

START ⬇

( ) C103: .047 μF Mylar.

NOTE: When you install ceramic capacitors, do not push the insulated portion of the leads into the circuit board holes. This could make it difficult to solder the leads to the foil.

INSULATION   INSULATION

( ) C102: .001 μF ceramic.

( ) C111: .05 μF ceramic.

( ) C101: .01 μF ceramic.

NOTE: The IC sockets supplied with your kit can be correctly installed in either direction. Be sure the socket pins are straight before you mount the sockets in the following steps. Solder the pins to the foil as you install each socket. Be sure all of the socket pins are through the circuit board before you begin to solder the pins to the foil.

( ) 16 pin IC socket at U101.

( ) C104: .05 μF ceramic.

( ) C106: .01 μF ceramic.

( ) C107: 400 pF Mica.

( ) Solder the leads to the foil and cut off the excess lead lengths.

( ) C108: 20 pF ceramic.

( ) C105: 100 pF ceramic.

( ) C113: .05 μF ceramic.

( ) C115: .05 μF ceramic.

( ) C129: 750 pF ceramic.

( ) C128: .0015 μF (1500 pF) ceramic.
NOTE: Board may be screened "1500."

( ) Solder the leads to the foil and cut off the excess lead lengths.

CONTINUE ⬇

( ) C123: .01 μF ceramic.

( ) C117: .05 μF ceramic.

( ) C109: .05 μF ceramic.

( ) C116: .05 μF ceramic.

FLAT
1/4"
E   C
B
FLAT

To mount a transistor as in the following steps, first bend the center lead toward the flat side of the transistor. Then line up the flat on the transistor with the outline of the flat on the circuit board and insert the leads into their corresponding holes. Solder the leads to the foil and cut off the excess lead lengths.

( ) Q101: MPSA20 transistor (#417-801).

( ) Q102: 2N3904 transistor (#417-875).

( ) Q104: MPSA13 transistor (#417-881).

( ) Q103: MPSA13 transistor (#417-881).

( ) C122:   .0027 μF Mylar.

( ) C124: .05 μF Mylar.

( ) C126: 150 pF mica.

( ) C127: .068 μF Mylar.

( ) Solder the leads to the foil and cut off the excess lead lengths.

FIGURE 18–28
Heat dissipator for microprocessor package. (Courtesy of International Electronic Research Corp.)

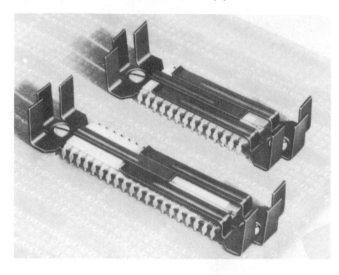

lists for their products. Most nonaxial components have standardized pad dimensions. When preprinted artwork patterns are used for multipad components, the artwork has highly toleranced pad locations. Multipad patterns are simply located on the layout at the required positions corresponding to the grid pattern whenever possible.

For determining lead spacing of axial lead components, IPC-CM-770B (proposed) suggests the lead should extend nominally .060 in. (1.52 mm) straight out from the component body before the start of the bend, Fig. 18–34 (A). It recommends a high-density packaging minimum of .030 in. (.76 mm) lead extension before the bend in accordance with MIL-STD-275D. Both the IPC and MIL-STD specifications suggest a minimum bend

FIGURE 18–29
Heat dissipator for PCB components. (Courtesy of International Electronic Research Corp.)

FIGURE 18–30
Axial lead component mounting: A = minimum straight part; B = component body length; C = total height; D = minimum spacing; E = insulate top; F = pad (lead) spacing.

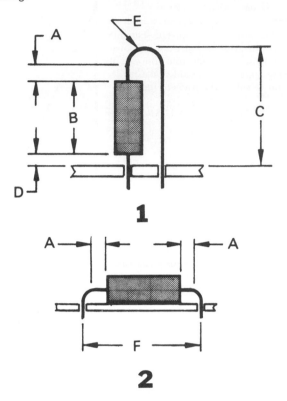

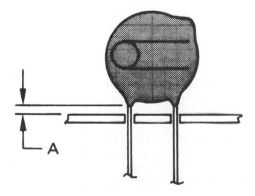

radius equal to one to two times (depending on lead diameter) the maximum lead diameter.

To determine the minimum lead spacing for axial lead components, the following formula can be used:

$$\text{LS(MIN)} = \text{CL(MAX)} + 2(\text{LE}) + 2\text{BR(MIN)} + \text{LD}$$

where LS(MIN) = minimum lead spacing (this should be rounded up to nearest standard grid increment)

FIGURE 18–31
Mounting ceramic capacitors; A = .060 in. (1.52 mm).

FIGURE 18–32
Transistor mounting. (Provided courtesy of Bishop Graphics, Inc.)

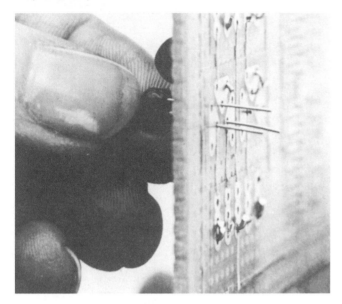

CL(MAX) = maximum component length (this includes coating meniscus, solder seal, solder or weld bead, or any other extension)

LE = lead extension (2× lead diameter) minimum .030 in. (.76 mm) preferred .060 in. (1.52 mm)

BR(MIN) = minimum bend radius

LD = lead diameter (2× lead radius)

Lead Diameter	Min. Bend Radius
Up to .027 in. (.69 mm)	1 × lead dia.
.028 in. (.70 mm) to 047 in. (1.19 mm)	1.5 × lead dia.
Over .047 in. (1.19 mm)	2 × lead dia.

For example,

$$CL(MAX) = .280 \text{ in. } (7.11 \text{ mm})$$
$$LE = .030 \text{ in. } (.76 \text{ mm}) \text{ (MIN)}$$
$$BR(MIN) = .026 \text{ in. } (.66 \text{ mm}) \text{ } (1 \times LD)$$
$$LD = .026 \text{ in. } (.66 \text{ mm})$$

In English units (inches),

$$LS(MIN) = .280 + 2(.030) + 2(.026) + .026 = .418 \text{ in.}$$

In metric units (millimeters),

LS(MIN) = 7.11 + 2(.76) + 2(.66) + .66 = 10.61 mm

Component Lead Spacing

.500 in. (12.70 mm)
.450 in. (11.43 mm)
.425 in. (10.80 mm)

FIGURE 18–33
Component mounting: (A) straight through; (B) offset; (C) spreader; (D) spacer; (E) clip or heat sink; (F) inverted; (G) flat pack mounting; (H) dual-in-line (DIP) package. (Provided courtesy of Bishop Graphics, Inc.)

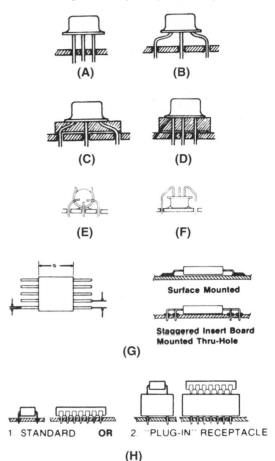

FIGURE 18–34
Pad (lead) spacing: (A) .060 in. (1.52 mm) for axial lead components, .030 in. (.76 mm) for high-density packaging; (B) component body length (max); (C) center-to-center pad spacing (lead and spacing).

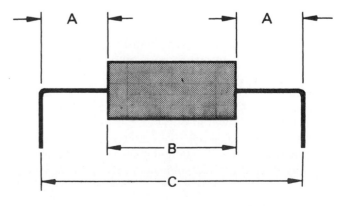

Grid System

.100 in. (2.54 mm)
.050 in. (1.27 mm)
.025 in. (.64 mm)

There are additional factors to consider when a board is being designed for automatic (NC) component insertion. Generally, a minimum of two .125 in. (3.18 mm) diameter tooling holes, accurately located and sized, are required within the longest straight board area.

Component mounting holes should be enlarged to accommodate additional tolerance factors (like locational accuracy of insertion equipment and tooling pin-to-tooling hole fit). For practical purposes, .008 in. (.20 mm) should be added to standard component mounting hole diameters to allow for these tolerance factors. (See "Component Terminal Holes.")

Component placement is more significant in designs with automatic insertion. It is preferable to have all components on a single axis with common lead spacing and orientation. Additional tooling changes, programming, and handling may be required for variances. Second in preference is to have all components on a single axis with common lead spacing. Next, in order of preference, is to vary the lead spacing and, finally, to use more than one axis. Axial lead spacing should be .300 in. (7.62 mm) minimum and 1.300 in. (33.02 mm) maximum, according to IPC recommendations.

Check with the manufacturer of the automatic insertion equipment being used for other specific requirements and limitations concerning board size, uninsertable areas, and component types, sizes, and clearances. IPC-CM-770B (proposed revision) is an excellent reference for automatic component insertion design data.

After the PC board has been assembled, a **conformal coating** may be required for protection of the board and its components. The PC board is dipped or sprayed with a thin layer of epoxy or resin and then dried with an air or baking process. Conformal coating protects the PC board from dust, dirt, moisture (and fungus growth), and minor mechanical damage.

Board Design

The first consideration of PC board design is the selection of the board design and basic outline. The rectangle is the most common shape for outlines and easiest to lay out. Round boards (Fig. 18–3) and other irregularly shaped boards have been designed for special situations. The PC board shown in Fig. 18–35 is an example of an irregularly shaped board. Even though this has irregular sides, it still *implies* a rectangular shape. The design was possibly started as a rectangle, and then after the layout

FIGURE 18–35
Irregularly shaped printed circuit board. (Courtesy of International Business Machines Corporation)

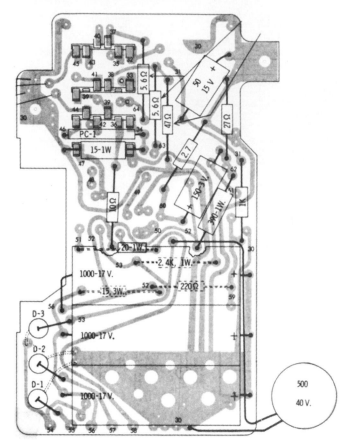

was completed, portions of the board were cut away to reduce space.

When the PC board package or module shape has been predetermined by the engineer, the board geometry is used to define subsequent board *working area* as well as limitations of conductor area and component location. When the board is designed first and packaged second, the board geometry is determined after the layout requirements are met. Common dimensions should be used when the board size and configuration are established.

Before board geometry is determined, the following considerations must be taken into account:

1. Type of housing into which the board will fit
2. Maximum amount of three-dimensional space available for the board
3. Mounting, fastening, and clamping method to be used
4. Electrical connection cable, connector, terminals, wiring requirements, and so on
5. Type, thickness, and material to be used for the board

6. Component mounting orientation (one side or both sides of the board)

Board Geometry

The board outline (physical configuration) is determined after the above needs are assessed and the assembly, fabrication, and manufacturing methods available are determined. In general, a simple rectangular board with connector strip is the easiest and most simple board geometry design. Boards designed with cutouts, curves, and irregular angles, as in Fig. 18–35, are more expensive to design, fabricate, and manufacture. The simple board geometry shown in Fig. 18–36 is the most common type of configuration.

The board outline has dimensions defining the configuration within specific toleranced limits. In Fig. 18–36 the board outline and the tooling holes are shown, but the actual dimensions are omitted. In Fig. 18–37 the tooling holes and the board outline are shown and dimensioned. The connector strip is also detailed, along with a notch or cutout located at one end of the strip to ensure proper installation. Connector strips are designed into boards that will require removal for testing, servicing, and other operational requirements. They should not be used where the board is to be securely and permanently mounted.

FIGURE 18–36
Board geometry.

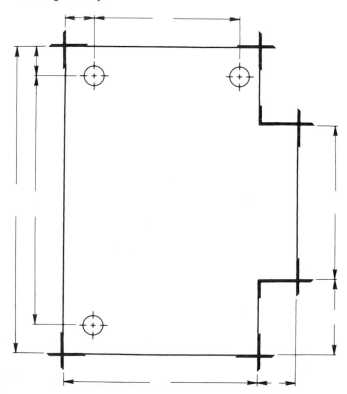

The usable *working* area of the board is determined by the available area after the board is manufactured. The smallest toleranced dimension is used to provide a basic size from which to calculate the working area. Fasteners, component sizes, conductor requirements, and possible heat sinks and ground planes must all be taken into account before the final board working area is established.

In general, a minimum of .025 in. (6.35 mm) should be provided between the closest trace and the board edge. Components are also restricted to a minimum distance from the edge of the board. Where the PC board design starts with the board itself and is not limited to a specific package, the designer can establish the board outline working area after laying out components and conductor traces. Where the board is to be restricted within a previously determined module size or package shape, the working area is determined from the given board geometry size.

Tooling Holes

In Fig. 18–37, four tooling holes are shown on the board. Tooling holes are used during the fabrication and manufacturing stages of the board's production. Tooling holes establish accurate reference points for dimensioning and locating other holes and fixtures required for machining the board. In most cases the tooling holes will be within the board outline. When the PC board is crowded and densely packed with components, tooling holes can be placed outside the board's outline and then trimmed off later. Tooling hole locations can also be used as register marks for artwork registration.

Registration Techniques

The relative position of one or more printed circuit patterns or parts of the patterns, with respect to their desired location on the printed circuit board, is called their *register*. The relationships of the elements of register and the techniques used to achieve register are called *registration*.

PCB register consists of three basic elements:

1. Overall pattern alignment from layer to layer
2. Alignment of individual terminal pads from layer to layer
3. Alignment of drilled holes to the circuit pattern

The first and third elements are primarily manufacturing registration concerns. If the photographic reduction of the **master pattern** is not carefully controlled, size differences between artwork layers can contribute to layer-to-layer misregistration of the circuit patterns. The same condition can occur if master artwork sheets are exposed to different environmental conditions. If

FIGURE 18–37
Tooling holes, board geometry, and working area.

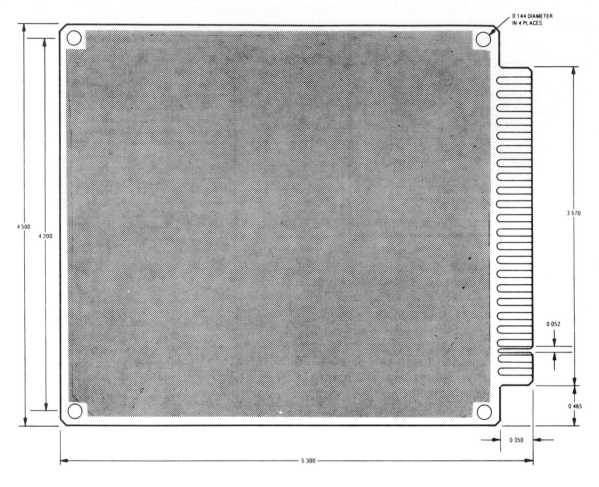

the mechanical alignment of the master patterns to the drilled hole pattern and to each other is not accurate during fabrication, holes will not align with terminal areas.

The second element is an artwork function. Precise control of photography and fabrication alignment cannot compensate for terminal areas individually misaligned on the artwork.

Artwork Markings. Printed circuit board artwork markings relating to registration include datum points, register marks, indexing holes, and photoreduction points. Other types of board markings are part numbers, reference designations, and corner markers.

Datum points are points assumed to be exact from which the location of printed wiring board features may be established by dimensional computation, as shown in Fig. 18–38. Three datum points establish the perpendicular Cartesian coordinate system of grid dimensioning. Three datum points are shown in Fig. 18–39. Note that they lie inside the area of the board. The

FIGURE 18–38
Pads for datum points and indexing holes. (Provided courtesy of Bishop Graphics, Inc.)

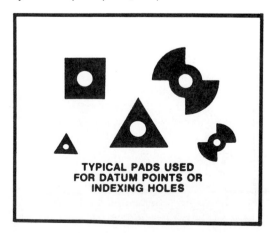

FIGURE 18–39
Two-color taping of a printed
circuit board. (Provided courtesy
of Bishop Graphics, Inc.)

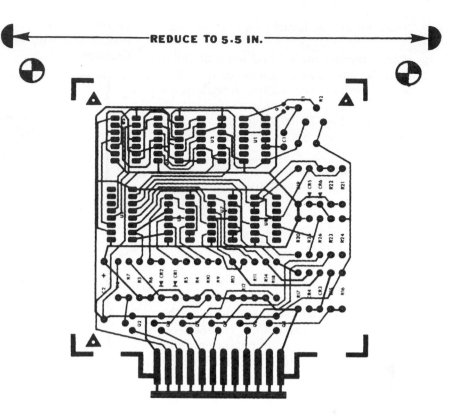

FIGURE 18–40
Register marks. Bishop Graphics' Universal Target® is
designed to achieve optimum visual register. (Provided
courtesy of Bishop Graphics, Inc.)

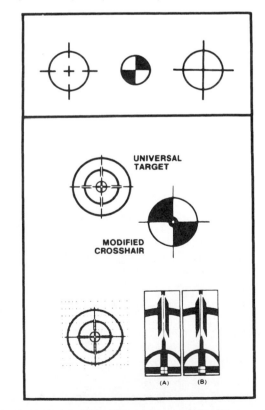

board geometry (shape) is delineated by board markings at the corners of the PCB, as shown here. Datum points are normally indicated by distinctively shaped pads when they are located directly on the printed circuit board. In Fig. 18–39 a triangular datum shape is used. All mechanical dimensioning of the master pattern should be made from datum points.

Register marks are bold geometric shapes used to establish registration, as shown in Fig. 18–40. Register marks are affixed to the artwork on all layers so that when the layers are overlaid the patterns are superimposed. Register marks may be located inside or outside the board outline, as appropriate. When located inside the board area, they may be drilled for mounting holes after they have served their original purpose. In Fig. 18–39 the register marks are positioned outside the board outline.

Indexing or *tooling holes* are placed in a printed circuit board so that the board can be accurately positioned during manufacturing processes. They may or may not appear within the finished board areas, but they must appear on the master pattern. Indexing or tooling holes can be established from datum points when the points are positioned within the area of the board outline (Fig. 18–39). A minimum of two indexing holes is required, and the holes should be diagonally located to encompass the greatest area. A third hole is sometimes used, and it should form a right triangle with the first two. They should always be located on grid intersections.

Datum points, register marks, and indexing holes may be combined in the same locations, or they may be completely independent of each other. A registration target such as Bishop Graphics' Universal Target® or Modified Crosshair® (Fig. 18–40) may be used as a combination datum point symbol, indexing hole symbol, and register mark.

Photoreduction points, shown in Fig. 18–41, are two points spaced an exact distance apart to visually indicate how much printed circuit artwork is to be photographically reduced (Fig. 18–39). Care should be used in selecting a mark that clearly indicates the precise measurement point. In Fig. 18–39 the photoreduction marks are at the top of the artwork and are specified with a reduction dimension between the two points: *reduce to 5.5 in.*

Use reduction marks or simple quadrants since the exact point of measurement cannot be confused—it must be taken from the edge. In addition, the points of intersection indicate proper photo exposure and development. *Do not use* cross hairs for reduction points, since the width of the line makes the exact measurement point ambiguous.

Pin Registration. Accurately registered printed circuit board artwork can be produced from pin-registered multisheet artwork techniques. With this method, a pad

FIGURE 18–41
Photoreduction marks. (Provided courtesy of Bishop Graphics, Inc.)

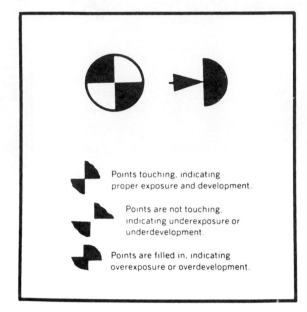

Points touching, indicating proper exposure and development.

Points are not touching, indicating underexposure or underdevelopment.

Points are filled in, indicating overexposure or overdevelopment.

master base sheet is created on transparent, dimensionally stable drafting film. An accurate pattern of precisely sized locating holes punched into the sheet is used for pinning successive overlays in precise registration, as shown in Fig. 18–42. A PC board designer uses a light

FIGURE 18–42
Pin registration. (Provided courtesy of Bishop Graphics, Inc.)

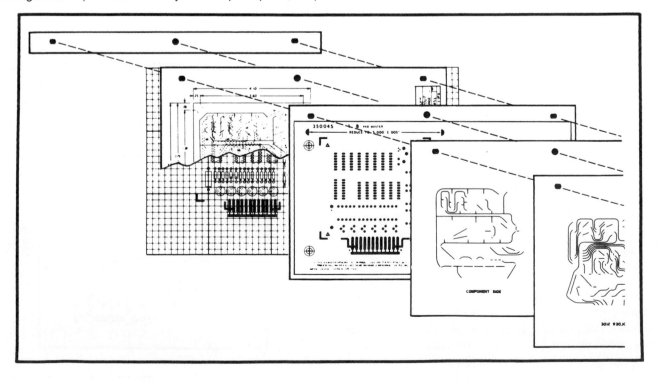

FIGURE 18–43

Designer using a light table and a pin registration to prepare PCB artwork. (Provided courtesy of Bishop Graphics, Inc.)

table and pin registration for laying out printed circuit artwork, as shown in Fig. 18–43.

When a multisheet artwork system is used, it is necessary to identify each sheet to ensure correct orientation when composite artwork is assembled for checking, rework, duplication, or photographic reduction. A sheet index chart, shown in Fig. 18–44, should be placed on the base sheet in an area that remains clear on each overlay. Beginning with the base sheet, all artwork layers should be clearly described in this chart, including artwork identification numbers, revision codes, sheet numbers, sheet or circuit layer descriptions, and other pertinent data.

Each artwork overlay sheet identification should be added while this sheet is superimposed over the base sheet. The information should be *right reading* and not interfere with any other data on the base sheet or other overlays in the set when composite artwork is assembled.

FIGURE 18–44

Sheet index chart for artwork identification. (Provided courtesy of Bishop Graphics, Inc.)

DWG. NO.		SH	REV.	DESCRIPTION
	350045	1	B	PAD MASTER
	350045	2	B	COMP SIDE CKT
	350045	3	B	SOLDER SIDE CKT
	350045	4	B	MARKING ARTW
	350045	5	B	SOLDER MASK

Auxiliary Documents.   Additional artwork and documentation have registration requirements, either in production or during fabrication, and can be created from pin registration techniques. Printed circuit board markings (silkscreen), ground plane, and solder mask artwork can all be prepared with appliqués, tape, cut and peel film, or reprographic methods while pin registered to the pad master. Assembly, outline, drill template, and fabrication drawings can also be made from overlay drafting or reprodrafting techniques while pinned to sheets of pin-registered composite artwork and layouts.

Pin Pattern.   Critical to the results of pin registration are hole shapes and locations, the accuracy and consistency of the hole pattern, and the closeness of fit between the pin and holes. Because all film materials tend to change size due to temperature and humidity variations, it is important to locate a single round hole, called a **pin pattern**, as close to the artwork center as practical. This round hole is punched with tolerances to ensure that virtually no changes can occur to affect its position or registration when the film is secured on its corresponding round metal pin, as shown in Fig. 18–45.

Additional registration holes are necessary to prevent the film from rotating around this round center hole. These additional holes should be elongated, with close-fitting edges on the axis radiating from the round hole (Fig. 18–45). The elongation of the holes allows for film expansion and contraction caused by environmental changes and keeps misregistration at the outer edges of the artwork sheets to a minimum. Use of more than

FIGURE 18–45

Pin registration for overlay drafting. (Provided courtesy of Bishop Graphics, Inc.)

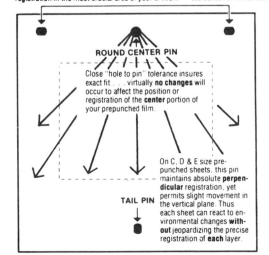

one restrictive round hole can cause buckling and distortion as sheets change size. The fit tolerance of the holes to the pins must be maintained with precise accuracy. With a pin registration system incorporating all these requirements, an artwork image registration accuracy of ±.002 in. (.051 mm) should be consistently attainable.

PRINTED CIRCUIT BOARD LAYOUT

A printed circuit design layout, shown in Fig. 18–46, is a preliminary drawing that describes the physical packaging design of an electronic circuit. It can be in the form of a rough sketch or a formal engineering drawing. The design layout is a necessary development aid, which is used as a reference to translate the electrical schematic or logic diagram into a master printed circuit artwork and mechanical documentation package.

Content

The layout should contain all the design information necessary to produce the printed circuit artwork and documentation package. All pertinent electronic component data should be represented, including shape, location, orientation, lead spacing, reference designations, and any special mounting requirements. It should also include form factor information such as board out-

FIGURE 18–46
PCB layout drawing.

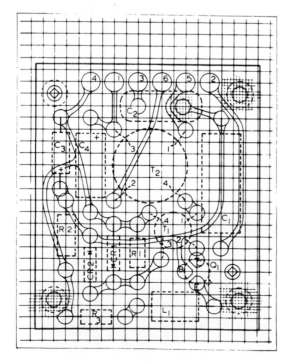

line, dimensions, tooling holes, and relationships with mating connectors and other external structures. All interconnection circuitry and scaled terminal areas should be depicted with general and local notations or keys for hole size, conductor width and clearance, terminal area size, layer designation, material, plating, and any other physical design requirements.

Preliminary Considerations

Before beginning a PC layout, you should have a schematic or logic diagram, a parts list, and all specifications that are applicable to the design, as shown in Fig. 18–47. From this information you will develop the board component layout and the master pattern. Figures 18–48 through 18–50 show the schematic drawing, parts list, board outline, component layout, and the circuit patterns for a double-sided PC board.

Compile information on all components, including physical size, lead pattern and spacing, special mounting data, required hole and terminal area sizes, and electrical and thermal limitations.

Determine the grid system, scale, nominal and minimum conductor width and spacing, board size, and number of layers before starting the layout. A 2:1 layout scale is the most common, although 4:1 is also used.

Be aware of any special circuit conditions such as test point requirements, high-voltage distribution and decoupling, or the thermal sensitivity of individual components that may affect the physical layout.

Give considerable thought to how the board is to be produced. Automated processes, such as artwork generation, drilling, routing, component insertion, wave soldering, and testing all have specific design requirements. Different etching and plating processes also affect the design layout requirements.

Preliminary List

The following items should be gathered, understood, and studied before the actual layout process begins:

1. Study the finalized schematic diagram.
2. Understand all symbols, reference designations, and component specifications.
3. Using the component specification sheet, determine component body configuration, size (length and width), mounting options, pad size (and pattern), lead spacing (pad spacing), and other requirements.
4. Note all polarity requirements for polarized components. Identify cathode end for diodes, emitter for transistors, and polarity for capacitors.
5. Group components according to common connections and determine component pad orientation for multilead components such as transistors.

FIGURE 18–47
Schematic diagram for PCB.

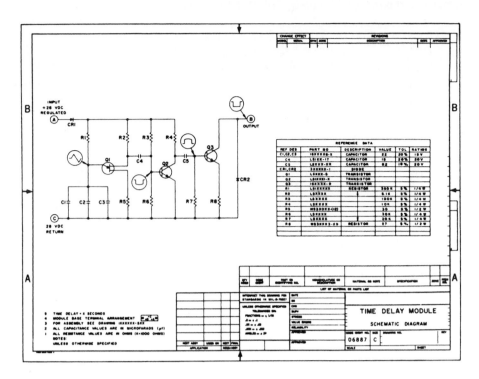

6. Establish layout scale and convert all component dimensions to the enlarged scale. A scaled PC layout template can be used for this purpose. Cutout dolls or preprinted component patterns that have been scaled to the layout size can also be used.

7. Determine grounds and voltage connections, including minimum width and thickness.

8. Establish conductor widths and thicknesses.

9. Obtain information on environmental considerations like heat, vibration, humidity, and other operating restrictions and conditions.

10. Understand connection requirements like terminal strips, cable connectors, and wiring needs.

11. Establish heat sinks, ground plane patterns, and other special conductor areas.

12. Establish the maximum module size restrictions, mounting method, and overall packaging design configuration in which the PC board is to be mounted, such as the chassis and cabinet.

13. Tentatively group components to establish overall space requirements.

Layout Sketch

Before beginning the PC design layout, make a list of components that are connected to a common point for use as a reference when making interconnections. Sometimes one or several rough freehand trial layout sketches are drawn to relate the schematic to the physical board. Components are usually represented as schematic symbols with the leads oriented in the same manner as the actual components, as shown in Fig. 18–51. Most circuit crossovers and junctions, along with component groupings, can be worked out in this stage.

Design Techniques

Design convention dictates that layouts be viewed from the component side of the board. Adhering to this standard eliminates possible confusion leading to costly errors. A designer should always create the layout and artwork at the same scale on or over a grid sheet. This aids in both preparing and checking the finished artwork. The original layout sketch is to be done freehand with grid paper. It is necessary to try a variety of layout options before a practical placement sequence can be established.

One method of layout starts with placing the *most important* component first, such as the transistor or DIP. If a particular component is connected to more components than any other component, then this is the one to build around. Component orientation, axis determination, and the physical shape of the grouped components should be taken into account at this time. If possible, lay out the components with one or two axes and with a rectangular or square overall grouping.

Design techniques vary considerably due to the nature and complexity of the circuit, as well as designer preference. Generally, component placement is the next step. Some designers, however, prefer to integrate interconnection routing with component placement, particularly on PC boards with many discrete components.

FIGURE 18–48
PCB schematic and parts list. (Courtesy TRW LSI Products)

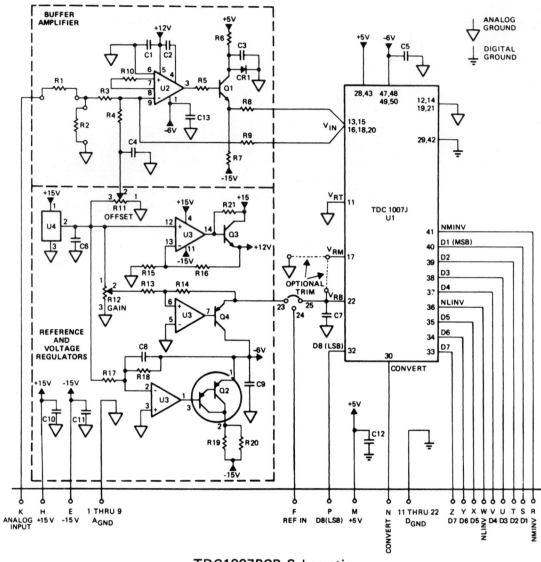

TDC1007PCB Schematic

PARTS LIST

RESISTORS

R1	0 Ω*	1/4W	2%
R2	80.6 Ω*	1/4W	2%
R3	1.0 KΩ	1/4W	2%
R4	4.2 KΩ	1/4W	2%
R5	10 Ω	1/4W	2%
R6	56 Ω	1/4W	5%
R7	240 Ω	2W	5%
R8	6.8 Ω	1/2W	5%
R9	2.0 KΩ	1/2W	2%
R10	†	1/4W	2%
R11	2.0 KΩ	1/4W	Multiturn Cermet Pot
R12	2.0 KΩ	1/4W	Multiturn Cermet Pot
R13	21.5 KΩ	1/4W	2%
R14	21.5 KΩ	1/4W	2%
R15	11.3 KΩ	1/4W	2%
R16	42.2 KΩ	1/4W	2%
R17	21.5 KΩ	1/4W	2%
R18	51.5 KΩ	1/4W	2%
R19	24 Ω	2W	10%
R20	24 Ω	2W	10%
R21	392 Ω	1/4W	2%

CAPACITORS

C1	0.1 μF	50V
C2	2.0 pF †	50V
C3	0.1 μF	50V
C4	0.1 μF	50V
C5	0.1 μF	50V
C6	1.0 μF	10V
C7	10.0 μF	10V
C8	0.001 μF	50V
C9	100.0 μF	10V
C10	10.0 μF	20V
C11	10.0 μF	20V
C12	10.0 μF	10V
C13	0.1 μF	50V

INTEGRATED CIRCUITS

U1	TRW TDC1007J	
U2	Plessey SL541C	
U3	Motorola MC4741	
U4	Motorola MC1403U	

TRANSISTORS

Q1	2N5836
Q2	2N6034
Q3	2N2222
Q4	2N2907

DIODE

CR1	1N4001

MISCELLANEOUS

A1 Cambion 64 pin socket
704-4064-01-04-12 for U1**

A2 Thermalloy heat sink
60738 FOR Q2

A3 TRW Cinch edge connector
251 22 30 160

A4 Printed circuit board
TRW TPC 1007

A5 Moore Systems stitch weld
pins 700508 for R1, R2
(4 Required)

FIGURE 18–49
PCB layout. (Courtesy TRW LSI Products)

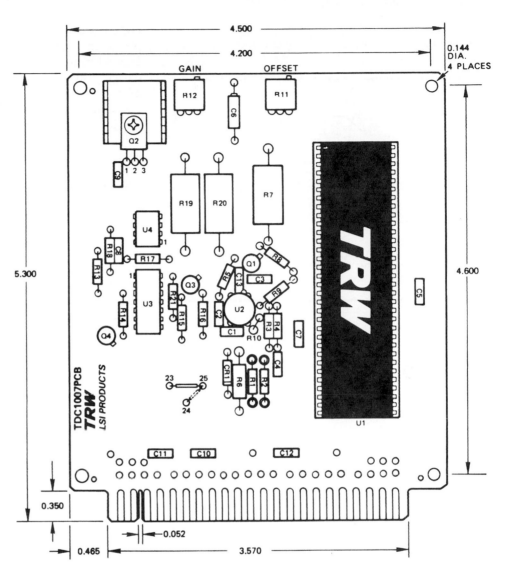

FIGURE 18–50
Evaluation printed circuit board—two sides. (Courtesy TRW SLI Products)

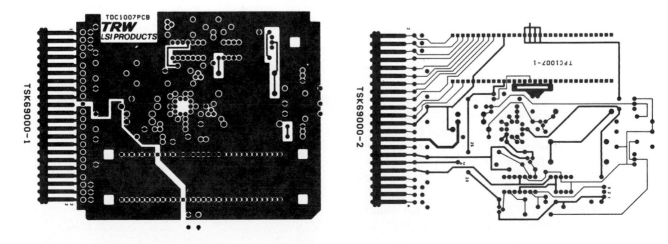

FIGURE 18–51
Trial layout sketch. (Provided courtesy of Bishop Graphics, Inc.)

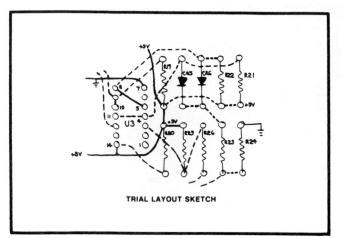

TRIAL LAYOUT SKETCH

Four basic component placement concepts may be used independently or in combination: schematic orientation, peripheral placement, central component placement, and fixed array.

The most basic concept, used primarily on medium- to low-density analog boards, is called *schematic orientation*. If the schematic has been drawn with a physical sense and has a minimum of interconnection crossovers, it is often possible to place components as they are physically drawn on the schematic diagram. This method works especially well if the signal input can be placed on one edge of the board and the output along the opposite edge.

The *peripheral placement* method is appropriate when board edge connectors or other components that require a specific fixed location, board edge placement, or off-board mounting are used. These components should be positioned first with any interconnecting components and then placed radiating inwardly from their locations.

The *central component placement* concept is applicable for boards that have one or more complex multiple lead devices, such as integrated circuits, relays, or modules with supporting peripheral components. In this case the predominant multilead components are centrally placed with the supporting components and then placed radiating outwardly from them. This technique is also used with fixed circuit group patterns such as semiconductor memory groupings. Figure 18–49 is an example of this method, except that here the IC has been placed to the right side of the board and all other components on the left.

The *fixed array* concept is typically used for straight digital logic boards comprised almost exclusively of inte-

grated circuits. With this method the ICs are logically placed in a fixed pattern, with a space allotment expressed in square inches per equivalent 14- or 16-lead device. This technique sometimes allows for the use of preprinted layout sheets, including board outlines, connector and component outlines, and terminal areas.

In general, when locating components on the layout, try to provide an orderly appearance. Component bodies should be parallel to a board edge and to each other, with the same orientation and lead spacing for like components (Fig. 18–46). Their orientation should provide for optimal interconnection routing. Note that many boards are designed with components at angles, such as the board in Fig. 18–52. This procedure is in no way incorrect and is frequently encountered when boards are hand assembled. When the board is to be produced by mechanized assembly processes, during the layout stage try to align components in rows with only one axis direction. Figure 18–46 is an example of one-axis orientation. The PC board in Fig. 18–27 has components grouped into mini groupings and aligned along two axes. Two-axis orientation is one of the most common arrangements found in industry.

Figures 18–49 and 18–52 show multiaxis layouts. Many books suggest that multiaxis arrangements are to be avoided or that they are unacceptable. From a purely theoretical point of view, this may be true. However, from a practical standpoint, multiaxis arrangements are

FIGURE 18–52
Component layout. (Courtesy Triplett Corp.)

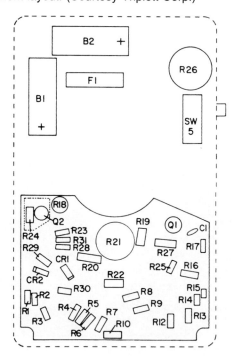

not only common but may actually reduce overall board size, shorten conductor length (by providing the shortest conductor path), and provide optimal component and conductor connection. The obvious drawbacks include problems in mechanized assembly, hole-to-grid alignment, and cluttered appearance.

Where all axial lead components are the same size and have the same lead spacing, their bodies should be oriented in the same direction (axis). This will ensure the most economical and efficient application of mechanical assembly.

It is good practice to follow automatic component insertion design guidelines when possible. Heavy components should always be placed near board supports. The direction of airflow and the requirements for isolation or heat sinking of heat-generating components must be considered during placement since they affect component location and orientation.

Polarity marks, pin patterns, and reference designations for like components should not be put on the layout until the interconnection stage. This allows greater flexibility when connections are routed.

Many designers prefer using color coding to facilitate circuit interconnection on the design layout. The most common color coding practice is to use different colors for connections on each board circuit layer. Red and blue lines work especially well when the artwork is prepared with the **red and blue taping method.** Color coding can also be used to designate power and ground connections, conductor widths, and terminal area sizes. For single-layer boards, the circuitry should always be placed on the side opposite the components. The board acts as an insulator between the components and circuitry, and this practice allows greater flexibility when connections are routed. If many unresolved crossovers remain after several attempts at rearrangement, try using a multilayer design. However, if only a few crossovers cannot be eliminated, wire jumpers can be used, as shown in Fig. 18–53.

Access to connector strips should be taken into account during the preliminary layout stage of the board, as shown in Fig. 18–54. Consideration of access at an early stage in the design layout process will reduce or eliminate the need for jumpers and crossovers.

On multilayer boards, interfacial connections (sometimes called vias or *feedthroughs*), consisting of plated-through holes with or without component leads, are used to transfer the circuit to the opposite side of the board to resolve crossovers. For maximum reliability, try to keep common connections on a single board surface. On digital logic board layouts it is a common practice to route all circuitry on one side of the board perpendicular to the circuitry on the opposite side. Although this practice requires the use of more vias, it

FIGURE 18–53
Jumpers. (Provided courtesy of Bishop Graphics, Inc.)

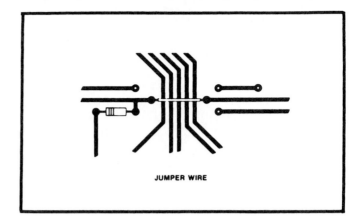

JUMPER WIRE

permits maximum routing flexibility, improves reliability, and typically increases circuit packing density.

When all interconnections have been successfully completed, the layout should be refined to enhance producibility. Excessive via holes, traces between pads, and long conductor runs should be eliminated wherever possible. Conductor widths and spaces should be enlarged wherever space permits. These refinements make the board easier to produce and therefore more reliable and less costly.

Drafting Methods

There are several techniques for doing the actual layout. The most primitive is to painstakingly draw each component from dimensions, along with the interconnecting circuitry. Due to the inherent trial-and-error techniques of PC layout, this is not very practical.

FIGURE 18–54
Access consideration for component layout. (Provided courtesy of Bishop Graphics, Inc.)

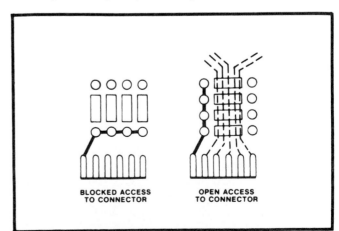

BLOCKED ACCESS
TO CONNECTOR

OPEN ACCESS
TO CONNECTOR

FIGURE 18–55
Layout with a template. (Provided courtesy of Bishop Graphics, Inc.)

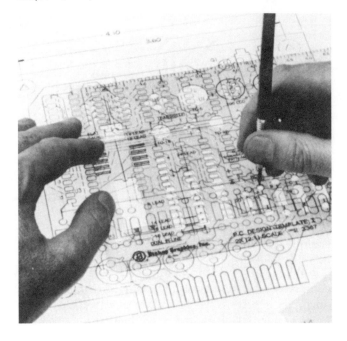

FIGURE 18–56
Printed circuit board templates. (Courtesy of Berol RapiDesign)

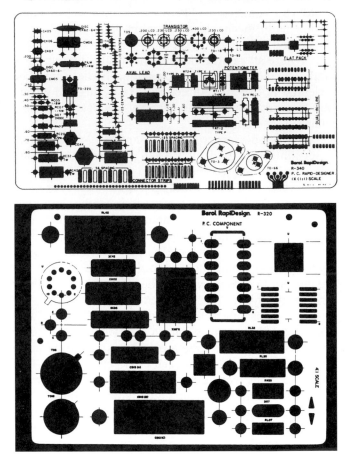

An alternative method is to use a template to draw the components, as shown in Fig. 18–55. Templates (as shown in Fig. 18–56) are available with most common types of component outlines and in a variety of scales. Changes and rearrangements are still time consuming, however, and often the particular layout must be discarded and begun anew.

Another method is to use component outline *dolls*. These are component shapes drawn on paper, cardboard, or drafting film and cut out with scissors. This method, used in conjunction with an interconnection overlay, allows convenient relocation of components, without redrawing the component outlines, until the design is final. There are several drawbacks to this technique. First, after the design is complete, the layout must still be drawn with a template to obtain a stable permanent record. Second, the dolls are difficult to handle and may be easily mislaid, damaged, or knocked askew.

One of the most common methods is to use preprinted component outlines such as Bishop Graphics' PUPPETS™, shown in Fig. 18–57. This is a printed circuit layout system based on the component dolls concepts. However, the traditional disadvantages have been eliminated and several benefits added. A wide assortment of preprinted and die-cut electronic component outline shapes and component type cross-reference charts is commercially available, as shown in Fig. 18–58.

Preprinted component outlines are positioned directly on the glossy surface of a precision grid or on a clear overlay, as shown in Fig. 18–59. They adhere to any shiny (nonmatte) drafting film surface without adhesive.

FIGURE 18–57
Layout artwork with PUPPETS™. (Provided courtesy of Bishop Graphics, Inc.)

FIGURE 18–58
Component assembly outlines.

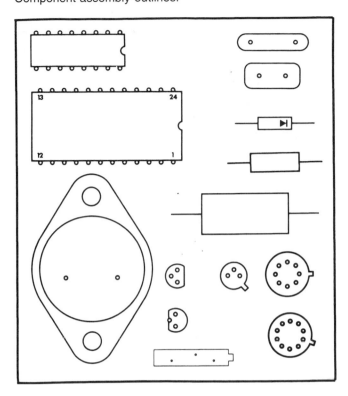

Interconnections are made on matte finish overlay sheets pin registered or taped over the component outlines and grid pattern.

Changes and refinements are easily accommodated by repositioning the PUPPETS℠ and revising or re-

FIGURE 18–59
Layout using dolls. (Provided courtesy of Bishop Graphics, Inc.)

drawing only the interconnection overlay. Permanent reproducible file copies, checkprints, and component assembly drawings can be prepared from the PUPPETS℠ layout using contact printers, diazo reproduction machines, and some office copiers.

Whichever basic drafting technique you use, it is a good idea to begin with the component shapes on one sheet and the interconnections on either overlays or copies of the component layout to minimize rework.

Layout Checklist

The four checklists that follow will help ensure that you have completed all necessary steps in designing a PCB layout.

General

1. Have layout grid, scale, and number of layers been indicated?
2. Have you checked electrical continuity using copies of the layout and the schematic or logic diagram by marking out each connection with a colored marker on both copies?
3. Have all mechanical dimensions pertaining to board size, mounting locations, cutouts, and clearances been indicated?
4. Is the board size compatible with photographic and fabrication equipment capacities?

Components

1. Have all components on the schematic been included, and are they properly designated?
2. Are all component shapes in the proper scale?
3. Have components been located in an orderly fashion?
4. Have standard lead spacings been used for like components?
5. Has orientation been indicated for multiple lead and polarized components?
6. Is there adequate clearance between components and board edges, mounting hardware, and other components?
7. Have special mounting requirements been accommodated (insulation, heat sinks, supports, hardware, etc.)?
8. Is there adequate access for components requiring replacement or adjustment after installation?

Holes

1. Are all holes located on a standard grid or dimensioned from a grid location?
2. Has a separate hole been provided for each component lead or terminal?

3. Have all hole sizes and types been indicated, and do they meet design requirements?
4. Does all hole-to-hole spacing meet requirements?
5. Are all via holes clear of component bodies or other obstructions?

Conductors

1. Have all conductor widths and spacing been properly indicated, and do they meet design requirements?
2. Have all terminal area sizes been properly indicated, and do they meet design requirements?
3. Have conductors been routed in the most efficient manner (smooth flow, short as practical, minimum quantity of jumpers, etc.)?
4. Have critical circuit points been accommodated (conductor length, shielding, isolation, ground planes, voltage planes, heat sinks, etc.)?
5. Has the difference between conductors on separate layers been indicated clearly?

Ground Planes

A *ground plane* is a continuous conductive area used either as a common reference point for circuit returns, signal potentials, and shielding or as a heat sink. Ground planes are tied to the ground circuit and are separated from other traces and pads. Also, high-frequency digital circuits have ICs and circuits that terminate in continuous copper conductor areas called planes.

Ground planes on the noncomponent side of a PCB larger than a .50-in. (12.7-mm) circle are sometimes broken up (relieved) into a striped or checkered pattern, as shown in Fig. 18–60. Here, the nonconductive area is approximately 50% of the conductive area. This prevents blistering and warping during soldering operations. The use of grid and strip relieved ground plane areas depends on the use of wave or hand soldering of the components. Where the PC board is to be hand soldered or the ground plane is on the component side of the board, the use of ground plane relieving is normally not necessary, as shown in Fig. 18–61. Ground plane relieving is used when wave soldering is used and the ground plane is on the noncomponent side of the board.

In some designs the ground plane is continuous, as shown in Fig. 18–61. Note that the conductor areas are numbered from 1 through 14. Instead of having taped conductors, the artwork for this PC board was created when the nonconductive areas of the board were removed after the entire layout was covered with a continuous ground plane. The conductor area is greater than the nonconductive area of the board.

Adequate clearance should be provided around

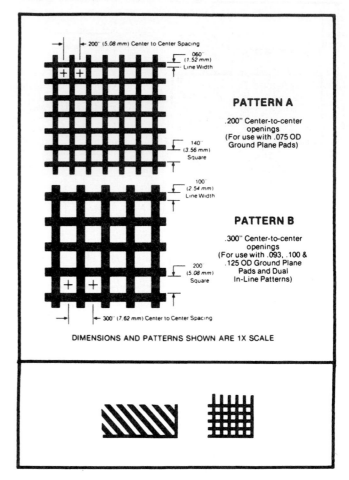

nonfunctional terminal areas in ground planes on external board layers, as shown in Fig. 18–62. On internal layers of multilayer sandwich boards, use of a nonfunctional terminal area is not necessary if a diameter (usually the size of the terminal area) is left clear around the hole, shown in Figs. 18–61 and 18–63.

Figure 18–63 shows the artwork for the fourth layer (as viewed from the component side) of a ten-layer PC board. The master artwork was generated by computer-aided design (CAD) and plotted on the Gerber Plotter #6240.

When a hole terminates in a ground plane and electrical continuity is required, a terminal pad should be used. Clearance should be provided between the terminal pad and ground plane for two to four connections to preserve circuit continuity, as in Fig. 18–64. This prevents *heat sinking* of the terminal area during soldering operations, which can result in an inferior solder joint.

FIGURE 18-61
Printed circuit board showing component outlines.

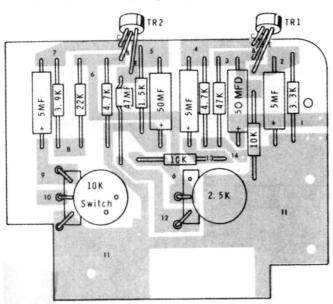

FIGURE 18-62
Clearance for nonfunctional terminal areas in ground planes. (Provided courtesy of Bishop Graphics, Inc.)

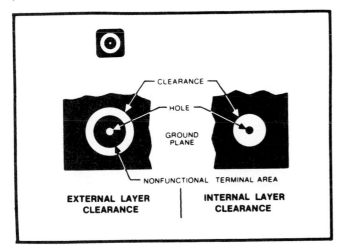

FIGURE 18-63
Front view, component side. (Courtesy of Motorola, Inc., Semiconductor Products Sector)

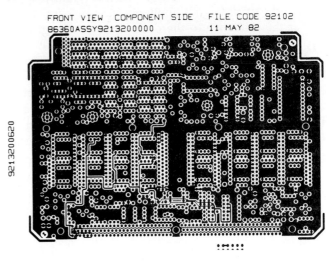

FIGURE 18-64
Hole termination in a ground plane. (Provided courtesy of Bishop Graphics, Inc.)

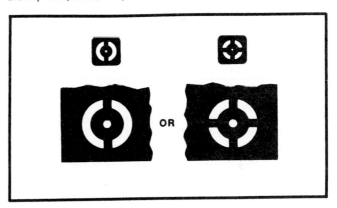

FIGURE 18-65
Ground planes from ground plane grid strips. (Provided courtesy of Bishop Graphics, Inc.)

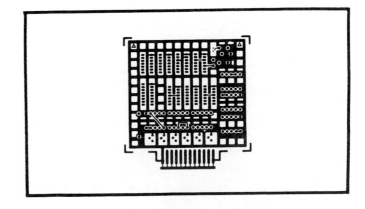

Applying Ground Plane Grid Strip and Stick-ons. Bishop Graphics' Ground Plane Grid Strips® or other commercially available graphic aids can be easily used to form PC ground plane areas wherever needed. Ground plane pads and other PC artwork patterns can be used with ground plane grid strips, as shown in Fig. 18-65.

The following steps show how to apply commercial ground plane artwork patterns:

1. Place artwork requiring a ground plane over a PC layout or padmaster (emulsion side down). Carefully

remove the backing and position the ground plane strip in desired areas on the artwork. Center pads in open grid areas as much as possible. Gently apply the grid strip. Do not apply pressure or burnish.

2. Cut the grid strip to the desired size of the ground plane area. Carefully trim away any portion of gridded ground plane that interferes with pads or circuit connections (this can be quickly accomplished with an X-Acto® knife or compass cutter). Be careful to cut only grid strips and not the artwork underneath.

3. Apply PC drafting aids where desired. Artwork is finished, complete with ground plane areas, as shown in Fig. 18–66.

For PC board layout using CAD, the artwork is photoplotted. The PC board shown in Fig. 18–66 was photoplotted. Note the use of continuous ground plane areas. This PC board artwork is one layer of a double-sided board.

Planes on the component side of a double-sided PCB are normally left with a continuous conductive area since wave soldering will not affect this area. The noncomponent side of the board should be broken into ground plane pattern areas as described above.

Ground planes can also be created photographically with red and blue tape. This method is faster and cheaper than the manual method described previously. The ground plane area is applied first and the red and blue conductor traces last, the opposite of the manual method.

Solder Masks. *Wave soldering* has become a common method of attaching components to printed circuit boards. With this technique a *wave* of molten solder is passed over the circuit (noncomponent or solder) side of the board, making all solder connections. The solder comes in contact with the component leads and the solder side of the board, thereby fusing them to the conductor foil. The board is actually passed through a wave of molten solder at a predetermined rate, with the board *floating* over the molten-hot solder.

To prevent *bridging* (shorting of adjacent conductors with excess solder), a polymer coating called a solder mask is applied to the board covering all conductors except terminal pads, connector lands, and test points. Bridging is a common problem on high-density boards with close spacing of conductors and thin conductor widths.

In addition to the prevention of bridging, solder masks provide physical circuit protection and insulation, cause less solder to be applied, and minimize metallic contamination of the run-off solder. The solder mask material is applied to the board by either screening liquid resist or imaging photosensitive coating or lamination film. To allow for processing tolerances, the terminal pad clearance areas should be .010 in. (.25 mm) to .040 in. (1.02 mm) larger than the terminal areas.

In Fig. 18–67, the pad areas of the layout drawing are being covered. This new drawing will be used to create a solder mask. The traces on the noncomponent side will be completely masked when the layout with the covered pad areas is used to create the solder mask drawing. A negative of the covered layout drawing is used as the solder mask.

PC ARTWORK STORAGE AND HANDLING

During the time that artwork is in progress, the temperature and humidity should be controlled to minimize any

FIGURE 18–66
Completed ground planes. (Courtesy TRW LSI Products)

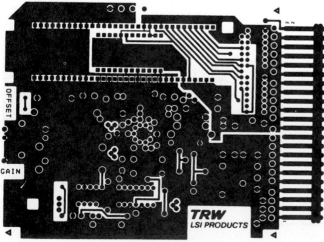

FIGURE 18–67
Solder masks. (Provided courtesy of Bishop Graphics, Inc.)

variation. If rapid or extreme changes take place, allow enough time to restabilize the material to the original conditions.

Observe good housekeeping practices—keep all liquids away from your artwork.

Use an antistatic film, and keep masters free of dust, lint, and erasure residue. Keep hands clean—grease keeps tape from sticking. Remember that the camera records everything—fingerprints, smudges, knife cuts, and scratches.

Always cover artwork with a protective sheet whenever it is not being worked on.

Never roll or hang master PC artwork. All plastic films flow and stretch. Always pack and store flat, and ship in flat containers.

Never store original taped artwork over extended periods of time. Contact copies should be made and the originals destroyed.

Never run the original taped-up PC artwork through any machine because you risk the destruction of the artwork! Suggestion: If you do not have a contact printer, make a *sun print* by putting the artwork and Diazo contact material on a flat surface and exposing it to direct sunlight.

Master Pattern

A master pattern is a one-to-one scale circuit pattern that is used to produce a printed circuit board within the accuracy specified on the master drawing. In cases where the artwork is produced at actual size, the master artwork and master pattern are identical. One of the most important steps in circuit board photofabrication is the artwork photography, since the quality of the finished part depends directly on the quality of the master pattern used to produce the part.

The master pattern should contain the following minimum information:

1. The master pattern should be identified with the master drawing number and the number of the layer. On multilayer boards, the patterns are identified as component side (front) and solder side (back). On multilayer sandwich boards, the layers are numbered progressively from the component side.
2. The applicable revision letter should appear on the master pattern and the PC board. Master patterns should also be serialized so that all parts can be easily traced whenever more than one copy exists.
3. The temperature and humidity at the time and place of creating the master pattern should be noted on the master pattern outside the board area.
4. A note should be included stating that reproductions are for reference only, unless the process and material used ensure the required dimensional stability.

Materials

Materials acceptable for use in preparing the master pattern should have the following characteristics:

1. Base material must be dimensionally stable and easy to handle and process, such as Accufilm® or photographic-quality glass.
2. The photosensitive coating should have high contrast and resolution.
3. The photosensitive coating should be stable.

Camera

PC artwork is photographically reduced on a graphic arts process camera. This type of camera is explicitly designed for copying line or halftone material. It must be capable of repeatedly producing the same photoreduction. The camera must be dimensionally stable and have distortion-free optics.

Master Pattern Inspection Checklist

The master pattern is a production tool. It should be inspected and controlled in the same manner as other precision tooling.

1. *Size:* Each master pattern should be measured at the photographic reduction marks and at least one other dimension at right angles to the first measurement to ensure proper reduction and size.
2. *Registration:* All patterns required to produce a printed circuit board must be inspected together to ensure that registration as stated on the master drawing has been achieved.
3. *Sharpness:* Image sharpness and resolution are necessary to prevent ragged conductor edge definition on the etched printed circuit board. Sample conductor line width and spacing measurements must be made to verify proper exposure and development.
4. *Density:* The image on the master pattern must be opaque enough to ensure that proper exposure of the photosensitive coating can be made.
5. *Quality:* All master patterns should be inspected for pinholes, scratches, fingerprints, and general cleanliness. All touch-up work should be made on the nonemulsion side of the film.

ARTMASTER DRAWING

The artmaster or *artwork* drawing, shown in Fig. 18–68, is used to produce the *to scale* master pattern artwork. Unless the artwork is produced on a CAD system, it is normally done at an enlarged scale of 2:1 or 4:1. Other scales, such as 10:1 or larger, are also used, depending on the eventual reduction size of the PC board. The smaller the finished product and the greater the need for holding extremely high tolerances, the larger the original artwork scale must be.

The artmaster shows all component pads and conductor patterns to be photographically reduced and printed on the copper clad board. The etching process removes all but the desired conductor paths and component pads (Fig. 18–68).

The required conductor paths, the component pad locations, and the board geometry are all determined before the actual tape-up of the artmaster takes place. Precision grid film is used for the placement of the component pads. The artmaster is laid out on a base material of .005- to .007-in.-thick film, which provides a dimensional- and temperature-stable base medium necessary for accurate artwork creation. The component pads are then connected as designated on the layout drawing and trace sketch with precision slit tape to represent the actual conductor paths between components. The engineer or designer must establish the mass of conductive material specified by calculating the total conductor area represented by the component pads, taped conductor traces, and other conductor areas such as ground planes.

Layout Methods

Four separate methods of artwork layout are commonly found in industry practice:

□ Separate-layer method
□ Three-layer method
□ Red and blue method
□ CAD method

FIGURE 18–68
Printed circuit master pattern.

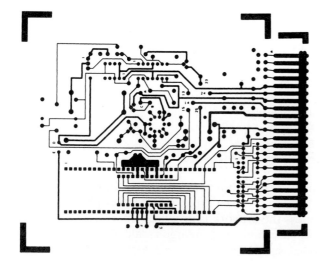

The *separate-layer method* uses a separate layer for each side of the PC board. This method requires that the component pads be spotted on each layer and then connected with the appropriate conductor traces on each side.

The *three-layer method* is similar to the separate-layer method, since separate sheets must be laid out for each side of the board. The difference lies in the use of a separate pad pattern. This method eliminates the need to spot pads on each layer of the artwork. This method is more accurate since the conductor paths for each layer connect to the same pad. Each layer of the board is laid over the pad master in a registration system as described previously.

In the *red and blue method,* different-colored tapes are used for opposite sides of the board. Only one artwork drawing needs to be created for this method. Different layers, or *sides,* are created photographically when first one color and then the other color are filtered out. This method produces perfectly registered sides for double-sided boards. The actual taping methods for the red and blue method require more attention to the taping process and accurate placement of the conductor traces, along with proper cutting and alignment of the pads and tapes, since two different colors of tape and two separate conductor paths are being laid out on the same film base.

The *CAD method* is described later in this chapter.

Regardless of the manual **taping method** used, separate layer, three layer, or red and blue, the actual taping techniques are identical.

Application Techniques for PC Artwork Drafting Aids

The introduction of pressure-sensitive drafting aids has had significant effect on PC artwork preparation. Use of precision manufactured, pressure-sensitive patterns makes it possible to produce extremely accurate photo tools while saving a considerable amount of drafting time and effort. The basic pressure-sensitive drafting aids used for printed circuit artwork are (1) precision slit, PC artwork tape; (2) precision die-cut, tape shape symbols; and (3) precision imprinted artwork pattern shapes. All these products produce photographically opaque images of superior quality.

The built-in accuracy of precision-manufactured symbols and tapes must be accompanied by accurate, professional application techniques.

Applying Multipad Artwork Patterns

The use of multipad component drafting patterns like the ones in Figs. 18–69 and 18–70 to produce master

FIGURE 18–69
Dual-in-line package patterns. (Provided courtesy of Bishop Graphics, Inc.)

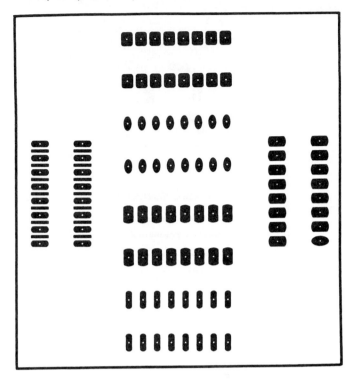

FIGURE 18–70
TO pad patterns. (Provided courtesy of Bishop Graphics, Inc.)

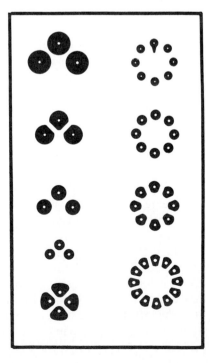

artwork is quicker and more accurate than to lay down individual pads for multipad configurations since they are manufactured to precisely toleranced specifications. Multipad configurations are recommended for all applications, particularly where numerically controlled drilling is indicated.

Multipad patterns are mounted on specially treated release liners. To remove a pattern, hold the backing paper and slip a knife blade under one edge of the pattern. While holding the pattern against the knife blade with a finger, gently peel the stick-on away from the backing, as shown in Fig. 18–71 (A).

For small symbols, use the knife blade as a holding tool and position the pattern over the artwork [Fig. 18–71 (B)].

Do not press the pattern so that, if repositioning is necessary, the pattern can be easily lifted and moved to the correct position.

FIGURE 18–71
Applying multipad stick-ons. (Provided courtesy of Bishop Graphics, Inc.)

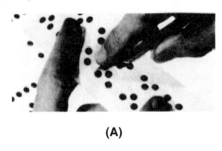

(A)

(B)

(C)

For large patterns, such as dual-in-line (DIP) patterns (Fig. 18–69), hold with one edge adhered to the knife blade, the opposite edge held in the fingers of the other hand. Line up the pattern on the side held with the fingers and gently apply, laying down the remainder of the pattern in a slow, smooth movement while removing the knife [Fig. 18–71 (C)].

When the pattern is properly positioned, hold its edge down with the forefinger of the free hand without pressing down hard and remove the knife.

To reposition a pattern, carefully slide a knife blade under it and peel the pattern gently up toward you. Realign as required.

Once the pattern is in the precise position desired, it can be affixed firmly and permanently to the surface by applying an even, gentle pressure over the entire surface of the pattern. No air bubbles should appear under the surface.

Applying Connector Strips and Prespaced Pads

Insertion connector patterns are available in a variety of standard configurations, as shown in Fig. 18–72. To apply the connector strip, first count the number of contacts or pads needed, cut off the excess, and save it for later use. Remove the release liner, and carefully position the strip over the layout.

Apply one end of the strip to the artwork, positioning precisely in the desired location (Fig. 18–72). After verifying the starting position, gently apply the remainder of the strip, working from the applied end. Low-tack adhesive allows repeated repositioning until exact placement is made.

Even the toughest material can be stretched. Remember to lay the connector patterns straight and flat without stretching in order to maintain the proper accuracy.

FIGURE 18–72
Applying connector strips. (Provided courtesy of Bishop Graphics, Inc.)

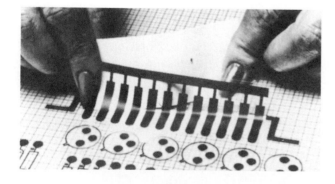

FIGURE 18–73
Conductor path tape aids: 90°, 45°, and 30° angles; radius; elbow; tee; and sample of tape widths. (Provided courtesy of Bishop Graphics, Inc.)

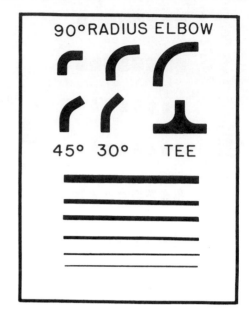

Applying Slit Tapes

Conductor paths are established with precision-cut tape. A wide variety of standard tape widths and tape aids, elbows, and tees is commercially available, as shown in Fig. 18–73.

When applying slit tape for pad-to-pad terminal interconnections, grasp the end of the tape between an X-Acto® knife and the forefinger and apply over the center of the pad area, as in Fig. 18–74 (A). While holding down the fixed end, begin unrolling the tape. With short conductors, run the tape over the center of the second pad area, and press down to establish the conductor. After cutting the roll end, press the tape down evenly with the finger along the entire length of the strip to assure an even surface adhesion, and affix it permanently.

With longer conductors, the finger should be run along the tape to affix it after the conductor is established but before it is cut at the pads [Fig. 18–74 (B)]. Whenever tape is being applied, care should be taken to ensure that stretching does not occur. Tape should be pressed into position without stress.

When cutting tape over a pad area, use a *fixed cut* to avoid cutting the pads. Hold the knife edge firmly in a fixed position in a straight line across the width of the tape. Pull the tape up and along the blade with the other hand, making sure the tape is pulled back at an angle to the knife edge to assure a clean cut, as in Fig. 18–75 (A).

FIGURE 18–74
Applying slit tape. (Provided courtesy of Bishop Graphics, Inc.)

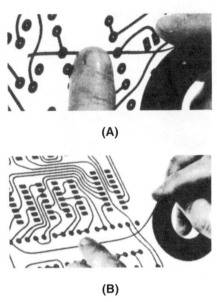

(A)

(B)

Curves.   When a narrow tape is used, the direction of conductor paths can be changed if the tape is formed into a smooth curve. When making a curve, press the tape down firmly up to the point where you want to change direction [Fig. 18–75 (B)]. Gently apply tension

FIGURE 18–75
Cutting tape and laying out curved corners. (Provided courtesy of Bishop Graphics, Inc.)

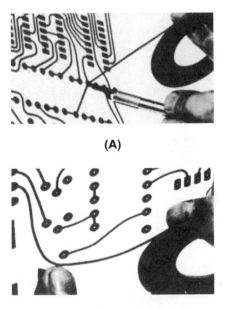

(A)

(B)

on the tape in the new direction while forming the curve with a rocking motion of the index finger of the other hand. For relatively small radius curves, this may be repeated several times before a smooth corner is formed.

Care should be taken to prevent excessive buckling on the inside edge or stretching on the outside edge, either of which can affect the width of the conductor or cause creeping of the tape on a curve. The minimum attainable bend radius varies with tape width and material. A minimum bend radius equal to approximately twice the tape width can usually be attained with practice and experience.

If you cannot make smooth curves by simply bending the straight tape during application, you may use either elbows with angles of 30°, 45°, 60°, or 90° or universal corners (Fig. 18–73). Universal corners (circular tape shapes) provide not only various bend radii, but also numerous angles when the appropriate circle is cut to the desired angle. Use commercial precut elbows and universal corners when a close or consistent radius is desired or when using the red and blue tape method.

Precision-slit artwork tapes may be squared or angled with a knife to form neat, accurate angled corners for conductor direction change, as in Fig. 18–76.

Consistent Conductor Spacing.   A quick way to maintain equal spacing between parallel conductors and terminals is to use *spacer* tape. Lay down a spacer tape run to exactly the width of your spacing increment. Then lay the conductor tape down next to the spacer, using it as a guide. Pull up the spacer and you have exact, consistent spacing. This works especially well with nested elbows and universal corners.

When spacing is at a premium, it is sometimes advantageous to narrow the width of a conductor from nominal down to minimal for a short distance to main-

FIGURE 18–76
Taping angled corners. (Provided courtesy of Bishop Graphics, Inc.)

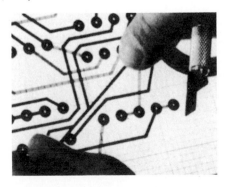

tain proper spacing. Remember—this reduces the current-carrying capacity of a conductor.

Preferred Taping Procedures. A variety of preferred taping procedures is shown in Fig. 18–77. Unacceptable procedures are also shown in this figure. Accuracy requirements for PC artwork are so high that the taping and pad locations are checked with precision scales, as in Figs. 18–78 and 18–79.

FIGURE 18–77
Artwork pattern configuration and taping techniques showing good and poor practice. (Provided courtesy of Bishop Graphics, Inc.)

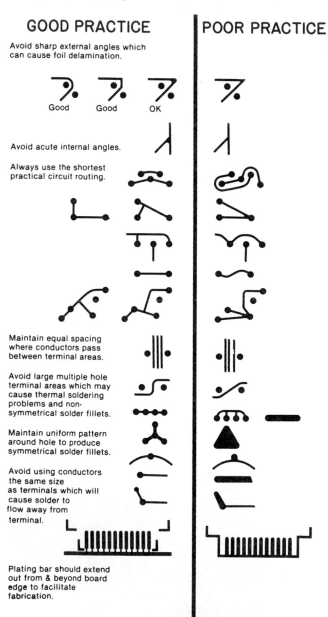

FIGURE 18–78
PC designer checking artwork for accuracy using an opto-scale. (Provided courtesy of Bishop Graphics, Inc.)

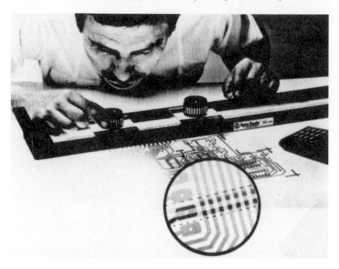

The following simplified standards are suggested for proper and accurate taping of conductor traces:

1. Conductor angles should be confined to standard angles of 45° or 90° whenever possible.
2. Traces running in the same general direction should

FIGURE 18–79
Accurate PCB taping. (Provided courtesy of Bishop Graphics, Inc.)

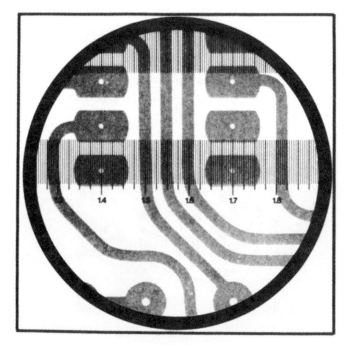

be laid out parallel (parallel and at the same angle when not horizontal or vertical).

3. To eliminate problems during the etching process, avoid taping conductors at an angle of less than 90° whenever possible. If internal angles are too sharp, the etching solution may build up and cause over-etching, thereby reducing the conductor width at the corner.

4. Avoid sharp external angles by using curves or double angles in the trace.

5. Keep conductor traces a minimum of .025 in. (6.3 mm) from the edge of the finished size board to ensure space for manufacturing processes, such as board shearing.

6. Avoid cutting pads. Avoid overlapping tape. Each conductor length between pads should be one continuous piece of tape.

7. Do not stretch tape since it may *creep* back to its original length and shape.

Master Fabrication Drawing

A printed circuit board master fabrication drawing presents a complete engineering description of all design features. Information recurring from board to board should always be placed in the same location on the drawing to facilitate communications between departments and between the company and its outside vendors. The master fabrication drawing is normally drawn at the same scale as the film photomaster. It is viewed from the noncomponent side of the board if only one side is to be shown. In Fig. 18–80, both sides of the PC board are detailed.

Content

The printed circuit board master drawing should always include the following information:

1. A reproduction of the conductor pattern image for the solder side of the board. It may also in-

FIGURE 18–80
PCB master drawing. (Provided courtesy of Bishop Graphics, Inc.)

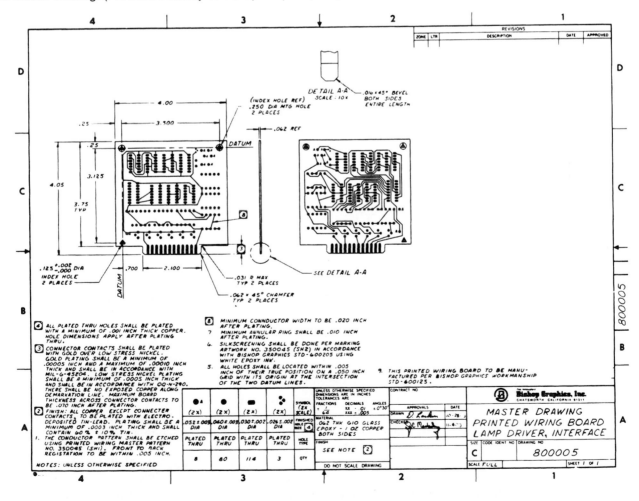

clude the circuit pattern images of all other layers.

2. The delineation of the complete board outline with all dimensions and tolerances necessary for fabrication, including any structural requirements. A complete description of the board geometry, along with dimensions and tolerances, is shown on the master drawing.

3. The complete designation of the material, process, and production specifications that apply to the board. This includes the plating and processing specifications needed for the board fabrication.

4. All hole identification information, including type, finished size, and location. A hole chart may be required. Hole sizes for the PC board are determined during the design and layout stage when the component type, size, and mounting process are decided on.

Layout of Master Drawings

There are several ways to delineate the circuit pattern on a printed circuit board master drawing. One of the simplest methods is to make a diazo or photoprint of the solder side artwork. This print should be made on a flat light table, vacuum frame, or copy board. Never process the original artwork through a machine that has a rolling action or one that emits fumes or excessive heat.

Board geometry along with the circuit pattern is transferred to the master drawing, and dimensions, notes, and specifications necessary for fabrication are included to complete the drawing.

Figure 18–80 shows a typical master drawing. Each PC board has a different set of specifications that must be clearly defined in the master drawing notes. Notes 1 through 8 describe the necessary specifications for the Lamp Driver Interface PC Board. Conductor pattern etching instructions, plating requirements, hole positional tolerances, silkscreen instructions, and minimum conductor spacing are included in these notes.

Material specification must appear on the master fabrication drawing (Fig. 18–80), where the title block has an area for material description:

.062″ THICK G10 GLASS
EPOXY–1 OZ. COPPER
BOTH SIDES

Note 2 (Fig. 18–80) provides specifications for the board's finish.

Hole Location.

There are three fundamental methods for locating holes on printed circuit boards: grid system, datum lines, and terminal pad centers.

Location by a *grid system* is often used to meet military requirements or to have all features on the board located to the grid pattern. This method is essential for automated drilling and component assembly techniques, since digitized grid coordinates are used for locating components. The drawback to this system is the precise accuracy required in

□ Layout grid
□ Location of terminal areas to the grid pattern
□ Photoreduction of the artwork
□ Layer-to-layer registration

Inaccuracy in any of these areas can result in serious pattern-to-hole misalignment.

Hole location by dimension from *datum lines* or X and Y axis coordinates is similar to the grid system method because terminal areas must be precisely placed in reference to the hole location dimensions. In a multilead component pattern, only one hole in the group should be located from the standard datum points or indexing holes. The other holes should be located dimensionally from this reference hole to preserve accuracy within the pattern. Besides the high risk of pattern-to-hole misalignment, this method can be very time consuming for the drafters and the PC board manufacturer.

The simplest method is to locate holes optically from the *terminal pad centers* on the master pattern or actual PC board. This method is usually least expensive and produces the most accurate pattern-to-hole alignment. It should be used whenever precise alignment to external parts or tooling is not a requirement.

A hole chart is also shown on Fig. 18–80. All holes for the PC board have been defined by symbol, size, type, and quantity. The size provides the tolerance parameters, and the hole type specifies if the hole is to be plated-through or not.

Location Tolerance.

The three basic methods of setting tolerances for hole locations are positional limitation tolerancing, coordinate tolerancing, and true-position tolerancing, as shown in Fig. 18–81.

Positional limitation tolerancing is applicable only when the optical alignment to terminal area center method is used for hole location. In defining the location tolerance, limitations are placed on the minimum annular ring of the terminal area. This method should only be used where precise alignment of holes to external parts or tooling is not a factor.

With the *coordinate tolerancing* method, the tolerance is applied directly to linear and angular dimensions, usually forming a rectangular area of allowable variation.

A *true-position tolerance* is expressed as a radius or diameter of allowable variation from the *true-position* center defined by a dimension or grid coordinate.

FIGURE 18–81
Tolerancing hole locations: (A) positional limitation; (B) coordinate; and (C) true position. (Provided courtesy of Bishop Graphics, Inc.)

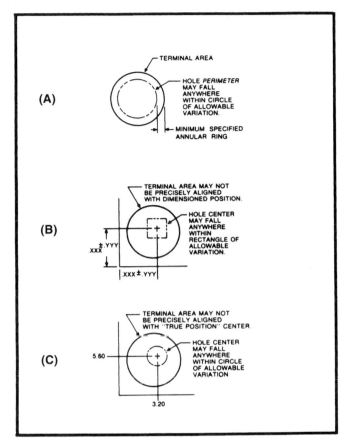

Master Drawing Checklist

The following list should be used to check the completed master fabrication drawing.

1. Has the circuit image been properly shown from the solder side?
2. Have hole types, sizes, and tolerances been properly indicated and referenced to the circuit image?
3. Has a hole location method been clearly defined?
4. Is the complete board outline clearly indicated and properly dimensioned?
5. Have all nonstandard holes (like board mounting holes, component hardware holes, etc.) been properly dimensioned?
6. Are all inside corners of notches and cutouts specified as maximum radii?
7. Has the base material been correctly specified?
8. Have all reference drawings, such as the master pattern and marking drawing, been properly referenced?
9. Are all applicable manufacturing and production specifications and standards referenced in the notes?
10. Has all plating been correctly specified and toleranced for circuitry, holes, and connector contacts?
11. Have processing variance allowances been indicated for registration, conductor width, and minimum annular ring?
12. Has the allowable amount and direction of base material warp been specified if critical?
13. Is the drawing scale clearly indicated?
14. Has the board identification number been specified as either etched or printed on the board?
15. Can the printed circuit board be completely and correctly fabricated from the master drawing without question?

Drill Master Drawing

The **drill master** drawing is used by the fabricator to locate all tooling and component mounting holes to be drilled in the board before assembly. A transparency of the artmaster is used for the drill master drawing, or one can be automatically prepared on a CAD system, as in Fig. 18–82. The drill master can be used to program NC drilling machines and generate NC magnetic or paper tapes, or it can be output directly to an on-line photoplotter. The drill drawing includes all material specifications, plating and baking instructions, board geometry specifications, and the toleranced diameters of all holes to be drilled.

The master artwork for a double-sided board is shown in Fig. 18–83. The drill drawing in Fig. 18–82 is used to locate the holes for this printed circuit board. Note that the Drill Data Table lists the hole symbol, diameter and tolerance, quantity, and plating for each size and type of hole. There are 1376 .029-in.-diameter lead holes in this board.

Marking Artwork

The circuit board artwork may have a separate sheet called a *marking* or *silkscreen* artwork containing component outlines and orientation symbols, as in Fig. 18–84. The marking artwork is viewed from the component side of the board. This drawing is used to mark the printed circuit board before assembly of the components. Component outlines, reference designations, and other necessary markings such as polarity terminations for diodes, are marked on the etched board with a silkscreen, stencil, or stamping process. The markings may be etched along with the circuit.

The marking drawing is prepared from the artmaster on the same enlarged scale, as in Fig. 18–85. Normally the marking artwork is prepared with commercial

FIGURE 18–82
Drill drawing and drill data table. (Courtesy of General Electric Co.)

DRILL DATA TABLE			
SYM	DIA OR SIZE	QTY	PLATING
+	.029 ± .003	1376	
✳	.033 ± .003	17	
⋈	.040 ± .003	—	YES
⊞	.055 ± .003	—	
⊖		—	
⊕	.111 ± .002	4	NO
✕	.125 ± .002	2	
◆	.067 ± .002	—	
▟	.096 ± .003	—	OPTIONAL
⊗	.140 ± .005	—	
⊠	.177 ± .005	—	

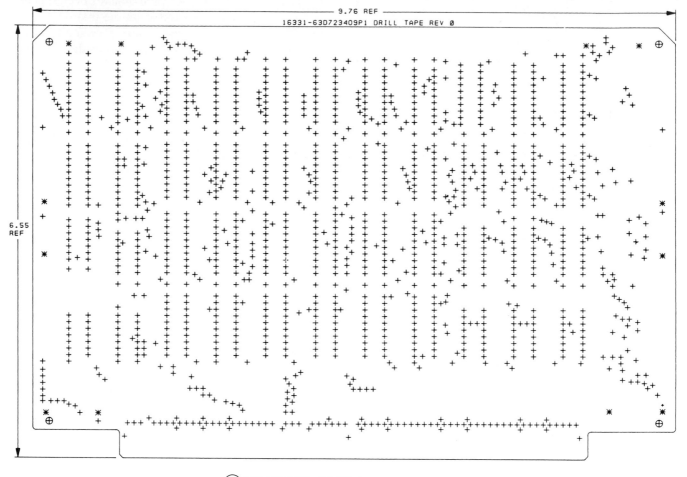

artwork patterns, although inking and Leroy lettering are also used. The marking drawing shown in Fig. 18–84 was generated on a CAD system. Artwork patterns and adhesive lettering are preferred for manually prepared marking patterns and reference designations, since the marking artwork must have clearly defined lineweights and clear, precise lettering for proper photographic processing.

After reduction, this pattern is printed, usually with the silkscreen method, using nonconductive ink on the component side of the board after it is etched and plated. The reference designations should be placed so that they are visible after assembly of components to the board (Fig. 18–84). They should all read from the same edge of the board. If this is not possible, the remaining designations should read from a second edge perpendicular to the first. Capitals must be used for all lettering on the marking drawing. Reference designations must not be smaller than $\frac{3}{32}$ in. (.09 mm) in height after reduction. Lettering must not fall on conductor traces,

FIGURE 18-83
Front and back of PCB artwork. (Courtesy of General
Electric Co.)

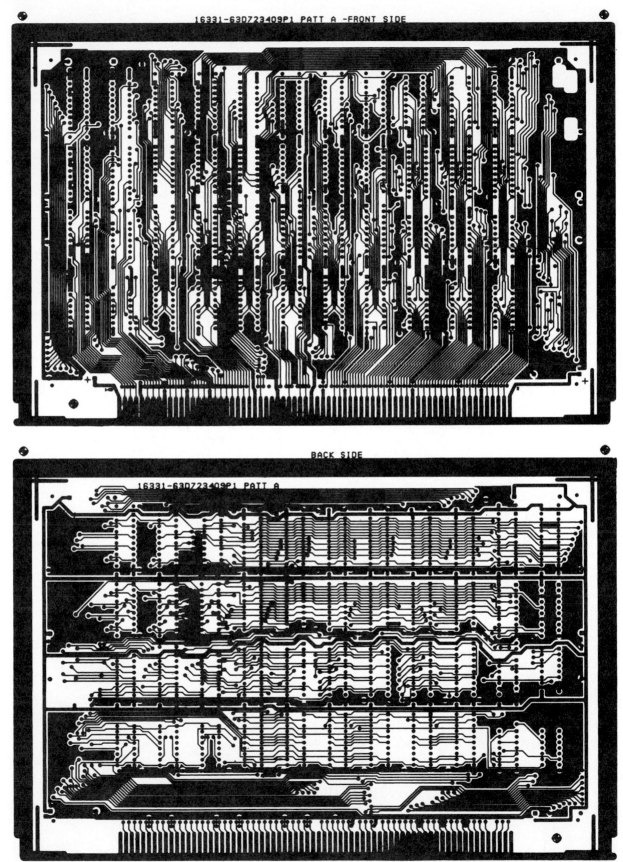

FIGURE 18–84
Marking drawing. (Courtesy of General Electric Co.)

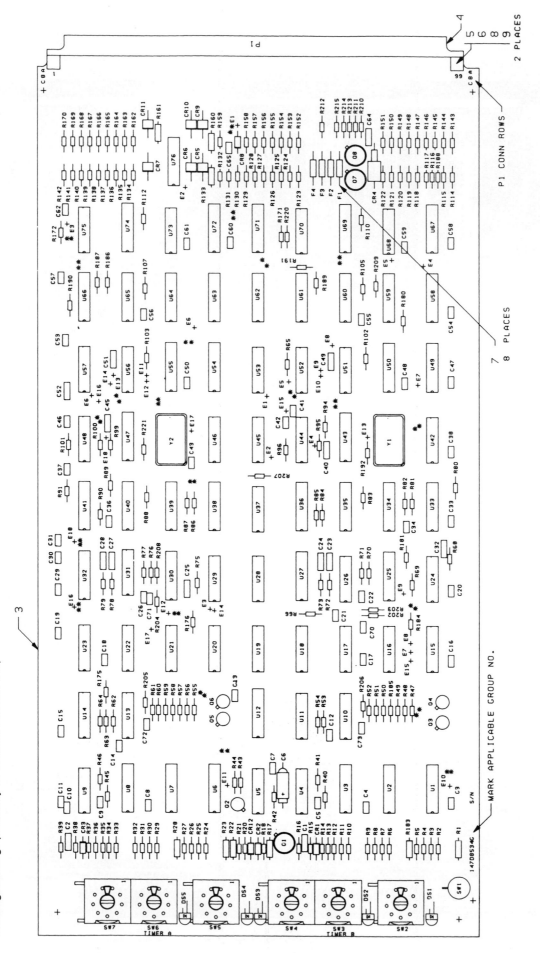

FIGURE 18–85
PCB master artwork. (Provided courtesy of Bishop Graphics, Inc.)

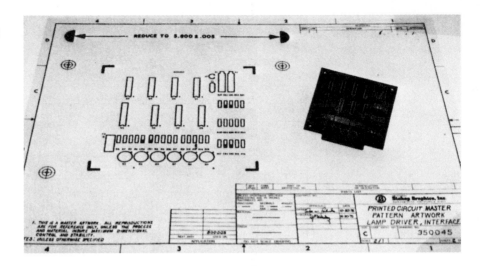

pads, terminals, or areas that require soldering connections. Lettering can be marked on nonconductive areas and on ground planes where they do not interfere with soldering.

Contact numbers and letters should be shown when the board uses board-edge contacts as the module connection method. Component orientation pins, marks, or tabs must be shown on the marking drawing to ensure proper mounting during assembly of the board.

The component outlines must be located in the same positions that the actual components will occupy after assembly, as shown in Fig. 18–86. No part of the component outlines or reference designations should cover a terminal area where solder is applied. Dimensions and other notes should not appear on this drawing. Reference, tooling, or production holes are usually not shown. Only the component outline, component designation, and registration marks are drawn on the marking artwork.

The marking artwork should also be prepared on dimensionally stable polyester film with reference designation, orientation, and component outline commercial artwork patterns. It should include registration marks that align with a part of the pattern that will appear within the board area after it is manufactured, as shown in Fig. 18–87.

Assembly Drawing

The printed circuit board assembly drawing is a complete engineering description of a printed circuit board assembly, including all components and parts mounted by fastening, soldering, or bolting, as in Fig. 18–13. A complete parts list should be included on the drawing or on a separate parts list sheet, as in Fig. 18–88. All parts, other than components, are *ballooned* on the assembly drawing (Fig. 18–89) and listed on the parts list. The assembly should show the board configuration, all components, reference designation markings, all mechanical parts (clamps, fasteners, and brackets), and the dimensions required for accurate assembly and component orientation.

FIGURE 18–86
Marking artwork.

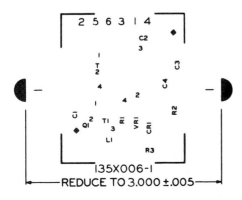

FIGURE 18–87
Photoreduction marks, crop marks, and registration targets. (Provided courtesy of Bishop Graphics, Inc.)

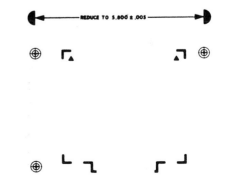

FIGURE 18–88
Board assembly interface parts
list.

	BOARD ASSEMBLY	
	INTERFACE PARTS LIST	
QTY.	DESCRIPTION	PART NO.
2	4-40 x 3/8" screw	250-4
3	#4 x 5/16" self-tapping screw	250-163
2	4-40 nut	252-2
2	#4 lockwasher	254-9
4	6-32 x 1/4" black screw	250-116
1	6-32 x 3/8" screw	250-9
1	6-32 x 3/8" flat head screw	250-32
10	#6 x 3/8" self-tapping screw	250-592
8	#6 x 1-1/8" self-tapping screw	250-1137
2	6-32 nut	252-3
2	6-32 brass insert nut	252-170
3	#6 lockwasher	254-1
1	Terminal strip	431-86
21	Connector block	432-874
2	14-pin IC socket	434-298
1	8-pin IC socket	434-230
1	Power transformer	54-893
4	Grommet	260-700
4	Retainer ring	260-701
4	Plastic foot	261-34

FIGURE 18–89
PCB assembly drawing.
(Provided courtesy of Bishop
Graphics, Inc.)

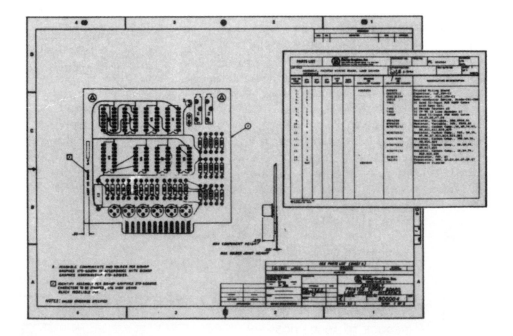

The assembly drawing is drawn as viewed from the component side of the board and usually at the same scale as the artwork drawing (Fig. 18–90).

Mounting procedures, location dimensions for any hardware, and maximum component height should be shown on the assembly drawing. Symbols and designations are to follow the schematic diagram. Jumpers are to be identified and noted clearly to provide easy assembly of the components and wiring, as in Fig. 18–91.

Content

The assembly drawing should include, as a minimum, the following information:

1. A view of the component side of the completed board
2. A complete parts list or bill of materials (often a separate document)
3. All general and specific notes to enable the board to be fabricated, inspected, and tested

The printed circuit board assembly drawing is usually drawn at the scale used in creating the master artwork. However, any easily readable scale may be used. The scale must be noted on the drawing.

The parts list should have the following information for each entry: part number, item description, number of parts required, and the parts list item numbers (Fig. 18–88). For electronic components, the reference designations are used as find numbers and should be included. Other information that sometimes is helpful or required by contract is the Federal Code Identification and drawing zone numbers indicating location.

Techniques

Printed circuit board assembly drawings are easily prepared from a number of contact reproduction methods using either wash-off or diazo photosensitive drafting films.

The artwork or master pattern of the component side of the board and a halftone screen (for example, 85 line, 40%) are used to produce a screened copy of the artwork configuration within a standard drawing format. The component outlines and reference designations are then added in solid black with a pencil and template or preprinted pressure-sensitive drafting aids. The screened artwork pattern provides a reference that allows components to be easily located during assembly and testing. The same basic technique may use an out-

FIGURE 18–90
Assembly drawing.

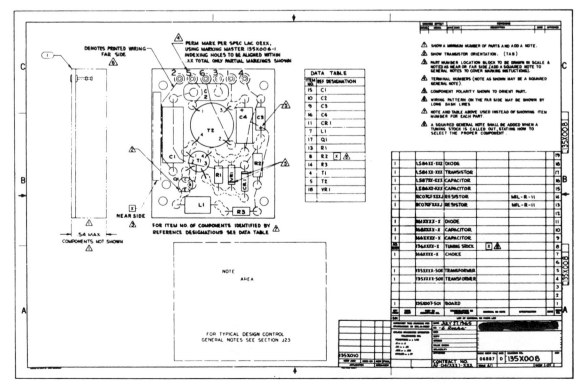

FIGURE 18–91
Board assembly interface.

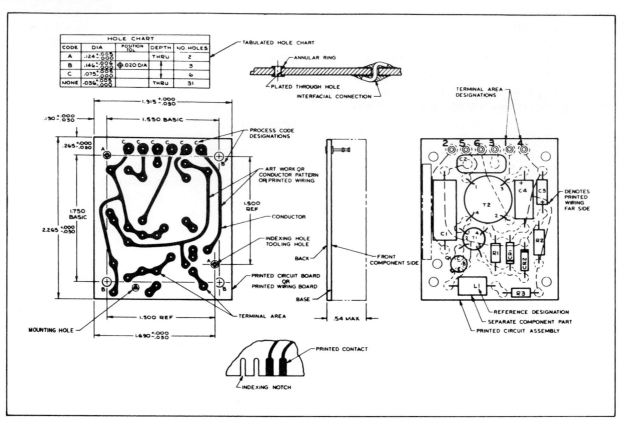

line artwork pattern instead of the screened pattern. This technique uses a spread negative made from a clear film spacer between the master negative and duplicating film.

The spread negative and the original artwork then are used to produce the assembly drawing. An assembly drawing made with the outline method is easier to change or revise.

Minimum documentation requires only that the component outlines be placed in their proper location within a blank board outline.

The quickest, easiest method of preparing assembly drawings is to make a reproducible copy of the component portion of a layout, draw in the board outline, and add the appropriate notes and details, as shown in Fig. 18–92.

Assembly Drawing Checklist

In general, the PC board assembly drawing is the same as any engineering assembly drawing. Parts, materials, and assembly instructions should appear on the drawing. The following list can be used to check the assembly drawing for proper view, procedures, and drawing specifications:

1. Printed circuit board assembly drawings should be drawn viewed from the component side. If components are mounted on both sides, then both front and back views should be shown.

FIGURE 18–92
Laying out an assembly drawing. (Provided courtesy of Bishop Graphics, Inc.)

2. Auxiliary views should be used to clarify any details, such as the precise location and orientation of mechanical parts.
3. Maximum component height or overall board thickness should be included.
4. Electronic components should be identified within the component outline with the reference designation used on the schematic diagram.
5. All parts, other than electronic components, should be identified by ballooned item numbers.
6. The orientation of components (pin number, polarity, and so on), when applicable should be clearly shown.
7. The assembly part number, revision level, and serial number (if applicable) should be clearly marked on the assembled board.

CAD/CAM PRINTED CIRCUIT BOARD DESIGN

This section discusses the use of a computer-aided design (CAD) system as a tool for the printed circuit board drafter and designer.

Lightweight, complex, and densely packed electronic equipment for the space program and for the military requires extremely accurate production of multilayer PC boards. In order to accomplish this task, software packages for automated drafting and design systems were created. A high-level CAD system can now entirely replace manual design and documentation of PC boards.

Registration of multiple layers and the packing of increasingly dense circuitry in smaller areas require very precise artwork. Placement of components for ease of routing and the routing of interconnections make the use of manual design and documentation a slow and sometimes inaccurate process. Only the integration of CAD throughout all phases of drafting, layout, and design can meet the demands of present PCB requirements.

PCB circuit density, multilayer construction (making layer-to-layer connections and registration difficult), and the many narrow interconnect paths have necessitated the use of CAD in the formative design stage and throughout the entire development and fabrication processes. CAD can eliminate the need for photoreduction of the artwork since computers can generate 1:1 art within specified tolerances, unlike hand-drawn artwork, which must be laid out at a 2:1 minimum scale, and normally at 4:1.

Designing a PCB involves a tremendous amount of repetitive work—reproducing the same components in many locations, annotating and dimensioning, component placement and routing, checking and updating de-

signs and documentation. CAD systems can make this work faster and easier.

System Description

Computer-aided design systems come in a variety of configurations. Typically, CAD systems used for PC board design and documentation consist of a central processing unit (CPU), a magnetic tape unit, a disk storage unit, and a keyboard device for data input. They also have at least one graphics display (CRT) console with a mouse or pen, a digitizing tablet with a pen plotter, and a photoplotter for making artwork plots.

CAD systems are available with a variety of capabilities depending on the hardware configuration and the software level. A mid- to low-end system, shown in Fig. 18–93, is an excellent tool for aiding the drafter and designer in the layout and design of a PCB. The PCB design process associated with a typical midlevel system is shown as a flow diagram in Fig. 18–94.

Normally, the design of a typical PCB starts with the creation of the schematic diagram, as shown in Fig. 18–95. Commands are normally available to speed up the layout of the schematic diagram. Commands such as draw junctions, buses, reference designators, and connect packages are usually available. Schematics created on the system can then be plotted and a material list generated, as shown in Fig. 18–96. The material list shows the parts in the schematic and their quantities. The typical materials list contains the reference designator, which is a multicharacter descriptor used to group like components on the schematic. In general, the first letter in the reference designator indicates the compo-

FIGURE 18–93
PCB design system. (Courtesy P-CAD)

FIGURE 18–94
PC board design process on CAD. (Courtesy of Hewlett–Packard Co.)

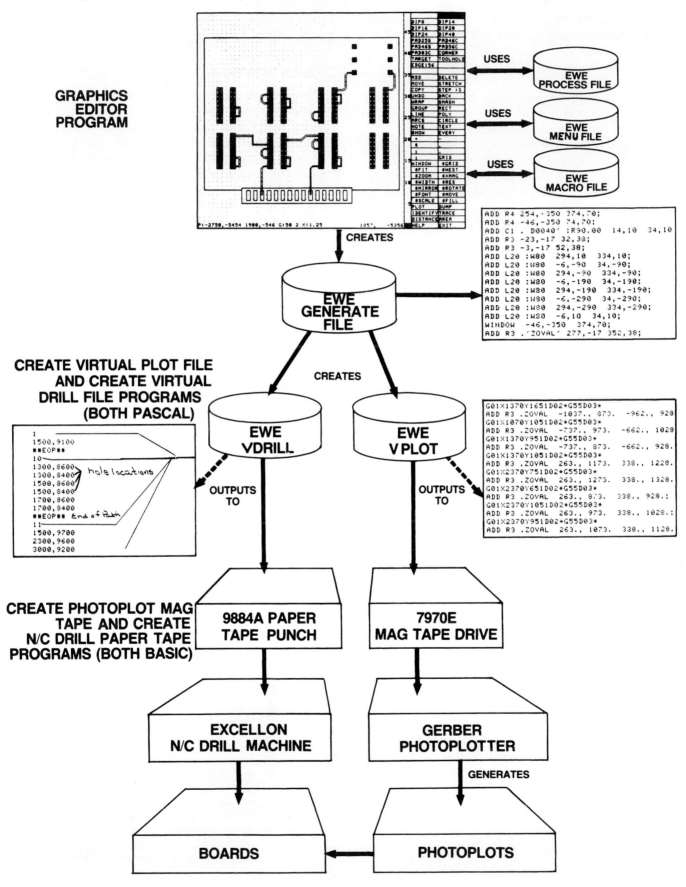

FIGURE 18–95
Schematic drawing of screen. (Courtesy of Hewlett–Packard Co.)

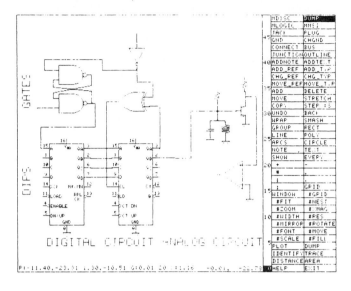

FIGURE 18–96
Example material list sorted both ways. (Courtesy of Hewlett–Packard Co.)

```
                        Material List
                           SCH1
ET12781
BOARD 1

                                      13-Oct-81    PAGE 1

   STOCK NO   QTY         DESCRIPTION           REFERENCE
  ----------  ----  ----------------------  ----------------------

  0180-0100    11   4.7UF 35V CAPACITOR     C1,C10,C11,C12,C14,
                                            C15,C2,C3,C4,C8,C9
  0180-2803     3   100UF 50V CAPACITOR     C5,C6,C7
  0683-1035     1   10K RESISTOR            R5
  0683-3025     1   3K RESISTOR             R2
  0698-6943     2   20K RESISTOR            R3,R4
  0764-0018     2   4.7K RESISTOR PACKAGE   RP1,RP2
  1820-0904     1   93L24  4 BIT MAGNITUDE COMPARATOR  U6
  1820-1195     2   74LS175 QUAD D-TYPE FLIP-FLOP      U10,U11
```

```
                        Material List
                           SCH1
ET12781
BOARD 1

                                      13-Oct-81    PAGE 1

   REFERENCE        STOCK NO           DESCRIPTION
  -----------     ----------     ----------------------

  C1              0180-0100      4.7UF 35V CAPACITOR
  C10             0180-0100      4.7UF 35V CAPACITOR
  C11             0180-0100      4.7UF 35V CAPACITOR
  C12             0180-0100      4.7UF 35V CAPACITOR
  C14             0180-0100      4.7UF 35V CAPACITOR
  C15             0180-0100      4.7UF 35V CAPACITOR
  C2              0180-0100      4.7UF 35V CAPACITOR
  C3              0180-0100      4.7UF 35V CAPACITOR
  C4              0180-0100      4.7UF 35V CAPACITOR
  C5              0180-2803      100UF 50V CAPACITOR
  C6              0180-2803      100UF 50V CAPACITOR
  C7              0180-2803      100UF 50V CAPACITOR
  C8              0180-0100      4.7UF 35V CAPACITOR
  C9              0180-0100      4.7UF 35V CAPACITOR
```

nent type. For instance, resistors may be R1, R2, ... R34; capacitors may be C1, C2, ... C25. The type and value of the component are also listed. Most companies like to assign their own internal inventory control stock number to electronic parts, as shown on the materials list in Fig. 18–96.

FIGURE 18–97
PC board under construction. (Courtesy of Hewlett–Packard Co.)

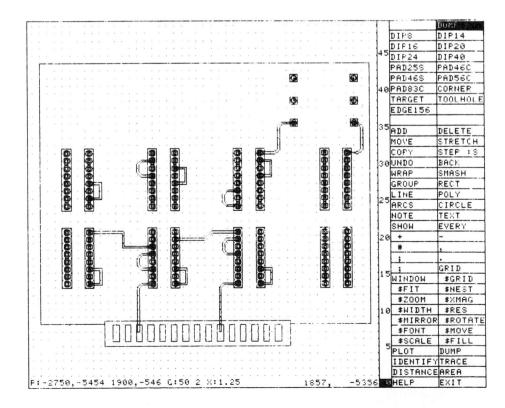

FIGURE 18–98
Example library parts. (Courtesy of Hewlett–Packard Co.)

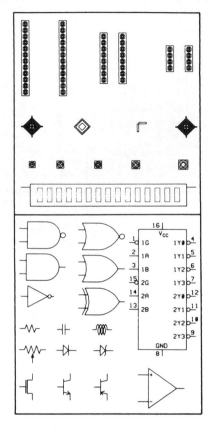

The PC board layout module diagrammed in Fig. 18–94 uses graphics generated from the general drafting software to create PC drawings and magnetic and paper tapes to drive production tooling. This system does not have an automatic component placement, automatic trace routing, or design rule verification option. Basically, this system is a drafting and layout system that replaces the time-consuming, tedious, error-prone methods associated with manual PC board layout procedures. The computer aids the drafter and designer in all areas of documentation and design from original PC board construction, as in Fig. 18–97, to artwork creation and plotting, to the plotting of drill drawings.

PC board software can be used throughout the layout and design process to draw and then generate a variety of documentation. The user can obtain plots to verify accuracy, paper tapes to feed PC board hole drilling NC machines, and magnetic tapes to drive photoplotters. Photoplotters in turn create the finished artwork, which is used for etching the traces on PC boards. Production, assembly, drilling, and other drawings are also generated from the original design input.

The general drawing software of a typical midlevel system allows the user to create an unlimited library of often repeated objects and symbols, as shown in Fig. 18–98. Library parts can be recalled by specific keyboard commands, or they can be inserted from a menu. The system shown in Figs. 18–93 through 18–99 has a screen menu accessed from a digitizer tablet for quick operator

FIGURE 18–99
Graphic editor screen. (Courtesy of Hewlett–Packard Co.)

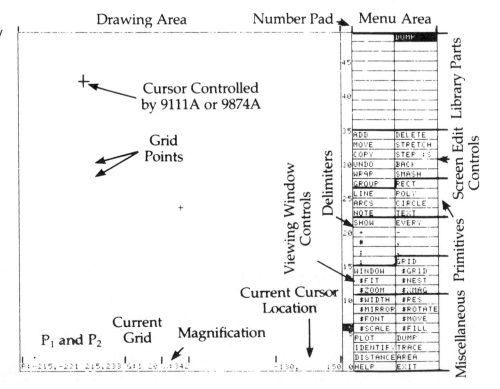

command processing. In Figs. 18–95 and 18–97, the right-hand column of the CRT displays the menu commands. Menu files can contain commands, library part names, and other text, which are displayed in the menu area of the CRT screen. The cursor, which is controlled by the digitizer or data tablet, can access any part of the screen, allowing the selection of commands from the menu area. The digitizer tablet may use overlays with PC components, electronic symbols or whole systems, as defined during the original library file creation.

The screen menu area of this system allows a series of editor commands (Fig. 18–99):

Move: Moves an object to another defined position on the screen

Copy: Duplicates any number of times an object already drawn on the screen, and positions it to the commanded location

Delete: Removes any object on the screen

Stretch: Stretches the vertex of an object on the screen

Rotate: Rotates any library part on the screen for proper orientation defined by the user

Width: Allows lines to be drawn with a specified line thickness

Scale: Adds a library part to the screen scaled up or down as specified

Wrap: Allows the user to enclose portions of objects on the screen and then store them as new library parts

The preceding sample of editor commands is typical to all CAD systems. The command name may be different, but its function is similar to other systems' commands.

PCB DESIGN USING STATE-OF-THE-ART CAD

High-level, sophisticated interactive graphics systems allow the designer almost unlimited freedom in design options and capabilities. The block diagram shown in Fig. 18–100 documents the flow and design sequence of one of these systems, Lockheed's CADAM®. Starting with the schematic diagram creation and ending with output hard copy, the system almost eliminates the noncreative manual part of PCB design, documentation, and manufacturing.

As has been stated, one of the most important and useful aspects of CAD system PCB design is the registration of multilayer boards. High-level systems can color code each layer. The system shown in Fig. 18–101 can differentiate up to eight layers simultaneously with seven different colors.

CAD systems are available with capabilities of 80 layers, hundreds of color options, and other design options. Artwork masters, shown in Fig. 18–102, solder masks, in Fig. 18–103, component drawings, in Fig. 18–104, silkscreen masters, in Fig. 18–105, and pad masters and drill drawings, in Fig. 18–106, can all be generated from the original PC board design data, with most aspects of the process automated.

The board can be drilled using the CAD database directly when a CNC drilling machine is integrated into the process (Fig. 18–107).

The Engineering Schematic Diagram

The first step in the process of designing PC boards is the design and creation of the engineering schematic diagram. The typical engineering schematic is a rough sketch, usually hand drawn by the design engineer, which must then be redrawn in a neat, understandable form by a designer or drafter before it can be used to create a printed circuit board design. This is not only a time-consuming task, but an unnecessary duplication of effort that requires constant interaction between the design engineer and the person responsible for redrawing the schematic. Creation of an engineering schematic is usually accomplished in the following manner: (1) placement of the schematic symbols, (2) interconnection of these symbols, (3) rearrangement of the symbols to form an even flow of data, and (4) annotation of the schematic.

Interactive CAD Approach

The designer sets the stage with the initial creation of a symbol library (Fig. 18–98). This small step is one key to the effectiveness of a CAD system. Storing frequently used symbols—major and minor, simple or complex—in the schematic or board component symbol library greatly simplifies preparation of the schematic and PC board. Once entered in the database, these symbols can be called up immediately and placed at the proper coordinate points on the schematic or board. Since frequently used symbols are now available at the touch of a button, the CAD operator creates the symbols only once and may then access these symbols as often as needed.

The operator can design *on the tube*, using an electronic light pen, menu, and keyboard to enter schematic diagram symbols, characters, and commands. The system responds—in a matter of seconds—to any of these commands. It stores the schematics and associated data while producing visual feedback on the CRT screen.

FIGURE 18–100
Printed circuit board design on CAD/CAM. (Courtesy of Lockheed-California Co.)

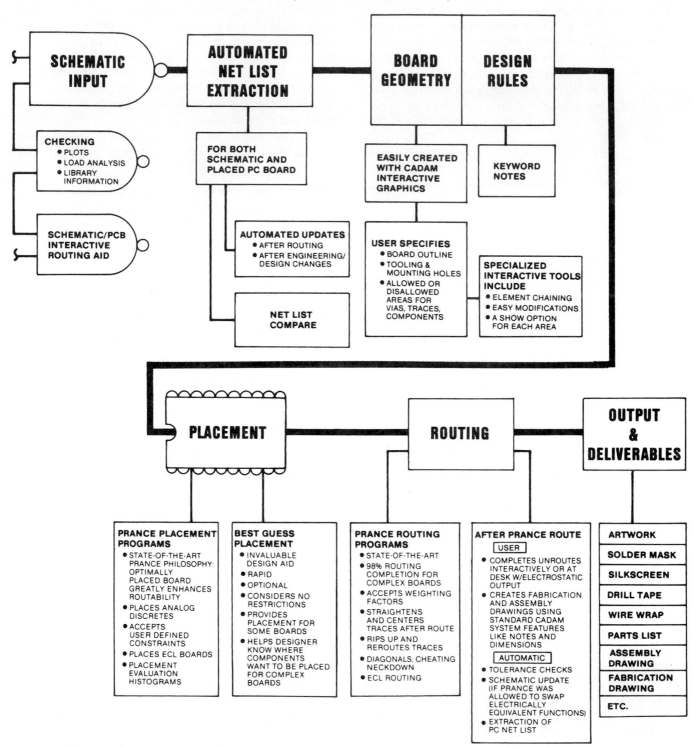

The combination of data tablet, electronic light pen, and graphic display allows the designer to enter schematics with a minimum of time consumed in detail drawing. The same electronic pen and tablet are used for entering symbol retrieval and drafting commands, as well as X–Y positional data. (CAD can also enter schematic data into the system through an on-line digitizer; however, digitizing has not proved to be as rapid a

FIGURE 18–101
GSI's task-oriented PCB CAD system includes a 19-in. color graphics display, which enables the PCB designer to display simultaneously or separately up to eight levels of data in seven colors. (Courtesy of Gerber Scientific Instrument Co.)

FIGURE 18–103
Solder masks generated from the same database as the artwork master. (Courtesy of Gerber Scientific Instrument Co.)

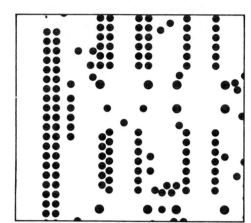

method of initiating schematic creation, although it is still widely used.)

One major time-saving feature of this method of computer-aided schematic preparation is faster annotation. Entering text manually—even with the help of a digitizer and keypunch—can be a slow process. With a CAD system, the designer can call up text associated with a symbol directly from the library. The designer also can enter additional text particular to an individual symbol usage on the terminal keyboard. The operator can then place the text in the appropriate locations by simply pointing to that location on the tablet using an

FIGURE 18–102
Artwork masters: 1:1 artwork masters are automatically generated from the database, eliminating costly photoreduction. (Courtesy of Gerber Scientific Instrument Co.)

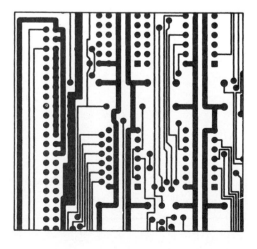

FIGURE 18–104
Component drawings for breadboarding prototypes or hand-inserting components; assembly drawings generated from the same database as the original design. (Courtesy of Gerber Scientific Instrument Co.)

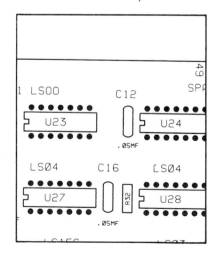

FIGURE 18–105
Silkscreen masters: the component layout and reference designators that are commonly created as part of the symbols, extracted to generate a silkscreen master. (Additional text can be added through the print function of the Digitize/Edit program.) (Courtesy of Gerber Scientific Instrument Co.)

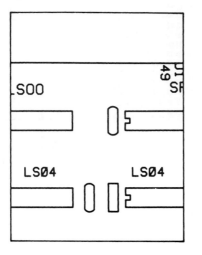

electronic pen. Text can be rearranged on command without the need to erase and reenter.

CAD Board Geometry

Before transferring any schematic information to the printed circuit design, the PC board geometry or physical PCB layout must be created. By manual methods, the designer must draw the PC board outline and any mechanical details by hand (mounting holes, cutouts, and other physical constraints and keep-out areas) before laying out the circuit interconnections.

With CAD, the designer may, after initial on-screen creation of the board geometry, rapidly change physical and manufacturing information on-line, as often as required, with immediate visual feedback. The designer can selectively stretch or shorten lines and dimensions and add or delete mechanical details such as mounting holes and cutouts—without ever using an eraser or pencil. If a family (similar group) of boards is to be designed, the board outline can be stored as a library item and be recalled and used repeatedly without the necessity to re-create it. Component package placement can be integrated into the menu, as shown in Fig. 18–108, allowing the designer quick and efficient access to package placement, orientation, and component configuration.

Some CAD systems offer an automatic dimensioning package. This is an extremely useful tool for the designer. With it, the designer and drafter can specify a single dimension of the particular part, and the system automatically dimensions the entire part using the specified dimension as a guide for all dimensions to follow.

Component Placement

When the schematic and board geometry have been completed, the PC components must be placed on the boards typically to minimize total wire length once the components are interconnected. This is one of the most difficult tasks facing the designer. He or she must take into account numerous gate or discrete component packaging alternatives, *keep-out* areas, and other physical constraints. Using manual layout methods, the designer must first arrive at a functional placement that will allow for an even flow of signals across the PC board to optimize subsequent routing results. This is a matter of trial and error and usually requires many insertions before a successful placement is achieved. After determining the proper placement, the designer must then individually draw each PCB component. Depending on the complexity of the design, this placement process can prove to be a substantial task.

CAD systems use computer power to eliminate many of the problems of component placement. By using the system, the designer can enter all the PCB parts needed for the particular design once, as in schematic parts creation, and then place these components on the PC board using interactive techniques to achieve the desired goal: minimum wire length with no areas of abnormal congestion. On some systems the component package, once selected, becomes the cursor itself, enabling the designer to freely move it about the board for optimal location placement. Once components are placed, a feature referred to as a **rat's nest** can be uti-

FIGURE 18–106
Pad masters generated from the same database as the artwork master and used to create a drill pattern. (Courtesy of Gerber Scientific Instrument Co.)

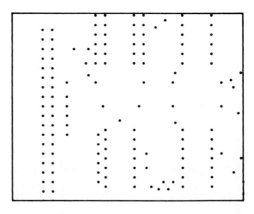

FIGURE 18–107
PCB microdrilling system.
(Courtesy Excellon Automation)

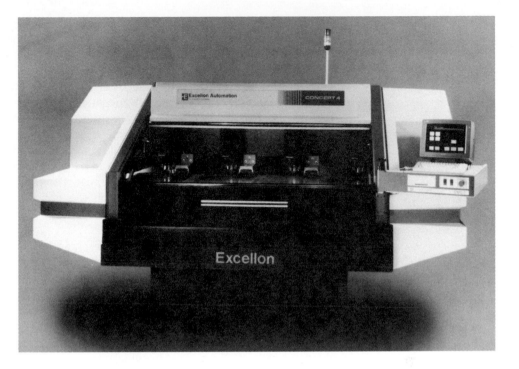

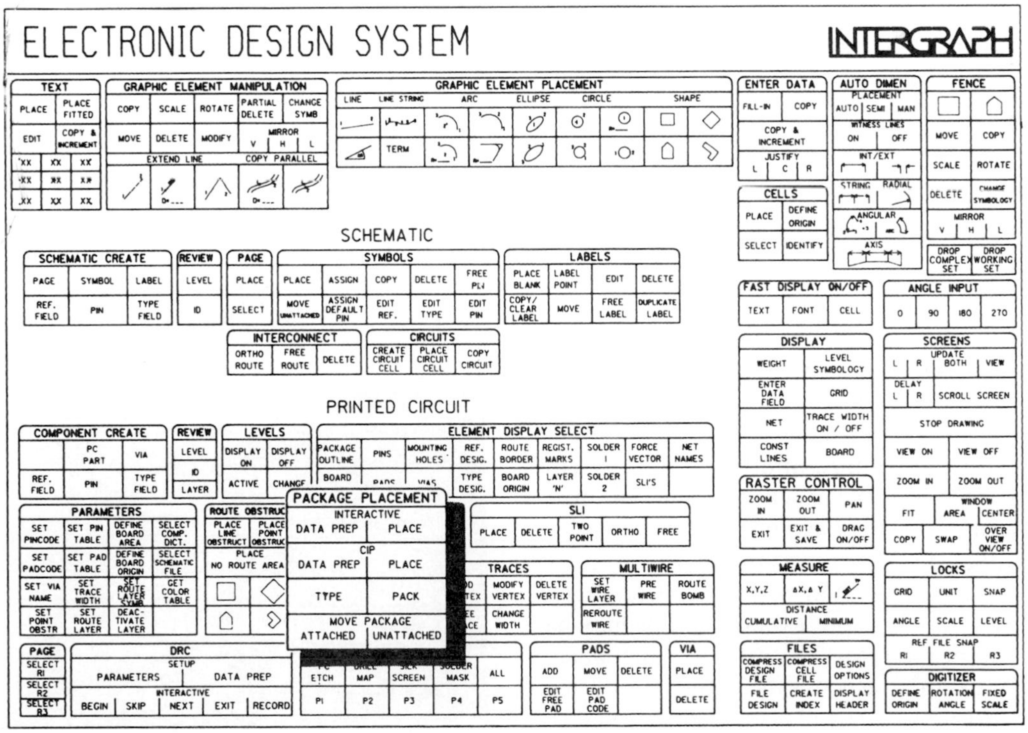

FIGURE 18–108
Electronic design menu for a CAD system PC package placement. (Courtesy of
Intergraph Corp.)

FIGURE 18–109
Rat's nest, a CAD feature that allows the user to view all computer-determined interconnections between components to see if they accommodate a good signal route. (Courtesy of Prime-Computervision Corp.)

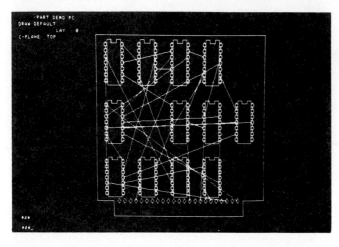

lized. This feature, available on many systems, allows the user to view all the computer-determined interconnections between each component and to determine whether particular components must be moved to accommodate a good signal route. An example of a *rat's nest* is shown in Fig. 18–109.

Another feature useful to the designer is *rubber banding.* This feature is most valuable when a designer wants to move components after initial placement. By using *rubber banding,* the designer may place a component into a *dynamic* mode, which allows the component to be *tracked* across the CRT screen to the desired location, as shown in Fig. 18–110. While *tracking,* the interconnections of the component being moved visually seem to stretch and bend as the component is moved, giving the designer an excellent opportunity to select the location where the component will fit best in the flow of the PC board. Although the rat's nest feature is available on most systems today, rubber banding is available only on systems that offer a raster or *refresh* graphics CRT.

By an interaction of gate assignment and IC placement on the board to minimize interconnection length and complexity, the operator achieves an optimum placement by defining (1) the location of all IC packages with gates assigned, (2) the location of all discrete components, (3) the interconnection wiring between them, and (4) the in and out pins and test points (this may be design-automated or manual). The results of the placement operation are then used as the basis for routing the interconnections.

Some systems offer a completely automated placement package. The board is optimally placed first; the routing program can normally route even complex boards to completion. A high-level placement program such as Lockheed's CADAM® Prance program will swap gates from component to component, or swap gates internal to a given component to reduce trace length in its search for optimum placement. If this does not meet the design specifications, the designer has the option to fix all gates and component assignments. The designer can also restrict certain device types to specific areas on the board. The program will then optimize placement of these components within the restricted area.

Several methods of automated placement are available. One procedure entails the system making the *best-guess* placement without restrictions and before classes have been assigned, allowing anything to land anywhere. For some boards this system provides optimal placement. In a matter of minutes, this first-pass automated placement appears on the screen, showing the designer which components want to be tied together or grouped. This feature is an excellent tool for inexperienced designers.

Components that tend to group together are assembled into classes. To establish classes, the designer instructs the system to fetch all component outlines from the schematic and display them on the screen. From this display the designer defines components that comprise a particular class. When the classes are established and

FIGURE 18–110
Rubber banding, a CAD feature that allows the user to move a component across the CRT screen while simultaneously stretching all related interconnections to maintain signal continuity. (Courtesy of Prime-Computervision Corp.)

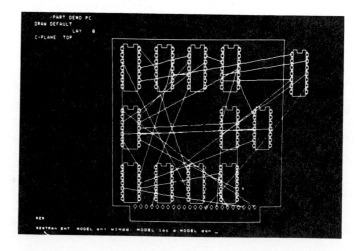

labeled, the designer can divide the board into sections and tell the system which class or classes of components are to be placed in a section. The program will consider discrete components just as it does integrated circuits, optimizing placement within the total board area. This contrasts with the more rudimentary placement programs that place ICs, line up the discrete components around the edge of the board, and rely on the designer to manually place them in the open areas.

When the designer is satisfied that an optimal placement has been obtained, routing of the board may be completed. Many systems have partial or complete routing programs available.

Signal Routing

Signal routing is by far the most tedious and time-consuming operation in the manual design process. After initial placement, the designer must hand draw or place all component interconnections individually. This again is a trial-and-error operation and takes a considerable amount of time to complete. Many times designers get 30% to 40% through the interconnection phase before they realize that the routing or placement must be changed in order to complete the design.

Computer-aided design systems offer many features to assist the designer in accomplishing signal interconnections more quickly and easily. The objectives may include maximum-percent completion of the interconnection task, minimization of the number of via holes, minimization of the total conductor run length, limited trace density or congestion, and, where completion is less than 100%, the option for leaving unused areas for manual completion. It is also necessary to avoid tooling holes, keep-out areas, and prerouted signals. The most powerful computer aid is the automatic router software program. The router is used after component placement to automatically interconnect all the signal runs between components on the PC board. Some routers can achieve results as high as 90% to 95% with little or no interaction by the designer.

Editing

At many times during the design process, it is necessary to edit functions, for example, in developing and refining a PCB design or in implementing engineering changes. These functions are not usually required during the initial stages of manual design, but become powerful tools for the designer during the computer-aided design procedures after completion of the automatic routing phase. The interactive editing functions of some CAD systems offer commands such as layer discrimination and signal highlighting to allow the designer to easily *stitch in* any signal runs not completed by the router. Some of these commands allow the designer to move, instead of erasing and redrawing, certain signal runs and *feed-throughs* to facilitate the addition of signal runs not automatically completed, such as changing, deleting, or adding text, components, and other items.

Checking

As a final step before production of manufacturing documentation, certain checks are made on-line. These checks for layout interconnections differ from the schematic or geometric design rule violations. During the *Compare Net* process, an automatic comparison is made between the routed printed circuit board net list and the schematic net list to check interconnections. Errors are flagged and noted, and they can be interactively corrected by the designer. After correction, signal nets are listed and back-annotated to the schematic. This ensures a consistent documentation package, including a schematic reflecting the final design.

Now that electrical considerations are satisfied, mechanical route quality is verified by a second checking process. This is an automatic geometric design rule check, executed to compare the routed board's physical characteristics to physical design rule constraints, such as line-to-line clearance, pad-to-line clearance, and so

FIGURE 18–111
Automatic drafting machine (photoplotter), which prepares up to eight layers for a single gate. (Courtesy of Motorola, Inc., Semiconductor Products Sector)

on. Errors are flagged and noted and can be interactively corrected.

Documentation

A variety of features for interactive graphics systems relieve the designer of many of the menial tasks associated with the design process. One feature allows the designer to automatically extract nongraphical information from the database and format it to appear in any form desired, such as bill of materials, reports, and point-to-point wire lists. This eliminates the need for a designer or drafter to hand prepare these documents.

Other features include programs to put certain data directly into numerical control format to be used on various NC drilling and insertion machines. Again, this eliminates the need for manual preparation of these items. By using the intelligent database of the CAD system, the designer can automatically generate such things as artwork photoplots, silkscreen drawings, assembly drawings, mechanical detail drawings, and NC drill masks simply by typing a few commands.

Drawings are generated on a plotter, such as the flatbed plotter. Artwork is normally plotted with a photoplotter, as in Fig. 18–111, because of the accuracy necessary to produce a high-quality printed circuit

FIGURE 18–112
Master artwork generated by a CAD system and plotted on a Gerber Plotter #6240, showing first three layers of a ten-layer PCB. (Courtesy of Motorola, Inc., Semiconductor Products Sector)

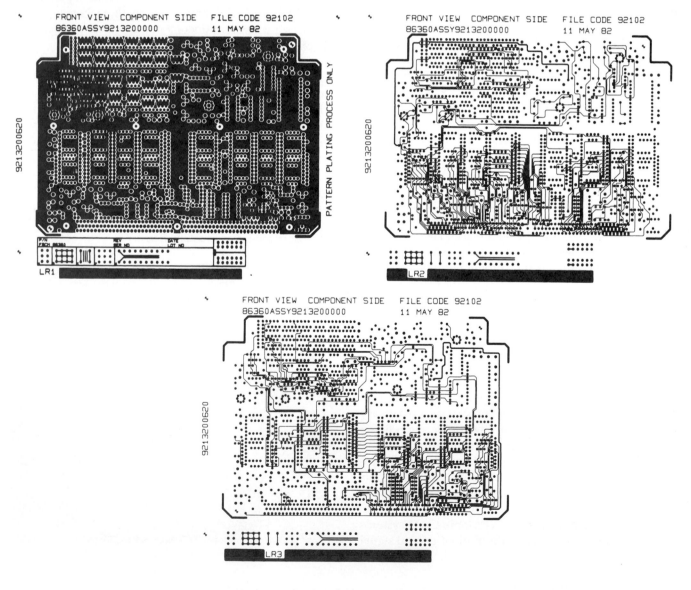

board. Photoplots are entered with a digitizer or by magnetic tape generated by a computer system. Photoplotting involves a complex set of operations involving the separation of the layers for the master pattern. Figure 18–112 is an example of three separate layers of a ten-layer board. The artwork was generated on a photoplotter. Besides the master conductor artwork generation, the photoplotter breaks down the design input into separate patterns necessary for production of the PC board. The master terminal pattern, component and solder side conductor pattern, marking drawing pattern, ground layer pattern, and solder mask pattern are all normally required in the design and fabrication of a typical PC board.

PCB DESIGN USING CAD: CASE STUDY

Designing a PCB using manual drawing or taping methods is a very tedious and time-consuming task. Having first prepared all the necessary information for the layout, a skilled drafter finds a suitable component-placement solution within spatial restrictions and design constraints. The drafter then creates a tracking pattern for hundreds, possibly thousands, of connections, all the time working to a high degree of accuracy to avoid dimensional spacing errors. The completed design is manually checked for spacing errors, for connectivity errors (by comparing it against the circuit diagram), and for artwork quality. Finally, when the board is ready for manufacture, drive tapes are programmed for any NC manufacturing and testing machines to be used.

When a CAD system is used, the schematic drawing package includes many aids to ensure fast schematic layout. The circuit layout is carried out at a design station. The interaction with the graphics display is via the keyboard, pen, or puck and the tablet. Options are displayed on the graphics screen, and the designer selects the appropriate command from the menu. A designer is provided with various capabilities to assist his or her efforts thoughout the layout stage of the project. The designer may ZOOM or WINDOW into selected areas of a drawing, set different items to different colors and layers, and make certain items invisible to reduce the amount of data displayed and to assist clarity. Automatic layout is available on some CAD systems; this allows for nongridded input of symbols and nodal lines directly from an engineer's sketch (Figure 18–113). Automatic layout aligns symbols and lines on the grid, producing the first document automatically (Figure 18–114).

Schematic design symbols are called from the library and placed on the screen interactively using the tablet menu and electronic pen or puck. The designer moves and rotates symbols as required. Subcircuits from previous work are stored and used in the same way, providing a good starting point for new schematics. The designer defines the point-to-point connection pattern of the symbols interactively while working at the graphics display. The interconnections are made with a choice of line widths and colors to enable easy differentiation between voltage, ground, and signal connections. Text in various sizes is added by typing. Once defined, the text is positioned, rotated, and mirrored on the graphics display as required. The designer reproduces repetitive areas of circuitry with ease, creating sections of circuitry only once, and then defines an area as a subdrawing or block. The subdrawing is then positioned and replicated on the main drawing as required.

Once the schematic has been verified by the circuit engineer, data are transferred to a PCB application package. Data transfer includes full details on components and is correlated with the corresponding symbol or symbols on the schematic. More powerful systems allow verification of the completed PCB against the schematic and automatically update the schematic drawing to ensure that both agree.

Routines are provided to move components interactively, rotate them, and fix them in position (Figs. 18–115 and 18–116). All associated connections move with the component to enable assessment of location and rotation. Text and component names are manipulated in a manner similar to components; they are mirrored so the text will appear on either side of the board.

Automatic component placement allows the designer to lay out integrated circuits and discrete components on grids. Routines are provided to swap components to improve the connection scheme. The designer has flexibility in using the placement routines; he or she fixes certain components in position and restricts the routines to operate on particular sections of the board. In addition, the designer interrupts the automatic routines to place components interactively as required.

The designer converts circuit connections to tracks with interactive routing and also modifies route paths with interactive routing. Connections are displayed as straight lines (Fig. 18–117), which are then converted to a series of orthogonal or angled track segments. Segments can be moved and swapped between layers. Via holes are created automatically where required.

The interactive placement and routing routines provide the capability to design from start to finish. However, all the decisions and the implementation of those decisions are done by the designer. The computer does most of the work required to do the placement and routing.

A pin-to-pin connection of all signals is used to verify connections or in conjunction with routing

FIGURE 18–113
Schematic input. (Courtesy
Prime-Computervision Corp.)

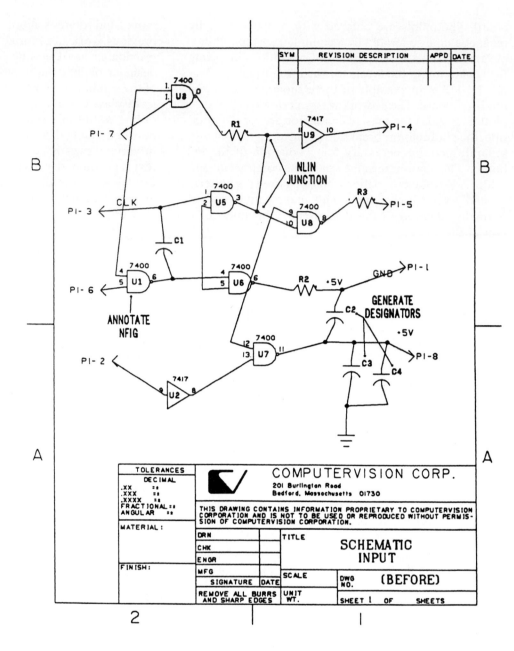

FIGURE 18–114
Schematic drawing of PCB.
(Courtesy Prime-Computervision
Corp.)

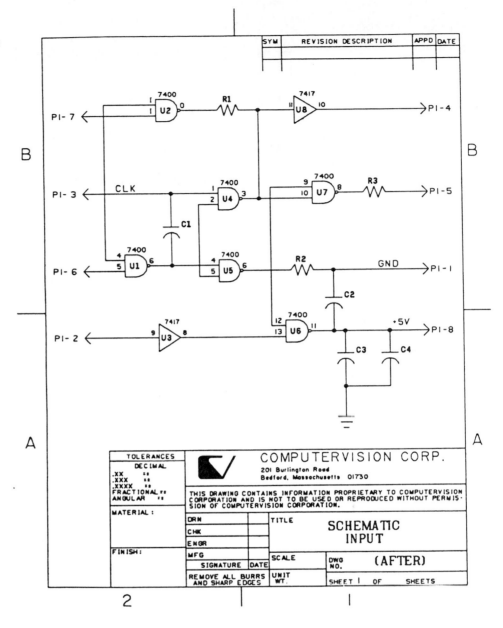

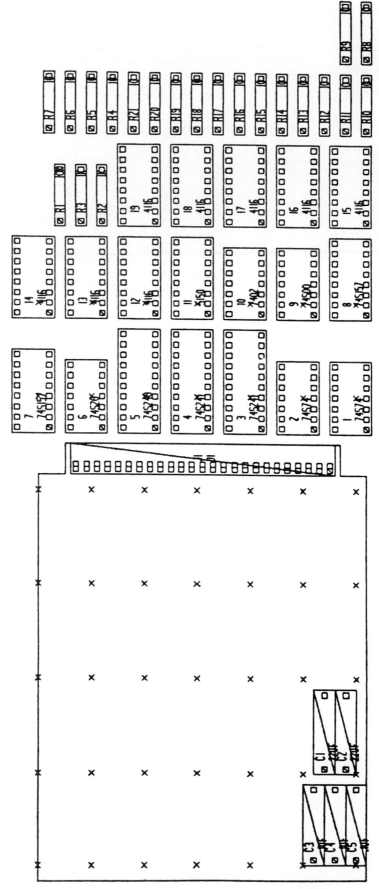

FIGURE 18–115
Initialized placement of components. (Courtesy Prime-Computervision Corp.)

480

FIGURE 18–116
Component placement with
keep-outs. (Courtesy Prime-
Computervision Corp.)

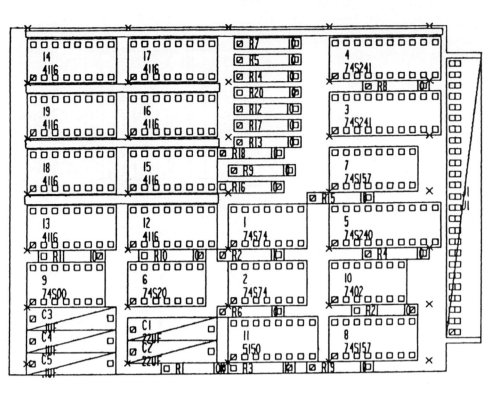

FIGURE 18–117
Component placement with rat's
nest. (Courtesy Prime-
Computervision Corp.)

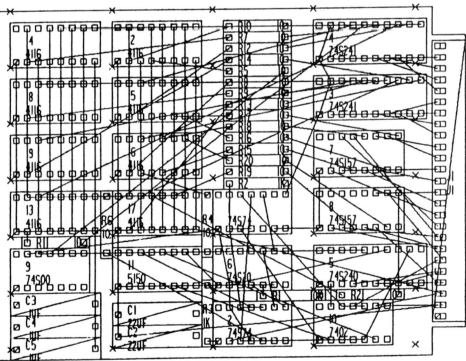

FIGURE 18–118
Display net capability. (Courtesy Prime-Computervision Corp.)

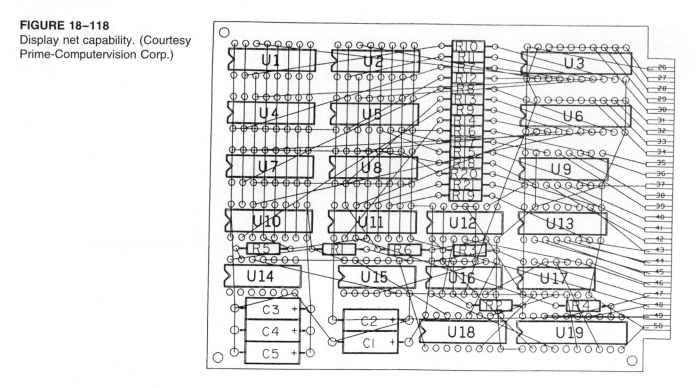

processes (Fig. 18–118). *Rubber banding* allows the designer to move components to see the relative location of traces on the board. This aids in the placement of components in congested areas (Fig. 18–119).

The majority of circuit connections can also be auto-matically converted to routes (Fig. 18–120). Automatic routing (Fig. 18–121) includes routines for routing power and ground connections, memory arrays, and the remaining signal connections. More powerful systems include routers specifically designed for multilayer

FIGURE 18–119
Rubber banding of components. (Courtesy Prime-Computervision Corp.)

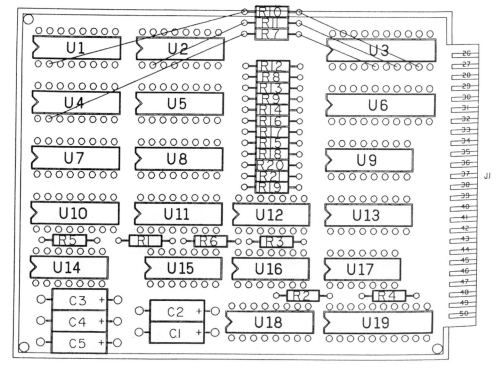

FIGURE 18–120
Autoroute. (Courtesy Prime-
Computervision Corp.)

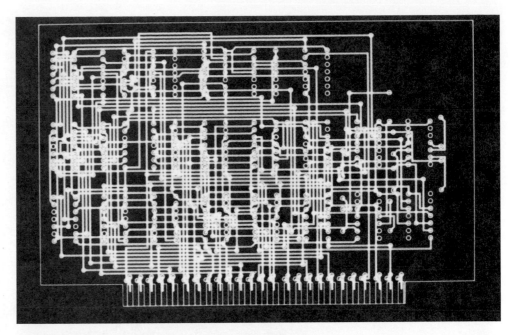

FIGURE 18–121
Autorouting. (Courtesy P-CAD)

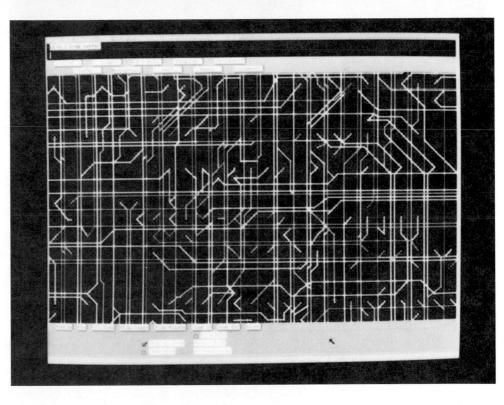

FIGURE 18–122
Multilayer board design using
DesignCAD. (Courtesy
American Small Business
Computers)

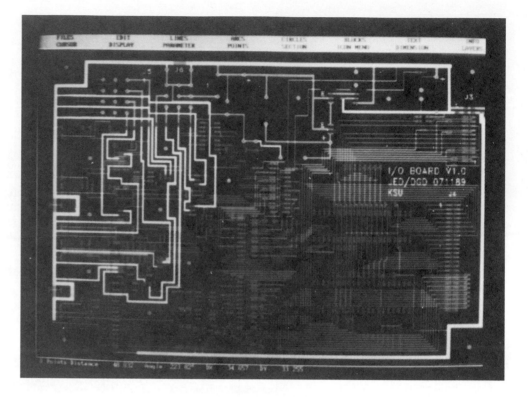

PCBs (Fig. 18–122). The router minimizes the number of via holes and prevents the insertion of vias underneath integrated circuits. Automatic routing is further enhanced by routines to do automatic gate and pen swapping. Finally, editing and checking can be performed directly on the system (Fig. 18–123).

Assembly drawings (Fig. 18–124) and a 3D model (Fig. 18–125) are extracted and plotted from the original

FIGURE 18–123
Editing and checking.
(Courtesy Prime-
Computervision Corp.)

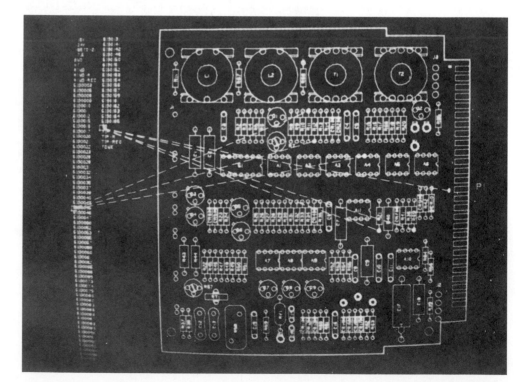

FIGURE 18–124
Drawing setup. (Courtesy Prime-Computervision Corp.)

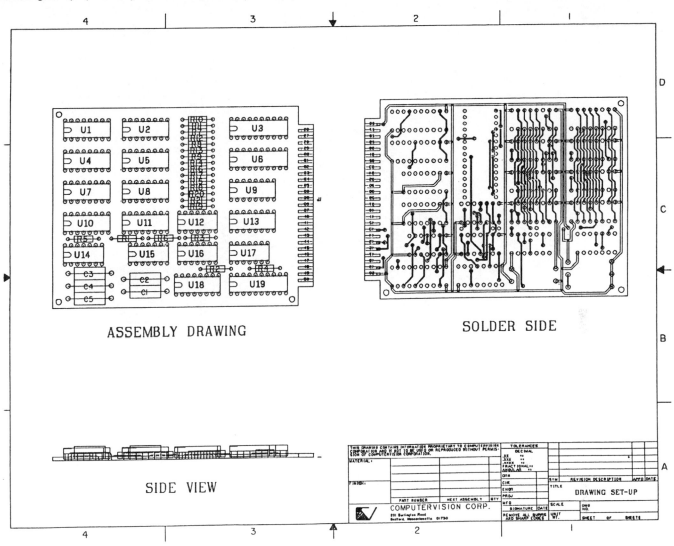

FIGURE 18–125
3D model of assembly. (Courtesy Prime-Computervision Corp.)

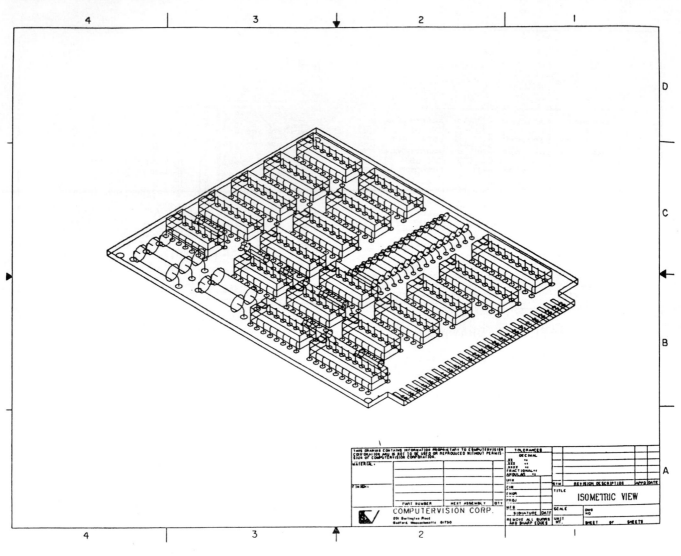

FIGURE 18–126
Pen plotter. (Courtesy Gerber Scientific Instrument Company)

FIGURE 18–127
Electrostatic plotter. (Courtesy Versatec)

FIGURE 18–128
Photoplotter. (Courtesy Gerber Scientific Instrument Company)

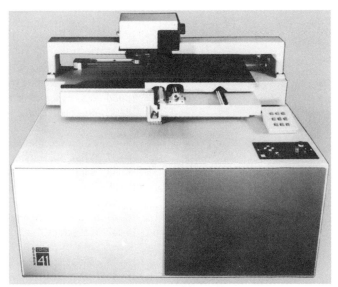

design database as the last step. Most systems include postprocessors to link to pen plotters (Fig. 18–126), hard-copy units, and photoplotters. Electrostatic plotters provide check plots quickly at any stage in the design (Fig. 18–127). Photoplotters provide high-quality, one-to-one, or scaled drawings for manufacturing artwork (Fig. 18–128).

Output from the system does not end with manufacturing artwork. Artwork masters, solder masks, component drawings, silkscreen masters, pad masters, drill drawings, and NC drill and NC automatic component-placement information are all generated from the original PCB design data, with much of the process automated. To complete the design cycle, the system provides design documentation with engineering reports and part listings.

REVIEW QUESTIONS

18–1 What types of documentation are normally required for a typical PC board assembly?

18–2 What are the three types of printed circuit boards and how do they differ?

18–3 What is a prepreg?

18–4 What are tooling holes used for?

18–5 Describe a ground plane; why it is sometimes relieved?

18–6 Why is registration necessary for printed circuit board documentation?

18–7 What is meant by conductor path and what does it take the place of?

18–8 Define the two major methods of PC board production and explain their differences.

18–9 Why is conductor spacing an important part of PC layout?

18–10 Why are the width and thickness of a printed circuit conductor important?

18–11 Name four methods or techniques of laying out a printed circuit board.

18–12 How does a CAD system improve PCB design?

18–13 What is the drill master drawing used for and what will appear on this drawing?

18–14 Which drawing does the parts list appear on?

18–15 What are the uses of an assembly drawing? What items normally appear on this drawing?

18–16 Name three types of board materials.

18–17 Name three types of terminal shapes used as pads. What is an offset terminal?

18–18 What are the advantages of a printed circuit over a circuit connected by wiring?

18–19 Define slit tape and its uses.

18–20 What enlarged scale is normally used for the artwork master?

18–21 What is shown on a marking master and what is this drawing used for?

18–22 How is a master pattern obtained when the manual method is used? When CAD is used?

18–23 Describe the difference between a plated-through hole and an unplated hole.

18–24 What is a component lead and how is it attached to a conductor pattern?

18–25 What is wave soldering and how should a PC board be designed to accommodate this process?

18–26 What is meant by board outline or board geometry?

18–27 What is a resist and what is it used for?

18–28 Define conformal coating and describe its uses.

18–29 Name and describe three types of electrical connections used for connecting the PC board or module to the equipment assembly.

18–30 Describe the *red and blue method* of creating artwork for a double-sided board.

18–31 Describe the subtractive (etching) method of PC board production.

18–32 What is the additive process and how is it used to create a conductor pattern?

18–33 How is copper foil specified?

18–34 Why is a precision-grid system used for PC board layout?

18–35 What conductor width and conductor spacing are recommended for low-voltage applications?

18–36 When are plated-through holes drilled?

18–37 What is a master layout drawing?

18–38 What is meant by *component side?*

18–39 How does a solder mask protect the conductor traces? Why is it used?

18–40 Describe the use of a pad master.

18–41 Name and define three types of artwork markings.

18–42 How does the board working area affect the layout of conductor traces and component location?

18–43 How should horizontally mounted axial lead components be mounted?

18–44 Name two methods of removing unwanted heat from a component on a portion of a board.

18–45 If an axial lead component is mounted vertically, how far above the board should it be?

18–46 How is the lead spacing determined for an axial lead component? For a radial lead component?

18–47 What is pad spacing?

18–48 Name three preliminary considerations before starting a PC board layout.

18–49 What is shown on a typical layout sketch?

18–50 What is automatic component placement and how will it affect the PC board layout stage?

18–51 What is the difference between one, two, and multiaxis component placement?

18–52 How should reference designations be oriented on the marking drawing?

18–53 How will NC drilling affect the design of a PC board?

18–54 What does a typical assembly drawing normally include?

18–55 Define rubber banding.

PROBLEMS

18–1 Using the board layout of the digital readout in the following figure, redesign the conductor pattern. Lay out the terminal holes as shown and tape the conductor traces as per the specifications in the chapter. Use an enlarged scale of 2 : 1 or 4 : 1 as assigned. The project is shown full scale in the text. Straighten the trace runs and attempt to shorten the conductor lengths by finding a more direct path between connections. As an example, the conductor trace for pin 18 could possibly be run around the right side of the board, making it shorter in total length. If possible, reduce the number of jumpers on the layout.

Use tape on this project; do not show the digital readout. After the board outline and terminal pads are located on the grid pattern, place a sheet of tracing paper or film over the layout and sketch in the conductor traces as per the specifications in the text. As much as possible, run the conductor traces on grid lines and make 45° angled corners, instead of the curved lines used on the original example.

PROBLEM 18–1
(Used with permission from Radio Shack, a division of Tandy Corporation, Fort Worth, TX 76102)

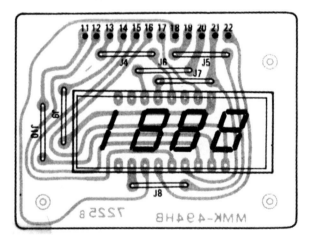

18–2 Follow the general instructions in problem 18–1 for the accompanying figure.

PROBLEM 18–2
(Used with permission from Radio Shack, a division of Tandy Corporation, Fort Worth, TX 76102)

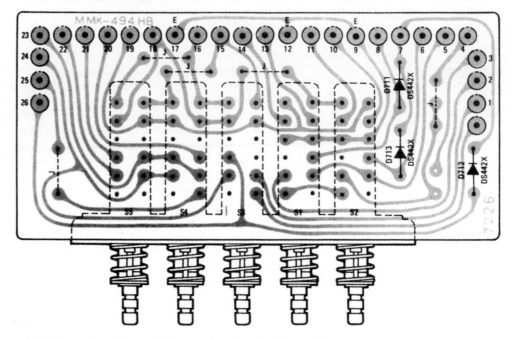

18–3 Draw the schematic diagram for the PCB shown here.

PROBLEM 18–3
(Courtesy of International Business Machines Corporation)

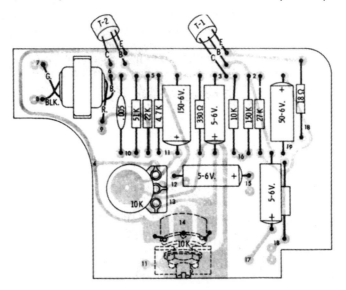

18–4 Follow the general instructions in problem 18–1 for the accompanying figure. Establish a rectangular board outline as required for the layout. Provide a minimum of .25 in. (13.1 mm) as an edge distance for both conductors and components. Electrical connection will be established by wiring as shown. Note that the project is shown full scale. Lay out problem at 1:1 or 2:1 scale as assigned. Also do a schematic diagram.

PROBLEM 18–4

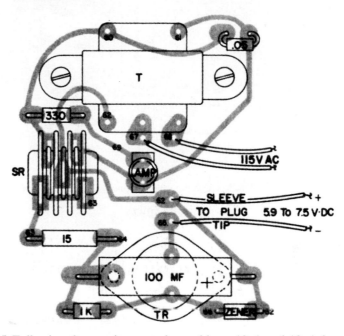

18–5 Following the requirements for problems 18–1 and 18–4, lay out the given project at four times the book size (this will provide a full 1:1 scale for the layout). Establish the board geometry as required and add a terminal connector strip along the bottom of the board. Use Fig. 18–49 as an example; the terminal strip should be designed with a notch to provide correct card insertion. Attempt to straighten and shorten conductor runs and keep jumpers to a minimum when connecting the given terminals to the connector strip. Use grid paper. Also draw a schematic diagram.

PROBLEM 18–5

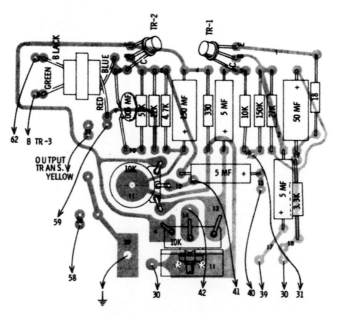

18–6 Repeat problem 18–4. Show all components and align them with the grid pattern. Draw the project at three times the book size. Draw a schematic diagram.

PROBLEM 18–6

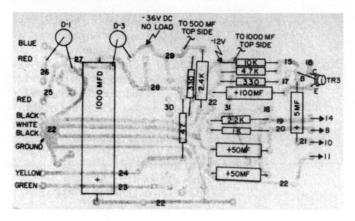

18–7 Tape the given artwork at three times the book scale.

PROBLEM 18–7

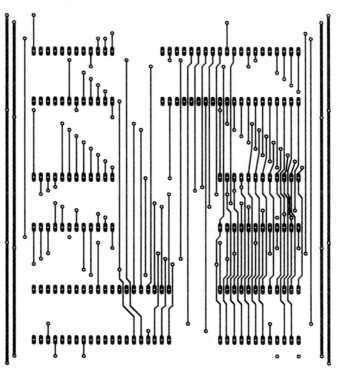

Layout and Design

For the following problems, provide a series of documentation drawings as assigned by your instructor. A complete documentation package will include the following drawings: layout sketch, layout master, marking master, artwork master, drill master/drill table, master drawing/parts list, and assembly drawing. Each project is to be completed at 2:1 or 4:1 scale as assigned. Use tape for all conductor traces. Conductor width and minimum spacing is to be .0312 in. (.793 mm).

Establish board geometry and use terminal connector strip as on Figs. 18–49 and 18–37. Adjust board size and terminal connections as per design. Provide four more connectors along the terminal strip than required by the schematic. Determine all component shapes and sizes from manufacturing tables. Use a gridded pad master and register all drawings. Calculate terminal holes, pad sizes, and pad spacing using tables in the appendix. Specify 2-oz copper conductor.

18–8 For this problem:
1. Resistors are $\frac{1}{4}$ W, ±10%.
2. Transistor is TO-18 case size.

PROBLEM 18–8
(Courtesy of Jerry Rye, Evergreen Valley College, San Jose, CA)

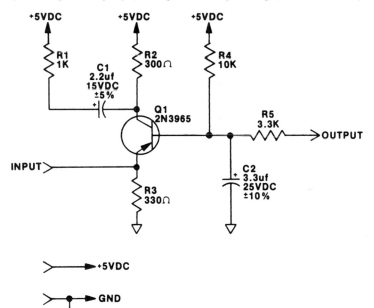

18–9 For this problem:
1. Resistors are ¼ W, ±5%.
2. Transistor is TO-18.
3. Capacitors are 50 V DC and CK06 case size.

PROBLEM 18–9
(Courtesy of Jerry Rye, Evergreen Valley College, San Jose, CA)

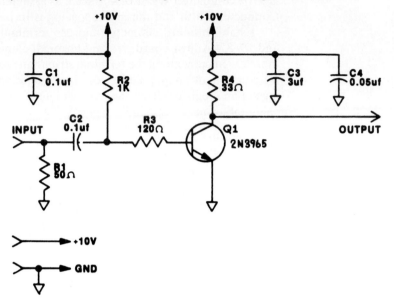

18–10 For this problem:
1. Resistors are RC-07 case size.
2. Diodes are DO-07 case size.
3. Transistors are TO-18.
4. Capacitors are CS13.

PROBLEM 18–10
(Courtesy of Jerry Rye, Evergreen Valley College, San Jose, CA)

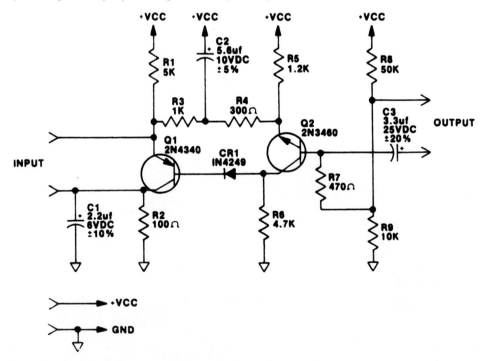

18–11 For this problem:
1. Resistors are $\frac{1}{4}$ W, ±5%.
2. Transistors: Q1 is TO-18, Q2 and Q3 are TO-5 case.
3. Capacitors are CS13A.

PROBLEM 18–11
(Courtesy of Jerry Rye, Evergreen Valley College, San Jose, CA)

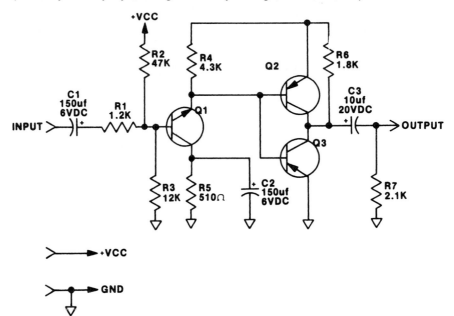

18–12 For this problem:
1. Resistors are RC type.
2. Transistors are 2N4249 and are TO-18 case size.
3. Diodes are DO-7 case size.
4. Polarized capacitors are CS13A case size and nonpolarized capacitors are CK06 case size.

PROBLEM 18–12
(Courtesy of Jerry Rye, Evergreen Valley College, San Jose, CA)

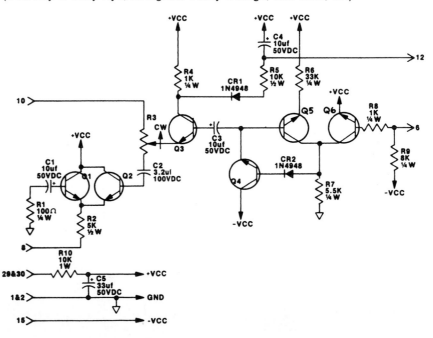

18–13, 18–14, and 18–15 Component size and type must be determined according to the voltage requirements provided on the schematic. In these projects the board geometry and the connector strip are to be designed by the student using the text as a reference.

PROBLEM 18–13

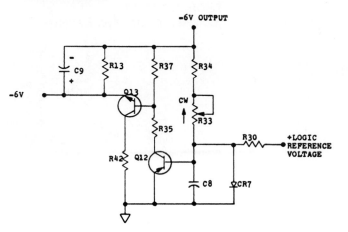

PROBLEM 18–14

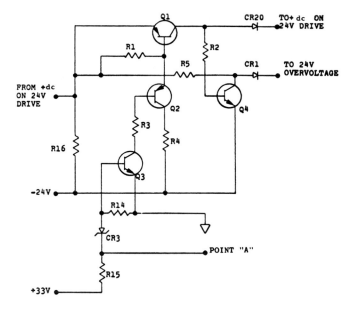

PROBLEM 18–15

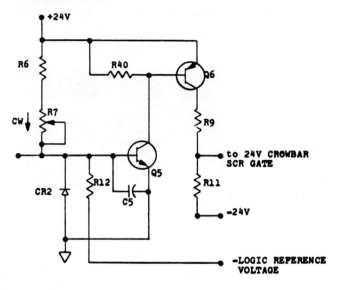

Double-Sided Boards

18–16 Redesign the double-sided PC board shown in Figs. 18–48, 18–49, and 18–50. Use a one- or two-axis layout for the components. Attempt to eliminate crossovers and jumpers. Spread components to eliminate crowding. Complete the project at 2 : 1 scale.

18–17 Redesign the double-sided board shown. Draw the artwork at three times the book size. Lay out conductor traces along grid lines, and use 45° angles for the corners.

PROBLEM 18–17
(Courtesy of Whirlpool Corp.)

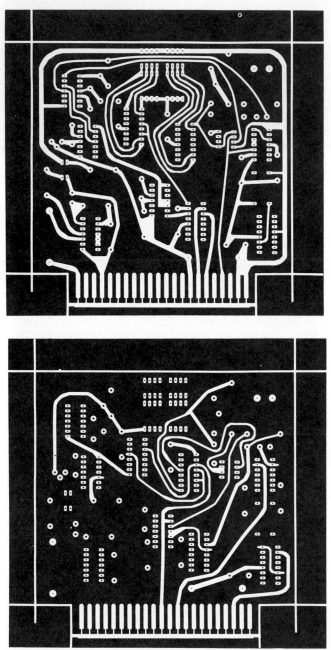

18–18 Lay out both sides of the double-sided board artwork shown in Fig. 18–39. Complete artwork at 2:1 scale.

18–19 Sketch a schematic of the simple PCB.

PROBLEM 18–19
PCB.

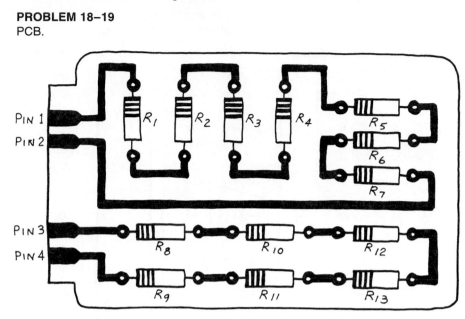

18–20 Draw a schematic of the PCB.

PROBLEM 18–20
PCB.

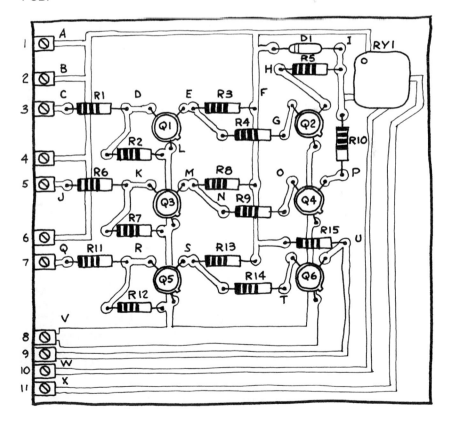

18–21 Draw a schematic of the double-sided PCB.

PROBLEM 18–21
Double-sided PCB.

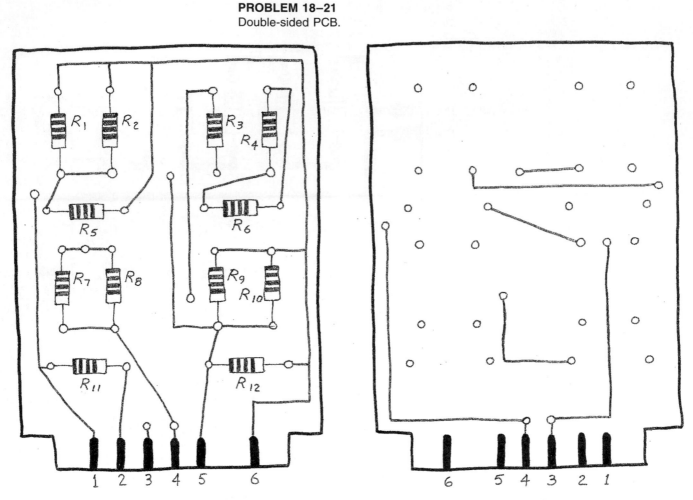

18–22 Lay out the PCB according to your instructor's specifications.

PROBLEM 18–22
PCB schematic.

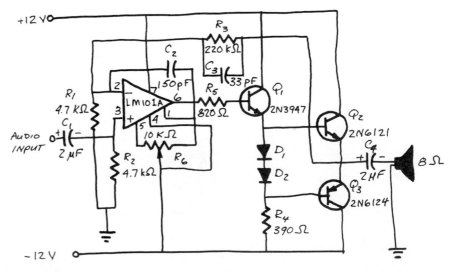

18–23 through 18–27 Lay out each of the analog printed circuit boards per the instructions following this section.

PROBLEM 18–23
A 2-kV negative power supply.

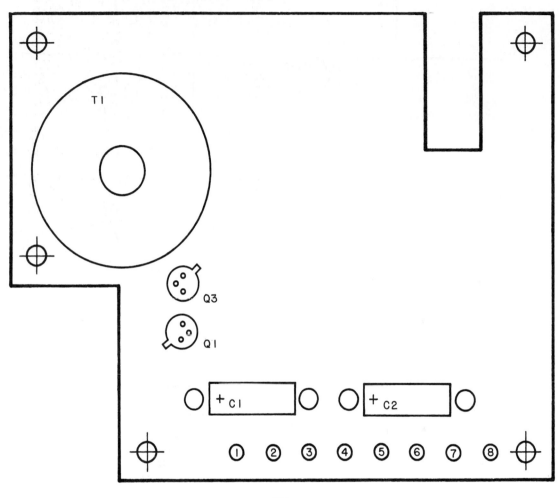

(A)

PROBLEM 18–23 (cont.)

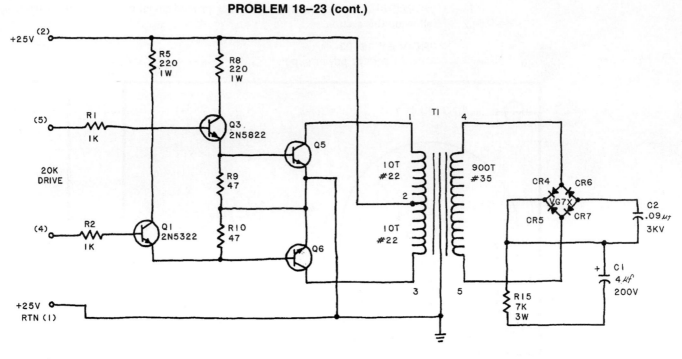

(B)

PROBLEM 18–24
Buffer amplifier.

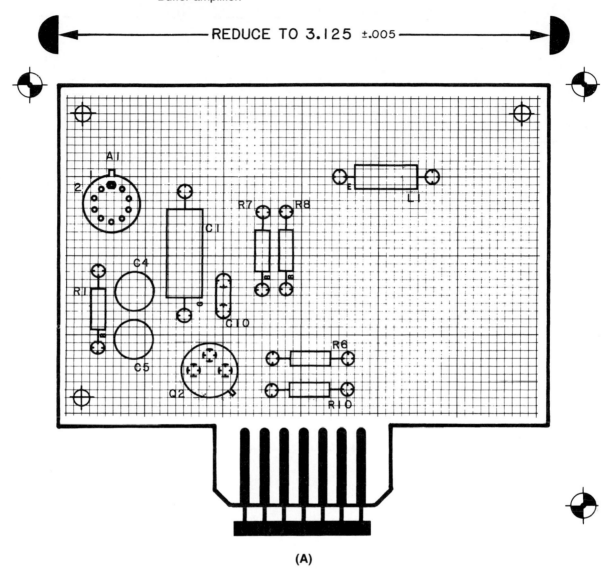

(A)

PROBLEM 18–24 (cont.)

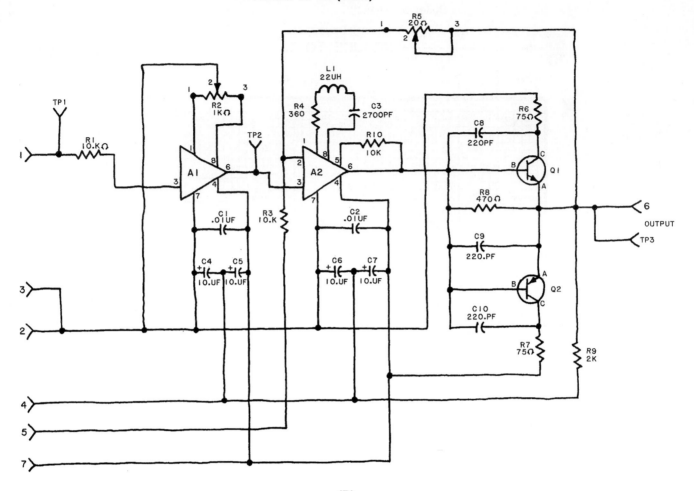

(B)

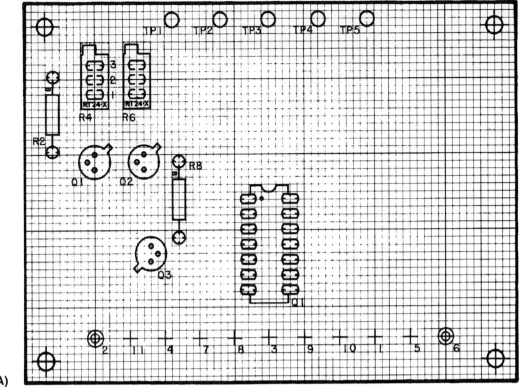

(A)

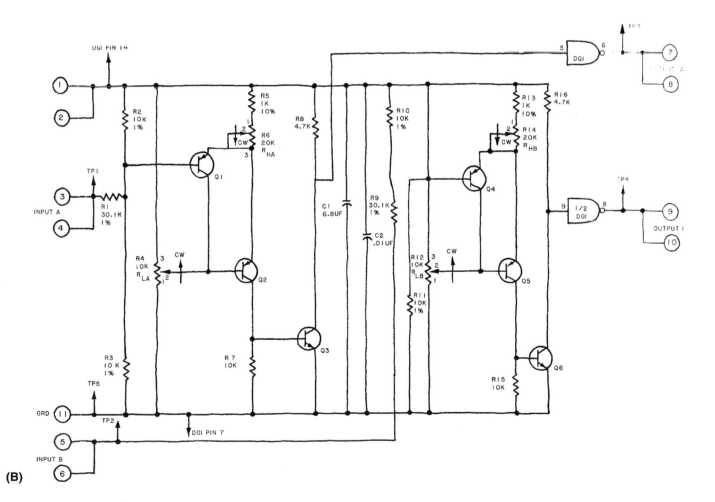

(B)

505

PROBLEM 18–26
High-speed analog comparator.

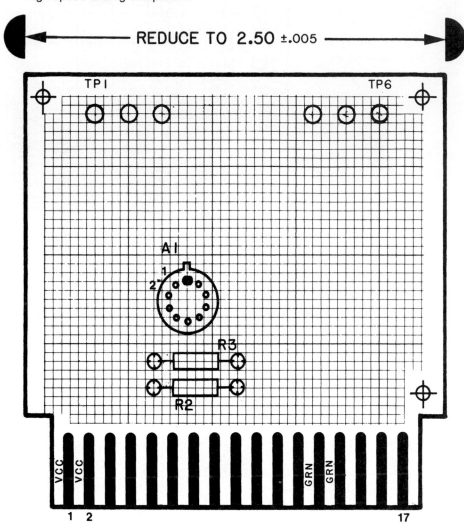

(A)

PROBLEM 18–26 (cont.)

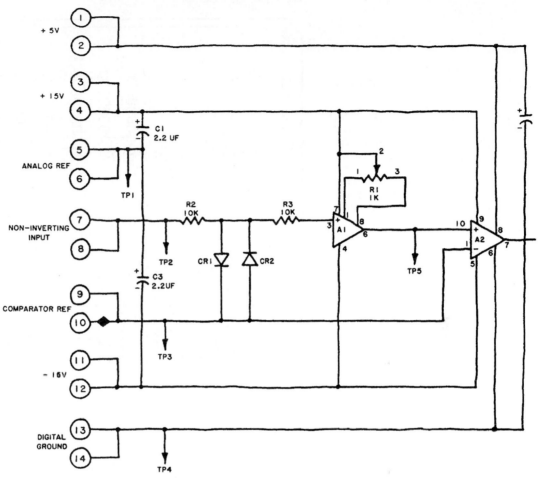

(B)

PROBLEM 18–27
Clock storage PCB.

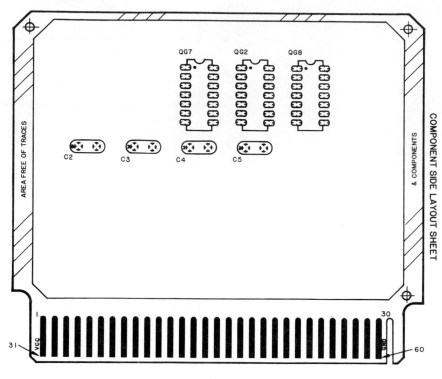

(A)

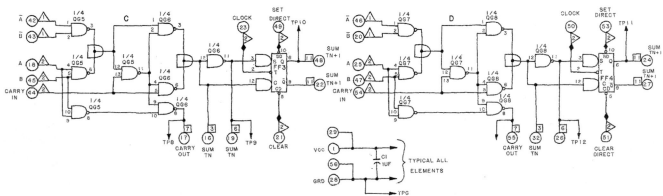

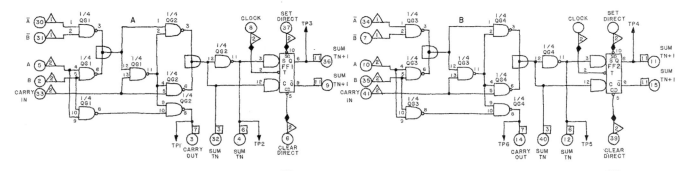

(B)

18–28 through 18–31 Lay out each of the digital boards per the instructions following this section.

PROBLEM 18–28
Input NAND gates—DTL.

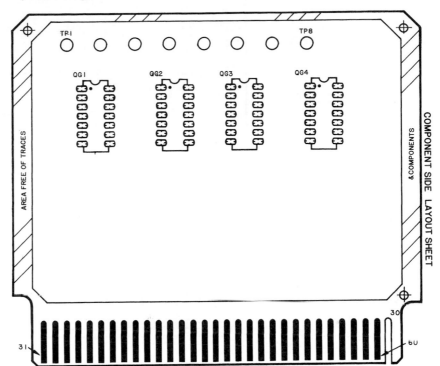

(A)

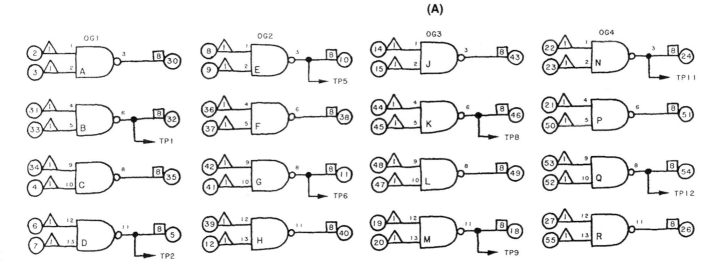

UNUSED CONNECTOR PINS 13,16,17,25

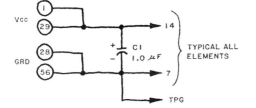

(B)

PROBLEM 18–29
Six-form A Reed Relays with three-input drivers—DTL.

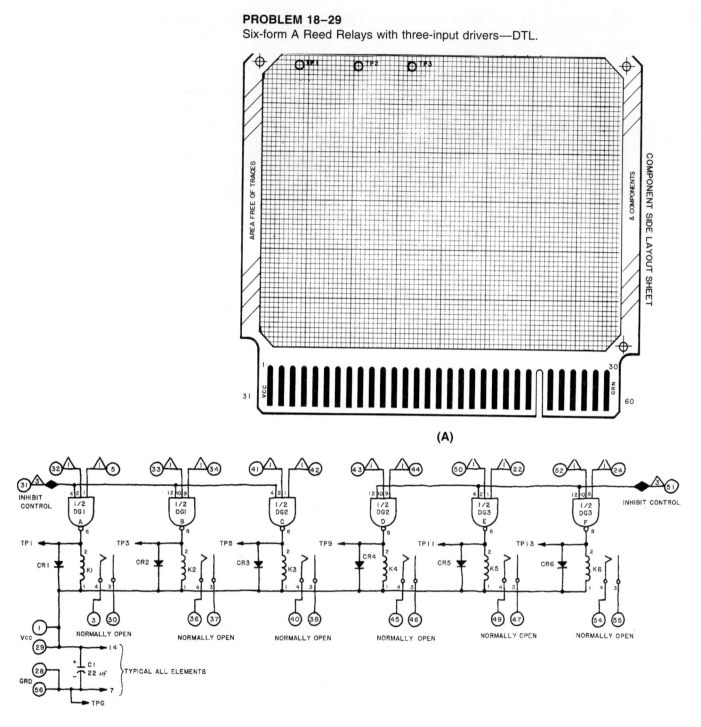

(A)

(B)

PROBLEM 18–30

A 12-bit shift register with gates parallel entry DTL.

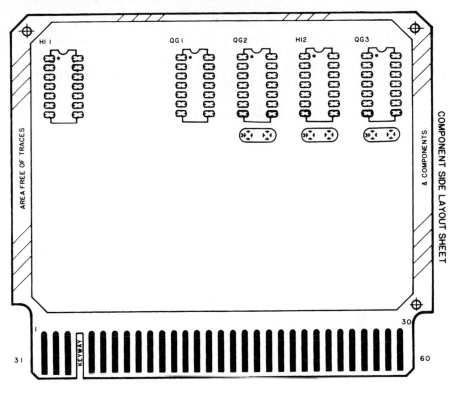

(A)

PROBLEM 18–30 (cont.)

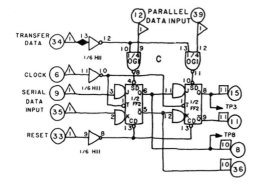

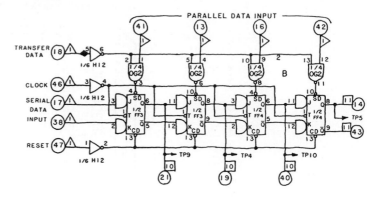

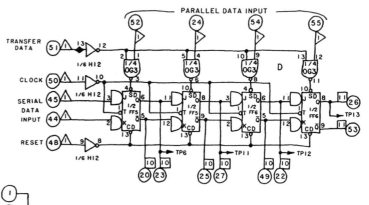

(B)

PROBLEM 18–31
Flasher circuit.

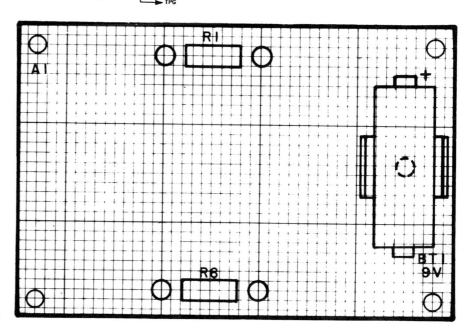

(A)

PROBLEM 18–31 (cont.)

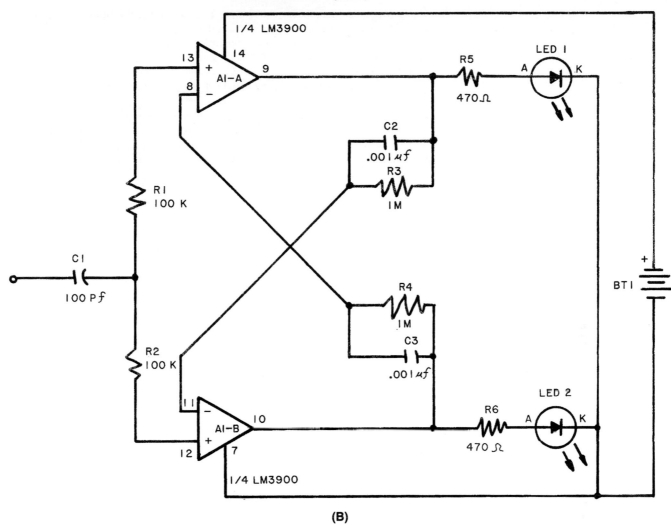

(B)

General Printed Circuit Board Instructions for Problems 18–23 through 18–31

Printed Circuit Boards Lay out the PCB represented by the LOGIC DIAGRAMS and SCHEMATIC DIAGRAMS. The board geometry has been provided for each PCB project in this section. For each of the boards to be designed, use the following specifications:

1. Scale: 2 : 1, 4 : 1, or 10 : 1 for manually laid out boards; 1 : 1 for CAD designs.
2. Medium: 10 × 10 grid (nonreproducible).
3. Color code: component side traces blue and solder side traces red.
4. Trace widths: signal lines, .015 in. wide; power/ground lines and primary bus, .100 in. wide; secondary bus lines, .050 in. wide, at 2 : 1 scale.
5. Pad diameters: .050, .062, .100, .200, and .250 in. at 2 : 1 scale.
6. Component placement: all on 10 × 10 grid cross hairs.
7. Feed-through pads: .050 in., at 1 : 1 scale.
8. Power pins or Vcc: normally, pins 1 and 31 for most digital PCBs, but instructor may wish to change the pin assignments.
9. Ground pins or Vdd: normally, pins 30 and 60 for most digital boards, but assignments can be changed.
10. Put three registration targets, reduction marks, and reduction size (at ±.005-in. tolerance) on all analog boards where these are not given.

Artwork Tape-up Master Rules

1. Using prepunch .003, .005, or .007 clear film, and align the film with registration pins. The film will be laid on top of the red and blue layout drawing. If registration pins and prepunched film are not available, use registration targets and register each layer of film to the layout. Place the correct-sized pads as required for the pad master.
2. *Registration:* All pads will be centered on the grid cross hairs. Apply reduction marks and dimensions as required. All reduction dimensions will have a ±.005-in. tolerance. Place three registration marks outside the board outline.
3. *Pad master overlay:* Put three tooling holes on each board when they are not already provided. For the following components use a square pad: (a) transistor emitters; (b) pin 1 and all DIP packages (unless the given layout marks pin 1 with a dot or a different shape pad); (c) capacitors that are polarized (square pad for the plus side); (d) cathode end for diodes; (e) wiper side for variable resistors; (f) pin 1 on SIPS resistor packs; and (g) jumper wires. Note that the pad master will have the board geometry shown on it. Use preprinted pads if available.
4. *Solder side:* This is the next overlay that will be placed on or under the pad master. Show all red traces on this overlay. Show the following on this layout: SOLDER SIDE in large letters, board part number, board revision letter, and serial number. These will normally be shown on this document but reversed. Use tape for conductor traces when available.
5. *Component side:* This is the next overlay that will be placed over the pad master. Show all blue traces that are on the layout. Show COMPONENT SIDE in large lettering on this document along with the assembly number of the board.
6. *Silkscreen:* This will be the last overlay placed on the pad master. It will show the location of all components and reference designations. Use template or Leroy lettering. Ink the drawing unless instructor requests otherwise. Preprinted component outlines may also be used.

Board Fabrication Rules

1. A board fabrication drawing shows board geometry, dimensions, and tolerances.
2. Note the board material; G10, G10-FR, or another material.
3. A table showing hole diameters, tolerance, and number of holes should be included on the drawing. Code the holes so they can be identified on the circuit side of the board.
4. Include the name, number, and any other specifications that may apply to the part.
5. The circuit-side pattern images should be shown with all traces and pads. Holes are drilled from the circuit side of the board.
6. If the board is multilayer, each layer should be indicated on a cross-sectional view of the board with thickness and tolerance given.

Assembly Drawing Rules

1. The assembly drawing should show component placement and reference designation markings and assembly and test specifications. Include a list of materials to be ordered.
2. The assembly drawing is viewed from the component side of the board.
3. Orientation and indexing of all transistor tabs, polarized capacitors, diodes, variable resistors, DIP packages, connector pin 1, and every tenth pin should be indicated.
4. All specifications should be included in the notes on this drawing.

19

ELECTRONIC PACKAGING

INTRODUCTION

The total packaging of an electronic or electrical item includes both the exterior form and the interior parts for securing electronic modules and other parts to the frame and enclosure. Therefore, the field of **electronic packaging** covers the functional and esthetic design of commercial computer products, as well as the mounting and structural configurations for modules and electronic subassemblies involving **sheet metal** parts. These areas can be referred to as exterior *enclosures* (shells) and interior securing configurations. This simple division, however, cannot always be applied to packages that incorporate both the exterior and interior package into one unit.

Exterior packaging includes the **cabinet** and envelope design for all types of electrical, electronic, and computer equipment. The computer products shown in Fig. 19–1 provide examples of the possible variations in packaging size and configuration.

Interior packaging includes the **mounting plates, panels, brackets, chassis, cage,** and other parts, and subunits that secure or enclose the actual electronic and electrical modules, assemblies, and subassemblies to the outer holding envelope or cabinet.

For small intrinsic units such as portable metering equipment, shown in Fig. 19–2, the exterior and interior division does not apply since the electronic components, circuit boards, and other parts may be at least partially secured directly to the outer shell or envelope.

Packaging can also be divided into functional categories; consumer product, private business or medical, industrial, public communications, power generation, military or space, and computer. Each category has its own design specifications based on environmental, functional, esthetic, and cost factors.

Products manufactured for sale in the consumer market include radio, stereo, video, television, calculator, personal computer, computer game, household, and transportation products. In the manufacture of military products and power generation or communications equipment, reliability and life length are the most important considerations. Private consumer products have the added requirements of a high degree of esthetic and cost engineering considerations.

Private sector business and medical applications include computers, communication equipment, and medical or dental equipment. Medical and dental equipment must be designed and manufactured according to strict government standards because of their life sustaining or threatening aspects. Medical equipment includes nuclear devices and life monitoring equipment.

FIGURE 19–1
Computer input, processing, and storage equipment.
(Courtesy of International Business Machines Corporation)

Industrial equipment includes electronic controls and monitoring equipment for robotics and computer-aided manufacturing equipment, along with traditional industrial machinery and production or assembly equipment. The artificial vision system shown in Fig. 19–3 is a programmable image sensing and processing system that inspects, identifies, counts, sorts, and positions parts. It can also be used as a stand-alone sensor for robots. Robotics equipment, CAD/CAM, and other automated electronic equipment are the fastest growing and most dynamic areas of electronic engineering and package design.

Power generation and communications equipment are also extremely important areas where electronic and electrical equipment are used extensively.

FIGURE 19–2
Logic analyzer. (Courtesy Gould, Inc.)

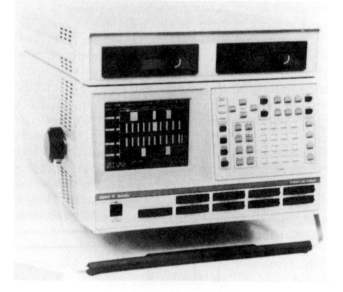

FIGURE 19–3
Artificial vision system and control station for an industrial robotic system. (Courtesy Automatix)

Military and space exploration applications for electronics include everything from simple hand-held enemy detection systems to missile and early warning systems. The military uses electronic equipment extensively on ships, aircraft, and land-based equipment. All military electronic equipment is governed by strict *Mil Specs.* Military electronic packaging must meet stringent design specifications, including tough requirements for strength, reliability, and operation, including human element engineering. Space exploration and satellite communications equipment, shown in Fig. 19–4, require the highest degree of electronic state-of-the-art design and packaging because of the extreme degrees of unknown and known variables encountered while in operation. Most state-of-the-art designs and inventions (including calculators and computers) are a direct result of the space program.

Computer equipment and products span the full range of categories: personal computers for private consumer tasks; military computers for radar, satellite, and military equipment; industrial computers to run CAM, CAD, and robotics operations; and communications and business operations computers for business tasks. CAD system hardware (see Chapter 6) is an example of electronics packaging in the computer industry. Note that a CAD system's packaging must take into account human factors of workstation design and esthetic values associ-

FIGURE 19-4
NASA training and test facility equipment for space flight networks. (Courtesy NASA)

ated with highly visible business and engineering office equipment.

DESIGN OF ELECTRONIC PACKAGING

The design of electronic packaging is such a varied and complex subject that this text is only able to introduce general design factors.

Package design starts with acquiring the limiting factors involved in the particular project. The size, material, and environmental and cost restrictions must be understood at the early stages of design.

The system's block diagram, schematic diagram, and wiring diagram should be readily available to the package designer. The block diagram provides essential information on the number of assemblies and the individual units that make up the system. The functional relationship between main components and assemblies is also defined in the block diagram.

Schematic diagrams describe in detail the functional relationship between modules, subassemblies, and components for a particular portion of the system. All component sizes, module dimensions, and mounting requirements need to be known before the actual system can be packaged. The parts list on the schematic describes each component type, size, and value designation.

The wiring diagram should be available when the packaging design starts. Wiring diagrams provide information regarding the interconnection of individual units, assemblies, and main components, along with the required connections to panel switches, lights, metering components, and other parts. The physical space requirements must take into account space taken up by wires, harnesses, cables, and connection items. Space for wiring must include extra areas for the long wire lengths needed to provide the unit with slide-out capabilities.

Cost restrictions on package design are influenced by the eventual use of the item and governing standards. Reliability and strength factors take precedence over cost when military packaging is being designed. For consumer and business products, esthetic and human engineering factors are as important as reliability. The computer terminal and microsystem emulator shown in Fig. 19-5 must meet high standards for appearance and human use to be attractive and salable. The printed circuit modules on the left are packaged in the emulator shown in the right of the photograph. Size, shape, ease of maintenance, and esthetic considerations were taken into account during the package design of this product.

Environmental conditions encountered during the use of the equipment must be considered in the early stages of the package design. Temperature variation, vibration, humidity, pressure, corrosion, and other variables are important in the selection of materials, finishes, fasteners, and the general design configuration of the package. The portability of the logic analyzer shown in Fig. 19-2 demands environmental design considerations different from those imposed on the designer when stationary equipment is involved.

Sophisticated CAD electronic packaging programs enable the designer to pack more electronics into less space. Designers can test and refine many component arrangements, materials, and packaging configurations on-line, clearly visualizing the effect of each on the emerging design. Two- and three-dimensional programs allow the designer unlimited freedom in creative and efficient design of even the most complex systems. The 3D option allows visualization and design in pictorial

FIGURE 19-5
CRT, computer, and printed circuit boards. (Courtesy Gould, Inc.)

views. At any point in the design process, visual interference checks can be run on even the most difficult packaging configurations. After the design is completed, the system automatically generates assembly drawings, NC tapes, or a flat pattern for metalworking.

Package materials range from traditional metals or metal alloys to sophisticated plastics. The printed circuit board and accompanying switches, battery, and fuse compartment of a VOM, shown in Fig. 19–6, use a molded plastic base unit. The computer equipment in Fig. 19–5 is packaged in a molded plastic enclosure, which requires no upkeep since the color and texture of the plastic meet all esthetic and functional requirements.

Metals are used where durability and toughness are more important than esthetics. The logic analyzer in Fig. 19–2 has a metal shell. A variety of sheet metal materials are used in package design for panels, chassis, brackets, mounts, and cabinets. Oxidizing metals must be anodized, alodined, or painted.

Exterior and Interior Parts of Electronic Packaging

Electronic packaging has been previously separated into exterior and interior packages with securing devices. Exterior configurations include all types, shapes, and sizes of cabinets, as well as simple enclosures and envelopes.

FIGURE 19–6
Printed circuit board with the function and range switches, the zero OHM pot, and the battery and fused compartment of the 260 VOM. (Courtesy Simpson Electric Co.)

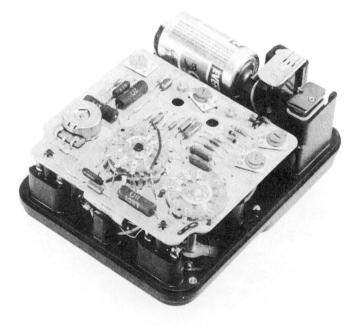

Cabinets range from standard units to custom-designed units for specific users. The maze of electronic equipment shown in Fig. 19–4 is an example of user-defined, custom-designed packaging. Electronic equipment is mounted in a cabinet designed for easy access and use by the operator. A standard cabinet as shown in Fig. 19–7 is a unit designed for consumer purchase to meet the needs of a variety of users. Standard cabinets are used for computer memory expansion units and a variety of electronic equipment.

Rack and panel assemblies are shown pictorially in Fig. 19–8, with the rear and side views shown in Fig. 19–9. They consist of separate electronic packages, units, assemblies, and modules mounted in their respective chassis or enclosures, secured to adjustable drawers or racks. Each unit is attached and wired to its respective front panel. Rear panels are not used on this type of open rack and panel assembly design. All wiring between units is confined to the wire duct, Fig. 19–9 (22), on this particular configuration. In most cases, the design and layout of electronic equipment on the rack and panel assembly must take into account space for wiring and cable connections. Each rack in this figure can be adjusted for the electronic equipment to be mounted on the structure. A detailed materials list for this rack and panel assembly is provided in Fig. 19–10.

Chassis

Chassis design includes modules and panels. Normally, a module is housed in some form of chassis arrangement as in Fig. 19–11, where the PC board module has been secured to a sheet metal chassis, which in turn is fastened to the molded plastic enclosure.

Modules are specific electronic packages providing one separate function to the total electronic system. Each module can be separately removed, serviced, maintained, and replaced individually without disturbing other parts of the system. Electronic modules may be a single PC board or a subassembly made up of multiple PC boards and individual components as long as they form a complete, separate individual function. In general, the term *module* is used to refer to any plug-in, printed circuit board, whether completely blank as in Fig. 19–12, provided with IC pads and connections, or fully assembled as a working entity (Fig. 19–11).

In Fig. 19–12 the plate, board, and panel are shown encased in a drawerlike assembly. The chassis assembly shown in Fig. 19–13 consists of a series of individual pull-out (drawerlike) modules installed in a cage. Each cage unit is designed to be mounted on a cabinet or rack and panel assembly.

The expansion mounting box shown in Fig. 19–14 has a built-in power supply, mounting frame, connector

FIGURE 19–7
Cabinet and parts list. (Courtesy Digital Equipment Corp.)

Item	Description	Part Number	Quantity
1	H9542-FA Basic Frame Kit	E-H9542-FB	1
2	Stabilizer and Leveler Kit	E-H9544-HA	1
3	Trim Kit	E-H9544-CA	1
4	Rear Door Assembly Kit	E-H9544-BA	1
5	Power Control 120V 12A	E-00872-A	1
6	Cable Assembly	E-70-08288-08	1
7	Expansion Cabinet Kit	E-H9544-JA	1
8	Filler Panel (5-1/4")	E-H9504-SC	2
9	Filler Panel (10-1/2")	E-H9504-UC	1
10	Blank Bezel (1.75")	E-H9544-DA	1
11	Bottom Cover (6")	E-H9544-DB	1
12	Hardware Kit (for assembling Frame Assembly, item 1)	E-22-00034-00	1
13	Hardware Kit	E-22-00035-00	1
14	Screw, Sems, Phl Truss Hd No. 10-32 × .50	E-90-09700-00	5
15	Retainer, "U" – Nut No. 10-32	E-90-07786-00	4
16	Cable Clamp, Screw Mount 3/8"	E-90-07083-00	1

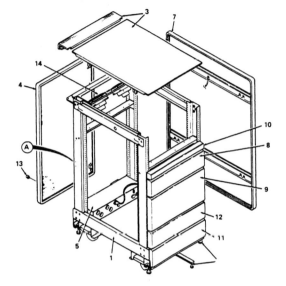

blocks, and examples of four possible module variations. This *system unit* is a set of module connecting blocks in a mounting frame, which in turn mounts to a cabinet or rack and panel assembly like the one shown in Fig. 19–8. The box is a drawerlike, rack-mountable, fully enclosed chassis for circuitry that cannot be housed within a computer's own enclosure. This box is also known as an *expansion mounting chassis*.

In Fig. 19–15, another example of an expansion enclosure (chassis) is shown along with its individual retainers for securing separate PCB modules that are positioned vertically or stacked. The front of a chassis such as this is attached to a functional panel. The chassis in Fig. 19–16 is a completely detailed custom enclosure. The finish, marking, and drilling requirements are specified in the notes. The chassis is a completely welded unit.

Panels

Panels are provided on most electronic packages. A panel serves as a mounting surface for indicators, dials, lights, switches, buttons, meters, and other controls. Panels can be as simple as the front panel of a logic analyzer (Fig. 19–2) or as complex as the multiple con-

trol panels for a NASA tracking station (Fig. 19–4). Panels for specific consumer or industrial products also have the brand name, logo, and other identifying callouts embossed, engraved, painted, labeled, or otherwise attached to their surface for easy identification. The small mounting plate (panel) in Figs. 19–17 and 19–18 shows the meter mounting cutout and securing holes and the current and voltage identifying markings.

Figure 19–19 shows the suggested panel and control station layout for industrial controls. Figure 19–20 shows a rear-panel assembly drawing of a data-acquisition computer.

A complete hard-drive assembly is shown in Fig. 19–21. Figure 19–22 shows a detailed RF panel.

SHEET METAL LAYOUT

Much of electronic packaging involves the use and fabrication of sheet metal parts to be used for chassis, panels, mounting plates, and a variety of enclosures and envelopes. Sheet metal parts are normally made from a *blank* of sheet metal cut to the required size. Panels, mounting plates, and other parts are normally flat sheets of metal

FIGURE 19–8
Pictorial of rack and panel
assembly (see Figs. 19–9 and
19–10). (Courtesy Interactive
Computer Systems, Inc.)

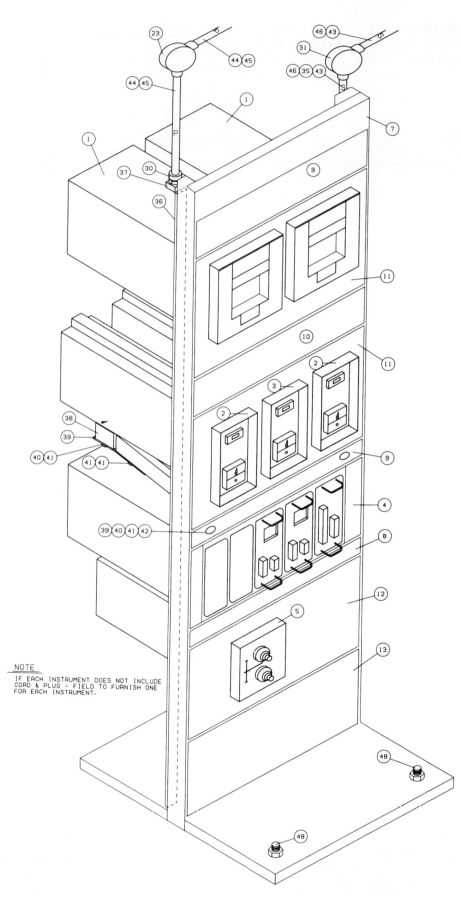

NOTE

IF EACH INSTRUMENT DOES NOT INCLUDE
CORD & PLUG – FIELD TO FURNISH ONE
FOR EACH INSTRUMENT.

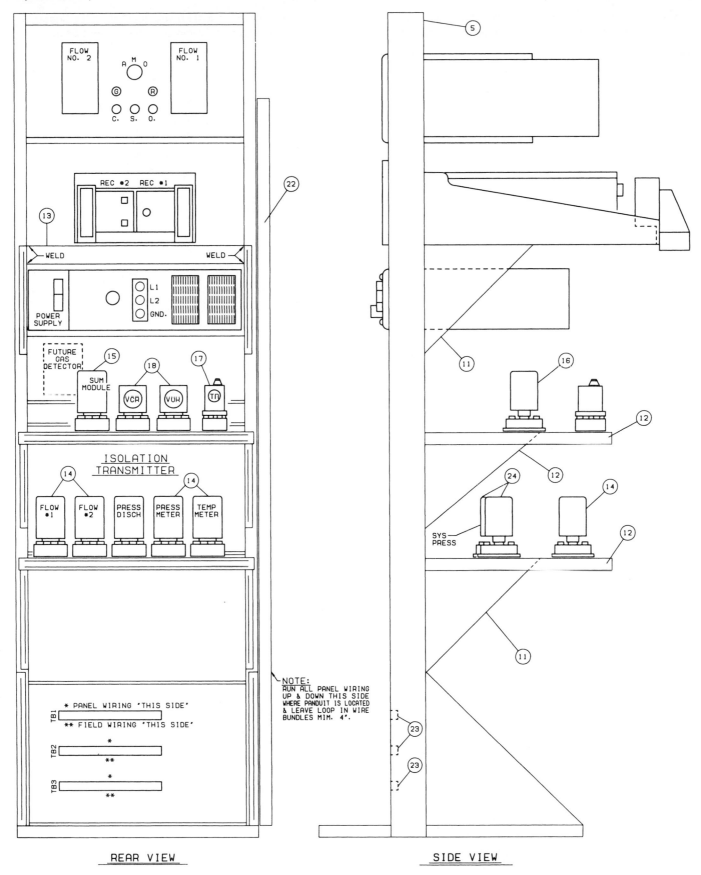

FIGURE 19–10
Parts list for rack and panel assembly shown in Figs. 19–8 and 19–9. (Courtesy Interactive Computer Systems, Inc.)

ITEM	QUAN	DESCRIPTION	REQ. NO.
		MATERIALS LIST	
1	2	DANIEL MOD 2239 TOTALIZER	S-1132 ('81)
2	2	FOXBORO MOD E-20S-V ELECT. CONSOTROL	S-1328 ('81)
		RECORDER	
3	1	FOXBORO MOD EH4D ELECT. SHELF	
		COMPLETE WITH TERMINAL BORAD ASSEMBLY	
4	1	THUNDERCO CSA-5 CONTROL SYSTEM	S-0006
5	1	BUD *RR-1264, OPEN TYPE RELAY RACK	S-0087
6	1	SFA-3167, 19"W X 12 1/4"H SURF. SHLD. PANEL	
7	2	SFA-3166, " " X 10 1/2"H " " "	
8	2	SFA-3165, " " X 8 3/4" H " " "	
9	1	SFA-3164, " " X 7"H " " "	
10	1	SFA-3161, " " X 1 3/4"H " " "	
11	6	MB-1268, 9" X 9" TRIANGULAR MTG BRACKET	
12	2	CB-1976, RACK SHELF, 19"W X 15"H	
13	1	2"W X 19" LG X 1/4"THK. STRAP IRON	
14	5	ACTION-PAK MOD AP4300113, ISOLATION TRANS.	
		COMPLETE WITH BT5-B BASE AND TRK1 CHANNEL	
15	1	ACTION-PAK MOD4402 SUMMING MODULE	S-1329 ('81)
		COMPLETE WITH BT5-B BASE AND TRK1 CHANNEL	
16	1	ACTION-PAK MOD 1021 CURRENT ALARM MODULE	
		COMPLETE WITH BTS-11 BASE AND TRK1 CHANNEL	
17	1	OMRON/AGASTAT MOD STPMHADA TIMER WITH 8 PIN	S-0146
		OCTAL BASE SOCKET	·
18	2	POTTER-BRIMFIELD MOD KRPN-11A DPDT RELAY	
		WITH 8 PIN OCTAL SOCKET *27E122	
19	1	G.E. MOD CR104B532 SELECTOR SWITCH	S-0146
20	2	" " MINIATURE INDICATING LAMP, ONE RED AND	·
		ONE GREEN. G.E. *CR104C	·
21	3	OILTIGHT PUSHBUTTON. G.E. *CR104A	·
22	1	PANDUIT, *E-1-5X2 LG 6' PLASTIC WIRE DUCT	
		COMPLETE WITH COVER	
23	3	TERMINAL BLOCK (40 PIECES)	S-0146
24	2	ROSEMOUNT *SPS2102P POWER SUPPLY, 3 WATT	S-1331 ('81)
		24VDC WITH SCREW DOWN SOCKET	

FIGURE 19–11
Close-up view of main logic board secured to the sheet metal chassis and molded plastic computer enclosure. (Courtesy Apple Computer, Inc.)

length. All surfaces of the object are connected along their common bend lines. The chassis enclosure shown in Fig. 19–23 has been developed as an inside-up pattern in Fig. 19–24. Both drawings are dimensioned. Note

FIGURE 19–12
Module for expansion cage. (Courtesy Vector Electronics Co.)

cut to the functional outline, with the proper slots and holes punched out or machined as required. These flat, thin pieces of metal can be a variety of thicknesses determined by the needs of the packaging design.

The maximum thickness of sheet metal parts that must be bent, formed, or shaped in some way is limited by how easily the material used can bend. Sheet metal configurations such as enclosures, chassis, cages, and some cabinets are laid out as developments of the original design. A development is the unfolding of the object onto one plane. The **flat pattern development** can then be machined as required, before the forming or bending process.

A variety of electronic parts and enclosures is made from flat sheet stock material. Parts designed to be produced of flat materials are cut from a pattern that is drawn as a development from the orthographic/multiview projection. The complete layout drawing of a part showing its total surface area in one view is then constructed with true length dimensions. This flat plane drawing shows each surface of the part as true shape; therefore, all lines in a development pattern are true

FIGURE 19–13
Fully assembled ten-module cage. (Courtesy Vector Electronics Co.)

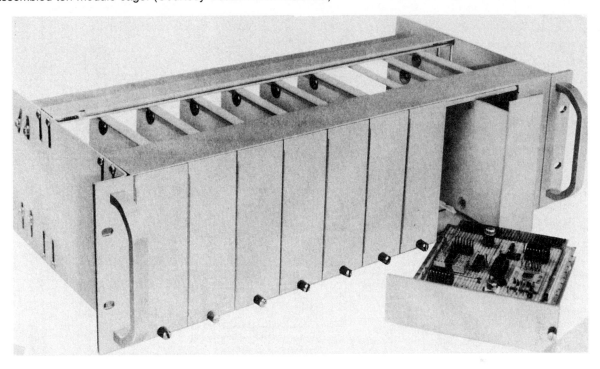

that the dashed lines on the pattern development are *bend lines,* lines along which the flat sheet metal will be bent.

A pattern is made from the original development drawing and used in the shop to scribe or set up the true shape configuration of a part to be produced. The actual flat sheet configuration is then cut according to its pattern. The final operations include bending, folding or rolling, and stretching the part to its required design. Welding, gluing, soldering, bolting, seaming, or riveting can be used to join the parts with *seam* edges.

The four most common shapes that can be developed are the prism, pyramid, cylinder, and cone in all of their variations. The surfaces of an object are normally

FIGURE 19–14
Expansion mounting box with built-in power supply. (Courtesy Digital Equipment Corp.)

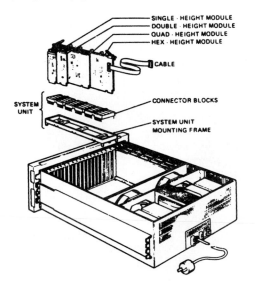

FIGURE 19–15
Circuit board retainer for enclosure. (Courtesy International Electronic Research Corp.)

FIGURE 19–16
Chassis detail. (Courtesy J. Higgins)

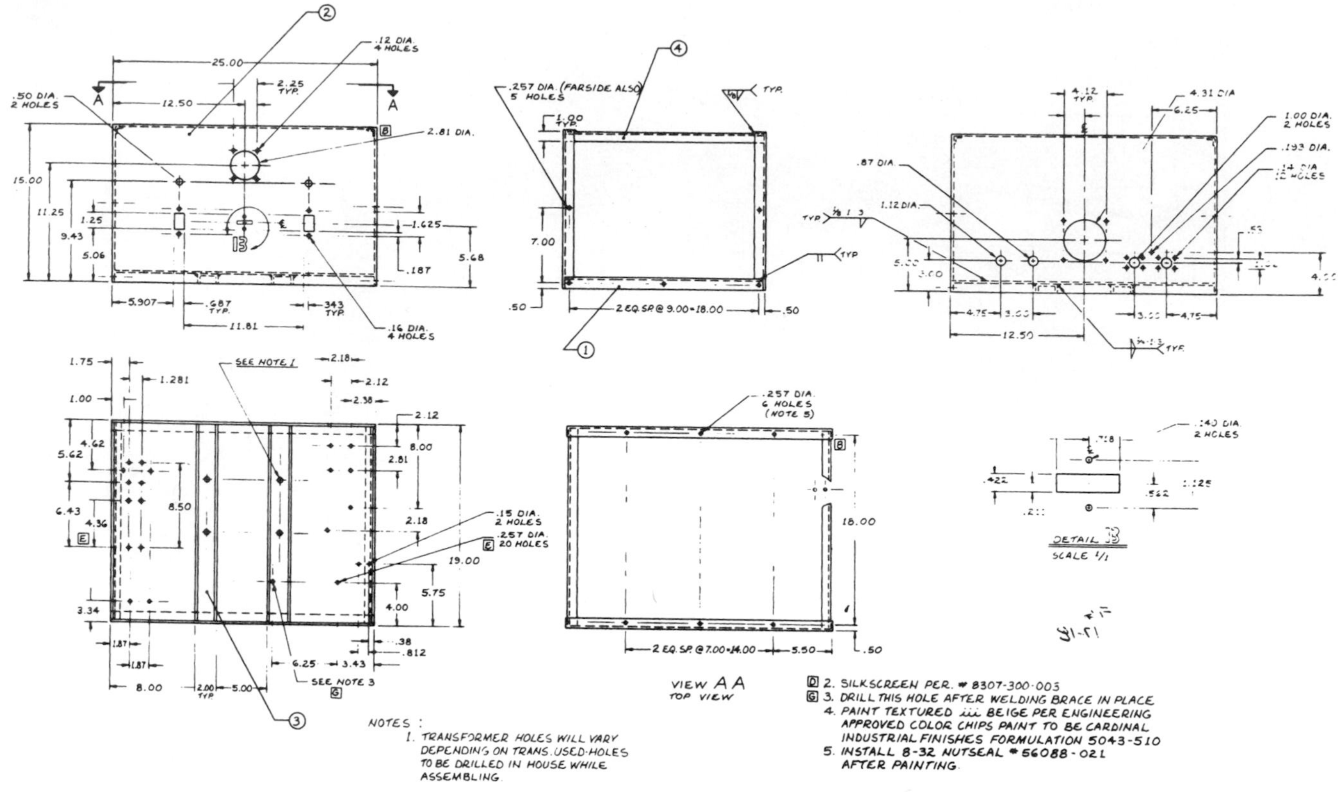

NOTES :

1. TRANSFORMER HOLES WILL VARY DEPENDING ON TRANS. USED. HOLES TO BE DRILLED IN HOUSE WHILE ASSEMBLING.

2. SILKSCREEN PER. # 8307-300-003
3. DRILL THIS HOLE AFTER WELDING BRACE IN PLACE.
4. PAINT TEXTURED iii BEIGE PER ENGINEERING APPROVED COLOR CHIPS PAINT TO BE CARDINAL INDUSTRIAL FINISHES FORMULATION 5043-510
5. INSTALL 8-32 NUTSEAL # 56088-021 AFTER PAINTING.

VIEW A A
TOP VIEW

FIGURE 19–17
Wattmeter. (Courtesy Triplett Corp.)

FIGURE 19–18
Meter panel and enclosure for wattmeter shown in Fig. 19–17. (Courtesy Triplett Corp.)

PANELS AND CONTROL STATION LAYOUT

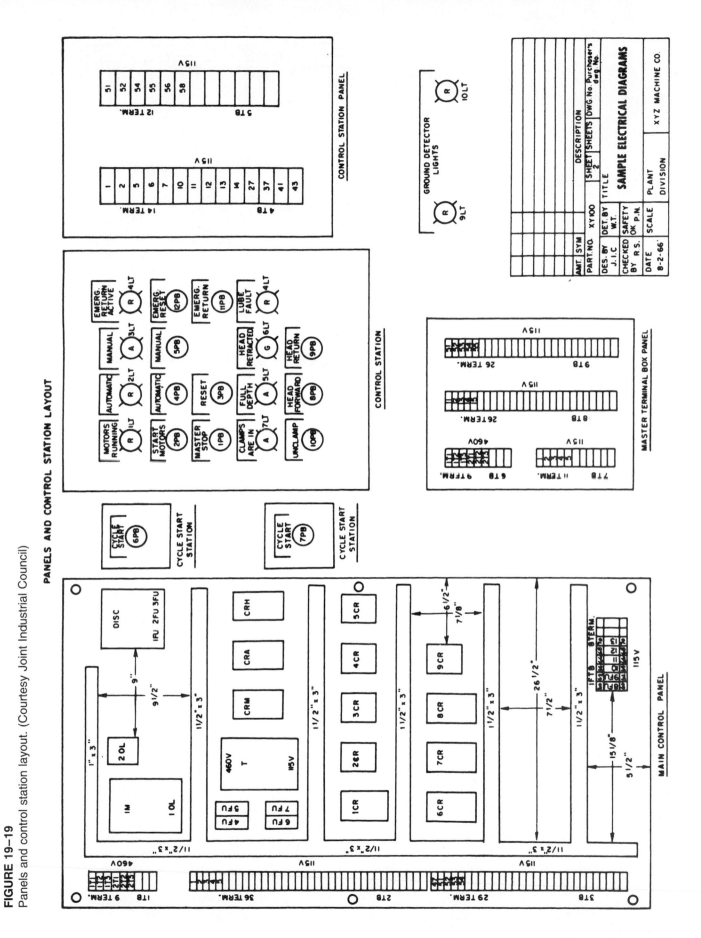

FIGURE 19-20
Rear panel assembly for data-acquisition computer.

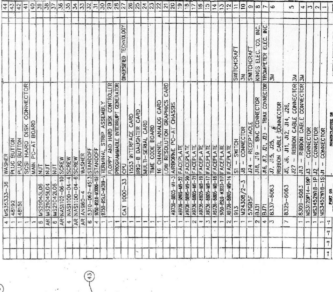

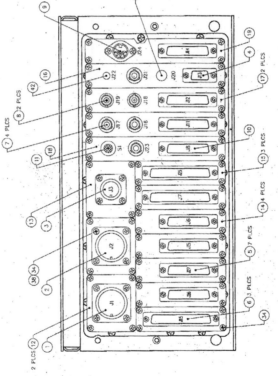

CONNECTOR LAYOUT

NOTES:

⚠ REAR PANEL, MOUNTING BAR AND PLATE REMOVED FOR INTERIOR VIEW

⚠ FRAME, FRONT PANEL OMITTED FOR CLARITY

⚠ HALF-LENGTH MULTI-SERIAL CARD, ITEM #24, OCCUPIES FORWARD SLOT – IRIG-B DAUGHTER CARD, ITEM #25, LOCATED IN REAR.

4. WIRE PER DWG 9750-RSLO-E3404

CONNECTOR FUNCTION TABLE

CONNECTOR	ITEM #	PURPOSE
J1	1	POWER IN
J2	2	POWER OUT
J3	3	POWER I/O
J4	5	SERIAL PORT #1
J5	5	SPARE
J6	6	ADC BUS
J7	10	SPARE
J8	5	SPARE
J11	5	SPARE
J12	4	VIDEO DISPLAY
J13	5	PRINTER
J14	8	FLOPPY DISK DRIVE
J15	7	1553 BUS A
J16	7	1553 BUS B
J17	8	IRIG-B IN
J18	8	IRIG-B OUT
J19	–	SPARE
J20	5	SPARE
J21	7	SPARE
J22	7	SPARE
J23	9	KEYBOARD
J24	4	ANALOG DATA
J25	5	SCSI BUS A
J26	5	SCSI BUS B
J27	11	CPU RESET
S1		

FIGURE 19–21
Data-acquisition hard-drive assembly.

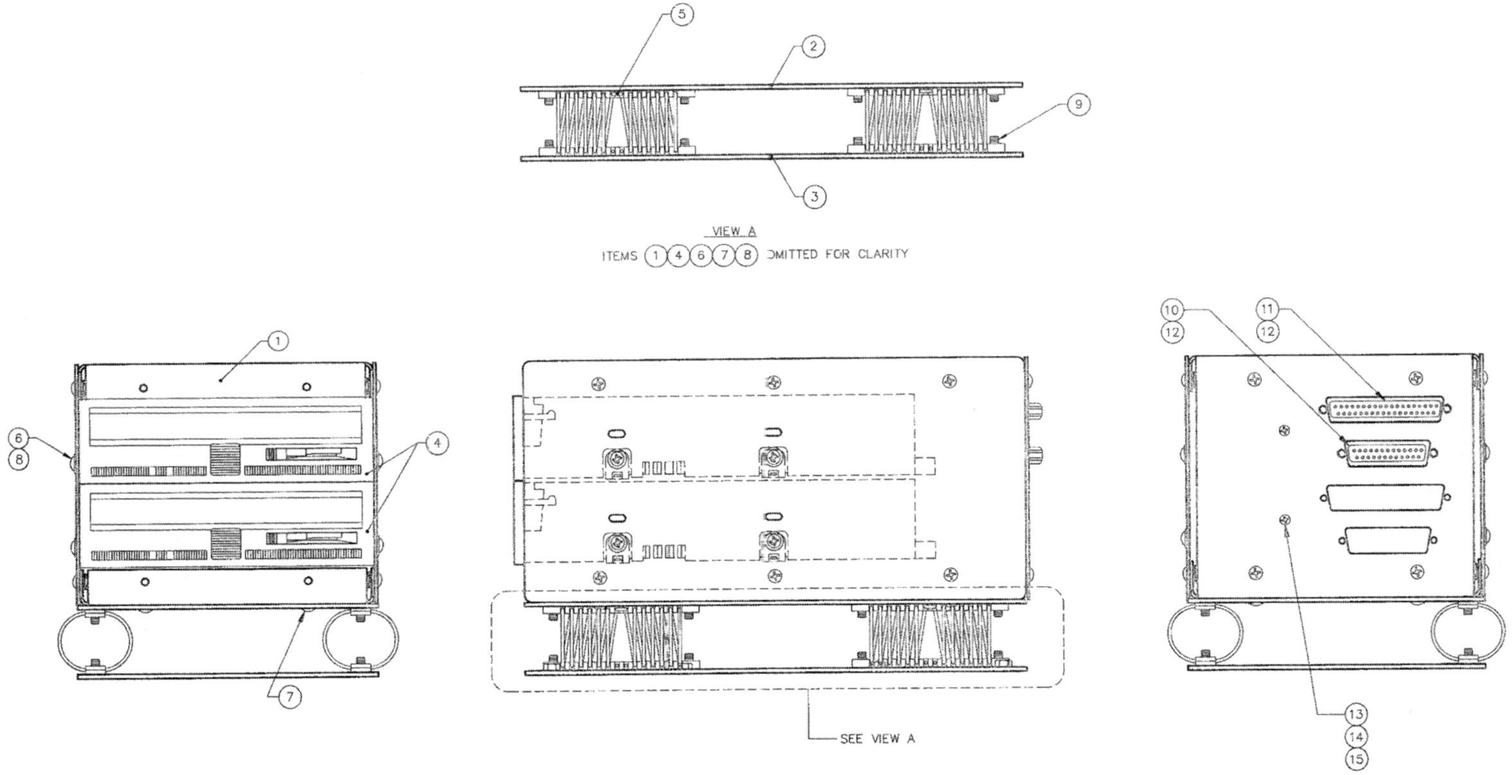

0.15 DIA. HOLE, 2 PLCS.

1.75

1.546

3.326

SEE VIEW A

VIEW A

ITEMS ① ④ ⑥ ⑦ ⑧ OMITTED FOR CLARITY

PARTS LIST

QTY REQD -4	QTY REQD -3	QTY REQD -2	QTY REQD -1	PART OR IDENTIFYING NO	NOMENCLATURE OR DESCRIPTION	MATERIAL/SPECIFICATION	ZONE	ITEM NO.
			2	MS21045L04	NUT			15
			2	NAS1100-04-6	SCREW			14
			4	9750-RSL0-WJ013-5	TERMINAL STRIP			13
			1	5935-01-052-9435	JACKSOCKET, ELECTRICAL CONNECTOR CANNON			12
			1	M24308/4-4	DCC-37P RECTANGULAR CONNECTOR CANNON			11
			1	M24308/4-3	DB-25P RECTANGULAR CONNECTOR CANNON			10
			16	NAS1102-08	SCREW			9
			8	NAS1100-06-8	SCREW			8
			6	NAS1100-3-8	SCREW			7
			AR	AN960-6	WASHER			6
			4	C2-11-B10-12	ISOLATOR	AEROFLEX		5
			1	SQ555	DISK DRIVE	SYQUEST		4
			1	9750-RSL0-WJ013-3	BASE PLATE			3
			1	9750-RSL0-WJ013-1	MOUNTING PLATE			2
			1	A9736-9007-W1	CHASSIS ASSEMBLY			1

527

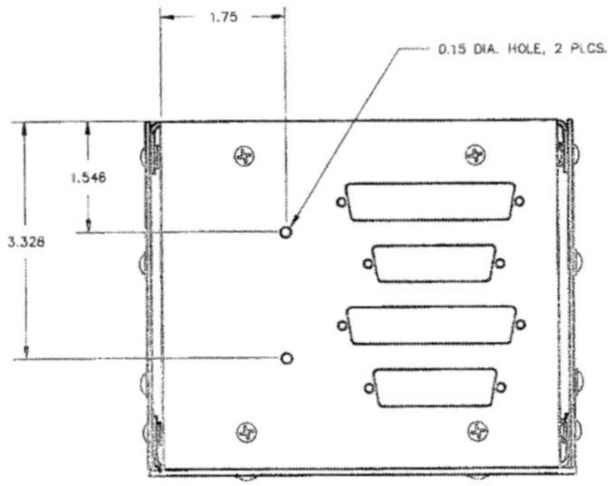

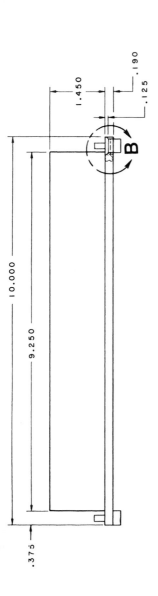

FIGURE 19–22
Panel detail.

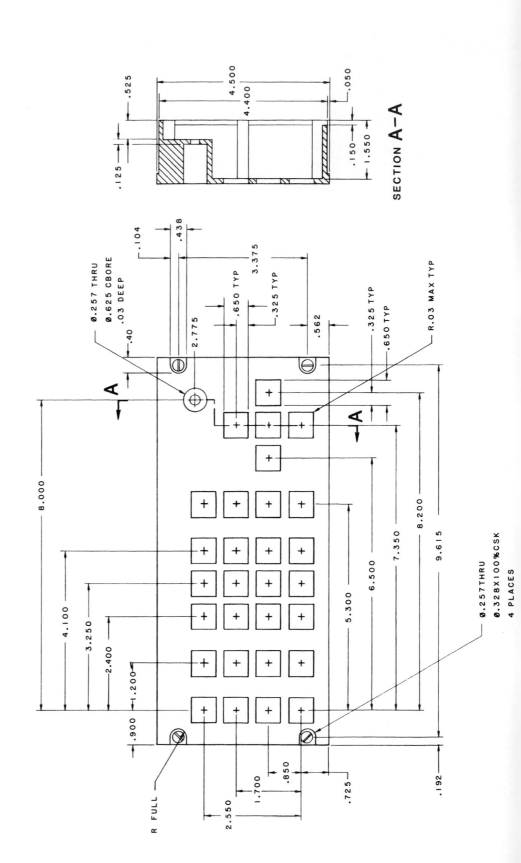

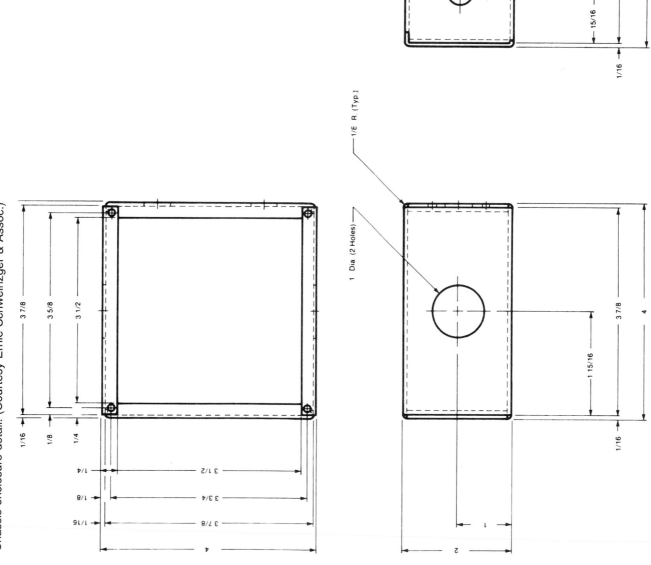
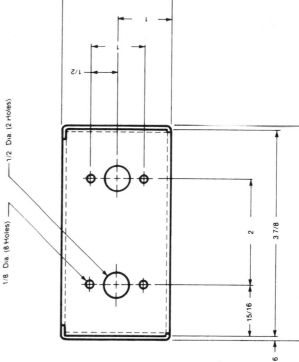

FIGURE 19–24
Development of chassis enclosure shown in Fig. 19–23. (Courtesy Ernie Schweinzger &
Assoc.)

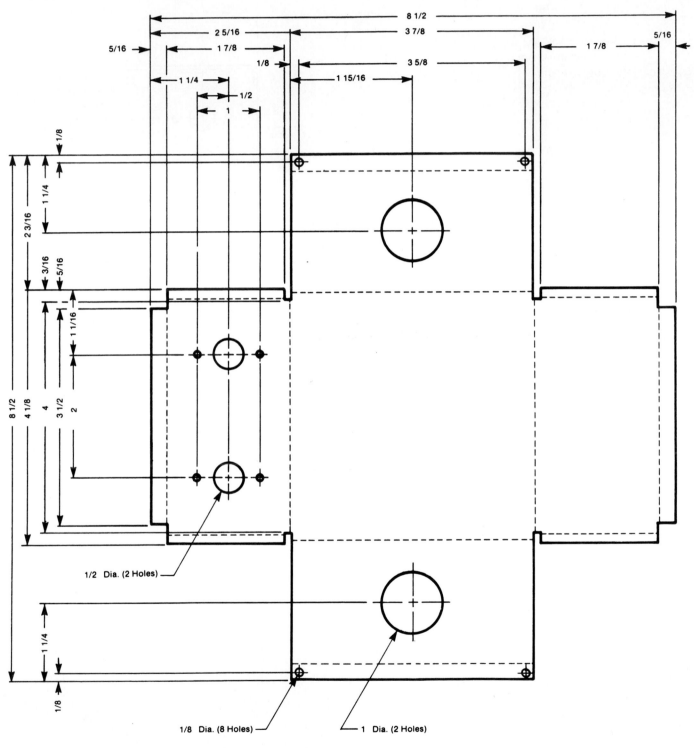

unfolded or unrolled onto the plane of the paper. The actual pattern drawing of the object consists of showing each successive surface as true shape, connected along common edges. On some designs where the part is to be *boxed* or completely enclosed, one of the edge lines serves as a seam for joining the plane surfaces.

Objects can be developed as an *inside-up* pattern drawing; the object is unfolded or unrolled so that the inside surface is face up. In some cases a pattern may be required to show as an *outside-up* development. Normally, patterns are inside-up since the bending is done along the inside bend lines.

Many complex three-dimensional shapes are fabricated from flat sheet materials. The shape to be formed is subdivided into its simplest elements or planes. In electronic packaging a majority of shapes to be fabricated from sheet metal are composed of flat plane surfaces, rectangular box shapes, or two-surface brackets. These shapes are formed from a sheet of material. The sheet is first cut to the proper pattern and then folded or rolled into its three-dimensional form. Aluminum and sheet metal are common materials for this type of part.

The fabricator, working from the design engineer's drawings and specifications, develops pattern drawings of each configuration to be produced in the fabrication shop. These patterns are made to the full size of the object and can only be made after the true lengths have been determined of all lines that lie on the pattern.

A pattern is a drawing composed entirely of true length lines. Therefore, all patterns are true shape and size, like the one in Fig. 19–24. Each development must be drawn accurately so that the final product has the correct shape within given tolerance limits. A bend allowance is usually provided on the pattern drawing to accommodate the space taken by the bending process. An extra tab or lap must be added to the pattern if adjoining edges are to be attached. This is called the *seam edge*. The width of this tab depends on the type of joining process. The length of the lap is normally established along the shortest edge to limit the amount and length of the seam.

Although a pattern is normally drawn full size, a reduced scale model can also be made by the designer during the refinement, analysis, and implementation stages of the design process. Small-scale, accurate models are constructed to aid in design analysis and to explain design variations to the fabricator or purchaser. Models have a distinct advantage over drawings because they can be viewed from any angle and are always seen in their true three-dimensional form. The multiview drawing is needed before a development pattern and model can be made.

A model pattern can be drawn on cardboard. Lightweight cardboard, the kind used for file folders, is an excellent material for making small models. The pattern outline and bend lines are easily transferred, and the cardboard folds well, making sharp corners. Note that tabs must be added along seam edges so that they can be joined with glue or tape. The pattern is transferred onto the cardboard from a carefully executed projection by small pin pricks at controlling points (endpoints of edge or bend lines) or by the use of carbon paper. The pattern is drawn on the cardboard and cut along the outline. The design model can then be analyzed to ensure acceptable function, size, and configuration.

CAD software programs are available for flat pattern developments. The sheet metal enclosure shown pictorially on the screen in Fig. 19–25 is an example of sheet metal design on a CAD system. Flat pattern development on a CAD system improves the speed and accuracy of transferring 3D part models into developed flat patterns. Such programs allow the designer to unfold the planes of the 3D part model on the screen of a graphics workstation. The series of drawings in Fig. 19–26 was generated by a CAD system. This typical CAD pattern software package allows the designer complete flexibility in design, dimensioning, and programmed manufacturing. As can be seen, a CAD system allows the designer and drafter unlimited freedom in the design, layout, detailing, and manufacturing processing. Fig. 19–26 shows the sheet metal form as a 3D part model during the first stage of unbending (originally a

FIGURE 19–25
Pictorial view of sheet metal enclosure. (Courtesy Computervision Corp.)

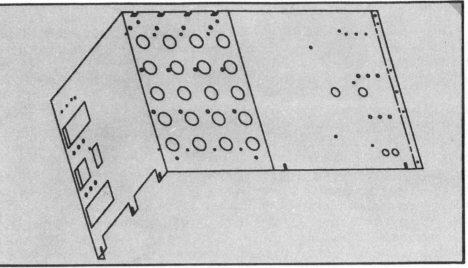

The first stage of unbending is complete.

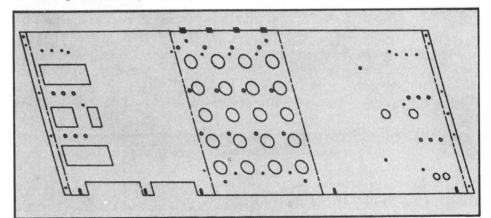

The finished flat pattern can be displayed in an auxilliary view.

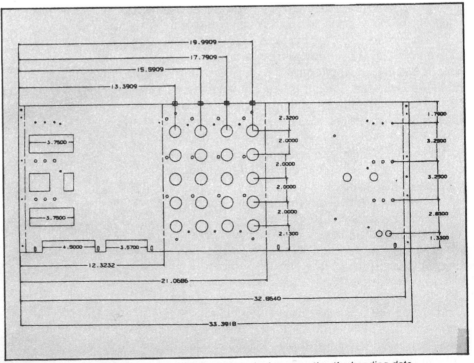

The system automatically dimensions the flat pattern by incorporating the bending data.

U-shape) both as a finished flat auxiliary view and as a dimensioned flat pattern shop drawing.

SHEET METAL DIMENSIONING

When a sophisticated CAD program is used to design, lay out, and detail a sheet metal part, dimensioning is completed automatically. The sheet metal heat sink in Fig. 19–27 was done with a CAD system.

In most cases, sheet metal parts are designed and detailed by the traditional manual drafting techniques. In Fig. 19–24, the sheet metal chassis was drawn and dimensioned with fractions. The use of fractions for sheet metal drawings is a common American practice where there is no need for close tolerance fabrication and machining. The dimensioned chassis drawing shown in Fig. 19–28 and the mounting plate detail in Fig. 19–29 were hand drawn. These drawings were dimensioned with datum lines and rectangular coordinate dimensioning methods. Figure 19–28 was dimensioned with a rectangular coordinate system without connecting dimension lines. The notes in the lower left-hand corner of the drawing give the material, finish, and bending radius for the part. The dimensions and text in Fig. 19–29 were completed with a Kroy lettering set.

The methods and procedures presented in Chapter 4 should be closely adhered to for sheet metal layout drawings and all other types of enclosure, chassis, and cabinet parts regardless of material type. Note that the use of rectangular coordinate dimensioning without dimensioning lines but with hole charts expedites the programming stage when numerical control is used for the manufacturing process. Holing charts and coordinate dimensioning eliminate the need for calculating the location of each hole, slot, or other feature, since the dimensions are taken from a predetermined datum or reference line established with a zero point and X–Y axes.

This process is especially beneficial when the part to be machined is a simple, flat (or thin), one-surface object with many holes and slot requirements, as in the mounting plate in Fig. 19–29. When the object is a typical rectangular, boxlike chassis with holes and slot cutouts on more than one surface, this process is also useful if the part is dimensioned and then machined as a flat pattern development. In Fig. 19–28, the part is projected as a multiview dimensioned drawing, not a flat pattern development. The dimensioning for the two sides is not datum line dimensioning as on the front surface. Possibly, the machining processes are to be completed after the part is laid out, developed, and bent to the final design configuration shown in the figure.

CHASSIS MARKINGS

In general, most chassis are marked for component placement and for general parts numbers. These markings are shown on the detail drawings where applicable. Silkscreening the numbers or other characters on the part is widely used. Figure 19–28 provides the silkscreen specifications in the drawing notes: Silkscreen per dwg. 18014-201. Here, only the part number is marked on the part, not the component locations.

FASTENERS AND CONNECTION PROCESSES

Electronic packaging uses a variety of methods to fasten, connect, and attach electronic components, modules, and subassemblies to their respective chassis, mounting plates, panels, brackets, enclosures, and cabinets. The various metal and plastic enclosure parts are also fastened and connected with one another to form functioning units. The type of fastening or connection method must be determined by the use, function, environment, and life-length requirements of the equipment package.

Each of the different classifications of electronic and electrical items has different design specifications. Military equipment must be more rugged, vibration and shock resistant, and noncorrosive than industrial or consumer products. Space and satellite communications electronic equipment must withstand special environmental characteristics, including excessive temperature, humidity, or pressure variations and weight restrictions. Commercial, military, and industrial products need to be easily disassembled for repair, servicing, and troubleshooting.

Fastening and connecting methods include *permanent connecting processes;* welding, brazing, soldering, riveting, bending/crimping. Fastening methods also include *nonpermanent fasteners:* pins, clips, spring locks, screws and bolts, nuts and washers. Methods that do not allow easy disassembly are considered *permanent* connections, and those that permit simple, quick disassembly are considered *nonpermanent.*

Welding may be any number of types, although spot welding and arc welding predominate. Spot welding is a commonly applied procedure for chassis and cabinetry assembly. In spot welding, pressure and heat are applied in the form of electric current to the two surfaces that need to be attached. Crimping may also be used in this process, or it may be used alone. Arc welding, soldering, and brazing are also used where the parts must be made into one unit. In Fig. 19–16, the chassis is to be formed from sheet metal and standard steel angle and channel shapes. The total unit is welded.

FIGURE 19-27
Heat sink.

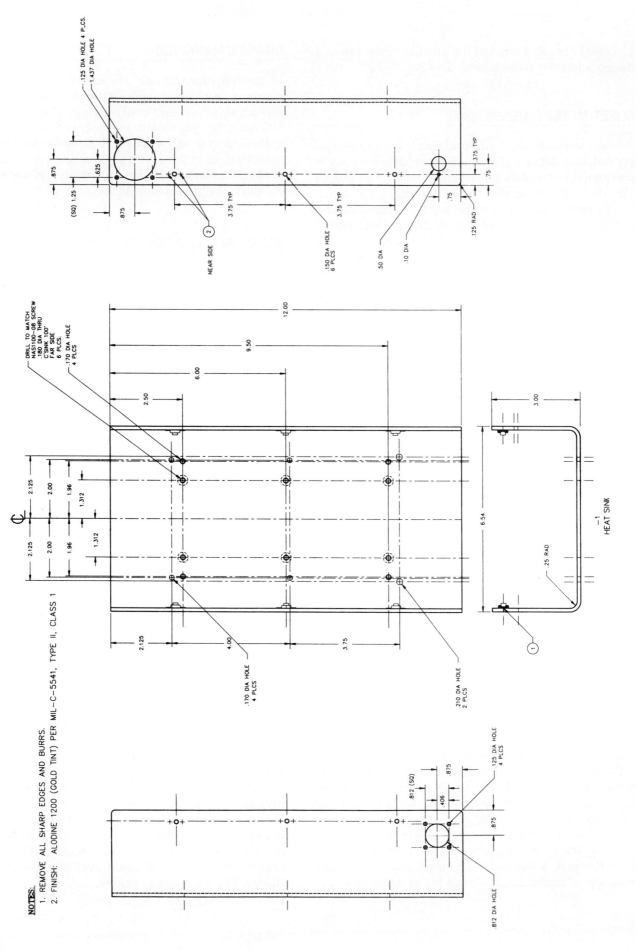

NOTES:
1. REMOVE ALL SHARP EDGES AND BURRS.
2. FINISH: ALODINE 1200 (GOLD TINT) PER MIL-C-5541, TYPE II, CLASS 1

FIGURE 19-28
Dimensioned chassis.

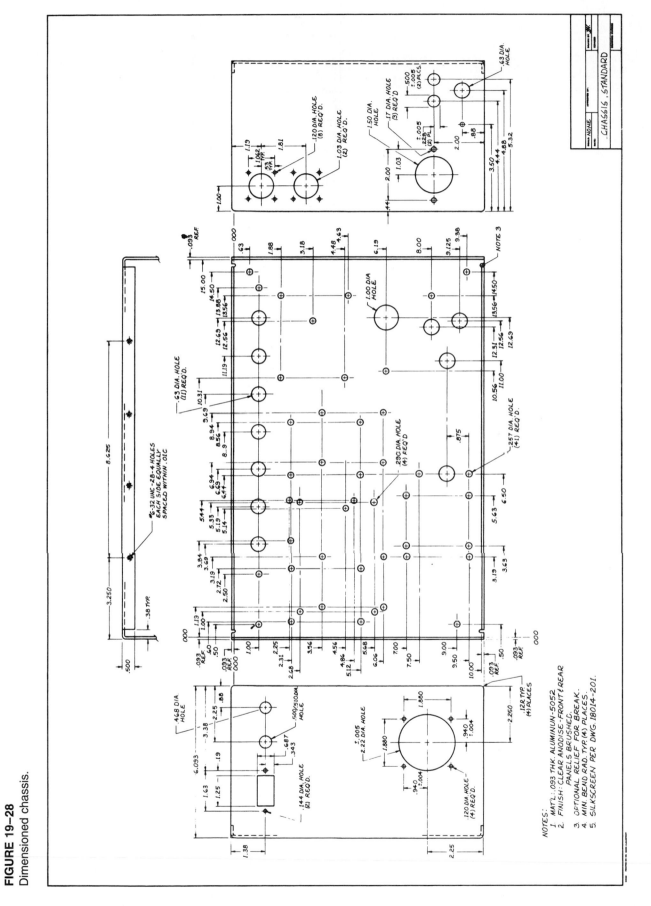

535

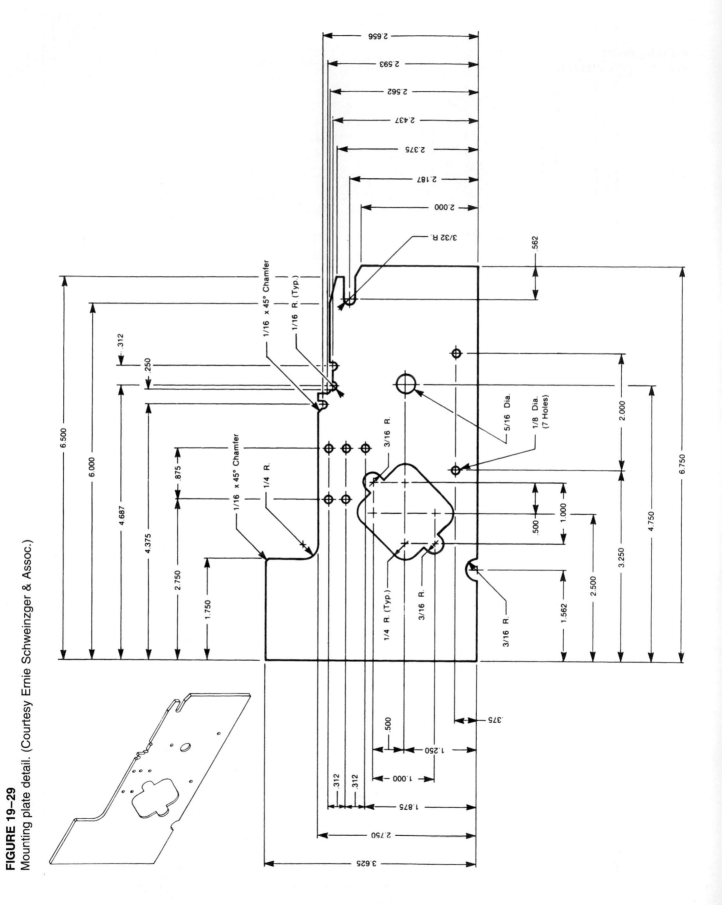

FIGURE 19-29
Mounting plate detail. (Courtesy Ernie Schweinzger & Assoc.)

FIGURE 19–30
Fasteners. (Courtesy Holo-Krome Co.)

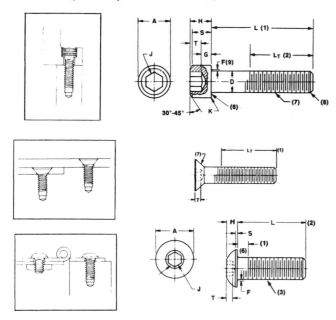

FIGURE 19–31
Mounting bracket and fasteners. (Courtesy Digital Equipment Corp.)

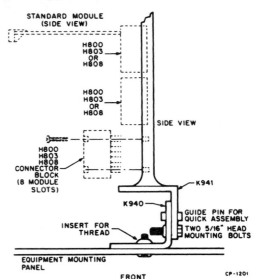

Rivets are a simple, quick, and inexpensive connection method used throughout the electronics industry. A wide variety of shapes and types of rivets is available. Rivets are ideal for joining aluminum or steel sheet metal parts. Rivet materials include aluminum, brass, and steel.

The type and variety of fastening methods found in the electronics industry are extensive. In Fig. 19–30, socket cap, flathead screws, and buttonhead screws are shown. The flathead and caphead varieties are shown

flush with the upper part surface. The mounting bracket assembly detail shown in Fig. 19–31 uses bolts and nuts to secure the bracket and mounting panel. In Fig. 19–32, the isometric exploded pictorial provides an assembly view and parts list. The panel is secured to the bracket with (Phillips) flathead screws and speed nuts. Nuts and washers of all kinds are used to assemble electronic equipment. In Fig. 19–33, fastenings for military electronic equipment are provided along with their associated military standard. The table includes washers, pins, screws, and nuts.

FIGURE 19–32
Panel assembly and fasteners.
(Courtesy Digital Equipment
Corp.)

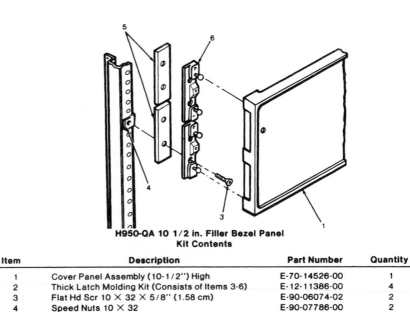

H950-QA 10 1/2 in. Filler Bezel Panel
Kit Contents

Item	Description	Part Number	Quantity
1	Cover Panel Assembly (10-1/2") High	E-70-14526-00	1
2	Thick Latch Molding Kit (Consists of Items 3-6)	E-12-11386-00	4
3	Flat Hd Scr 10 × 32 × 5/8" (1.58 cm)	E-90-06074-02	2
4	Speed Nuts 10 × 32	E-90-07786-00	2
5	Spacer	E-12-14780-00	1
6	Latch Molding	E-12-09224-00	1

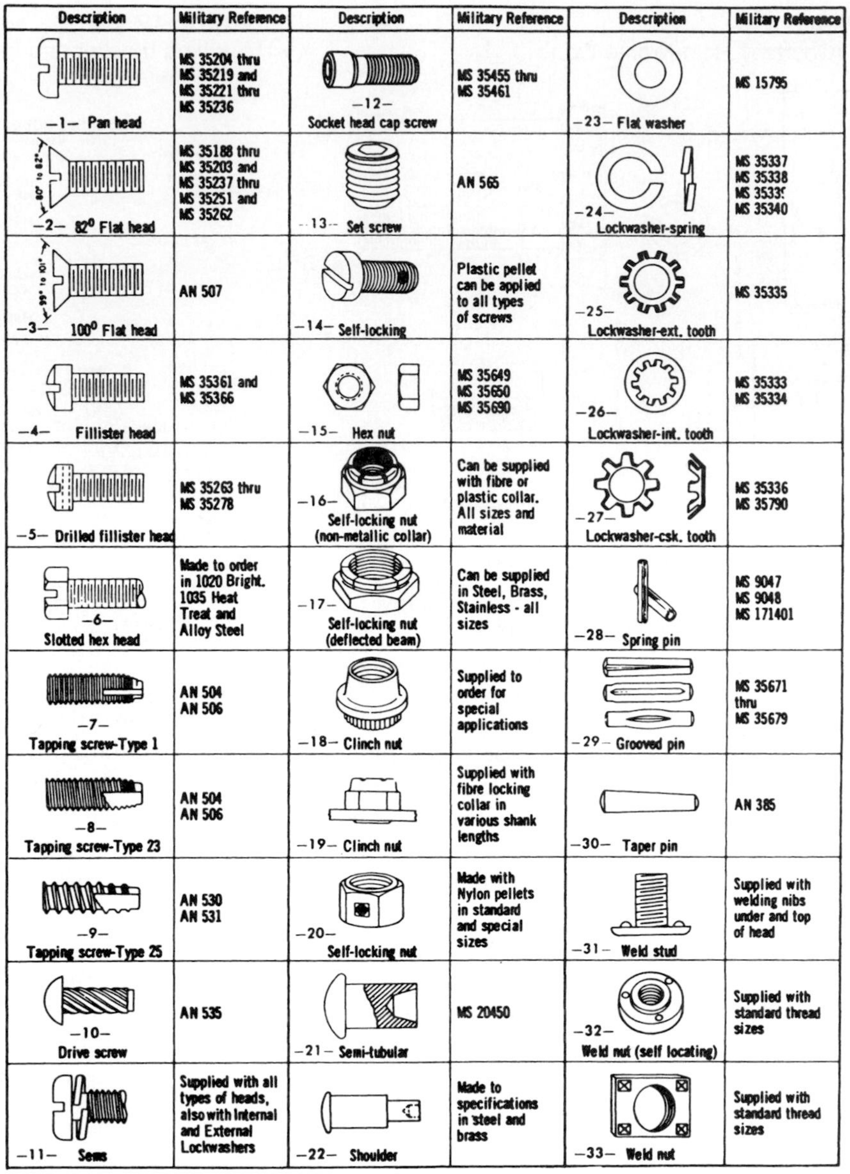

FIGURE 19–33
Fasteners for military electronic equipment. (Courtesy of Federal Screw Products, Inc.)

Description	Military Reference	Description	Military Reference	Description	Military Reference
—1— Pan head	MS 35204 thru MS 35219 and MS 35221 thru MS 35236	—12— Socket head cap screw	MS 35455 thru MS 35461	—23— Flat washer	MS 15795
—2— 82° Flat head	MS 35188 thru MS 35203 and MS 35237 thru MS 35251 and MS 35262	—13— Set screw	AN 565	—24— Lockwasher-spring	MS 35337 MS 35338 MS 35339 MS 35340
—3— 100° Flat head	AN 507	—14— Self-locking	Plastic pellet can be applied to all types of screws	—25— Lockwasher-ext. tooth	MS 35335
—4— Fillister head	MS 35361 and MS 35366	—15— Hex nut	MS 35649 MS 35650 MS 35690	—26— Lockwasher-int. tooth	MS 35333 MS 35334
—5— Drilled fillister head	MS 35263 thru MS 35278	—16— Self-locking nut (non-metallic collar)	Can be supplied with fibre or plastic collar. All sizes and material	—27— Lockwasher-csk. tooth	MS 35336 MS 35790
—6— Slotted hex head	Made to order in 1020 Bright. 1035 Heat Treat and Alloy Steel	—17— Self-locking nut (deflected beam)	Can be supplied in Steel, Brass, Stainless - all sizes	—28— Spring pin	MS 9047 MS 9048 MS 171401
—7— Tapping screw-Type 1	AN 504 AN 506	—18— Clinch nut	Supplied to order for special applications	—29— Grooved pin	MS 35671 thru MS 35679
—8— Tapping screw-Type 23	AN 504 AN 506	—19— Clinch nut	Supplied with fibre locking collar in various shank lengths	—30— Taper pin	AN 385
—9— Tapping screw-Type 25	AN 530 AN 531	—20— Self-locking nut	Made with Nylon pellets in standard and special sizes	—31— Weld stud	Supplied with welding nibs under and top of head
—10— Drive screw	AN 535	—21— Semi-tubular	MS 20450	—32— Weld nut (self locating)	Supplied with standard thread sizes
—11— Sems	Supplied with all types of heads, also with Internal and External Lockwashers	—22— Shoulder	Made to specifications in steel and brass	—33— Weld nut	Supplied with standard thread sizes

REVIEW QUESTIONS

19–1 Name four considerations in the design of electronic packaging.

19–2 Electronic products produced for the private sector are called:
    a. Industrial   c. Commercial
    b. Computer   d. Military

19–3 Name four environmental considerations in the design of electronic packaging.

19–4 Cabinets for computer products are normally made of (   ), whereas military and communications cabinets are made of (   ).

19–5 *True or false:* A chassis is a sheet metal part on which electronic components and modules are mounted.

19–6 Name four items that are normally mounted on the front panel.

19–7 Flathead screws and socket head cap screws are used where the fastener must be (   ) with the mounting surface.

19–8 Name four methods of fastening or connecting.

19–9 Which fastening methods are considered permanent?

19–10 Sheet metal flat pattern developments are normally formed into their final design shape by (   ).

PROBLEMS

19–1 Redraw the chassis box shown in the accompanying figure. Develop the part and dimension completely. Scale the drawing from the text and assume that it is shown at ¼ scale.

PROBLEM 19–1
Chassis box.
(Courtesy of J. Higgins)

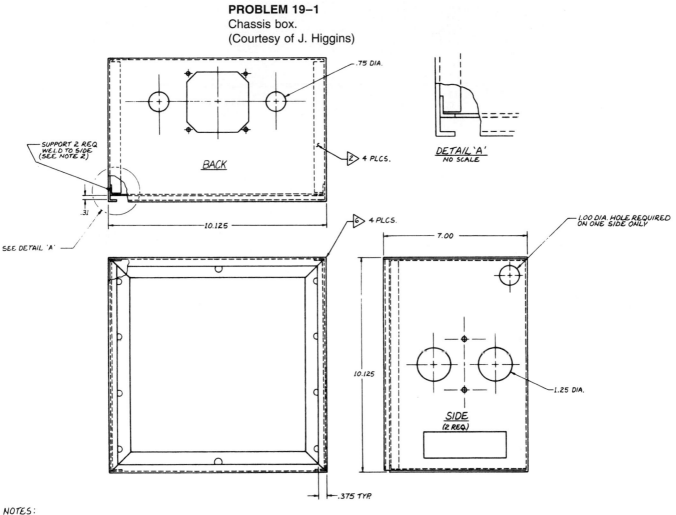

NOTES:
1. MATERIAL 16 GA. SHEET STEEL.
2. SPOT WELD SEAMS, MAX. SPACING 1.5 INCHES
3. BEND RADIUS .06 MAX. ALL BENDS.
4. PAINT WITH BLACK TEXTURED, FED. STD. 595 A #27038.
5. FILL AND SMOOTH CORNER SEAMS BEFORE PAINT
6. DIMENSIONS GIVEN ARE FINISHED DIMENSIONS.

SCALE ¾ : 1

19–2 Draw and dimension the base/tray in the accompanying figure.

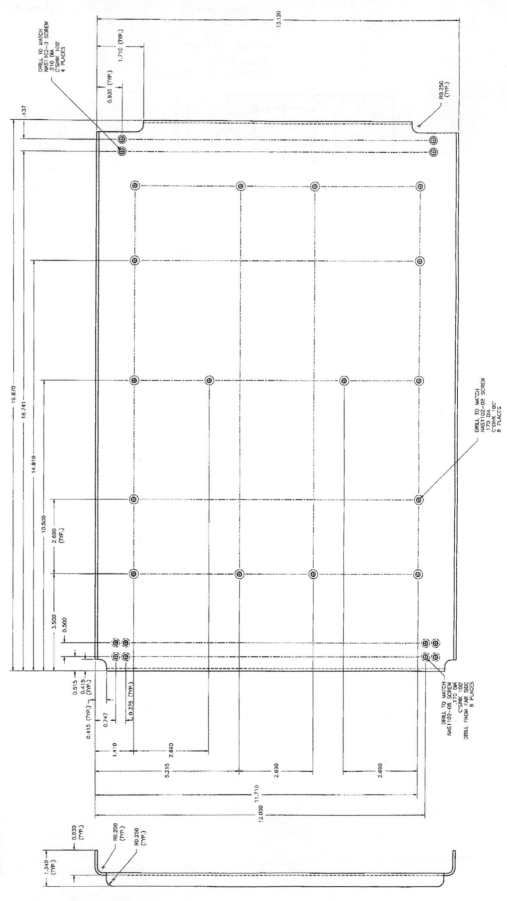

PROBLEM 19–2
Base/tray.

19–3 Draw, develop, and dimension the mounting bracket shown.

PROBLEM 19–3
Mounting bracket.

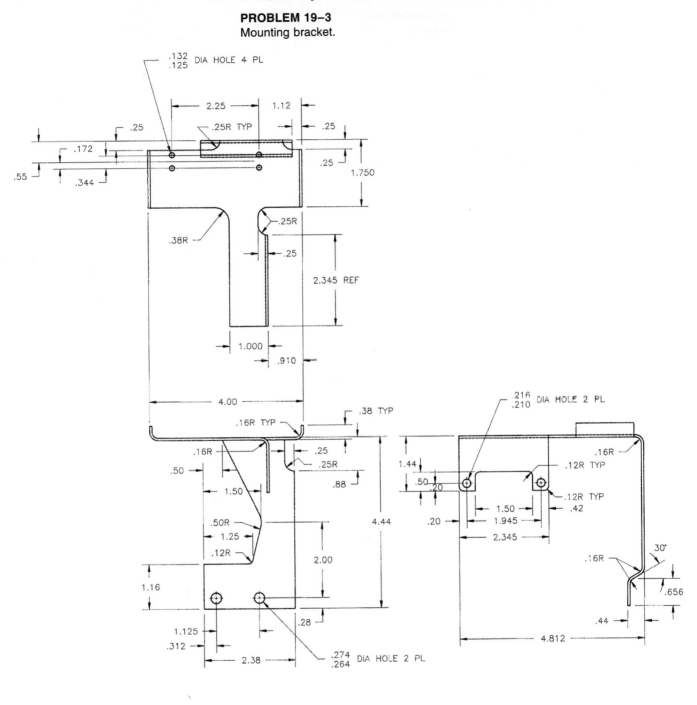

−105 POT MOUNTING BRACKET

MAT'L: 6061−T6 AL ALY .063 THICK

19–4 Draw, develop, and dimension the sheet metal part in the accompanying figure.

PROBLEM 19–4
Sheet metal part.

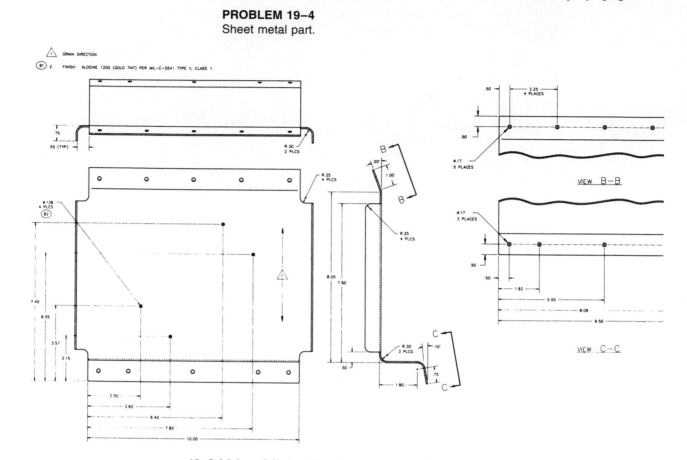

19–5 Make a full-sized development pattern for the chassis enclosure shown in Figs. 19–23 and 19–24. Use lightweight cardboard.

19–6 Draw the chassis shown in Fig. 19–28 and dimension it using a hole chart. Establish the location of holes based on view (right side, front, and left side) and datum line dimension. Use specifications from Chapter 4 for this project.

19–7 Draw the mounting plate shown in Fig. 19–29. Dimension the part with rectangular coordinate dimensions without dimension lines. Use a hole identification list. Use dual dimensioning.

19–8 Draw and dimension the heat sink in Fig. 19–27.

19–9 Draw the chassis shown in Fig. 19–16. Establish an X and Y axis and dimension using them as datums. For clarity, dimension lines may be eliminated.

19–10 Develop the sheet metal enclosure shown in Fig. 19–28 and dimension completely. Use datum lines and dual dimensioning.

19–11 Design a complete electronic package for the circuit selected by your instructor.

19–12 Design a chassis for the PC board selected by your instructor.

Appendix A

TABLE A–1
Inch/Metric Unit Equivalents

Fraction	DECIMAL EQUIVALENT Customary (in.)	DECIMAL EQUIVALENT Metric (mm)	Fraction	DECIMAL EQUIVALENT Customary (in.)	DECIMAL EQUIVALENT Metric (mm)
1/64	.015625	.3969	33/64	.515625	13.0969
1/32	.03125	.7938	17/32	.53125	13.4938
3/64	.046875	1.1906	35/64	.546875	13.8906
1/16	.0625	1.5875	9/16	.5625	14.2875
5/64	.078125	1.9844	37/64	.578125	14.6844
3/32	.09375	2.3813	19/32	.59375	15.0813
7/64	.109375	2.7781	39/64	.609375	15.4781
1/8	.1250	3.1750	5/8	.6250	15.8750
9/64	.140625	3.5719	41/64	.640625	16.2719
5/32	.15625	3.9688	21/32	.65625	16.6688
11/64	.171875	4.3656	43/64	.671875	17.0656
3/16	.1875	4.7625	11/16	.6875	17.4625
13/64	.203125	5.1594	45/64	.703125	17.8594
7/32	.21875	5.5563	23/32	.71875	18.2563
15/64	.234375	5.9531	47/64	.734375	18.6531
1/4	.250	6.3500	3/4	.750	19.0500
17/64	.265625	6.7469	49/64	.765625	19.4469
9/32	.28125	7.1438	25/32	.78125	19.8438
19/64	.296875	7.5406	51/64	.796875	20.2406
5/16	.3125	7.9375	13/16	.8125	20.6375
21/64	.328125	8.3384	53/64	.828125	21.0344
11/32	.34375	8.7313	27/32	.84375	21.4313
23/64	.359375	9.1281	55/64	.859375	21.8281
3/8	.3750	9.5250	7/8	.8750	22.2250
25/64	.390625	9.9219	57/64	.890625	22.6219
13/32	.40625	10.3188	29/32	.90625	23.0188
27/64	.421875	10.7156	59/64	.921875	23.4156
7/16	.4375	11.1125	15/16	.9375	23.8125
29/64	.453125	11.5094	61/64	.953125	24.2094
15/32	.46875	11.9063	31/32	.96875	24.6063
31/64	.484375	12.3031	63/64	.984375	25.0031
1/2	.500	12.7000	1	1.000	25.4000

TABLE A–2
Fraction, Decimal, and
Millimeter Conversions

Millimeters to Inches		Inches to Millimeters		Inches (Decimals) to Millimeters	
mm	in.	in.	mm	in.	mm
1 =	.0394	$\frac{1}{32}$ =	.794	.01 =	.254
2 =	.0787	$\frac{1}{16}$ =	1.587	.02 =	.508
3 =	.1181	$\frac{3}{32}$ =	2.381	.03 =	.762
4 =	.1575	$\frac{1}{8}$ =	3.175	.04 =	1.016
5 =	.1969	$\frac{5}{32}$ =	3.968	.05 =	1.270
6 =	.2362	$\frac{3}{16}$ =	4.762	.06 =	1.524
7 =	.2756	$\frac{7}{32}$ =	5.556	.07 =	1.778
8 =	.3150	$\frac{1}{4}$ =	6.350	.08 =	2.032
9 =	.3543	$\frac{9}{32}$ =	7.144	.09 =	2.286
10 =	.3937	$\frac{5}{16}$ =	7.937	.10 =	2.540
11 =	.4331	$\frac{11}{32}$ =	8.731	.12 =	3.048
12 =	.4724	$\frac{3}{8}$ =	9.525	.14 =	3.556
13 =	.5118	$\frac{13}{32}$ =	10.319	.16 =	4.064
14 =	.5512	$\frac{7}{16}$ =	11.112	.18 =	4.572
15 =	.5906	$\frac{15}{32}$ =	11.906	.20 =	5.080
16 =	.6299	$\frac{1}{2}$ =	12.699	.22 =	5.588
17 =	.6693	$\frac{17}{32}$ =	13.493	.25 =	6.350
18 =	.7087	$\frac{9}{16}$ =	14.287	.26 =	6.604
19 =	.7480	$\frac{19}{32}$ =	15.081	.28 =	7.112
20 =	.7874	$\frac{5}{8}$ =	15.875	.30 =	7.620
21 =	.8268	$\frac{21}{32}$ =	16.668	.32 =	8.128
22 =	.8662	$\frac{11}{16}$ =	17.462	.34 =	8.636
23 =	.9055	$\frac{23}{32}$ =	18.256	.36 =	9.144
24 =	.9449	$\frac{3}{4}$ =	19.050	.38 =	9.652
25 =	.9843	$\frac{25}{32}$ =	19.843	.40 =	10.160
26 =	1.0236	$\frac{13}{16}$ =	20.637	.50 =	12.699
27 =	1.0630	$\frac{27}{32}$ =	21.431	.60 =	15.240
28 =	1.1024	$\frac{7}{8}$ =	22.225	.70 =	17.780
29 =	1.1418	$\frac{29}{32}$ =	23.018	.80 =	20.320
30 =	1.1811	$\frac{15}{16}$ =	23.812	.90 =	22.860
50 =	1.9685	1 =	25.400	1.00 =	25.400
100 =	3.9370	12 =	304.800	2.00 =	50.800

TABLE A–3
Standard Unified Thread Series[a]

Inch		Metric Equiv.	COARSE (NC) (UNC)		FINE (NF) (UNF)		EXTRA-FINE (NEF) (UNEF)	
PRESENT UNIFIED THREAD NOMINAL SIZE—DIAMETER			Threads Per Inch	Tap Drill[b]	Threads Per Inch	Tap Drill[b]	Threads Per Inch	Tap Drill[b]
.060	0	1.52	—	—	80	$\frac{3}{64}$	—	—
.073	1	1.85	64	No. 53	72	No. 53	—	—
.086	2	2.18	56	No. 50	64	No. 50	—	—
.099	3	2.51	48	No. 47	56	No. 45	—	—
.112	4	2.84	**40**	No. 43	48	No. 42	—	—
.125	5	3.17	40	No. 38	44	No. 37	—	—
.138	6	3.50	**32**	No. 36	40	No. 33	—	—
.164	8	4.16	**32**	No. 29	36	No. 29	—	—
.190	10	4.83	**24**	No. 25	**32**	No. 21	—	—
.216	12	5.49	24	No. 16	28	No. 14	32	No. 13
.250	$\frac{1}{4}$	6.35	**20**	No. 7	**28**	No. 3	32	No. 2
.3125	$\frac{5}{16}$	7.94	**18**	F	24	I	32	K
.375	$\frac{3}{8}$	9.52	**16**	$\frac{5}{16}$	24	Q	32	S
.4375	$\frac{7}{16}$	11.11	**14**	U	20	$\frac{25}{64}$	**28**	Y
.500	$\frac{1}{2}$	12.70	**13**	$\frac{27}{64}$	20	$\frac{29}{64}$	**28**	$\frac{15}{32}$
.5625	$\frac{9}{16}$	14.29	**12**	$\frac{31}{64}$	18	$\frac{33}{64}$	24	$\frac{17}{32}$
.625	$\frac{5}{8}$	15.87	**11**	$\frac{17}{32}$	18	$\frac{37}{64}$	24	$\frac{19}{32}$
.6875	$\frac{11}{16}$	17.46	—	—	—	—	24	$\frac{41}{64}$
.750	$\frac{3}{4}$	19.05	**10**	$\frac{21}{32}$	16	$\frac{11}{16}$	**20**	$\frac{45}{64}$
.8125	$\frac{13}{16}$	20.64	—	—	—	—	**20**	$\frac{49}{64}$
.875	$\frac{7}{8}$	22.22	**9**	$\frac{49}{64}$	14	$\frac{13}{16}$	**20**	$\frac{53}{64}$
.9375	$\frac{15}{16}$	23.81	—	—	—	—	**20**	$\frac{57}{64}$
1.000	1	25.40	**8**	$\frac{7}{8}$	12	$\frac{59}{64}$	**20**	$\frac{61}{64}$
1.0625	$1\frac{1}{16}$	26.99	—	—	—	—	18	1
1.125	$1\frac{1}{8}$	28.57	**7**	$\frac{63}{64}$	12	$1\frac{3}{64}$	18	$1\frac{5}{64}$
1.1875	$1\frac{3}{16}$	30.16	—	—	—	—	18	$1\frac{9}{64}$
1.250	$1\frac{1}{4}$	31.75	**7**	$1\frac{7}{64}$	12	$1\frac{11}{64}$	18	$1\frac{13}{64}$
1.3125	$1\frac{5}{16}$	33.34	—	—	—	—	18	$1\frac{17}{64}$
1.375	$1\frac{3}{8}$	34.92	**6**	$1\frac{13}{64}$	12	$1\frac{19}{64}$	18	$1\frac{5}{16}$
1.4375	$1\frac{7}{16}$	36.51	—	—	—	—	18	$1\frac{3}{8}$
1.500	$1\frac{1}{2}$	38.10	**6**	$1\frac{21}{64}$	12	$1\frac{27}{64}$	18	$1\frac{29}{64}$
1.5625	$1\frac{9}{16}$	39.69	—	—	—	—	18	$1\frac{1}{2}$
1.625	$1\frac{5}{8}$	41.27	—	—	—	—	18	$1\frac{9}{16}$
1.6875	$1\frac{11}{16}$	42.86	—	—	—	—	18	$1\frac{5}{8}$
1.750	$1\frac{3}{4}$	44.45	**5**	$1\frac{35}{64}$	—	—	**16**	$1\frac{11}{16}$
2.000	2	50.80	$4\frac{1}{2}$	$1\frac{25}{32}$	—	—	**16**	$1\frac{15}{16}$
2.250	$2\frac{1}{4}$	57.15	$4\frac{1}{2}$	$2\frac{1}{32}$	—	—	—	—
2.500	$2\frac{1}{2}$	63.50	4	$2\frac{1}{4}$	—	—	—	—
2.750	$2\frac{3}{4}$	69.85	4	$2\frac{1}{2}$	—	—	—	—
3.000	3	76.20	4	$2\frac{3}{4}$	—	—	—	—
3.250	$3\frac{1}{4}$	82.55	4	3	—	—	—	—
3.500	$3\frac{1}{2}$	88.90	4	$3\frac{1}{4}$	—	—	—	—
3.750	$3\frac{3}{4}$	95.25	4	$3\frac{1}{2}$	—	—	—	—
4.000	4	101.60	4	$3\frac{3}{4}$	—	—	—	—

[a]Adapted from ANSI B1.1-1960.
Bold type indicates Unified threads. To be designated UNC or UNF.
Unified Standard—Classes 1A, 2A, 3A, 1B, 2B, and 3B.
For recommended hole-size limits before threading, see Tables 38 and 39, ANSI B1.1-1960.
[b]Tap drill for a 75% thread (not Unified—American Standard).
Bold-type sizes smaller than $\frac{1}{4}$ in. are accepted for limited applications by the British, but the symbols NC or NF, as applicable, are retained.

TABLE A–4
Thread Sizes and Dimensions: Fraction/Decimal/Metric

NOMINAL SIZE		DIAMETER (MAJOR)		DIAMETER (MINOR)		TAP DRILL (FOR 75% TH'D.)			THREADS PER INCH		PITCH (mm)		T.P.I. (APPROX.)	
Inch	mm	Inch	mm	Inch	mm	Drill	Inch	mm	UNC	UNF	Coarse	Fine	Coarse	Fine
—	M1.4	.055	1.397	—	—	—	—	—	—	—	.3	.2	85	127
0	—	.060	1.524	.0438	1.092	3/64	.0469	1.168	—	80	—	—	—	—
—	M1.6	.063	1.600	—	—	—	—	—	—	—	.35	.2	74	127
1	—	.073	1.854	.0527	1.320	53	.0595	1.499	64	—	—	—	—	—
1	—	.073	1.854	.0550	1.397	53	.0595	1.499	—	72	—	—	—	—
—	M.2	.079	2.006	—	—	—	—	—	—	—	.4	.25	64	101
2	—	.086	2.184	.0628	1.587	50	.0700	1.778	56	—	—	—	—	—
2	—	.086	2.184	.0657	1.651	50	.0700	1.778	—	64	—	—	—	—
—	M2.5	.098	2.489	—	—	—	—	—	—	—	.45	.35	56	74
3	—	.099	2.515	.0719	1.828	47	.0785	1.981	48	—	—	—	—	—
3	—	.099	2.515	.0758	1.905	46	.0810	2.057	—	58	—	—	—	—
4	—	.112	2.845	.0795	2.006	43	.0890	2.261	40	—	—	—	—	—
4	—	.112	2.845	.0849	2.134	42	.0935	2.380	—	48	—	—	—	—
—	M3	.118	2.997	—	—	—	—	—	—	—	.5	.35	51	74
5	—	.125	3.175	.0925	2.336	38	.1015	2.565	40	—	—	—	—	—
5	—	.125	3.175	.0955	2.413	37	.1040	2.641	—	44	—	—	—	—
6	—	.138	3.505	.0975	2.464	36	.1065	2.692	32	—	—	—	—	—
6	—	.138	3.505	.1055	2.667	33	.1130	2.870	—	40	—	—	—	—
—	M4	.157	3.988	—	—	—	—	—	—	—	.7	.35	36	51
8	—	.164	4.166	.1234	3.124	29	.1360	3.454	32	—	—	—	—	—
8	—	.164	4.166	.1279	3.225	29	.1360	3.454	—	36	—	—	—	—
10	—	.190	4.826	.1359	3.429	26	.1470	3.733	24	—	—	—	—	—
10	—	.190	4.826	.1494	3.785	21	.1590	4.038	—	32	—	—	—	—
—	M5	.196	4.978	—	—	—	—	—	—	—	.8	.5	32	51
12	—	.216	5.486	.1619	4.089	16	.1770	4.496	24	—	—	—	—	—
12	—	.216	5.486	.1696	4.293	15	.1800	4.572	—	28	—	—	—	—
—	M6	.236	5.994	—	—	—	—	—	—	—	1.0	.75	25	34
1/4	—	.250	6.350	.1850	4.699	7	.2010	5.105	20	—	—	—	—	—
1/4	—	.250	6.350	.2036	5.156	3	.2130	5.410	—	28	—	—	—	—
5/16	—	.312	7.938	.2403	6.096	F	.2570	6.527	18	—	—	—	—	—
5/16	—	.312	7.938	.2584	6.553	I	.2720	6.908	—	24	—	—	—	—
—	M8	.315	8.001	—	—	—	—	—	—	—	1.25	1.0	20	25
3/8	—	.375	9.525	.2938	7.442	5/16	.3125	7.937	16	—	—	—	—	—
3/8	—	.375	9.525	.3209	8.153	Q	.3320	8.432	—	24	—	—	—	—
—	M10	.393	9.982	—	—	—	—	—	—	—	1.5	1.25	17	20
7/16	—	.437	11.113	.3447	8.738	U	.3680	9.347	14	—	—	—	—	—
7/16	—	.437	11.113	.3726	9.448	25/64	.3906	9.921	—	20	—	—	—	—
—	M12	.471	11.963	—	—	—	—	—	—	—	1.75	1.25	14.5	20
1/2	—	.500	12.700	.4001	10.162	27/64	.4219	10.715	13	—	—	—	—	—
1/2	—	.500	12.700	.4351	11.049	29/64	.4531	11.509	—	20	—	—	—	—
—	M14	.551	13.995	—	—	—	—	—	—	—	2	1.5	12.5	17
9/16	—	.562	14.288	.4542	11.531	31/64	.4844	12.3031	12	—	—	—	—	—
9/16	—	.562	14.288	.4903	12.446	33/64	.5156	13.096	—	18	—	—	—	—
5/8	—	.625	15.875	.5069	12.852	17/32	.5312	13.493	11	—	—	—	—	—
5/8	—	.625	15.875	.5528	14.020	37/64	.5781	14.684	—	18	—	—	—	—
—	M16	.630	16.002	—	—	—	—	—	—	—	2	1,5	12.5	17
—	M18	.709	18.008	—	—	—	—	—	—	—	2.5	1.5	10	17
3/4	—	.750	19.050	.6201	15.748	21/32	.6562	16.668	10	—	—	—	—	—
3/4	—	.750	19.050	.6688	16.967	11/16	.6875	17.462	—	16	—	—	—	—
—	M20	.787	19.990	—	—	—	—	—	—	—	2.5	1.5	10	17
—	M22	.866	21.996	—	—	—	—	—	—	—	2.5	1.5	10	17
7/8	—	.875	22.225	.7307	18.542	49/64	.7656	19.446	9	—	—	—	—	—
7/8	—	.875	22.225	.7822	19.863	13/16	.8125	20.637	—	14	—	—	—	—
—	M24	.945	24.003	—	—	—	—	—	—	—	3	2	8.5	12.5
1	—	1.000	25.400	.8376	21.2598	7/8	.8750	22.225	8	—	—	—	—	—
1	—	1.000	25.400	.8917	22.632	59/64	.9219	23.415	—	12	—	—	—	—
—	M27	1.063	27.000	—	—	—	—	—	—	—	3	2	8.5	12.5

TABLE A–5
Twist Drill Sizes: Decimal/Metric

	NUMBER SIZES								LETTER SIZES		
No. Size	Decimal Equivalent	Metric Equivalent	Closest Metric Drill (mm)	No. Size	Decimal Equivalent	Metric Equivalent	Closest Metric Drill (mm)	Size Letter	Decimal Equivalent	Metric Equivalent	Closest Metric Drill (mm)
1	.2280	5.791	5.80	41	.0960	2.438	2.45	A	.234	5.944	5.90
2	.2210	5.613	5.60	42	.0935	2.362	2.35	B	.238	6.045	6.00
3	.2130	5.410	5.40	43	.0890	2.261	2.25	C	.242	6.147	6.10
4	.2090	5.309	5.30	44	.0860	2.184	2.20	D	.246	6.248	6.25
5	.2055	5.220	5.20	45	.0820	2.083	2.10	E	.250	6.350	6.40
6	.2040	5.182	5.20	46	.0810	2.057	2.05	F	.257	6.528	6.50
7	.2010	5.105	5.10	47	.0785	1.994	2.00	G	.261	6.629	6.60
8	.1990	5.055	5.10	48	.0760	1.930	1.95	H	.266	6.756	6.75
9	.1960	4.978	5.00	49	.0730	1.854	1.85	I	.272	6.909	6.90
10	.1935	4.915	4.90	50	.0700	1.778	1.80	J	.277	7.036	7.00
11	.1910	4.851	4.90	51	.0670	1.702	1.70	K	.281	7.137	7.10
12	.1890	4.801	4.80	52	.0635	1.613	1.60	L	.290	7.366	7.40
13	.1850	4.699	4.70	53	.0595	1.511	1.50	M	.295	7.493	7.50
14	.1820	4.623	4.60	54	.0550	1.397	1.40	N	.302	7.671	7.70
15	.1800	4.572	4.60	55	.0520	1.321	1.30	O	.316	8.026	8.00
16	.1770	4.496	4.50	56	.0465	1.181	1.20	P	.323	8.204	8.20
17	.1730	4.394	4.40	57	.0430	1.092	1.10	Q	.332	8.433	8.40
18	.1695	4.305	4.30	58	.0420	1.067	1.05	R	.339	8.611	8.60
19	.1660	4.216	4.20	59	.0410	1.041	1.05	S	.348	8.839	8.80
20	.1610	4.089	4.10	60	.0400	1.016	1.00	T	.358	9.093	9.10
21	.1590	4.039	4.00	61	.0390	.991	1.00	U	.368	9.347	9.30
22	.1570	3.988	4.00	62	.0380	.965	.95	V	.377	9.576	9.60
23	.1540	3.912	3.90	63	.0370	.940	.95	W	.386	9.804	9.80
24	.1520	3.861	3.90	64	.0360	.914	.90	X	.397	10.084	10.00
25	.1495	3.797	3.80	65	.0350	.889	.90	Y	.404	10.262	10.50
26	.1470	3.734	3.75	66	.0330	.838	.85	Z	.413	10.491	10.50
27	.1440	3.658	3.70	67	.0320	.813	.80				
28	.1405	3.569	3.60	68	.0310	.787	.80				
29	.1360	3.454	3.50	69	.0292	.742	.75				
30	.1285	3.264	3.25	70	.0280	.711	.70				
31	.1200	3.048	3.00	71	.0260	.660	.65				
32	.1160	2.946	2.90	72	.0250	.635	.65				
33	.1130	2.870	2.90	73	.0240	.610	.60				
34	.1110	2.819	2.80	74	.0225	.572	.55				
35	.1100	2.794	2.80	75	.0210	.533	.55				
36	.1065	2.705	2.70	76	.0200	.508	.50				
37	.1040	2.642	2.60	77	.0180	.457	.45				
38	.1015	2.578	2.60	78	.0160	.406	.40				
39	.0995	2.527	2.50	79	.0145	.368	.35				
40	.0980	2.489	2.50	80	.0135	.343	.35				

Fraction-size drills range in size from one sixteenth—4 in. and over in diameter—by sixty-fourths.

TABLE A–6
U.S. Standard Sheet Metal Gauges

Gauge	Thickness		Wt. Per Sq. Ft.		Gauge
10	.1406 in.	3.571 mm	5.625 lb	2.551 kg	10
11	.1250 in.	3.175 mm	5.000 lb	2.267 kg	11
12	.1094 in.	2.778 mm	4.375 lb	1.984 kg	12
13	.0938 in.	2.383 mm	3.750 lb	1.700 kg	13
14	.0781 in.	1.983 mm	3.125 lb	1.417 kg	14
15	.0703 in.	1.786 mm	2.813 lb	1.276 kg	15
16	.0625 in.	1.588 mm	2.510 lb	1.134 kg	16
17	.0563 in.	1.430 mm	2.250 lb	1.021 kg	17
18	.0500 in.	1.270 mm	2.000 lb	.907 kg	18
19	.0438 in.	1.111 mm	1.750 lb	.794 kg	19
20	.0375 in.	.953 mm	1.500 lb	.680 kg	20
21	.0344 in.	.877 mm	1.375 lb	.624 kg	21
22	.0313 in.	.795 mm	1.250 lb	.567 kg	22
23	.0280 in.	.714 mm	1.125 lb	.510 kg	23
24	.0250 in.	.635 mm	1.000 lb	.454 kg	24
25	.0219 in.	.556 mm	.875 lb	.397 kg	25
26	.0188 in.	.478 mm	.750 lb	.340 kg	26
27	.0172 in.	.437 mm	.687 lb	.312 kg	27
28	.0156 in.	.396 mm	.625 lb	.283 kg	28
29	.0141 in.	.358 mm	.563 lb	.255 kg	29
30	.0120 in.	.318 mm	.500 lb	.227 kg	30

TABLE A–7
Chassis Wiring Color Code

Color	Abbrev.	Numerical Code	Circuit
Black	BK	0	Grounds, grounded elements, and returns
Brown	BR	1	Heaters of filaments off ground
Red	R	2	Power supply B-plus
Orange	O	3	Screen grids
Yellow	Y	4	Cathodes, emitters
Green	GN	5	Control grids, base
Blue	BL	6	Plates (anodes), collectors
Violet	V	7	Power supply, minus
(or Purple)	PR		
Gray	GY	8	AC power lines
White	W	9	Miscellaneous, returns above or below ground, AVC, etc.

TABLE A–8
Circuit-Identification Color Code for Industrial Control Wiring

Color	Circuit
Black	Line, load, and control circuit at line voltage
Red	AC control circuit
Blue	DC control circuit
Yellow	Interlock panel control when energized from external force
Green	Equipment grounding conductor
White	Grounded neutral conductor

TABLE A–9
Wire Numbers and
Dimensions

No. AWG[a] MCM	Stranding[b]	Diameter (in.)	Area (cmil)	Area[c] (sq in.)
40	Solid	.0031	10	.000008
38	Solid	.0040	16	.000013
36	Solid	.0050	25	.00002
34	Solid	.0063	40	.00003
32	Solid	.0080	64	.00005
30	Solid	.0100	100	.00008
28	Solid	.0126	159	.00012
26	Solid	.0159	253	.00020
24	Solid	.0201	404	.00032
22	Solid	.0253	640	.00050
20	Solid	.0320	1023	.00080
18	Solid	.0403	1620	.0013
16	Solid	.0508	2580	.0020
14	Solid	.0641	4110	.0032
12	Solid	.0808	6530	.0051
10	Solid	.1019	10,380	.0081
8	Solid	.1285	16,510	.0130
6	7	.184	26,240	.027
4	7	.232	41,740	.042
3	7	.260	52,620	.053
2	7	.292	66,360	.067
1	19	.332	83,690	.087
0	19	.372	105,600	.109
00	19	.418	133,100	.137
000	19	.470	167,800	.173
0000	19	.528	211,600	.219
250	37	.575	250,000	.260
300	37	.630	300,000	.312
350	37	.681	350,000	.364
400	37	.728	400,000	.416
500	37	.813	500,000	.519
600	61	.893	600,000	.626
700	61	.964	700,000	.730
750	61	.998	750,000	.782
800	61	1.030	800,000	.833
900	61	1.090	900,000	.933
1000	61	1.150	1,000,000	1.039
1250	91	1.289	1,250,000	1.305
1500	91	1.410	1,500,000	1.561
1750	127	1.526	1,750,000	1.829
2000	127	1.630	2,000,000	2.087

[a]Wire numbers from 40 to 0000 are AWG, from 250 to 2000 are MCM.
MCM ≅ 1000-circular-mil area.
[b]Solid conductors listed can be procured in stranded configurations.
[c]Area given is for a circle equal to the overall diameter of a stranded conductor.

TABLE A–10
Conductor Spacing
Minimums

Voltage between Conductors DC or AC Peak (Volts)	Uncoated 0–10,000 Ft. Alt. IPC-ML-910A	Uncoated above 10,000 Ft. Alt. IPC-ML-910A	Coated[a] and Internal Layers MIL-STD-275D and IPC-ML-910A	Voltage between Conductors DC or AC Peak (Volts)
0 15	.015 in. (.38 mm)	.025 in. (0.64 mm)	.005 in. (.13 mm)	0 15
16 30	.015 in. (.38 mm)	.025 in. (0.64 mm)	.010 in. (.25 mm)	16 30
31 50			.015 in. (.38 mm)	31 50
51 100	.025 in. (.64 mm)	.060 in. (1.52 mm)	.020 in. (.51 mm)	51 100
101 150		.125 in. (3.18 mm)		101 150
151 170	.050 in. (1.27 mm)		.030 in. (.76 mm)	151 170
171 250		.250 in. (6.35 mm)		171 250
251 300		.500 in. (12.70 mm)		251 300
301 500	.100 in. (2.54 mm)		.060 in. (1.52 mm)	301 500
500+	.0002 in./V (.0051 mm/V)	.0010 in./V (.0030 mm/V)	.00012 in./V (.00305 mm/V)	500+

[a]Coating per MIL-I-46058 (Provided Courtesy of Bishop Graphics, Inc.)

FIGURE A–1
Conductor thickness and width for use in determining current-carrying capacity and sizes of etched copper conductors for various temperature rises above ambient. (Provided Courtesy of Bishop Graphics, Inc.)

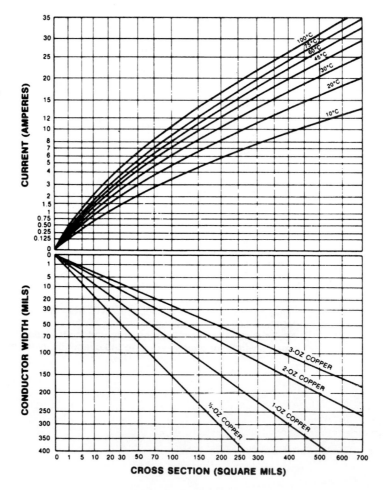

TABLE A–11
Recommended Terminal Areas for Unplated Holes

AWG	Lead Size Dia. in./mm	Finished Hole Dia. in./mm	Max. Allow Tol.	PREFERRED Term. Dia.	PREFERRED Pad 4X	PREFERRED Pad 2X	PREFERRED Pad 1X	STANDARD Term. Dia.	STANDARD Pad 4X	STANDARD Pad 2X	STANDARD Pad 1X	MINIMUM Term. Dia.	MINIMUM Pad 4X	MINIMUM Pad 2X	MINIMUM Pad 1X
34	.0063 in. 0.160														
33	.0071 in. 0.180														
32	.0080 in. 0.203	.020 in. 0.51	±.005 in. 0.127	.105 in. 2.67	D214 .416 in. 10.57	D168 .208 in. 5.28	D277 .106 in. 2.69	.065 in. 1.65	D358 .260 in. 6.60	D276 .130 in. 3.30	D370 .070 in. 1.78	.050 in. 1.27	D204 .200 in. 5.08	D101 .100 in. 2.54	D135 .050 in. 1.27
31	.0089 in. 0.226														
30	.0100 in. 0.254														
29	.0113 in. 0.287														
28	.0126 in. 0.320														
27	.0142 in. 0.361														
26	.0159 in. 0.404	.030 in. 0.76	±.005 in. 0.127	.115 in. 2.92	D224 .468 in. 11.89	D298 .230 in. 5.84	D265 .115 in. 2.92	.075 in. 1.91	D111 .300 in. 7.62	D203 .150 in. 3.81	D311 .075 in. 1.91	.060 in. 1.52	D170 .240 in. 6.10	D239 .120 in. 3.05	D136 .062 in. 1.57
25	.0179 in. 0.455														
24	.0201 in. 0.511														
23	.0226 in. 0.574														
22	.0253 in. 0.634	.040 in. 1.02	±.006 in. 0.152	.125 in. 3.18	D119 .500 in. 12.70	D109 .250 in. 6.35	D102 .125 in. 3.18	.090 in. 2.29	D377 .360 in. 9.14	D248 .180 in. 4.57	D372 .090 in. 2.29	.070 in. 1.78	D375 .280 in. 7.11	D280 .140 in. 3.56	D370 .070 in. 1.78
21	.0258 in. 0.724														
20	.0320 in. 0.813														
19	.0359 in. 0.912														
18	.0403 in. 1.024	.052 in. 1.32	±.006 in. 0.152	.140 in. 3.56	D121 .562 in. 14.27	D375 .280 in. 7.11	D280 .140 in. 3.56	.100 in. 2.54	D115 .400 in. 10.16	D204 .200 in. 5.08	D101 .100 in. 2.54	.085 in. 2.16	D376 .340 in. 8.63	D307 .170 in. 4.32	D371 .085 in. 2.16
17	.0453 in. 1.151														
16	.0508 in. 1.290	.067 in. 1.70	±.007 in. 0.178	.155 in. 3.93	D124 .625 in. 15.88	D113 .312 in. 7.92	D103 .156 in. 3.96	.115 in. 2.92	D224 .468 in. 11.89	D298 .230 in. 5.84	D265 .115 in. 2.92	.100 in. 2.54	D115 .400 in. 10.16	D204 .200 in. 5.08	D101 .100 in. 2.54
15	.0571 in. 1.450														

NOTE: This table was prepared on the basis of 2 oz./ft² copper with tin lead plate, double sided. The plated-thru holes have a minimum of .001 in. (.025 mm) copper plating and .001 in. (.025 mm) tin lead plating. (Provided Courtesy of Bishop Graphics, Inc.)

TABLE A-12
Recommended Terminal Areas for Plated-Through Holes

COMPONENT LEAD SIZE		RECOMMENDED FINISHED HOLE			RECOMMENDED TERMINAL AREA											
					PREFERRED				STANDARD				MINIMUM			
					Term. Dia.	Bishop Donut Pad/Dia.			Term. Dia.	Bishop Donut Pad/Dia.			Term. Dia.	Bishop Donut Pad/Dia.		
AWG	Dia. in. mm	Dia. in. mm	Max. Allow Tol.	Drill		4X	2X	1X		4X	2X	1X		4X	2X	1X
34	.0063 in. 0.160															
33	.0071 in. 0.180															
32	.0080 in. 0.203															
31	.0089 in. 0.226	.0145 in. .368	±.003 in. .076	#79	.100 in. 2.54	D115 .400 in. 10.16	D204 .200 in. 5.08	D101 .100 in. 2.54	.075 in. 1.91	D111 .300 in. 7.62	D203 .150 in. 3.81	D311 .075 in. 1.91	.0625 in. 1.59	D109 .250 in. 6.35	D102 .125 in. 3.18	D136 .062 in. 1.57
30	.0100 in. 0.254															
29	.0113 in. 0.287															
28	.0126 in. 0.320															
27	.0142 in. 0.361															
26	.0159 in. 0.404															
25	.0179 in. 0.455	.028 in. .71	±.003 in. .076	#70	.115 in. 2.92	D224 .468 in. 11.89	D298 .230 in. 5.84	D265 .115 in. 2.92	.085 in. 2.16	D376 .340 in. 8.64	D307 .170 in. 4.32	D371 .085 in. 1.91	.075 in. 1.91	D111 .300 in. 7.62	D203 .150 in. 3.81	D311 .075 in. 1.91
24	.0201 in. 0.511															
23	.0226 in. 0.574															
22	.0253 in. 0.634															
21	.0258 in. 0.724	.040 in. 1.02	±.004 in. .102	#60	.100 in. 3.18	D119 .500 in. 12.70	D109 .250 in. 6.35	D102 .125 in. 3.18	.100 in. 2.54	D115 .400 in. 10.16	D204 .200 in. 5.08	D101 .100 in. 2.54	.090 in. 2.29	D377 .360 in. 9.14	D248 .180 in. 4.57	D372 .090 in. 2.29
20	.0320 in. 0.813															
19	.0359 in. 0.912															
18	.0403 in. 1.024	.052 in. 1.32	±.004 in. .102	#55	.140 in. 3.56	D121 .562 in. 14.27	D375 .280 in. 7.11	D280 .140 in. 3.56	.110 in. 2.79	D117 .437 in. 11.10	D185 .220 in. 5.59	D373 .110 in. 2.79	.100 in. 2.54	D115 .400 in. 10.16	D204 .200 in. 5.08	D101 .100 in. 2.54
17	.0453 in. 1.151															
16	.0508 in. 1.290	.0625 in. 1.59	±.004 in. .102	1/16	.150 in. 3.81	D149 .600 in. 15.24	D111 .300 in. 7.62	D144 .150 in. 3.81	.120 in. 3.05	D188 .475 in. 12.07	D170 .240 in. 6.10	D239 .120 in. 3.05	.110 in. 2.79	D117 .437 in. 11.10	D185 .220 in. 5.59	D373 .110 in. 2.79
15	.0571 in. 1.450															

NOTE: This table was prepared on the basis of 2 oz./ft² copper with tin lead plate, double sided. A .015 in. (.38 mm) minimum annular ring is assumed. (Provided Courtesy of Bishop Graphics, Inc.)

552

PRINTED CIRCUIT DESIGN CRITERIA
DIMENSION AND TOLERANCE CONSIDERATIONS

The information presented in the chart below is for reference only. It is a compilation of various Military Standards (MIL-STD) and Institute for Interconnecting and Packaging Electronic Circuits (IPC) specifications. For more detailed information, consult the referenced specifications.

The classes of Materials and Processes on the right of the chart indicate progressive degrees of sophistication. The use of one class for a specific characteristic does not mean that the class must be used for all other characteristics and tolerances. Whenever possible, dimensions and tolerances should be selected from the preferred data since this will typically result in the most producible part. Design requirements should always be discussed with individual circuit board manufacturers as process capabilities may vary considerably.

In some instances where a conflict between MIL-STD and IPC data for a specific feature or tolerance exists, both sets of data are listed. Note that the MIL-STD specifications generally represent a printed circuit board *user's* point of view while the IPC specifications typically emphasize the printed circuit board *manufacturer's* viewpoint.

Printed Wiring Board Dimensional Features

- GLASS RESIN LAMINATE
- COPPER
- PLATING

PLATED-THRU HOLE   UNPLATED HOLE

DESIGN CRITERIA & TOLERANCES			DATA FOR CLASSES OF MATERIALS & PROCESSES				
CODE LETTER	DESCRIPTION QUALIFIERS SUB-QUALIFIERS	SOURCE DOCUMENT	PREFERRED CLASS 1	CLASS 2	STANDARD CLASS 3	REDUCED PRODUCIBILITY CLASS 4	CLASS 5
T	BOARD THICKNESS NOMINAL (SINGLE OR DOUBLE)		.062" (1,57mm)			.031" (0,79mm)	
	MAXIMUM	MIL-STD-275D	.100" (2,54mm)		.150" (3,91mm)	.200" (5,08mm)	
T Tol.	BOARD THICKNESS TOLERANCE (1) SINGLE OR DOUBLE SIDED LAMINATE .031" (0,79mm) THK .062" (1,57mm) THK .093" (2,36mm) THK .125" (3,18mm) THK .250" (6,35mm) THK	MIL-P-13949E	± .0065" (0,17mm) ± .0075" (0,19mm) ± .0090" (0,23mm) ± .0120" (0,30mm) ± .0220" (0,56mm)		± .004" (0,10mm) ± .005" (0,13mm) ± .007" (0,18mm) ± .009" (0,23mm) ± .012" (0,30mm)	± .003" (0,08mm) ± .003" (0,08mm) ± .004" (0,10mm) ± .005" (0,13mm) ± .006" (0,15mm)	
	FINISHED BOARD	MIL-STD-275D IPC-ML-910A	± 10% OF "T" NOMINAL OR ± .007" (0,18mm) MINIMUM				
MDT	MINIMUM THICKNESS MULTILAYER DIELECTRIC LAYERS	MIL-STD-275D	.008" (0,20mm)		.006" (0,15mm)	.0035" (0,09mm)	
L	MAXIMUM NUMBER OF MULTI-LAYER CONDUCTOR LAYERS (2)	MIL-STD-275D	6		12	20	
W	CONDUCTOR WIDTH (MINIMUM) (3) MIL-STD INTERNAL EXTERNAL	MIL-STD-275D	.015" (0,38mm) .020" (0,51mm)		.010" (0,25mm) .015" (0,38mm)	.008" (0,20mm) .008" (0,20mm)	
	IPC (MULTILAYER)	IPC-ML-910A	.020" (0,51mm)		.015" (0,38mm)	.010" (0,25mm)	
W Tol.	CONDUCTOR WIDTH TOLERANCE (4) MIL-STD & IPC (MULTILAYER) WITHOUT PLATING WITH PLATING	MIL-STD-275D IPC-ML-910A	+ .004" (0,10mm) − .006" (0,15mm) + .008" (0,20mm) − .006" (0,15mm)		+ .002" (0,05mm) − .005" (0,13mm) + .004" (0,10mm) − .004" (0,10mm)	+ .001" (0,03mm) − .003" (0,08mm) + .002" (0,05mm) − .002" (0,05mm)	
	IPC (SINGLE OR DOUBLE SIDED) WITHOUT PLATING WITH PLATING	IPC-D-300F	+ .007" (0,18mm) − .011" (0,28mm) + .016" (0,41mm) − .018" (0,28mm)	+ .005" (0,10mm) − .006" (0,15mm) + .009" (0,23mm) − .006" (0,15mm)	+ .003" (0,08mm) − .005" (0,10mm) + .005" (0,10mm) − .005" (0,10mm)	+ .002" (0,05mm) − .003" (0,08mm) + .004" (0,10mm) − .004" (0,10mm)	+ .002" (0,05mm) − .002" (0,05mm) + .003" (0,08mm) − .003" (0,08mm)
S	CONDUCTOR SPACING (MINIMUM) (5) MIL-STD & IPC (MULTILAYER)	MIL-STD-275D IPC-ML-910A	.020" (0,51mm)		.010" (0,25mm)	.005" (0,13mm)	
	IPC (SINGLE & DOUBLE SIDED)	IPC-D-320A(6)	.060" (1,52mm)	.030" (0,76mm)	.015" (0,38mm)	.010" (0,25mm)	.005" (0,10mm)
SE	CONDUCTOR SPACING (MIN) TO EDGE OF BOARD (5) MIL-STD & IPC (MULTILAYER) INTERNAL EXTERNAL	MIL-STD-275D IPC-ML-910A	.100" (2,54mm) .100" (2,54mm)		.050" (1,27mm) .100" (2,54mm)	.025" (0,64mm) .100" (2,54mm)	
SL	MINIMUM INTRALAYER CONDUC-TOR SPACING (MULTILAYER)	IPC-ML-910A	.007" (0,18mm)		.005" (0,13mm)	.003" (0,08mm)	

FIGURE A-2 (cont.)

CODE LETTER	DESCRIPTION QUALIFIERS SUB-QUALIFIERS	SOURCE DOCUMENT	PREFERRED CLASS 1	CLASS 2	STANDARD CLASS 3	REDUCED PRODUCIBILITY CLASS 4	CLASS 5
H	HOLE DIAMETER (UNPLATED) MINIMUM	IPC-D-300F	2/3T (MAX) (66%)	1/2T (MAX) (50%)	3/10T (MAX) (30%)	1/4T (MAX) (25%)	1/5T (MAX) (20%)
	MAXIMUM	MIL-STD-275D	NOMINAL LEAD DIA. + .020" (0,51mm) NOMINAL FLAT LEAD THICKNESS + .028" (0,71mm) MINIMUM EYELET BARREL DIA. + .010" (0,25mm)				
H Tol.	HOLE TOLERANCE (AS MACHINED) (7) MIL-STD & IPC (MULTILAYER) (TOTAL VARIANCE) Up to .032" (0,81mm) DIA. .033" (0,82mm)-.063" (1,60mm) DIA. .064" (1,61mm)-.188" (4,78mm) DIA.	MIL-STD-275D IPC-ML-910A	.004" (0,10mm) .006" (0,15mm) .008" (0,20mm)		.003" (0,08mm) .004" (0,10mm) .006" (0,15mm)	.002" (0,05mm) .003" (0,08mm) .004" (0,10mm)	
	IPC (SINGLE & DOUBLE SIDED) Up to .032" (0,81mm) DIA.		± .003" (0,08mm)	± .002" (0,05mm)	+ .001" (0,03mm) − .002" (0,05mm)	± .001" (0,03mm)	± .001" (0,03mm)
	.033" (0,82mm)-.063" (1,60mm) DIA.	IPC-D-300F	± .004" (0,10mm)	± .003" (0,08mm)	± .002" (0,05mm)	+ .001" (0,03mm) − .002" (0,05mm)	± .001" (0,03mm)
	.064" (1,61mm)-.188" (4,78mm) DIA.		± .005" (0,13mm)	± .004" (0,10mm)	± .003" (0,08mm)	± .002" (0,05mm)	+ .001" (0,03mm) − .002" (0,05mm)
PH	PLATED-THRU HOLE DIA. (8) MINIMUM	MIL-STD-275D IPC-ML-910A	1/3T (MAX) (33%)		1/4T (MAX) (25%)	1/5T (MAX) (20%)	
	MAXIMUM	MIL-STD-275D	MINIMUM LEAD DIA + .028" (0,71mm) NOMINAL FLAT LEAD THICKNESS + .028" (0,71mm)				
PH Tol.	PLATED-THRU HOLE DIA. TOLERANCE (8) (9) MIL-STD & IPC (MULTILAYER) (TOTAL VARIANCE) .015" (0,38mm)-.030" (0,76mm) DIA. .031" (0,77mm)-.061" (1,55mm) DIA. .062" (1,56mm)-.186" (4,72mm) DIA.	MIL-STD-275D IPC-ML-910A	.008" (0,20mm) .010" (0,25mm) .012" (0,30mm)		.005" (0,13mm) .006" (0,15mm) .008" (0,20mm)	.004" (0,10mm) .004" (0,10mm) .006" (0,15mm)	
	IPC (SINGLE & DOUBLE SIDED) Up to .032" (0,81mm) DIA. .033" (0,82mm)-.063" (1,60mm) DIA. .064" (1,61mm)-.188" (4,78mm) DIA.	IPC-D-300F	± .005" (0,13mm) ± .006" (0,15mm) ± .007" (0,18mm)	± .003" (0,08mm) ± .004" (0,10mm) ± .005" (0,13mm)	± .002" (0,05mm) ± .003" (0,08mm) ± .004" (0,10mm)	± .001" (0,03mm) ± .002" (0,05mm) ± .004" (0,10mm)	± .001" (0,03mm) ± .001" (0,03mm) ± .002" (0,05mm)
HL Tol.	HOLE LOCATION TOLERANCE (RADIUS TRUE POSITION VARIANCE) MIL-STD & IPC (MULTILAYER) Boards up to 12.000" (30,5cm) Boards over 12.000" (30,5cm)	MIL-STD-275D IPC-ML-910A	.005" (0,13mm) .007" (0,18mm)		.003" (0,08mm) .005" (0,13mm)	.002" (0,05mm) .003" (0,08mm)	
	IPC (SINGLE & DOUBLE SIDED) <6.00" (15,2cm) FROM DATUM >6.00" (15,2cm) FROM DATUM	IPC-D-300F	.010" (0,25mm) .014" (0,36mm)	.007" (0,18mm) .010" (0,25mm)	.005" (0,13mm) .007" (0,18mm)	.003" (0,08mm) .005" (0,13mm)	.002" (0,05mm) .003" (0,08mm)
AR	MINIMUM ANNULAR RING UNPLATED PLATED-THRU HOLE	MIL-STD-275D	.015" (0,38mm)				
	INTERNAL (MIL-STD) EXTERNAL (MIL-STD)	MIL-STD-275D	.008" (0,20mm) .010" (0,25mm)		.005" (0,13mm) .008" (0,20mm)	.002" (0,05mm) .005" (0,13mm)	
	INTERNAL (IPC) EXTERNAL (IPC)	IPC-ML-910A	.005" (0,13mm) .010" (0,25mm)		.002" (0,05mm) .005" (0,13mm)	.001" (0,03mm) .002" (0,05mm)	
TA	MINIMUM TERMINAL AREA (10)	MIL-STD-275D	MAX. EYELET OR TERMINAL FLANGE DIA. + .020" (0,51mm) OR MAX. DIA. OF UNSUPPORTED HOLE + .040" (1,02mm)				
FL Tol.	FEATURE LOCATION TOLERANCE (RADIUS TRUE POSITION VARIANCE) MIL-STD & IPC <12.000" (30,5cm) (MULTILAYER) >12.000" (30,5cm)	MIL-STD-275D IPC-ML-910A	.008" (0,20mm) .010" (0,25mm)		.007" (0,18mm) .009" (0,23mm)	.006" (0,15mm) .008" (0,20mm)	
	IPC (SINGLE & <6.000" (15,2cm) (11) DOUBLE SIDED) >6.000" (15,2cm) (11)	IPC-D-300F	.021" (0,53mm) .025" (0,64mm)	.014" (0,36mm) .018" (0,46mm)	.010" (0,25mm) .014" (0,36mm)	.007" (0,18mm) .010" (0,25mm)	.0045" (0,11mm) .007" (0,18mm)

NOTES:

(1) Board thickness tolerances apply only to copper-clad laminated glass/resin based materials.

(2) The number of conductor layers should be the optimum for the required board function and good producibility.

(3) The thickness and width of the conductors should be determined on the basis of the current-carrying capacity required. The allowable temperature rise should be determined in accordance with Figure I. Maximum size, consistent with minimum spacing requirements, should be maintained for ease of manufacture and durability in usage.

(4) The conductor width tolerances represent process tolerances that can be expected with normal processing. (Specific process tolerances should be discussed with the supplier.) These tolerances are based on 2 oz/ft² copper. For 1 oz/ft² copper, a .001" (0,025mm) reduction in variation can be expected per conductor edge. Final product drawings and specifications should specify only minimums for conductor width and spacing.

(5) The minimum spacing dimensions listed are limited by voltage rating, altitude and coatings. See Table IV for more detailed information.

(6) Revision A proposed.

(7) The tolerances listed are for nominal base material thicknesses up to and including 0.0625" (1,59 mm). For nominal base material thickness over 0.0625" (1,59 mm), add ±0.001" (0,025 mm) to each tolerance. Tolerances indicate total spread and may be varied from the nominal to satisfy design requirements.

(8) All references to plated-thru-hole diameters indicate finished hole diameters (after plating).

(9) The tolerances listed are for holes no less than one-third of the nominal overall board thickness. Tolerances indicate total spread, and may be varied from the nominal to satisfy design requirements. For hole diameter to nominal board thickness ratios greater than ⅓ add ±0.001" (0,025 mm) to these tolerances. For ratios greater than ¼ add ±0.002" (0,05 mm).

(10) MIL-STD minimum terminal areas are listed for reference only. For a discussion and examples on how to determine terminal area, see section on Design "Terminal Areas (Pads)."

(11) Tolerances apply when a registration datum is used.

Appendix B
Electronic and Electrical Drafting Glossary*

AC–DC A device that will operate on alternating or direct current; an active device; a component that contains voltage or other current sources.

Actuator The cam, arm, or similar mechanical device used to trip limit switches.

Air capacitor A capacitor that uses air as a dielectric.

Air inductor An inductor without a magnetic core.

Air gap The nonconductive air space between conductors, traces, pads, and so on.

Airline wiring diagram Connection diagram with a single horizontal or vertical line representing the cable location. Feeder lines branch off the main line.

Alternating current A current of electrons that moves in one direction and then reverses and flows in the opposite direction at specific time intervals. The current has alternating positive and negative values.

Ampere Unit of current.

Amplifier A device that uses electron tubes or transistors to increase voltage, current, or power.

Analog A type of computer that uses numbers representing directly measurable quantities.

Analog circuit A circuit composed primarily of discrete components that produce data for physical variables such as resistance, voltage, and so on.

Annular ring The conductor width surrounding a hole in a PC board.

Anode A positive terminal capable of attracting negative charges.

Antiplugging protection The effect of a control function or a device that prevents application of countertorque by the motor until the motor speed has been reduced to an acceptable value.

Apparatus (control) A set of control devices that accomplish the intended control functions.

Appliqués Electronic symbols, other graphical shapes, and lettering that are preprinted on sheets. They can be easily separated from mounting sheets and positioned on a drawing.

Armature The revolving part of a DC motor or generator. Also, the vibrating or moving part of a buzzer or relay.

Artwork A precise scaled drawing used to produce the finished master pattern for a PC board, created by hand taping or plotting with a CAD system and photoplotter.

Assembly drawing A drawing representing a group of parts constituting a major subdivision of the final product.

Auxiliary contacts In a switching device, supplements to the main-circuit contacts that function with movement.

Auxiliary device Any electrical device (other than motors and motor starters) necessary to operate the machine or equipment completely.

AWG American Wire Gauge. A standardized method for specifying a wire diameter. The larger the number, the smaller the corresponding wire diameter.

Axial leads Wire leads extending from the ends of various components, capacitors, resistors, and so on.

Back annotation A process to extract information from a completed printed circuit board and insert it in the logic elements of the schematic describing the board.

Base The thin section of a transistor between the emitter and the collector.

Bias Voltage applied to an electronic device, such as the grid of an electron tube, to establish a benchmark level for operation.

Bipolar A method of fabricating one type of IC by layering silicon that has two different electrical characteristics.

Block diagram A diagram showing the relationship of separate subunits (blocks) in the system.

Bonding conductor A transmitter that connects exposed metal surfaces.

Branch circuit The portion of a wiring system extending beyond the final overcurrent device protecting the circuit. (A device not approved for branch-circuit protection, such as a thermal cutout or motor overload protective device, is not considered the overcurrent device protecting the circuit.)

Breadboard Laying out an electronic circuit with components and wiring on a board for experimentation, testing, and designing.

Bridging When excess solder builds up during wave soldering, shorting out adjacent conductors by "bridging" the area between them.

Bus A conductive metal strip or trace used to distribute voltage, grounds, and so forth, to smaller branch traces on a PC board.

Bus bar The main power distribution point of a circuit. The bus bar is connected to the primary power source.

Bypass capacitor A component that provides a comparatively low impedance AC path around a circuit.

Capacitance A property of an electric circuit to oppose a change in voltage; also, an electric circuit's ability to store energy in an electrostatic field.

---

* Portions of the following were extracted from the JIC Standard Glossary, courtesy of the Joint Industrial Council.

Capacitor A fixed or variable device providing capacitance. A simplified capacitor comprises two metal plates separated by an insulator.

Captive screw A screw-type fastener that is retained when unscrewed and cannot easily be separated from the part it secures.

Cathode The emitter of an electron tube, diode, semiconductor, and so on.

Chassis A sheet metal box, frame, or simple plate on which electronic components and their associated circuitry can be mounted.

Circuit breaker A device to open and close a circuit nonautomatically, and to open the circuit automatically on a predetermined overload of current without injuring itself when properly applied within its rating.

Circuit interrupter A manually operated device to open, under abnormal conditions, a current-carrying circuit without injuring itself.

CMOS (complementary metal oxide silicon) A logic family using n- or p-channel enhancement-mode transistors fabricated compatibly on a silicon chip and connected into a push–pull complementary digital circuit.

Collector A conductor that maintains contact between moving and stationary parts of an electric circuit. Also, the portion of a transistor that attracts and collects electrons.

Component The smallest element of a circuit (i.e., resistor, capacitor, transistor, or integrated circuit package).

Conduit, flexible metal A flexible raceway of circular cross section specially constructed to pull in or withdraw wires or cables after the conduit and its fittings are in place.

Conduit, flexible nonmetallic A flexible raceway of circular cross section especially for pulling in or withdrawing wires or cables after the conduit and its fittings are in place.

Conduit, rigid metal A raceway specially constructed to pull in or wind wires or cables after the conduit is in place. Made of metal pipes, its standard weight and thickness permit cutting standard threads.

Conformal coating A sprayed, dipped, or brushed coating to protect a PC board or other electronic component from environmental damage such as moisture, fungus, or dust.

Connector A plug or receptacle for electrically interconnecting one or more cables or electronic circuits.

Connector tongue The edge of a PC board designed to mate with a receptacle that bridges (mechanically or electrically) the board and other circuitry.

Contactor A device for repeatedly establishing and interrupting an electric power circuit.

Control circuit The circuit of the control apparatus or system that carries the signals directing the controller's performance.

Control circuit transformer A voltage transformer that supplies a voltage suitable to operate control devices.

Control circuit voltage The voltage provided to operate shunt coil magnetic devices.

Control panel (See "panel.")

Control station (See "operator's control station.")

Controller, electronic A device or group of devices that governs in some predetermined manner the electronic signals delivered to the apparatus to which it is connected.

Cross-hatching Filling in an outline with a series of symbols to highlight part of a design.

Current The flow of an electrical charge measured in amperes; also, the rate of that flow.

Device, input A device that initiates a signal that is a condition of the system.

Device, output A device that accepts a signal and executes a control function.

Dielectric A nonconductor of a direct electric current; also, the insulator material between plates of a capacitor.

Diffusion A step in fabricating electronic components, transistors, diodes, ICs, and so forth, that uses high temperatures to move impurities into a slice of silicon, thus changing its electrical properties.

Digital A discrete representation of a physical quantity, for example, devices, elements, or circuits that respond in discrete steps (i.e., pulses or on–off operation).

Digital circuit A circuit composed primarily of ICs that operates like a switch (i.e., it is either on or off).

Disconnecting means A device that disconnects the current-carrying conductors of a circuit from their supply source.

Disconnect switch (motor circuit switch) A switch intended for use in a motor branch circuit. Rated in horsepower, it is also capable of interrupting the maximum operating overload current of a motor of the same rating at the rated voltage.

Discrete component A component fabricated prior to its installation, such as diodes, transistors, capacitors, and resistors.

Electrode A conductor that establishes electrical contact with a nonmetallic part of a circuit.

Electromechanical The term applied to any device in which electrical energy magnetically causes mechanical movement.

Electron An elementary particle of matter consisting of a charge of negative electricity.

Electronic control The term applied to electronic, static, precision, and associated equipment.

Elementary (schematic) diagram A wiring diagram that uses symbols and a plan of connections to illustrate the scheme of control simply. Also, an electrical diagram containing components, logic elements, wire nets, bullets, miscellaneous graphic and nongraphic information, and text annotation.

Enclosure The case, box, or structure surrounding the electronic equipment that protects it from contamination. The degree of tightness is usually specified.

ES (electrical schematic) A diagram of a detailed arrangement of hardware, using conventional component symbols.

External control devices All control devices mounted externally on the control panel.

Eyelet The mechanism on printed circuit boards that makes electrical connections from one side of the board to the other side.

Farad The unit of measurement for capacitance.

Feedback Transferring voltage from the output of a circuit back to its input point.

Feeder The circuit conductors between the service equipment, or the generator switchboard of an isolated plant, and the branch circuit overcurrent device.

Feed-through   On a printed circuit board, a plated-through hole used to provide an electrical connection between a trace on one side of a PC board and a trace on the other side. Normally, the hole is small with a small pad size, since it is not used to mount a component lead.

Filament   A conductor made incandescent by the passage of electric current.

Filter   A circuit, device, or material designed to suppress or minimize waves or oscillations of certain electrical frequencies.

Frequency   The number of complete cycles per second (cps).

Fuse   A safety device that opens an electric circuit when the circuit overloads. A current above the fuse rating will melt or break the fuse and open the circuit.

Gain   The ratio of the output power, current, or voltage to the input power, current, or voltage.

Gate   A circuit having two or more inputs and one output, the output depending on the combination of logic signals of the inputs.

Glass epoxy   Material used for the base of a PC board. Fiber glass is impregnated with an epoxy filler that is then laminated with a conductor material, usually copper.

Ground   The common voltage reference point in a circuit. Also, a connection to earth using plates or rods. The chassis of electronic equipment is sometimes used as the ground.

Grounded   Connected to the ground or to some conducting body that substitutes for the ground.

Grounded circuit   A circuit in which one conductor or point (usually the neutral or neutral point of the transformer or generator windings) is intentionally grounded (earthed), either solidly or through a grounding device.

Grounding conductor   A conductor that under normal conditions carries no current, but serves to connect exposed metal surfaces to an earth ground, to prevent hazards in case of a breakdown between current-carrying parts and exposed surfaces. The conductor, if insulated, is colored green, with or without a yellow stripe.

Ground plane   A condition on a PC board in which whole areas of the conductor material are left unetched and tied to the ground circuit throughout the board.

Henry   Unit of measurement of inductance.

Hole   The space left by a dislodged electron; also, the space left by a removed atom (in semiconductors).

IC (integrated circuit)   A tiny complex of electronic components and their connections on a slice of material such as silicon. A combination of inseparably interconnected active and passive circuit elements on or within a continuous substrate.

Impedance   The total resistance to the flow of an alternating current.

Inductance   A circuit's ability to oppose a change in current and to store energy in a magnetic field.

Inductor   An apparatus such as a coil that acts upon itself or another by induction.

Inrush current   The inrush current of a solenoid or coil is the steady-state current taken from the line with the armature blocked in the maximum rated open position.

Insulating (isolating) transformer   (See "isolating transformer.")

Interconnecting diagram   A diagram showing all terminal blocks in the complete system and identifying each terminal.

Interconnecting wire   A term referring to connections between subassemblies, panels, chassis, and remotely mounted devices; it does not necessarily apply to internal connections of these units.

Interconnection   Anything that connects one item to another. On PC boards, interconnections consist of copper runs connecting pads. On schematic drawings, interconnections are lines connecting elements.

Interlock   Actuated by the operation of some other device with which it is directly associated, this device governs succeeding operations of the same or allied devices. May be either electrical or mechanical.

Interrupting capacity   Interrupting capacity is the highest current at rated voltage that the device can interrupt.

Isolating transformer   A transformer that electrically isolates one circuit from another.

Joint   A connection between two or more conductors.

Lands   An enlarged portion of conductor material surrounding a component mounting hole.

LED (light-emitting diode)   A pn junction that emits light when biased in the forward direction.

Limit switch   A switch operated by some part or motion of a power-driven equipment to alter the equipment's electric or electronic circuits.

Logic control panel layout   The physical arrangement of the devices on a chassis or panel.

Logic design   Specifying the functions of various parts of a system in symbolic logic.

Logic diagram   A diagram showing the relationship of standard logic elements in a control system. No internal detail of the logic elements need be shown.

Logic element   A symbol with logical meaning; may also be called a logic symbol (e.g., gates or flip-flops).

Magnetic device   A device actuated electromagnetically.

Magnetic starter   A starter actuated electromagnetically.

Master pattern   A highly accurate scaled pattern that produces a PC board within a specified tolerance limit defined on the master drawing.

MOS (metal oxide silicon)   A logic family utilizing N- or P-channel transistors fabricated on a silicon chip connected to form digital circuits.

Mother board   A relatively large PC board on which other PC boards, modules, connectors, and subassemblies are mounted and interconnected by traces.

Motor junction (conduit) box   An enclosure on a motor that terminates conduit runs and joins the motor to power conductors.

Net   A collection of parts and connector pins that must be connected; also known as a signal.

Netlist   The list of names, symbols, and their connection points that are logically connected in a net.

Node Also called a junction point or branch point, a terminal of any branch of a network or a terminal common to two or more branches.

Nominal voltage The utilization voltage. See the appropriate NEMA standard for device voltage ratings.

Normally open and normally closed Terms signifying the position that a magnetically operated switching device, such as a contactor or relay, or the contacts thereof, takes when the operating magnet is de-energized. They apply only to nonlatching-type devices.

Open circuit A circuit that does not provide a complete path for signal flow.

Operator's control station (push-button station) A unit assembly of one or more externally operable push-button switches, sometimes including other pilot devices such as indicating lights or selector switches, in a suitable enclosure.

Oscilloscope A test instrument that uses a CRT to observe a circuit signal.

Outline drawing A drawing showing approximate overall shape with no detail.

Overcurrent The current in an electric or electronic circuit that will cause an excessive or dangerous temperature in the conductor or conductor insulation.

Overcurrent protective device A device operative on excessive current that causes and maintains power interruption in the circuit.

Overlapping contacts Combinations of two sets of contacts, actuated by a common means, each closing in one or two positions and arranged so that the contacts of one set open after the contacts of the other set have been closed.

Overload relay A device that provides overload protection for electrical equipment.

Panel A subplate on which the control devices are mounted inside the control compartment or enclosure.

Panel layout The physical position or arrangement of the components on a panel or chassis.

PCB (printed circuit board) Insulated substrate (often plastic) on which interconnection wiring has been applied by photographic techniques.

Photoconductive The ability of certain materials to increase in conductivity when exposed to a light source.

Photodiode A two-terminal junction diode that conducts upon exposure to light energy.

Plate The anode of an electron tube; also, a conductive electrode in a battery or capacitor.

Plated-through hole A hole in the PC board in which metal is deposited on the wall to connect conductors electrically on each side of the board.

Plating Application of a uniform coating of conductive material (e.g., tin, silver, gold) on the base metal of a PC board.

Plug A connecting device that can be pushed or screwed into a receptacle to make electrical connections.

Plug-in device A component or group of components and their circuitry that can be easily installed or removed from the equipment.

Polarized plug A plug arranged so that it may be inserted in its receptacle only in a predetermined position.

Potting A method of securing a component or group of components by encapsulation.

Power The rate of doing work or expending energy. The watt is the unit of electric power.

Pressure connector A conductor terminal applied with pressure to secure the connection mechanically and electrically.

Pulse An abrupt change in voltage, either positive or negative, that conveys information to a circuit.

Raceway Any channel for holding wires, cables, or bus bars designed and used solely for this purpose.

Radial lead Lead of a component that extends from its side instead of its end.

Receptacle A connecting device into which a plug can be pushed or screwed to make electrical connections.

Rectifiers A component or device that converts AC into a pulsating DC current (unidirectional current).

Registration Alignment of a PC board pad with its mate pad on the opposite side. Alignment of graphic documentation when designing and laying out a PC board, using a CAD or manual layout system.

Relay A device that a variation in the conditions of one electric circuit triggers to effect the operation of other devices in the same or another electrical circuit.

Resistance The quality of an electrical circuit that opposes the flow of current passing through it.

Resistor A component resisting or opposing flow of an electrical current.

Routing Placement of interconnections on a PC board. Also, the sequence of steps in producing a part or assembly.

Schematic diagram (See "elementary diagram.")

Semiconductor A device that can function either as a conductor or a nonconductor, depending on the polarity of the applied voltage (e.g., a rectifier or transistor with a variable conductance depending on the control signal applied).

Shielded cable Shielded cable is single- or multiple-conductor cable surrounded by a separate conductor (the "shield") to minimize the effects of adjacent electrical circuits.

Short An abnormal connection of relatively low resistance between two points of a circuit.

Signal The name associated with a net.

Signal highlighting Identifying the connection points of a net in a PC board.

Solenoid An electromagnetically energized, approximately cylindrical coil and an armature whose motion reciprocates within and along the coil axis.

Starter An electric controller that accelerates a motor from rest to normal speed. (Note: A device that starts a motor in either direction of rotation should be designated a controller).

Static circuit The behavior of a circuit under fixed, not changing, conditions.

Static device For electronic and other control or information-handling circuits, the term refers to devices with switching functions that have no moving parts.

Stepping relay (switches) A multiposition relay in which moving wiper contacts mate with successive sets of fixed contacts in a series of steps representing successive operations of the relay.

Subassembly   A portion of an assembly of electrical or electronic components, mounted on a panel or chassis, that forms a functional unit by itself.

Symbol   A sign, mark, or drawing representing an electrical or electronic device or component of it.

Temperature control   A control device responsive to temperature.

Terminal   A point of connection in an electronic circuit.

Terminal block   An insulating base or slab equipped with one or more terminal connectors to which electrical connections are made.

Threshold   The level of input voltage at which a binary logic circuit changes from one logic state to another.

Tie point   A distributing point in circuit wiring, other than a terminal connection, where junctions of leads are made.

Tooling hole   Hole drilled through a PC board to aid in setting dimension, positioning, and manufacturing. Dimensions for the board geometry are normally established from tooling holes.

Transducer   A device to transfer one form of energy to another or one type of input to another.

Transformer   A device that uses electromagnetic induction to transfer energy from one circuit to another.

Triode   A three-electrode vacuum tube consisting of a grid, plate, and cathode.

Truth table   A tabulation that shows the relation of all output logic levels of a digital circuit to all possible combinations of input logic levels to characterize the circuit functions completely.

TTL (transistor logic)   A logic family that uses transistors at the input to perform the logic function and drive the output transistors.

Voltage   The force that causes free electrons to move in a conductor. The volt is the unit of measure.

Watt   The unit of power measurement.

Wave soldering   Also called flow soldering. A method of soldering a PC board by moving the board over a wave of flowing molten solder in a solder bath. Eliminates the need to hand solder each individual component lead.

Wire bond   The method by which very fine wires are attached to semiconductor chips for interconnection with package leads.

Wire list   Wire run list containing only two connections in each wire; also called a from–to list.

Wire net   Subset of electrical connections in a logical net having the same characteristics and common identifiers. No physical order of connection is implied.

Wireway   Sheet metal troughs with hinged covers for housing and protecting electrical conductors and cable. Conductors are laid in place after the wireway has been installed as a complete system.

Wire wrapping   A technique to terminate conductors.

Wiring diagram   Diagram containing components, wire runs, wires, miscellaneous graphic and nongraphic information, and text annotation.

Appendix C
CAD/CAM Glossary

Absolute coordinates   The values of the X, Y, or Z coordinates with respect to the origin of the coordinate system. Contrast with incremental coordinates.

Absolute data   Values representing the absolute coordinates of a point or other geometric entity on a display surface. The values may be expressed in linear units of the display or of the engineering drawing.

Access time   The interval between the instant a computer instruction requests that data be extracted or placed on a storage device and the instant at which the data actually begin moving to or from the device.

Aiming device   A pattern of light activated by a light pen on the display surface to assist positioning of the pen and to describe the pen's field of view. (See "Cursor.")

Alphanumeric display   Device consisting of a typewriter-style keyboard and a display (CRT) screen on which text is viewed.

Annotation   The process of inserting text on a drawing.

APT (automatically programmed tools)   A program language used to prepare numerical control tapes.

Assembler   A computer program that converts user-written symbolic instructions into equivalent machine-executable instructions.

Associative dimensioning system   A package that allows automatic updating of dimensions as the entities to which they are linked change.

Automatic dimensioning   The means by which a CAD dimensioning program automatically measures distances and places extension lines, dimension lines, and arrowheads.

Autoplacement   A software option that automatically packages IC elements and optimizes the layout of components on a PC board.

Autoroute   A software option that automatically determines the placement of copper on the printed circuit board to connect part pins of the same signal.

Back annotation   A process of extracting information from a completed PC board and inserting it in the logic elements of the schematic that describes the board.

Baud rate   Unit of signaling (in bits per second) in measuring serial data flow between a computer and/or communication devices.

Benchmark   A set of standards used in testing a software or hardware product or system from which a measurement can be made. Benchmarks are often run on a system to verify that it performs according to specifications.

Binary code   A system of representing all numbers using combinations of 0 and 1.

Bit (binary digit)   The smallest unit of information on a binary system of notation.

BOM (bill of materials)   A listing of all the subassemblies, parts, materials, and quantities required to make one assembled product.

BPI (bits per inch)   The number of bits of binary data that one inch of magnetic tape can store.

Byte   A sequence of eight adjacent bits that operate as a unit.

CAD (computer-aided design)   A process whereby a computer assists in creating or modifying a design.

CAM (computer-aided manufacturing)   A process employing computer technology to manage and control the operations of a manufacturing facility.

Character   An alphabetic, numeric, or special graphic symbol.

CL file (cutter location file)   Output of an APT or graphics system that provides X, Y, and Z coordinates and NC information for target machine tool processing.

Code   A system of symbols and rules for representing information; it usually refers to instructions executed by a computer.

Compiler   A program that translates user-written PEP or FORTRAN instructions into binary machine-level code.

Computer graphics   Methods and techniques for converting any form of information to or from graphic displays using a computer.

Computer output microfilm (COM)   Microfilm containing computer-generated data; also, to place computer-generated data on microfilm.

Configuration   A combination of computer and peripheral devices at a single installation.

Construction plane   A predefined or operator-defined plane on which digitizers are projected.

Conversational mode   A mode of operation for a data processing system in which each unit of input entered by the user elicits a prompt response from the computer.

CPS (characters per second)   A measure of the speed with which data are input or output.

CPU (central processing unit)   The unit of the processor that includes the circuits that interpret and execute instructions.

Cross hairs   On a cursor, a horizontal line intersected by a vertical line to indicate a point on the display whose coordinates are desired.

CRT (cathode-ray tube)   A device that presents data visually using controlled electron beams.

Cursor   A manually movable marker that indicates the location of the next action.

Cut plane Defining and intersecting a plane with a three-dimensional object to derive a sectional view.

Database A comprehensive collection of information having predetermined structure and organization suitable for communication, interpretation, or processing.

Database relations Linkages within a database that logically bind two or more elements in it. For example, a nodal line (interconnect) is related to its terminal connection nodes (pins) because they all belong to the same electrical net.

Data extract Capability to obtain printed reports from the database.

Data tablet An input device with a writing surface corresponding directly between positions on the tablet and addressable points on the display surface of a display device.

Digitize Convert visual information into digital form.

Digitizer A device that converts coordinate information into numeric form readable by a digital computer.

Disk A device on which information is stored.

Display Representation of data for viewing on an output device.

Display device A device capable of presenting, on a viewing surface or image area, the display elements that visually represent data (e.g., CRT).

Display elements The basic building symbols for an application used to construct display images, (e.g , points, line segments, and characters).

Display foreground The collection of display elements or a display image subject to change by a computer program or by the operator of a display terminal in an interactive mode.

Display group A collection of display elements that can be manipulated as a unit and that may be further combined to form larger groups.

Display image The collection of display elements and display groups visually represented together on the viewing surface of a display device.

Display menu An option listed on a display allowing an operator to select the next action by indicating one or more choices with an input device.

Display parameters Data that control the appearance of graphics (e.g., line fonting).

Display space The usable area of the display surface that includes all addressable points.

Display surface The medium upon which a display image is produced (e.g., CRT screen, film, paper).

Drum plotter A plotter that draws an image on paper or film mounted on a drum.

Dynamic rotation When a display element, display group, or display image is rotated continuously at a fixed rate until stopped by the operator.

Element The lowest-level design entity with an identifiable logical, electrical, or mechanical function.

Entity The fundamental building blocks that a designer uses to represent a product (e.g., arc, circle, line, text, point, line, figure, nodal line).

Firmware Sets of instructions cast into user-modifiable hardware.

Flatbed plotter A plotter that draws an image on paper, glass, or film mounted on a flat table.

Flip The same as mirror-image projection.

Fonts, line A repetitive pattern used to give meaning to a line (e.g., solid, dashed, dotted, etc.).

Fonts, text A complete set of one character type.

Format The arrangement of data on a data medium.

Form flash To project a constant pattern such as a report form, grid, or map as background for a display. Synonymous with form overlay.

FORTRAN A high-level programming language used primarily for scientific or engineering applications.

Full frame A display image scaled to maximize use of the viewing surface or the area of a display device.

Function key A specific key that requests a predefined function whenever it is depressed.

Function keyboard An input device for an interactive display terminal consisting of a number of function keys.

Geometric entity description A sequence of data, in a computer-readable format, that completely specifies a geometric element (e.g., point, line, or circle).

Graphic tablet A surface through which coordinate points can be transmitted using a cursor or stylus.

Grid A network of uniformly spaced points on an input device used for locating position.

Hard copy Printed copy of machine output (e.g., drawings, printed reports, listings, summaries).

Hardware The physical components of a computer system (e.g., mechanical, magnetic, electrical, or electronic devices).

Hidden line removal In computer graphics, a technique for removing lines or line segments that a displayed object would normally obscure.

Hidden lines Line segments that the display of a solid three-dimensional object would obscure.

Host computer The computer attached to a network providing services such as computation, database management, and special programs.

Intelligent terminal A terminal with local processing power whose characteristics can be changed under program control.

Interactive display terminal A terminal consisting of one or more display devices and one or more input devices such as tablets, control balls, light pens, alphanumeric keyboards, function keys, and tape readers.

Interactive graphics Capability to perform graphics operations directly on the computer with immediate feedback.

I/O device Input/output equipment used to communicate with a system.

Joystick A data-entry device to enter coordinates manually in specific X, Y, and Z registers.

Keypunch A keyboard-actuated device that punches holes in cards.

Layer A logical concept used to distinguish subdivided group(s) of data within a given drawing; it may be thought of as a series of transparencies (overlaid) in any order, yet having no depth. The operator may specify display elements (layers) to be visible.

Layout A visual representation of a complete physical entity, usually to scale.

Library Collection of often-used parts.

Light pen A photosensitive device that detects light on the face of a CRT to provide a signal that may be interpreted by the display control program to determine either positional or display element information.

LIS (large interactive surface) An automated drafting table used to plot and/or digitize drawings (e.g., Interact IV). Also called a digitizer table.

Loop A sequence of instructions repeatedly executed, with or without modifications during each iteration, until a terminating condition is satisfied.

Machine instruction Machine-recognizable and executable instructions.

Machine language The actual language the computer uses when it performs operations, usually binary code.

Macro A combination of commands executed as a single command.

Mainframe In general, a central processing unit of a large-scale computer configuration.

Main storage The general-purpose storage of a computer, program addressable, from which instructions can be executed and from which data can be loaded directly into registers.

Mass storage Auxiliary or bulk memory that can store large amounts of data readily accessible to the computer, for example, a disk or magnetic tape.

Menu An input device consisting of command squares on a digitizing surface. It eliminates the need for an input keyboard for common comands.

Microcomputer A computer whose basic element is a single integrated circuit and that has a limited basic instruction set.

Minicomputer Generally, a 16-bit computer with limited memory addressability.

Mirror-image projection In computer graphics, the reflection of display elements or display groups with respect to a specific straight line or plane. Also, to reflect display elements or display groups with respect to a specific straight line or plane; synonymous with flip or reflect.

Mirroring A graphic construction aid; the ability to create a mirror image of a graphic entity.

Model A geometrically accurate, complete representation of a real object stored in a CAD/CAM database.

Multiprocessor Computer architecture that can execute one or more computer programs using two or more processing units simultaneously.

NC (numerical control) Prerecorded instructions for automatic computer control of machine tools, drafting machines, and other operations.

Nesting Imbedding data in levels of other data so that certain routines or data can be executed or accessed continuously, in loops.

Network Two or more central processing facilities interconnected.

Numerical control Automatic control of a process performed by a device that makes use of all or part of numerical data, usually introduced during the operation.

Off-line Equipment or devices in a system that are not under direct control of the system.

On-line Equipment or devices in a system that are directly connected to and under the control of the computer.

Operating system Software that controls the execution of computer programs and the flow of data to and from peripheral devices.

Order To place in sequence according to rules or standards.

Pad An area of plated copper on a PC board to which leads of components are soldered.

Partitioning A logical grouping of electrical functions within a given set of hardware components.

Passive graphics The use of a display terminal in a noninteractive mode, usually through such items as plotters and microfilm viewers.

Passive mode A method of operating a display device that does not allow any on-line interaction or alteration.

Photoplotter A device used to generate artwork photographically for PC boards.

Pin A connection point for electrical components and logical elements; may be referred to as a terminal on components.

Plotter A device that makes a permanent copy of a display image.

Postprocessor A software program or procedure that interprets graphical data and formats them for use by an NC machine or by other computer programs.

Preprocessor A method of converting data into computer-usable form for processing and output.

Program A set of machine instructions or symbolic statements combined to perform a task.

Prompt Any message or symbol from the computer system informing the user of possible actions or operations; a guide to the operator.

Properties User-established attributes that dictate the significance of an entity or subfigure within the model.

Puck Manually operated directional control device used to input coordinate data.

RAM (random access memory) Memory from which data can be retrieved regardless of input sequence.

Raster The geometric coordinate grid dividing the display area of a display device.

Raster display A display whose entire surface is scanned at a constant refresh rate.

Raster scan A line-by-line sweep across the entire display surface to generate elements of a display image.

Reflect The same as mirror-image projection.

Refresh CRT display technology requiring continuous restroking of the display image.

Repaint Redraw a display image on a CRT to reflect its updated status.

Repeatability (of display device) A measure of the hardware accuracy or the coincidence of successive retraces of a display element.

Resolution The smallest spacing between points on a graphic device at which the points can be detected as distinct.

Response time The time elapsed between a specified or specifiable signal sent by an entity and a related specified or specifiable signal received by the sending entity.

ROM (read-only memory) A storage device (memory) generally used for control programs, the content of which is not alterable.

Rotate To turn a display element, display group, or display image through an angle.

Router A program that automatically determines the routing path for the component connections on a PC board.

Routine A set of instructions arranged in proper sequence to cause a computer to perform a desired operation.

Rubber banding A technique for displaying a straight line with one end fixed and the other end following a stylus to some input device.

Scale The ratio of the current displayed image with respect to the database.

Scissor To apportion a drawing into segments that can be viewed on a CRT screen.

Section To construct and position a bounded or unbounded intersecting plane with respect to the object part(s) and then request generation and display of the total intersection geometry on a display surface.

Shape fill The solid painting-in within the boundaries of a shape using graphics.

Software A set of programs, procedures, rules, and associated documentation that directs the operation of a computer.

Source User-written instruction statements prior to translation by the computer into machine-executable form.

Storage tube A CRT that retains an image for a considerable period of time without redrawing.

Stylus A hand-held object that provides coordinate input to the display device.

Surface machining The ability to output three-, four-, and five-axis NC toolpaths using three-dimensional surface definition capabilities, (e.g., ruled surfaces, tabulated cylinders, and surfaces of revolution).

Surface of revolution Rotation of a curve around an axis through a specified angle.

Symbol Representation of something by relationship, association, or convention.

Tablet An input device that digitizes coordinate data indicated by stylus position.

Telewriter A typewriter-style keyboard device used to enter commands or to print out system messages.

Terminal The data entry or exit point in a system or communication network.

Time sharing Use of the same computer memory for two or more simultaneous tasks.

Toolpath Centerline of an NC cutting or drilling tool in a specific cutting operation, such as milling or boring.

Track In computer graphics, to cause the display device to follow and display or determine the position of a moving input device, such as the writing tip of a stylus.

Tracking symbol In computer graphics, a symbol such as a cross, dot, angle, or square used for indicating the position of a stylus.

Turnkey Pertaining to a computer system sold in a ready-to-use state.

Vector generation The process that determines all intermediate points between two endpoints of a line segment.

Verification In computer graphics, the message feedback to a display device acknowledging that an input was detected (e.g., the brightening of a display element selected by a light pen).

Via Means of passing from one layer or side of a PC board to the other.

Window A rectangular area on the display screen selected by the operator.

Z clipping The ability to specify depth parameters for a three-dimensional drawing such that all elements above or below the specified depth(s) become invisible. No change is made to the database of the part or drawing. Useful in viewing cluttered or complex part geometry.

Zoom To enlarge proportionately or decrease the size of the display entities by rescaling.

Appendix D
Comparative Glossary*

	Manual Definition	**CADD Definition**
Accuracy, design	Positional accuracy in critically tight areas determined graphically on large-scale study work plot or manual calculations.	Position accuracy in tight areas can be viewed graphically at any infinitely large scale and checked by requesting dimensional readout; used in interference checking.
Accuracy, drafting	Depending on scale of drawing, measure of positional accuracy of drawing elements. May be .01 in. (.254 mm) at best, if on dimensionally stable material.	Regardless of drawing scale, positional accuracy is plotted at .001 in. (.0254 mm), while database coordinates are accurate to 14 decimal places.
Add	Arithmetic or design/drafting function.	To sum by computer, like items or parts; augment a design or drafting image with further graphic, dimensional or alphanumeric information.
Add text	Letter or type additional alphanumeric notation.	Type at workstation keyboard additional alphanumeric notation that is electronically placeable at preselected text nodes.
Align	Place in line, as in parallel to a reference or object line.	Automatically line up design features, shapes, symbols, text, etc., parallel to a reference or object line.
Algorithm	A predefined sequence of steps to be taken to solve problems of a particular type; a procedure attributed to an Uzbek mathematician, Al-Khwarizmi, in ninth century.	A predefined program of steps to be followed by computer to solve problems of software furnished by CADD supplier to solve typical problems.
Annotate	Complement dimensioned drawing with explanatory text, labels, general notes, special notes, reference notes, subtitles, and titles.	Add explanatory text, labels, general notes, etc.; electronically copy and place repetitious notation common to many symbols, drawing segments, or drawings.
Array	In drafting, the alignment in X and Y of similar design entities or tabulated data.	Electronic alignment in X, Y, and Z of similar design entities or tabulated data.
Assemble	Place related units or parts into predetermined positions.	Electronically place related units or parts into predetermined positions as in arrangement and assembly drawings. Units or parts may be electronically copied from other drawings in the database.
Assign	Schedule or reserve a position for an activity or a drawing segment.	Schedule or reserve a position for an activity, data, symbol, or drawing segment to a layer, drawing coordinate, model, or disk memory space.
Auto, revise		Revise all drawing segments, subtitles, titles, drawing numbers at each occurrence of faulty data in database; effect change with one command on one or all drawings or documents of a set (see "Revise").
Auxiliary view	(See "View, auxiliary.")	(See "View, auxiliary.")

---

* From Lamit and Paige, *Computer-Aided Design and Drafting: CADD,* Columbus, Ohio: Merrill, 1987.

	Manual Definition	**CADD Definition**
Axis	One of a set of three lines intersecting at a common point in space in such a way that each axis is perpendicular to the plane containing the other two.	Same as manual definition.
BASIC		A computer language: Beginner's All-purpose Symbolic Instruction Code.
Batch process		Without benefit of interactivity by CRT; digitizer; tablet or keyboard; a means for creating alphanumeric and/or graphic output from data processed by any computer.
Baud		A unit of signaling speed equal to the number of discrete conditions or signal events per second.
Bill of material	A listing of parts or items required to fabricate, assemble, or erect an engineering design. Also BOM or BOM list.	A computer listing (lettered on drawing or tabulated on printer) of parts or items represented on an engineering design, automatically derived from the database.
Blank, line	Erase line from drawing	Electronically erase from view on screen, but leave in database for ultimate reuse or elimination.
Blank, model	Erase all views of a design feature on all drawings.	Electronically erase from view on screen, but leave in database for ultimate reuse, modification, or elimination.
Blank, screen		Electronically erase all data from view on screen, but leave in database for ultimate reuse, modification, or elimination.
Blank, submodel	Erase all views of a design on all drawings.	Electronically erase from view on screen, but leave in database for ultimate reuse, modification, or elimination.
CAD		Interchangeably: computer-aided design or computer-aided drafting; a generic term used in the United States, Europe, and Japan.
COM		Computer-output-microfilm: an electromechanical system for transforming the digital version of an engineering design directly onto 35-mm microfilm or 105-mm microfiche images.
CPU		Central processing unit; that section of the computer that contains the control unit, the arithmetic unit, and memory.
Cartesian coordinates	A reference system similar to engineering LEFT–RIGHT, UP–DOWN, BACK–FORWARD system for defining a point in three-dimensional space. Compares to projection planes: FRONTAL, HORIZONTAL, PROFILE.	A reference system along X(LEFT–RIGHT), Y(UP–DOWN), and Z(BACK–FORWARD); René Descartes' X–Y–Z system for defining a point in three-dimensional space.
Catalog	Compilation, in a printed book, of vendor's standard offerings; a reference document for designers.	Portion of the design database containing often-used vendor's or trade association's standard reference information, graphics, and dimensions.
Cathode-ray tube, refresh		Specialized type of CRT in which screen image is formed by continuously panning electron beam over phosphor-coated tube face, *rasterization.*
Cathode-ray tube storage		Specialized type of CRT in which screen image is formed by electron beam "stroking" phosphor-coated tube face.

	Manual Definition	**CADD Definition**
Checking	Inspection or recalculation of engineering drawing data for compliance with original design, vendor's drawings, or catalog information.	Same as manual definition.
Class 1 drawing	A designation of drawing type: NONDIMENSIONED, NOT TO SCALE; usually schematic or diagrammatic.	Same as manual definition.
Class 2 drawing	A designation of drawing type: DIMENSIONED, NOT TO SCALE; usually with tabulated dimensions, but may be isometric (piping, etc.)	Same as manual definition, except designer has choice of electronically "rubber banding" drawing to accurately reflect tabular dimensions.
Class 3 drawing	A designation of drawing type: DIMENSIONED, TO SCALE.	Same as manual definition, except the three-dimensional model or submodel in database is source for two-dimensional (orthographic) views needed for working drawings.
Composite	A multilayered drawing; a series of special overlays viewable and reproducible with a base drawing to make a composite print.	Separable layers or overlays of graphic and/or alphanumeric data in the database; viewable on screen or plotted in any combination of base drawings or overlays.
Database	A collection of interrelated data items that must be assembled by each application, thereby causing the "reinvention of the wheel" each time the database is needed.	A collection of interrelated data items organized by a consistent scheme that allows one or more applications to process the items without regard to physical storage locations.
Delete	Remove from document.	Remove electronically from database; selectively remove portion of symbol or drawing segment, linework, or text and automatically remove all such occurrences on any drawing or document in the database.
Designer	Degreed or paraengineer who creates preliminary or working drawings and documentation from engineer's notes and sketches.	Operates input design workstations with interactive CADD processing system.
Digitize	Laborious scaling and recording X–Y–Z coordinates or drawing or map elements from an origin point.	Automatically recording by pointing; X–Y–Z coordinates as in manual definition; enters graphical data into a CADD system.
Digitizer		Old name for CADD drafters'/designers' input workstation; a computer-oriented device for automating graphic data reduction; process called *digitizing*.
Dimension	Annotate drawings with lettered dimensions to denote sizes of elemental shapes and areas as well as locations.	Automated placement of dimension lines; witness lines with typed or computer-supplied dimensions to denote size of elemental shapes and areas as well as locations.
Display	Show.	Command drawing or document to appear from database on the workstation screen.
Drafter	U.S. government-mandated term for man or woman performing engineering drafting; formerly draftsman, draughtsman, draftswoman, draftsperson.	An ungraded drafter; uses CADD input or output workstations interactively to conduct all drafting functions.
Drawing	Graphic representation with annotation of an engineered physical object; sketch.	Digital version of graphic and alphanumeric representation of an engineered physical object stored in database; electromechanically plotted graphic and alphanumeric representation of the object.
Drawing, layer	(See "Composite.")	(See "Composite.")

	Manual Definition	CADD Definition
Drawing segment	A portion of complete drawing; may contain symbols and/or text.	A portion of complete drawing; a repeatable and electronically copied, reduced, or enlarged drawing segment for use on other drawings, with or without modification; a combination of symbols and/or text.
Drawing standards manual	Prepared by design/drafting management.	Same as manual definition, except for some special standards for CADD.
Drawings security	Method for protecting original documents (tracings) from theft or unauthorized alteration; usually by controlled access.	Method for protecting digital database from theft or unauthorized copying, alteration, or accidental erasure (see "Password").
Edit	Review or proofread for possible revision.	Review or proofread on screen for possible revision; perform revision electronically.
Edit station (CADD)		Workstation for inputting design or drafting changes to existing CADD drawing or document. Changes may be redesign, revision, construction ECO, or as-built data.
Enlarge	Make bigger than before.	Electronically cause the indicated portion of the picture to enlarge to fill screen. Operator can enlarge infinitely, until a decimal point, for example, is made to fill full 19-in. screen.
Erase	Rub out with abrasive material (erasure); remove with moist cotton swabs, as with water-erasable ink on plastic film; remove with No. 1 and No. 2 solution as on sepia print.	Remove electronically from database.
Font	A drafter's or printer's definition for unique sets of alphanumeric characters (e.g., Leroy lettering font, Futura type font).	Identifying name for lettering style or line characteristic (e.g., Leroy lettering font, dashed-line font).
FORTRAN		FORmula TRANslation: a computer language universally used in engineering.
Graphics command language	May be instructions in design/drafting manual spelling out graphic procedures, drawing composition usage, text sizes, etc.	A CADD designer's shorthand for communicating desired graphic actions and responses to the system.
Grid	A spaced array (graphical).	A design/drafting aid available as a placement background on the CRT screen. Grids may be square or stretch in X, Y, or Z.
Hidden line	A line on a drawing, usually shown dashed, representing an edge, contour, or surface that could be seen only if the object being drafted were transparent. May be omitted as well; must be constructed.	In three-dimensional work, all hidden lines appear at first as object lines. Designer has option to "touch" each line with cursor, ask computer to make it dashed, or eliminate it from view.
Host computer or host CPU		Term that used to be given to only very large mainframe computers, but now includes minicomputers for data management and number crunching.
Input/output		Communication with processing system (computer). Facilities for "talking" to computer; give data, coordinates, instructions; receive answers, listings, drawings in readable form.
Interactive		A technique of designer communication in which the system immediately acknowledges and acts on requests entered by the designer at a workstation.

	Manual Definition	**CADD Definition**
Isometric	A form of drawing projection in which three faces of an object or feature are shown on three major axes 120° apart and in which the angle the front edge makes with the vertical is 35°. True isometrics are to-scale drawings.	Same as manual definition.
Item	A component of a larger grouping or assembly; part of a design; a purchased unit under a single engineering specification used in fabrication, assembly, or erection.	A two- or three-dimensional symbol (cell), or submodel; an engineering database unit under a single specification; part; an elemental portion of a cell, submodel, or model.
Item select		A technique for selecting a symbol (cell) or submodel, including text, preparatory to executing a MOVE, COPY, or MIRROR command at the workstation.
Label	Hand-lettered note or callout identifying a drawing feature.	Computer-generated note or callout identifying a drawing feature.
Layer	(See "Composite.")	(See "Composite.")
Lettering, Leroy	(See "Font.")	(See "Font.")
Library	A cataloged collection of data.	Same as manual definition.
Line	A visible connection between two points in space.	Same as manual definition.
List		A command to request that a list of items be printed by the system printer.
Macro		Directions that generate a known set of instructions. Used to eliminate the need to write a set of instructions that are used repeatedly.
Menu		An area of the digitizing tablet reserved for an array of commands. Allows choosing the commands with the stylus.
Metrication	The act of replacing, relettering, or recalculating nonmetric dimensions or values into SI units.	Automatic assignment and conversion of nonmetric dimensions or values into SI units, including roundoff.
Mirror	Create, by tracing back of drawing; an opposite-hand view of a portion of a design.	Electronically command, display, or plot an opposite-hand view of a portion of a design, wherein "mirrored" text remains right-reading.
Mirror, about line	Create opposite-hand view similar to or symmetrical about a line.	Electronically display or plot opposite-hand view; used in two-dimensional design only.
Mirror, about plane	Impossible to create without redrawing.	Electronically display or plot opposite-hand view; used in three-dimensional design only.
Model	A three-dimensional object used to obtain physical data for drawing information.	The three-dimensional object that is being constructed electronically in the computer.
Modem	MOdulator–DEModulator; device used to send and receive data in high-speed bulk mode over telephone lines.	Same as manual definition.
Move	Completely erase and redraw or trace in a different location.	Electronically move a group of items without redrawing.
Orient	Line up with a known axis such as X, Y, or Z.	Line up working coordinating system to base coordinate system; in three-dimensional work, line up item relative to submodel or submodel to model.
Orthogonal	At right angles to each other.	Same as manual definition.
Orthographic projection	The projection of a point from one plane to another.	Same as manual definition.
Paint		Electronic drawing or lettering on the screen of a design.

	Manual Definition	**CADD Definition**
Part	Elemental physical object or its symbolic representation; also, elemental portion of a drawing (i.e., point or line).	Same as manual definition; sometimes called *item*.
Password		A word or code required to gain access to the system.
Picture	Any graphic representation.	The flat-plane (two-dimensional) view of a submodel or model, part of an engineering drawing to which dimensions, text, and titling will be added; a "window" portion of the complete design in database.
Plot		To get hard-copy output on a variety of plotters.
Plotter, belt		Upright high-speed pen (ballpoint or ink) plotter capable of handling drawings up to E size.
Plotter, electrostatic		A very high speed plotter in which lines are formed by a matrix of dots.
Plotter, flatbed		Large tablelike pen (ballpoint or ink) plotter capable of handling very long drawings.
Plotter, microfilm		Computer-output microfilm (COM) recorder that converts magnetic tape version of a drawing or document from CADD database and records a miniature version on 35-mm microfilm or 105-mm microfiche.
Point	A visible dot to represent some coordinate in space.	Same as manual definition.
Process (n.)	An orderly method for attaining a given result.	A predesigned procedure for computer to aid in attaining a given design, drafting, or documentation result.
Process (v.)	To proceed, step by step, for attaining a given result.	Activate a predesigned software command(s) procedure to create a given result from data furnished by computer.
Repaint		Electronic redraw or relettering of latest status of a design, used right after a revision or edit has been executed, for designer to view the effects of the revision.
Revise	Alter existing tracing or document by erase–replace or redraw–retype.	Alter database; electronic erase and replace, seldom redraw.
Rubber band		Electronically "stretch" space between design components and have interconnected lines "stretch" also.
Ruler	Scale. A tool for linear measurement. May be architect's, engineer's, mechanical (machine), or metric.	A scalelike image with the ability to be placed electronically on screen at any location or orientation as an aid for design layout at workstation. May be marked off as architect's, engineer's, mechanical (machine), or SI metric. English and metric units are interchangeable on command.
Save	Keep on file.	Same as manual definition.
Scale, architect's, USA	Measuring device marked off in units varying from $\frac{1}{32}$ to $\frac{1}{2}$ of an English inch.	Same as manual definition.
Scale, engineer's, USA	Measuring device marked off in units varying from $\frac{1}{10}$ to $\frac{1}{100}$ of an English inch.	Same as manual definition.
Scale, machine, USA	Measuring device marked off in equal units varying from $\frac{1}{4}$ size to full size.	Same as manual definition.
Scale, metric, SI	Measuring device marked off in tenths of a millimeter or centimeter.	Same as manual definition.

	Manual Definition	**CADD Definition**
Screen	(Sometimes used to describe printers' Benday screen of dot patterns created photographically on a background site or structure drawings.)	TV-like picture tube; cathode-ray tube (CRT) may be storage, vector, or raster type.
Select	To choose.	Identify to the computer the portion of drawing to be acted on next.
Shift	To move.	Same as manual definition.
Skew	Place at an angle off plumb or level line.	Place at an angle off plumb, level, or depth line.
Smooth	Drafter's refinement of plotted-point spline using French curves.	Electronic refinement of known value point curve.
Space	Scale off on drawing positions for repetitive symbols, drawing segments.	Electronic scaling, positioning of repetitive points, symbols, or drawing segments by typing only the overall dimension and the number of units to be repetitively spaced and placed on the screen.
Spline curve	Cumbersome pliable metal strip used to draft varying radius curves through preplotted points on drawing or template (as in Lofting).	Automatic curve-fit generator, with calculated offsets, through predetermined points for drawing or template plot (as in lofting, shipbuilding, or sheet metal).
Submodel	Portion of a three-dimensional object used to obtain physical data for drawing information.	Portion of a three-dimensional object that is being constructed geometrically.
Symbol library	Legend of symbols, names, and uses; key to symbols; may be plastic templates.	Digital version of symbols, by engineering discipline, instantly callable from symbol library database. Each symbol may contain associated text or preassigned space for varying text (see "Symbol, transfer").
Symbol template	Standard or custom, symbol cutout plastic template for tracing onto drawing; available by disciplines (e.g., piping, iso, electrical).	Not used (see "Symbol library").
Symbol, transfer	Rub-on version of preprinted standard drafting symbols; used in photodrafting.	Digital version of standard or special symbol, library database. Electronically placeable, spaceable, or copyable anywhere on drawing in any orientation.
Tabular dimensioning	Listing of numerical values keyed to dimension letters. Used on typical details.	Same as manual definition.
Text node	Space reserved on drawing for later addition of alphanumeric characters, dimensions, etc.	A preselected space reserved on symbol (cell), drawing segment, or drawing for later addition by keyboard typing of alphanumeric characters, dimensions, subtitles, and titles.
3D	Three-dimensional; relates to measurable volumes in engineering.	Three-dimensional; relates to measurable volumes. X–Y–Z coordinates used to make orthographic views automatically.
2D	Two-dimensional; relates to measurable areas in engineering.	Two-dimensional; relates to measurable areas and symbols used for drawings.
Trace	Recopy portion of repetitive drawing by tracing.	Electronically copy portion of repetitive drawing already in database.
View, auxiliary	Planar projection of a drawing portion not perpendicular to standard orthographic projection.	Same as manual definition.
Window		Portion of larger design area, filling the screen vertically and horizontally; the "distance" from which operator sees drawings or model; related to "scale."
Zoom	A photo technique for enlarging or reducing a portion of a drawing; a TV camera technique for enlarging or reducing field of view.	An electronic enlarge or reduce technique for changing scale on screen infinitely, up or down.

Appendix E
Reference Designations†

Reference designations distinguish one graphic symbol from another and correlate these identifications with actual components on the parts lists and assembly drawings.

Reference designations identify electronic, electrical, and mechanical components, as well as assemblies and subassemblies. These designations consist of a combination of letters and numbers identifying the component's class. The numerical designation following the class letter or letters always falls on the same line without a space or hyphen. When letters follow a numerical designation, they indicate a function separate from the designated component as a whole.

Reference designations may be placed above, below, or on either side of a graphic symbol. Other pertinent information, such as component value, tolerances or rating, terminal numbering, and other functional descriptions may be placed around the graphic symbols.

It is standard practice to number each class of components from left to right beginning with the upper left corner of the diagram and following horizontally and vertically across the sheet. This facilitates locating the symbols and related components on the schematic diagram.

When circuit changes are made, the schematic diagram is also changed by adding or deleting graphic symbols. The remaining symbols are not renumbered when a graphic symbol is deleted. Because of these changes, a reference designation table is often placed on the schematic diagram to avoid searching for a designation that may have been removed.

CLASS DESIGNATION LETTERS

This alphabetical listing of class designation letters is used in assigning reference designations for electrical and electronics parts and equipment as described in ANSI Y32.16–1968 and Y32.16a–1970.

Parts not specifically included in this list should be assigned a letter or class most similar in function.

Designations for general classes of parts are marked with an asterisk (*) to facilitate designation of parts not specifically included in this list.

A

accelerometer*
(1) assembly, separable or reparable (2)
circuit element, general
computer
divider, electronic
facsimile set
generator, electronic function
integrator
modulator
multiplier, electronic
recorder, sound
recording unit
reproducer, sound
sensor (transducer to electric power)
servomechanism, positional
subassembly, separable or reparable
telephone set
telephone station
teleprinter
teletypewriter

AR

amplifier (magnetic, operational, or summing)
repeater, telephone

AT

attenuator (fixed or variable)
bolometer
capacitive termination
inductive termination
isolator (nonreciprocal device)
pad
resistive termination

B

blower
fan
motor
synchro

BT

barrier photocell
battery

---

† Provided courtesy of Bishop Graphics, Inc.

571

blocking layer cell
cell, battery
cell, solar
transducer, photovoltaic

C

capacitor
capacitor bushing

CB

circuit breaker

CP

adapter, connector
coupling (aperture, loop, or probe)
junction (coaxial or waveguide)

CR

absorber, overvoltage
current regulator (semiconductor device)
demodulator, diode-type ring
detector, crystal
diode (capacitive, storage, or tunnel)
modulator, diode-type ring
photodiode
rectifier (metallic or diode)
selenium cell (rectifier)
semiconductor device, diode
thyristor (semiconductor diode-type)
transducer, photoconductive
varactor
varistor, asymmetrical

DC

coupler, directional

DL

delay function
delay line
slow-wave structure

DS

alarm, visual
annunciator
audible signaling device
bell, electrical
buzzer
device, indicating (excluding meter or thermometer)
flasher (circuit interrupter)
indicator (excluding meter or thermometer)
lamp (cold cathode, fluorescent, glow, incandescent,
    indicating, pilot, signal, neon)
light source, general
ringer, telephone
signal light
siren
sounder, telegraph
vibrator, indicating
visual signaling device

E*

antenna, loop or radar
arrestor, lighting
bimetallic strip
brush, electrical contact
carbon block
cell, aluminum or electrolytic
cell, conductivity
contact, electrical
core (adjustable tuning, electromagnetic, inductor, memory,
    transformer)
counterpoise, antenna
dipole antenna
ferrite bead rings
film element
gap (horn, protective, or sphere)
Hall element or generator
insulator
magnet, permanent
part, miscellaneous electrical
post, binding
protector (network, gap, telephone)
rotary joint (microwave)
shield (electrical or optical)
short (coaxial transmission)
spark gap
splice
terminal (individual)
terminal, circuit
termination, cable
valve element

EQ

equalizer
network, equalizing

F

fuse
fuse cutout
limiter, current (for power cable)

FL

filter

G

amplifier, rotating (regulating generator)
chopper, electronic
exciter (rotating machine)
frequency changer (rotating)
generator
magneto (ignition or telephone)
regulating generator
vibrator, interrupter

H*

hardware (common fasteners, etc.)

HP*

hydraulic part

HR

heater
lamp (heating or infrared)
resistor heating

HS

handset
operator's set

HT

earphone
headset, electrical
receiver (hearing-aid or telephone)

HY

circulator
hybrid coil (telephone usage)
junction, hybrid
magic T
network, hybrid circuit

J

connector, receptacle, electrical
disconnecting device (connector receptacle jack)
jack
receptacle (connector, stationary portion)
waveguide flange (choke)

K

contactor (magnetically operated)
relay (armature, solenoid, reed, thermal)

L

choke coil
coil (all not classified as transformers)
electromagnetic actuator
field (generator or motor)
inductor
inductor, shunt
reactor
saturable reactor
solenoid, electrical
winding

LS

horn, electrical
loudspeaker
loudspeaker–microphone
reproducer, sound
transducer, underwater sound

M

clock
coulomb accumulator
counter, electrical
gage
instrument
meter
oscillograph
oscilloscope

outdoor metering device
recorder, elapsed-time
strain gage
thermometer
timer, electric

MG

converter (rotating machine)
dynamotor
inverter (motor–generator)
motor–generator

MK

hydrophone
microphone
transmitter, telephone

MP*

brake
clutch
frame
gyroscope
interlock, mechanical
mechanical part
mounting (not electrical circuit, not a socket)
part, miscellaneous mechanical (bearing, coupling, gear,
    shaft)
part, structural
reed, vibration
tuning fork

MT

detector, primary
transducer (measuring or mode)

N(3)

subdivision, equipment

P

connector plug, electrical
disconnecting device (connector, plug)
plug (connector, movable portion)
waveguide flange (plain)

PS

inverter, static (DC to AC)
power supply
rectifier (complete power supply assembly)
thermogenerator

PU

eraser, magnetic
erasing head
head (with various modifiers)
pickup
recording head

Q

rectifier, semiconductor controlled
switch, semiconductor controlled

transistor
thyratron (semiconductor device)
thyristor (semiconductor triode)

R

magnetoresistor
potentiometer
resistor (adjustable, nonlinear, variable)
rheostat
shunt (instrument or relay)

RE

receiver, radio

RT

ballast (lamp or tube)
lamp, resistance
resistor (current-regulating or thermal)
temperature-sensing element
thermistor

RV

resistor, voltage-sensitive
varistor, symmetrical

S

contactor (manually, mechanically, or thermally operated)
dial, telephone
disconnecting device (switch)
governor (electrical contact type)
interlock, safety, electrical
key switch (telephone usage)
key, telegraph
speed regulator (electrical contact type)
switch
switch (hook, interlock, reed)
thermal cutout (circuit interrupter)
thermostat

SQ

link (fusible or sensing)
squib (electric, explosive, igniter)

SR

slip ring
ring, electrical contact
rotating contact

T

autotransformer
coil (telephone induction or repeating)
coupler, linear
taper (coaxial or waveguide)
transformer (current or potential)

TB

block, connecting
strip, terminal
terminal board
test block

TC

thermocouple
thermopile

TP

(4) test point

TR

transmitter, radio

U*(1)

integrated circuit package
microcircuit
micromodule
nonreparable assembly
photon-coupled isolator

V

cell (light sensitive, photoemissive, photosensitive)
counter tube (Geiger–Muller or proportional)
detector, nuclear-radiation (gas filled)
electron tube
ion-diffusion device
ionization chamber
klystron
magnetron
photoelectric cell
phototube
resonator tube (cavity-type)
thyratron (electron tube)
traveline wave tube
voltage regulator (electron tube)

VR

diode, breakdown
regulator, voltage (excluding electron tube)
stabistor
voltage regulator (semiconductor device)

W

bus bar
cable
cable assembly (with connectors)
cable, coaxial
conductor
dielectric path
distribution line
Goubau line
transmission line
transmission line, strip-type
waveguide
wire

WT (5)

tiepoint, wiring

X

fuseholder
lampholder
socket

Y

crystal unit (piezoelectric or quartz)
oscillator (excluding electron tube)
oscillator, magnetostriction
resonator, tuning-fork

Z

artificial line (other than delay line)
balun
carrier-line trap
cavity, tuned
discontinuity (usually coaxial or waveguide transmission
  use)
gyrator
mode suppressor
network, general (where specific class letters do not fit)
network, phase-changing
resonator (tuned cavity)
shifter, directional phase (nonreciprocal)
shifter, phase

tuned circuit
tuner (E–H, multistub, slide-screw)

1. The class letter A indicates that the item is separable or reparable. The class letter U indicates that the item is inseparable or nonreparable.
2. For economic reasons, fundamentally separable or reparable assemblies may not be so provisioned but may be supplied as complete assemblies. However, the class letter A shall be retained.
3. Not a class letter, but used to identify a subdivision or equipment in the location numbering method.
4. Not a class letter, but commonly used to designate test points for maintenance purposes. See American National Standard Y14.15–1966.
5. Not a class letter, but commonly used to designate a tie-point on connection diagrams. See American National Standard Y14.15–1966.

Appendix F
Quick Reference to Symbols*

*Courtesy of American National Standards Institute, ANSI.

576

Quick Reference to Symbols

1. Qualifying Symbols

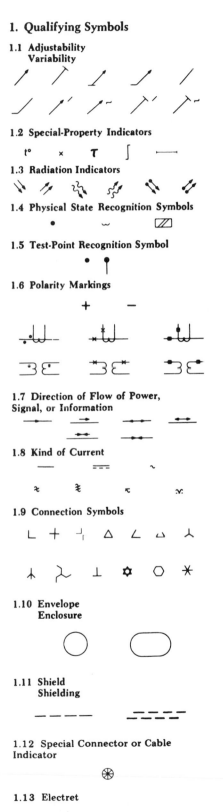

1.1 Adjustability
Variability

1.2 Special-Property Indicators

t°   x τ ∫   ⊢

1.3 Radiation Indicators

1.4 Physical State Recognition Symbols

1.5 Test-Point Recognition Symbol

1.6 Polarity Markings

\+   −

1.7 Direction of Flow of Power, Signal, or Information

1.8 Kind of Current

1.9 Connection Symbols

1.10 Envelope
Enclosure

1.11 Shield
Shielding

1.12 Special Connector or Cable Indicator

1.13 Electret

2. Fundamental Items

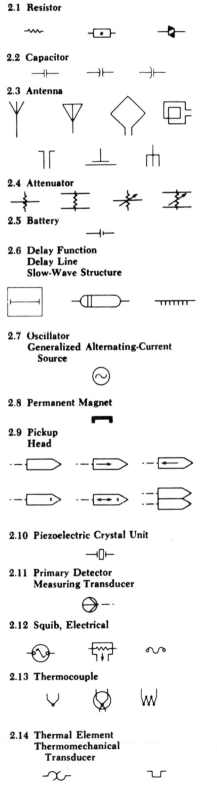

2.1 Resistor

2.2 Capacitor

2.3 Antenna

2.4 Attenuator

2.5 Battery

2.6 Delay Function
Delay Line
Slow-Wave Structure

2.7 Oscillator
Generalized Alternating-Current Source

2.8 Permanent Magnet

2.9 Pickup
Head

2.10 Piezoelectric Crystal Unit

2.11 Primary Detector
Measuring Transducer

2.12 Squib, Electrical

2.13 Thermocouple

2.14 Thermal Element
Thermomechanical Transducer

2.15 Spark gap
Igniter gap

2.16 Continuous Loop Fire Detector (temperature sensor)

2.17 Ignitor Plug

3. Transmission Path

3.1 Transmission Path
Conductor
Cable
Wiring

3.2 Distribution lines
Transmission lines

F   S   T   V

3.3 Alternative or Conditioned Wiring

3.4 Associated or Future

3.5 Intentional Isolation of Direct-Current Path in Coaxial or Waveguide Applications

Quick Reference to Symbols

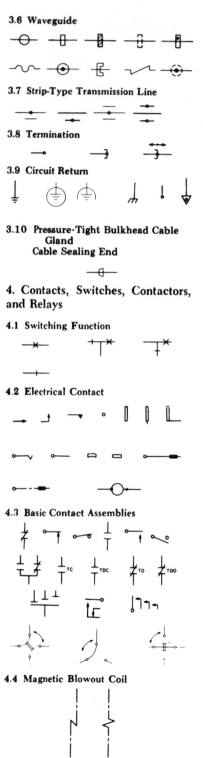

3.6 Waveguide

3.7 Strip-Type Transmission Line

3.8 Termination

3.9 Circuit Return

3.10 Pressure-Tight Bulkhead Cable Gland
Cable Sealing End

4. Contacts, Switches, Contactors, and Relays

4.1 Switching Function

4.2 Electrical Contact

4.3 Basic Contact Assemblies

4.4 Magnetic Blowout Coil

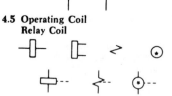

4.5 Operating Coil
Relay Coil

4.6 Switch

4.7 Pushbutton, Momentary or Spring-Return

4.8 Two-Circuit, Maintained or Not Spring-Return

4.9 Nonlocking Switch, Momentary or Spring-Return

4.10 Locking Switch

4.11 Combination Locking and Non-locking Switch

4.12 Key-Type Switch
Lever Switch

4.13 Selector or Multiposition Switch

4.14 Limit Switch
Sensitive Switch

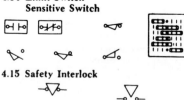

4.15 Safety Interlock

4.16 Switches with Time-Delay Feature

4.17 Flow-Actuated Switch

4.18 Liquid-Level-Actuated Switch

4.19 Pressure- or Vacuum-Actuated Switch

4.20 Temperature-Actuated Switch

4.21 Thermostat

4.22 Flasher
Self-interrupting switch

4.23 Foot-Operated Switch
Foot Switch

4.24 Switch Operated by Shaft Rotation and Responsive to Speed or Direction

4.25 Switches with Specific Features

4.26 Telegraph Key

4.27 Governor
Speed Regulator

4.28 Vibrator
Interrupter

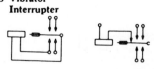

Quick Reference to Symbols

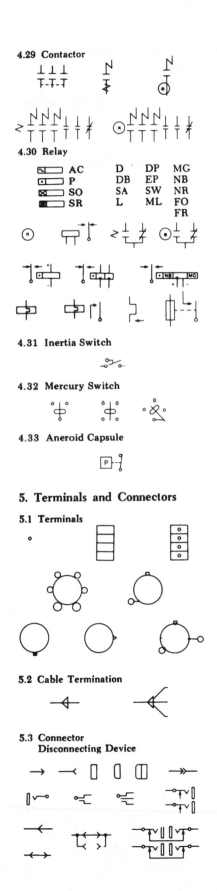

4.29 Contactor

4.30 Relay

AC	D
P	DB
SO	SA
SR	L

DP	MG
EP	NB
SW	NR
ML	FO
	FR

4.31 Inertia Switch

4.32 Mercury Switch

4.33 Aneroid Capsule

5. Terminals and Connectors

5.1 Terminals

5.2 Cable Termination

5.3 Connector
Disconnecting Device

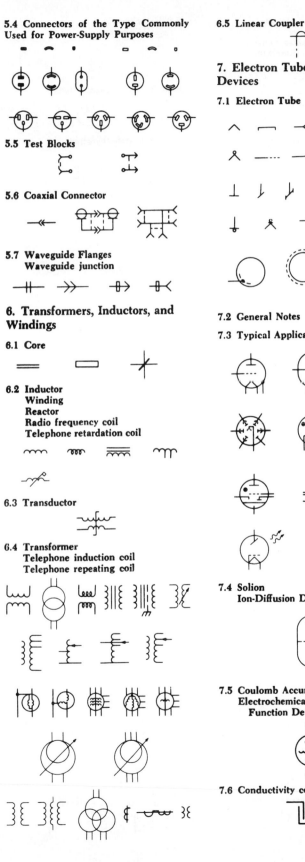

5.4 Connectors of the Type Commonly Used for Power-Supply Purposes

5.5 Test Blocks

5.6 Coaxial Connector

5.7 Waveguide Flanges
Waveguide junction

6. Transformers, Inductors, and Windings

6.1 Core

6.2 Inductor
Winding
Reactor
Radio frequency coil
Telephone retardation coil

6.3 Transductor

6.4 Transformer
Telephone induction coil
Telephone repeating coil

6.5 Linear Coupler

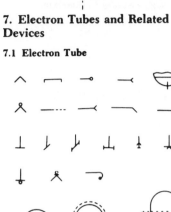

7. Electron Tubes and Related Devices

7.1 Electron Tube

7.2 General Notes

7.3 Typical Applications

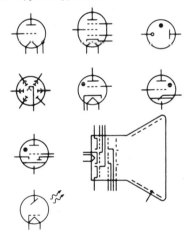

7.4 Solion
Ion-Diffusion Device

7.5 Coulomb Accumulator
Electrochemical Step-Function Device

7.6 Conductivity cell

Quick Reference to Symbols

7.7 Nuclear-Radiation Detector
Ionization Chamber
Proportional Counter Tube
Geiger-Müller Counter Tube

8. Semiconductor Devices

8.1 Semiconductor Device
Transistor
Diode

8.2 Element Symbols

8.3 Special Property Indicators

8.4 Rules for Drawing Style 1 Symbols

8.5 Typical Applications: Two-Terminal Devices

8.6 Typical Applications: Three- (or More) Terminal Devices

8.7 Photosensitive Cell

8.8 Semiconductor Thermocouple

8.9 Hall Element
Hall Generator

8.10 Photon-coupled isolator

8.11 Solid-state-thyratron

9. Circuit Protectors

9.1 Fuse

9.2 Current Arrester

9.3 Lightning Arrester
Arrester
Gap

9.4 Circuit Breaker

9.5 Protective Relay

C   F   φ   S   V
Z   GP W   T

10. Acoustic Devices

10.1 Audible-Signaling Device

10.2 Microphone

10.3 Handset
Operator's Set

10.4 Telephone Receiver
Earphone
Hearing-Aid Receivers

11. Lamps and Visual-Signaling Devices

11.1 Lamp

11.2 Visual-Signaling Device

12. Readout Devices

12.1 Meter
Instrument

A	DB	I	OP	RF	VA
AH	DBM	INT	OSCG	SY	VAR
C	DM	μA	PH	TLM	VARH
CMA	DTR	UA	PI	t°	VI
CMC	F	MA	PF	THC	VU
CMV	G	NM	RD	TT	W
CRO	GD	OHM	REC	V	WH

12.2 Electromagnetically Operated Counter
Message Register

13. Rotating Machinery

13.1 Rotating Machine

13.2 Field, Generator or Motor

13.3 Winding Connection Symbols

13.4 Applications: Direct-Current Machines

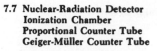

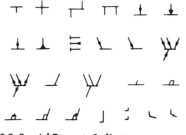
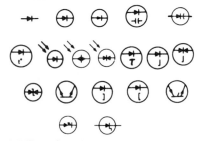
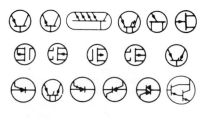
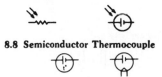

Quick Reference to Symbols

13.5 Applications: Alternating-Current Machines

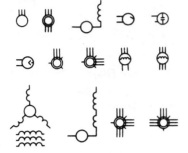

13.6 Applications: Alternating-Current Machines with Direct-Current Field Excitation

13.7 Applications: Alternating- and Direct-Current Composite

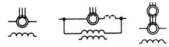

13.8 Synchro

CDX	TDX
CT	TR
CX	TX
TDR	RS

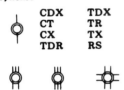

14. Mechanical Functions

14.1 Mechanical Connection
Mechanical Interlock

14.2 Mechanical Motion

14.3 Clutch
Brake

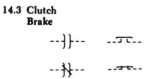

14.4 Manual Control

15. Commonly Used in Connection with VHF, UHF, SHF Circuits

15.1 Discontinuity

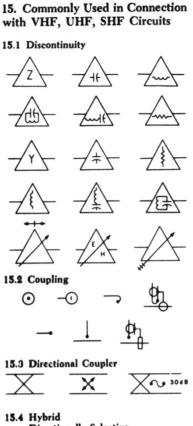

15.2 Coupling

15.3 Directional Coupler

15.4 Hybrid
Directionally Selective Transmission Devices

15.5 Mode Transducer

15.6 Mode Suppression

15.7 Rotary Joint

15.8 Non-reciprocal devices

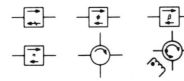

15.9 Resonator
Tuned Cavity

15.10 Resonator (Cavity Type) Tube

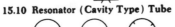

15.11 Magnetron

15.12 Velocity-Modulation (Velocity-Variation) Tube

15.13 Transmit-Receive (TR) Tube

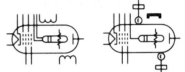

15.14 Traveling-Wave-Tube

15.15 Balun

15.16 Filter

15.17 Phase shifter

15.18 Ferrite bead rings

15.19 Line stretcher

16. Composite Assemblies

16.1 Circuit assembly
Circuit subassembly
Circuit element

EQ	FL-BP	RG	TPR
FAX	FL-HP	RU	TTY
FL	FL-LP	DIAL	CLK
FL-BE	PS	TEL	IND
ST-INV			

16.2 Amplifier

BDG	EXP	PRE
BST	LIM	PWR
CMP	MON	TRQ
DC	PGM	

Quick Reference to Symbols

16.3 Rectifier

16.4 Repeater

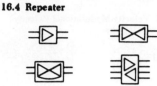

16.5 Network

16.6 Phase Shifter
Phase-Changing Network

16.7 Chopper

16.8 Diode-type ring demodulator
Diode-type ring modulator

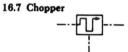

16.9 Gyro
Gyroscope
Gyrocompass

16.10 Position Indicator

16.11 Position Transmitter

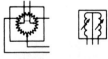

16.12 Fire Extinguisher Actuator Head

17. Analog Functions

17.1 Operational Amplifier

17.2 Summing Amplifier

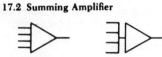

17.3 Integrator

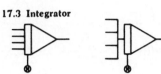

17.4 Electronic Multiplier

17.5 Electronic Divider

17.6 Electronic Function Generator

17.7 Generalized Integrator

17.8 Positional Servo-mechanism

17.9 Function Potentiometer

18. Digital Logic Functions

18.1 Digital Logic Functions
(See cross references)

19. Special Purpose Maintenance Diagrams

19.1 Data flow code signals

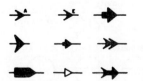

19.2 Functional Circuits

20. System Diagrams, Maps and Charts

20.1 Radio station

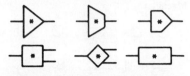

20.2 Space station

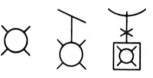

20.3 Exchange equipment

20.4 Telegraph repeater

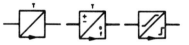

20.5 Telegraph equipment

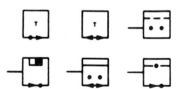

20.6 Telephone set

21. System Diagrams, Maps and Charts

21.1 Generating station

Quick Reference to Symbols

21.2 Hydroelectric generating station

21.3 Thermoelectric generating station

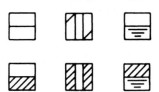

21.4 Prime mover

21.5 Substation

22. Class Designation Letters

A	DS	J	PU	TP
AR	E	K	Q	TR
AT	EQ	L	R	U
B	F	LS	RE	V
BT	FL	M	RT	VR
C	G	MG	RV	W
CB	H	MK	S	WT
CP	HP	MP	SQ	X
CR	HR	MT	SR	Y
D	HS	N	T	Z
DC	HT	P	TB	
DL	HY	PS	TC	

Appendix G
Typical Graphic Symbols for Electrical Diagrams*

*Courtesy of Joint Industrial Council

Typical Graphic Symbols for Electrical Diagrams

Typical Graphic Symbols for Electrical Diagrams

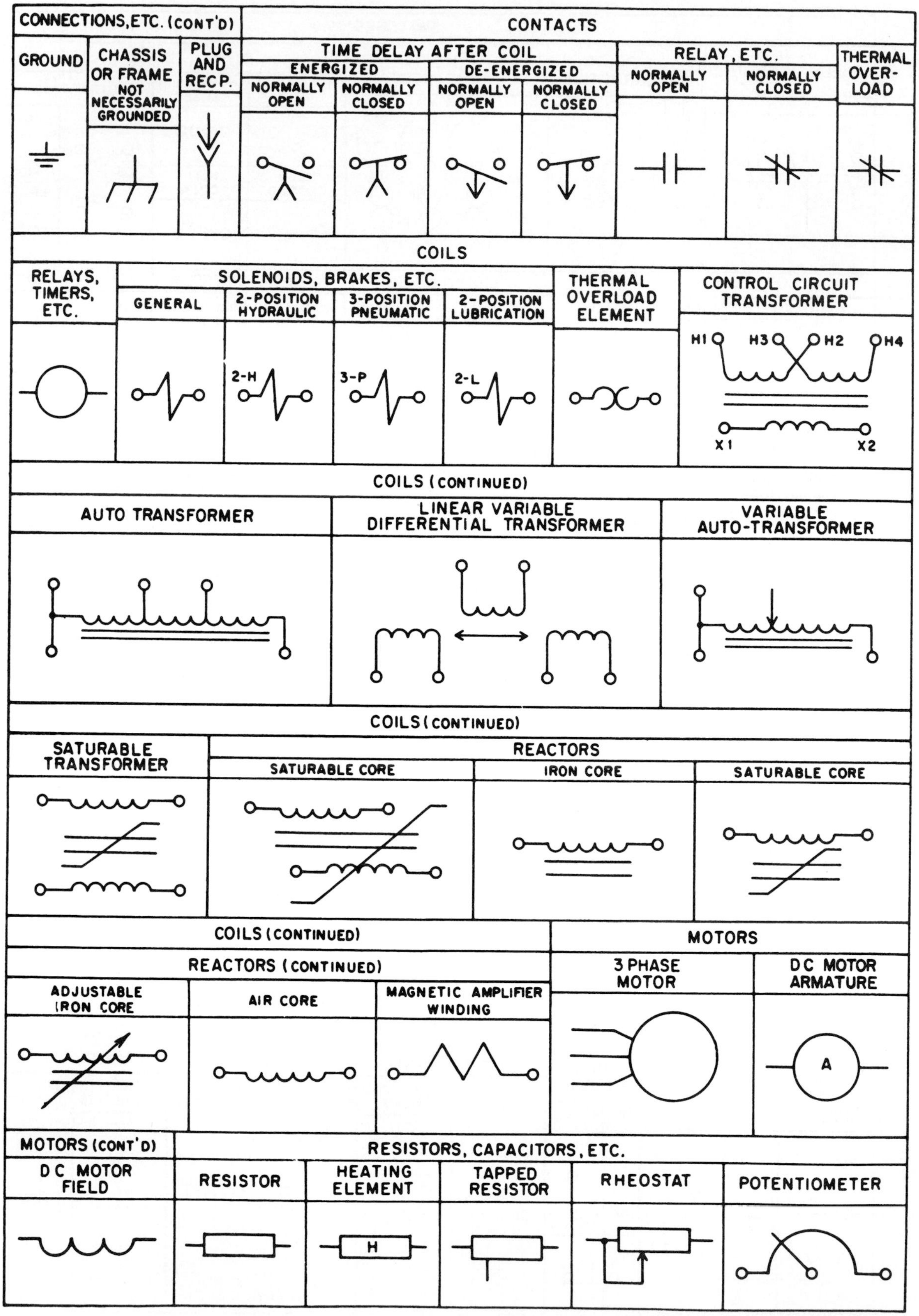
CONNECTIONS, ETC. (CONT'D)
GROUND
CHASSIS OR FRAME NOT NECESSARILY GROUNDED
PLUG AND RECP.
CONTACTS
TIME DELAY AFTER COIL
ENERGIZED
NORMALLY OPEN
NORMALLY CLOSED
DE-ENERGIZED
NORMALLY OPEN
NORMALLY CLOSED
RELAY, ETC.
NORMALLY OPEN
NORMALLY CLOSED
THERMAL OVER-LOAD
COILS
RELAYS, TIMERS, ETC.
SOLENOIDS, BRAKES, ETC.
GENERAL
2-POSITION HYDRAULIC
2-H
3-POSITION PNEUMATIC
3-P
2-POSITION LUBRICATION
2-L
THERMAL OVERLOAD ELEMENT
CONTROL CIRCUIT TRANSFORMER
H1
H3
H2
H4
X1
X2
COILS (CONTINUED)
AUTO TRANSFORMER
LINEAR VARIABLE DIFFERENTIAL TRANSFORMER
VARIABLE AUTO-TRANSFORMER
COILS (CONTINUED)
SATURABLE TRANSFORMER
REACTORS
SATURABLE CORE
IRON CORE
SATURABLE CORE
COILS (CONTINUED)
REACTORS (CONTINUED)
ADJUSTABLE IRON CORE
AIR CORE
MAGNETIC AMPLIFIER WINDING
MOTORS
3 PHASE MOTOR
DC MOTOR ARMATURE
A
MOTORS (CONT'D)
DC MOTOR FIELD
RESISTORS, CAPACITORS, ETC.
RESISTOR
HEATING ELEMENT
H
TAPPED RESISTOR
RHEOSTAT
POTENTIOMETER

Typical Graphic Symbols for Electronic Diagrams with Basic Device Designations

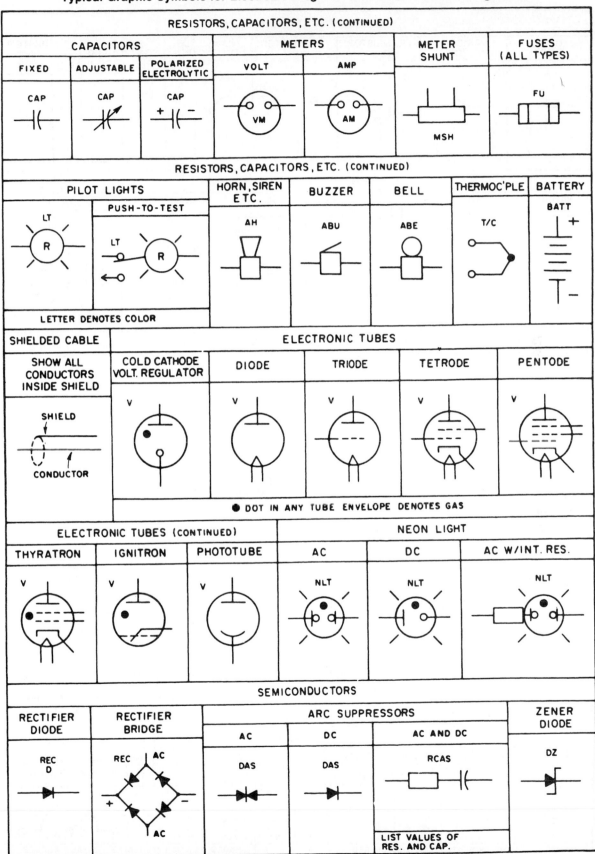

Typical Graphic Symbols for Electronic Diagrams with Basic Device Designations

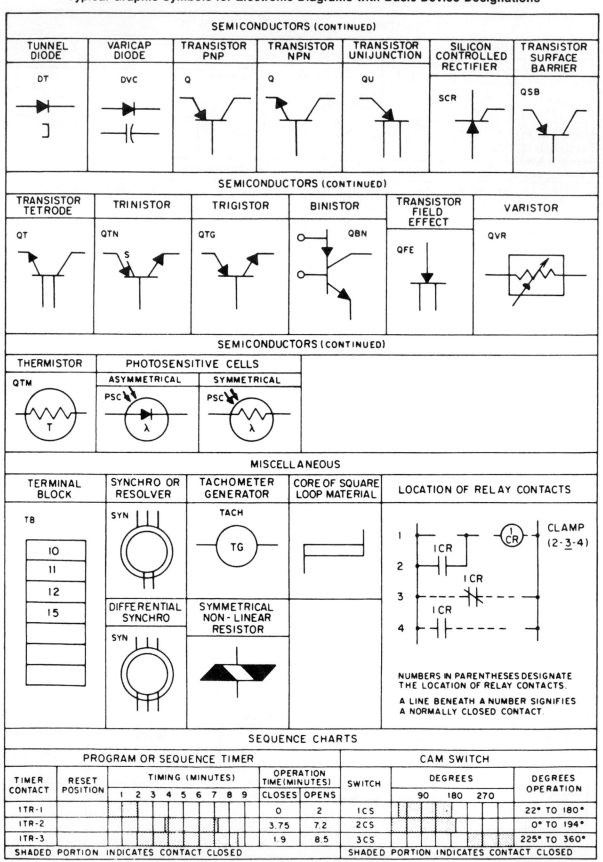

TIMER CONTACT	RESET POSITION	TIMING (MINUTES) 1 2 3 4 5 6 7 8 9	OPERATION TIME (MINUTES) CLOSES	OPENS	SWITCH	DEGREES 90 180 270	DEGREES OPERATION
1TR-1			0	2	1CS		22° TO 180°
1TR-2			3.75	7.2	2CS		0° TO 194°
1TR-3			1.9	8.5	3CS		225° TO 360°
SHADED PORTION INDICATES CONTACT CLOSED					SHADED PORTION INDICATES CONTACT CLOSED		

Index